Complex and Dusty Plasmas

Series in Plasma Physics

Series Editor:
Steve Cowley, Imperial College, UK and UCLA, USA

Other recent books in the series:

Applications of Laser–Plasma Interactions
Shalom Eliezer and Kunioki Mima

Plasma Kinetic Theory
D G Swanson

High Power Microwaves, *Second Edition*
J Benford, J A Swegle and E Schamiloglu

Controlled Fusion and Plasma Physics
K Miyamoto

Plasma Electronics: Applications in Microelectronic Device Fabrication
T Makabe and Z Petrović

An Introduction to Inertial Confinement Fusion
S Pfalzner

Aspects of Anomalous Transport in Plasmas
R Balescu

Non-Equilibrium Air Plasmas at Atmospheric Pressure
K H Becker, R J Barker and K H Schoenbach (Eds)

Magnetohydrodynamic Waves in Geospace: The Theory of ULF Waves and their Interaction with Energetic Particles in the Solar-Terrestrial Environment
A D M Walker

Plasma Physics via Computer Simulation (paperback edition)
C K Birdsall and A B Langdon

Plasma Waves, Second Edition
D G Swanson

Microscopic Dynamics of Plasmas and Chaos
Y Elskens and D Escande

Plasma and Fluid Turbulence: Theory and Modelling
A Yoshizawa, S-I Itoh and K Itoh

The Interaction of High-Power Lasers with Plasmas
S Eliezer

Introduction to Dusty Plasma Physics
P K Shukla and A A Mamun

Series in Plasma Physics

Complex and Dusty Plasmas

From Laboratory to Space

Edited by
Vladimir E. Fortov
Joint Institute for High Temperatures
Russian Academy of Sciences, Moscow, Russia

Gregor E. Morfill
Max-Planck-Institute of Extraterrestrial Physics
Garching, Germany

CRC Press
Taylor & Francis Group
Boca Raton London New York

CRC Press is an imprint of the
Taylor & Francis Group, an **informa** business
A TAYLOR & FRANCIS BOOK

CRC Press
Taylor & Francis Group
6000 Broken Sound Parkway NW, Suite 300
Boca Raton, FL 33487-2742

Printed in the United States of America on acid-free paper
10 9 8 7 6 5 4 3 2 1

International Standard Book Number: 978-1-4200-8311-8 (Hardback)

Library of Congress Cataloging-in-Publication Data

Complex and dusty plasmas : from laboratory to space / editors, Vladimir E. Fortov and Gregor E. Morfill.
p. cm. -- (Series in plasma physics ; 25)
Includes bibliographical references and index.
ISBN 978-1-4200-8311-8
1. Dusty plasmas. I. Fortov, V. E. II. Morfill, G. E. III. Title. IV. Series.

QC718.5.D84C66 2010
530.4'4--dc22 2009033249

Visit the Taylor & Francis Web site at
http://www.taylorandfrancis.com

and the CRC Press Web site at
http://www.crcpress.com

For Product Safety Concerns and Information please contact our EU representative:
GPSR@taylorandfrancis.com
Taylor & Francis Verlag GmbH, Kaufingerstraße 24, 80331 München, Germany.

Contents

Contributors

Vladimir E. Fortov
Joint Institute for High Temperatures, Russian Academy of Sciences, Moscow, Russia

Ove Havnes
University of Tromsø, Tromsø, Norway

Mihály Horányi
University of Colorado, Boulder, USA

Alexey V. Ivlev
Max-Planck-Institut für extraterrestrische Physik, Garching, Germany

Alexey G. Khrapak
Joint Institute for High Temperatures, Russian Academy of Sciences, Moscow, Russia

Sergey A. Khrapak
Max-Planck-Institut für extraterrestrische Physik, Garching, Germany

Boris A. Klumov
Max-Planck-Institut für extraterrestrische Physik, Garching, Germany

Vladimir I. Molotkov
Joint Institute for High Temperatures, Russian Academy of Sciences, Moscow, Russia

Gregor E. Morfill
Max-Planck-Institut für extraterrestrische Physik, Garching, Germany

Oleg F. Petrov
Joint Institute for High Temperatures, Russian Academy of Sciences, Moscow, Russia

Hubertus M. Thomas
Max-Planck-Institut für extraterrestrische Physik, Garching, Germany

Olga S. Vaulina
Joint Institute for High Temperatures, Russian Academy of Sciences, Moscow, Russia

Sergey V. Vladimirov
School of Physics, The University of Sydney, Sydney, Australia

Introduction

"Dusty", or "complex" plasmas are plasmas containing solid or liquid particles (dust) which are charged. The charges can be negative or positive, depending on the charging mechanisms operating in the plasmas. Dust and dusty plasmas are quite natural in space. They are present in planetary rings, comet tails, interplanetary and interstellar clouds (Goertz 1989; Northrop 1992; Tsytovich 1997), found in the vicinity of artificial satellites and space stations (Whipple 1981; Robinson and Coakley 1992), etc. Also, dusty plasmas are actively investigated in laboratories (Fortov *et al.* 2005; Thomas 2009). Currently, the term "complex plasmas" is widely used in the literature to distinguish dusty plasmas specially "designed" for such investigations.

The presence of massive charged particles in complex plasmas is essential for the collective processes. Ensembles of microparticles give rise to new very low-frequency wave modes which represent the oscillations of particles against the quasi-equilibrium background of electrons and ions. Overall dynamical time scales associated with the dust component are in the range 10–100 Hz. The particles themselves are large enough to be visualized individually, and hence, their motion can be easily tracked. This makes it possible to investigate phenomena occurring in complex plasmas at the most fundamental kinetic level.

Micron size particles embedded in a plasma not only change the charge composition, they also introduce new physical processes into the system, e.g., effects associated with dissipation and plasma recombination on the particle surface, variation of the particle charges. These processes imply new mechanisms of the energy influx into the system. Therefore, complex plasmas are a new type of non-Hamiltonian systems with the properties which can be completely different from those of usual multicomponent plasmas.

Dust plays an exceptionally important role in technological plasma applications, associated with the utilization of plasma deposition and etching technologies in microelectronics, as well as with production of thin films and nanoparticles (Selwyn et al. 1989; Bouchoule 1999; Kersten *et al.* 2001). To control these processes, it is necessary to understand the basic determining mechanisms, e.g., transport of dust particles, influence of dust on plasma parameters.

Due to large charges carried by the grains (typically, on the order of thousand elementary charges for a micron-size particle), the electrostatic energy of the mutual interaction is remarkably high. Hence, the strong electrostatic coupling in the dust subsystem can be achieved much more easily than in the electron-ion subsystem. In complex plasmas, one can observe transitions from a disordered gaseous-like phase to a liquid-like phase and the formation of ordered structures of dust particles-plasma crystals. The first experimental observation of the ordered (quasicrystalline) structures of charged microparticles obtained in a modified Paul's trap was reported by Wuerker *et al.* (1959).

The enormous increase of interest in complex plasmas was triggered in the mid 1990's by the laboratory discovery of plasma crystals. The possibility of dust subsystem crystallization in a nonequilibrium gas discharge plasma was predicted by Ikezi (1986). The first experimental observations of ordered particle structures were reported in 1994 in radio frequency (rf) discharges (Chu *et al.* 1994; Hayashi and Tachibana 1994; Melzer *et al.* 1994; Thomas *et al.* 1994). Later on, plasma crystals were found in direct current (dc) discharges (Fortov *et al.* 1996b), thermal plasmas at atmospheric pressure (Fortov *et al.* 1996a), and even nuclear-induced dusty plasmas (Fortov *et al.* 1999). Today, the physics of complex plasmas is a rapidly growing field of research, which covers various fundamental aspects of the plasma physics, hydrodynamics, kinetics of phase transitions, nonlinear physics, and solid states, as well as the industrial applications, engineering, and astrophysics. More and more research groups throughout the world have become involved in the field, and the number of scientific publications is growing exponentially.

In this book, we have made an attempt to provide a balanced and consistent picture of the current status of the field, by covering the latest development in the most important directions of the experimental and theoretical research, and have outlined the perspective issues to pursue in future. The major types of complex plasmas in ground-based and microgravity experiments are considered. Properties of the magnetized, thermal, cryogenic, ultraviolet, nuclear-induced complex plasmas and plasmas with nonspherical particles are discussed. Various basic plasma-particle interactions, including grain charging in different regimes, momentum exchange between different complex plasma species, electric potential distribution around particles in isotropic and anisotropic plasmas, and interactions between charged grains are investigated. The major forces acting on the particles and features of the particle dynamics in complex plasmas are highlighted. An overview of the wave properties in different phase states is given. Recent results on the phase transitions between different crystalline and liquid complex plasma states are presented. Possible existence of the liquid-vapor critical point in complex plasmas is briefly discussed. Fluid behavior of complex plasmas and the onset of cooperative phenomena are considered. Particular attention is given to astrophysical aspects of dusty plasmas and numerical simulation of their properties. Diagnostics of complex plasmas is discussed in detail. Possible applications of complex plasmas, interdisciplinary aspects, and perspectives are also considered. An important feature of this book is detailed discussion of unique experimental and theoretical aspects of complex plasmas related to the experimental investigations under microgravity conditions performed onboard Mir and ISS space stations. Here an inestimable contribution came from the expertise of the members of the Russian-German research team who are among the authors of this book.

References

Bouchoule, A. (1999). Technological impacts of dusty plasmas. In *Dusty plasmas: Physics, chemistry and technological impacts in plasma processing*, Bouchoule, A. (ed.), pp. 305–396. Wiley, Chichester.

Chu, J. H. and I, L. (1994). Direct observation of Coulomb crystals and liquids in strongly coupled rf dusty plasmas. *Phys. Rev. Lett.*, **72**, 4009–4012.

Fortov, V. E., Ivlev, A. V., Khrapak, S. A., Khrapak A. G., and Morfill, G. E. (2005). Complex (dusty) plasmas: Current status, open issues, perspectives. *Phys. Rep.*, **421**, 1–103.

Fortov, V. E., Nefedov, A. P., Petrov, O. F., Samarian, A. A., Chernyschev, A. V., and Lipaev, A. M. (1996a). Experimental observation of Coulomb ordered structure in sprays of thermal dusty plasmas. *JETP Lett.*, **63**, 187–192.

Fortov, V. E., Nefedov, A. P., Torchinskii, V. M., Molotkov, V. I., Khrapak, A. G., Petrov, O. F., and Volykhin, K. F. (1996b). Crystallization of a dusty plasma in the positive column of a glow discharge. *JETP Lett.*, **64**, 92–98.

Fortov, V. E., Vladimirov, V. I., Deputatova, L. V., Molotkov, V. I., Nefedov, A. P., Rykov, V. A., Torchinskii, V. M., and Khudyakov, A. V. (1999). Ordered dusty structures in plasma produced by nuclear particles. *Dokl. Phys.*, **44**, 279–282.

Goertz, C. K. (1989). Dusty plasmas in the solar system. *Rev. Geophys.*, **27**, 271–292.

Hayashi, Y. and Tachibana, S. (1994). Observation of Coulomb-crystal formation from carbon particles grown in a methane plasma. *Jpn. J. Appl. Phys.*, **33**, L804–L806.

Ikezi, H. (1986). Coulomb solid of small particles in plasmas. *Phys. Fluids*, **29**, 1764–1766.

Kersten, H., Deutsch, H., Stoffels, E., Stoffels, W. W., Kroesen, G. M. W., and Hippler, R. (2001). Micro-disperse particles in plasmas: From disturbing side effects to new applications. *Contrib. Plasma Phys.*, **41**, 598–609.

Melzer, A., Trottenberg, T., and Piel, A. (1994). Experimental determination of the charge on dust particles forming Coulomb lattices. *Phys. Lett. A*, **191**, 301–307.

Northrop, T. G. (1992). Dusty plasmas. *Phys. Scripta*, **45**, 475–490.

Robinson, P. A. and Coakley, P. (1992). Spacecraft charging-progress in the study of dielectrics and plasmas. *IEEE Trans. Electr. Insul.*, **27**, 944–960.

Selwyn, G. S., Singh, J., and Bennett, R. S. (1989). In situ laser diagnostic studies of plasma-generated particulate contamination. *J. Vac. Sci. Technol. A*, **7**, 2758–2765.

Thomas, Jr., E. (2009). Dust clouds in dc-generated dusty plasmas: Transport, waves, and three-dimensional effects. *Contrib. Plasma Phys.*, **49**, 316–345.

Thomas, H., Morfill, G. E., Demmel, V., Goree, J., Feuerbacher, B., and Mohlmann, D. (1994). Plasma crystal: Coulomb crystallization in a dusty plasma. *Phys. Rev. Lett.*, **73**, 652–655.

Tsytovich, V. N. (1997). Dust plasma crystals, drops, and clouds. *Phys. Usp.*, **40**, 53–94.

Whipple, E. C. (1981). Potentials of surfaces in space. *Rep. Prog. Phys.*, **44**, 1197–1250.

Wuerker, R. F., Shelton, H., and Langmuir, R. V. (1959). Electrodynamic containment of charged particles. *J. Appl. Phys.*, **30**, 342–349.

1

Types of experimental complex plasmas

Vladimir E. Fortov, Alexey G. Khrapak, Vladimir I. Molotkov, Gregor E. Morfill, Oleg F. Petrov, Hubertus M. Thomas, Olga S. Vaulina, Sergey V. Vladimirov

Plasmas have been investigated in the laboratory since the beginning of the last century. Already in his pioneering work, Langmuir found dust particles appearing in his discharges (Langmuir *et al.* 1924). At that time dust in a plasma was just seen as a dirt effect, so the physics of the dusty plasma was not a topic of research. For quite a while the topic of dusty plasmas was interesting mainly for astrophysicists, and theory was developed for the charging of dust grains, their interaction, transport, etc., in cometary and planetary atmospheres, interstellar matter, planet formation, etc. (Grün *et al.* 1984; Goertz 1989; Hartquist *et al.* 1992). In the late 1980s, laboratory research on dusty plasmas became important, again, especially for industry. In the fabrication of chips and microelectronics using plasma processes, the dust particles were always found after the manufacturing was finished. Claiming that the dust must have come from outside, before the vacuum chamber was closed, industry built the devices in the best clean rooms at that time. Nevertheless, the amount of dust was not decreasing with the cleanliness of the surrounding laboratory. In 1988 G. Selwyn at IBM recognized that the particles grew during the plasma process (see Figure 1.1) (Selwyn *et al.* 1989). Using laser light scattering he could show, in a series of experiments, that starting from molecules the dust grew via the nanometer scale up to micrometer sizes. The latter were causing the problems on the processed wafers because of contamination or short cutting circuits. After that discovery the research on dusty plasmas started in industry and application-oriented institutes. But, at that time, the research was not concentrated on the physics of this new topic; it was more related to avoiding the growth of dust. Nowadays, main trends in this application research have changed, because in times of nanomaterials the possibility to grow nanoparticles of different structure and composition opens up new possibilities in this growing field.

Parallel to the rediscovery of dusty plasmas in the laboratory, a fundamental research topic in dusty plasmas was formed. Ikezi (1986) proposed that Coulomb crystallization might occur in dusty plasmas for typical plasma conditions and particle sizes of micrometers. This prediction was based on the one-component plasma (OCP) model. He argued that similarly to this system a transition from fluid to crystalline states should occur in a multi-component plasma containing electrons, ions and charged dust particles, provided the electric coupling between the particles is strong enough. This paper led to discussions among scientists from the theoretical space dusty plasma community. They supposed, that gravity would be a hindering

FIGURE 1.1
Laser light scattering of a particle distribution above a wafer (Selwyn *et al.* 1989)

factor in forming large 3-dimensional crystalline structures from micrometer sized particles and proposed the experiment "Plasma Crystal" to ESA for the so-called Columbus Precursor Flights in 1991. They claimed that under microgravity conditions, Ikezi's forecast could become reality. The proposal was evaluated in a peer process and received the highest rankings. Nevertheless it was rejected, because based on theory only, it was considered too risky. Instead, the Principal Investigator, Gregor Morfill, was asked to start experimental work under gravity conditions first, before resubmitting the proposal. The latter followed the guidance of the board and started the work in the laboratory with a PhD student. Shortly after the assembly of the lab, it was found that plasma crystals could be formed in the laboratory, too. The charged microparticles could be levitated in the sheath electric field of a capacitively coupled rf discharge and could form, under special conditions, a 2-dimensional plasma crystal (see Figure 1.2). The main reason, that this could happen was that the microparticles were spherical and had a monodisperse size distribution. The result was that all the particles received the same high charge and were levitated at the same height in the sheath electric field. Another necessary condition was fulfilled by the neutral gas damping, which was responsible for cooling down the particle component. Finally, electric interaction between the microparticles was strong enough to initiate the transition to a crystalline structure. The discovery of plasma crystals was first presented during the International Conference on Phenomena in Ionized Gases (ICPIG) in Bochum in 1993 and published the following year by Thomas *et al.* (1994).

Interestingly, the so-called plasma crystals were discovered by two more groups independently at nearly the same time, which clearly showed that the discovery was just a matter of time (Chu and I 1994; Hayashi and Tachibana 1994). The discovery of plasma crystals can be regarded as the trigger for many plasma physicists to start investigations of this fascinating new state of matter, which shows so many interest-

FIGURE 1.2
First plasma crystal, an ordered structure of microparticles, charged by the surrounding plasma and levitated in the sheath electric field. The picture shows a top view of the reflected laser light by a single-layer plasma crystal (Thomas *et al.* 1994).

ing properties, first of all, the possibility to study processes on the most fundamental level – the kinetic level.

In the beginning low-temperature rf discharges were mostly used for the research. In the following years, dc discharges were also employed, providing new possibilities, like large 3-dimensional clouds levitated in striations. In the meantime, even fusion plasmas have become the topic of complex plasma research. In the next sections we introduce two most important discharges, the rf-capacitively or inductively coupled and the dc discharges, used in complex plasma research.

1.1 Complex plasmas in rf discharges

Most complex plasma experiments have been and still are performed in capacitively coupled rf discharges, as mentioned above. The reason for this is manyfold (Raizer 1991; Liberman and Lichtenberg 1994): (i) the capacitive sheath produces a strong self bias of a couple of 10 V, which can be used to levitate the charged microparticles; (ii) the rf-frequency is so high, that neither the dust particles nor the ions can respond to the changing electric field, and therefore they are (usually) cold, damped by the neutral component to the room temperature; (iii) the electron temperature, responsible for the value of the charge on the particles, is reasonably high, on the order of few eV.

Usually, the smaller the particle size is, the more layers in the vertical direction can be formed. It is even possible to fill the plasma bulk with particles, if their size is much below $1\,\mu$m – since then, gravity is no longer the dominating force in the system.

On the other side, to use the special property of complex plasmas – the kinetic observation of single particles in a many particle system – it is necessary to use particles with sizes above $1\,\mu$. Gravity restricts such a system to a few layers in the vertical direction. To perform experiments in the bulk with fully 3-dimensional complex plasma systems it is necessary to use additional levitating forces, e.g. gas flow or thermophoresis, or to remove gravity. The latter is possible on parabolic flights, on sounding rockets or in experiments onboard the International Space Station (ISS). For such microgravity experiments a special design of the rf discharge is mandatory, since the asymmetric discharge would destroy the homogeneity and symmetry gained through microgravity. A highly symmetrical plasma chamber, a symmetrically coupled parallel plate rf discharge, fulfills the special needs for a setup for microgravity.

The two different setup-types will be described in the following paragraphs.

1.1.1 The GEC-RF-Reference Cell

A typical set-up, a modified Gaseous Electronics Conference (GEC)-RF-Reference Cell (Hargis *et al.* 1994), will be explained in the following, as an example for a laboratory study of 2-dimensional dusty plasma systems. The electrode system consists of a driven rf electrode in the bottom of the apparatus and a grounded or floating counter electrode at the top (see Figure 1.3). If the latter is removed the metal vacuum chamber surrounding the electrode system, which is providing the vacuum conditions for the low-pressure discharge, acts as the counter electrode. The lower electrode is connected to an rf generator via a Pi-type matching unit. This provides the matching of the 50 Ohm signal from the generator to the high Ohmic plasma device. These directly applied rf currents and voltages create the high voltage capacitive sheath between the electrode and the bulk plasma and lead to stochastic or collisionless heating in the sheath and the ohmic heating in the bulk. The plasma can be excited and controlled in inert gas, typically argon, in a pressure range from ≈ 0.1 to ≈ 1000 Pa (depending on the gas). Other molecular gases can be used also, but inert gases do not form negative ions and provide a pure electropositive plasma.

The electrode is driven at a frequency of 13.56 MHz, producing a so-called low-temperature plasma. Since the ion plasma frequency is much lower than the excitation frequency, the ions are not affected by the alternating electric field and stay at room temperature due to collisions with the much higher number of neutral atoms or molecules (typical ionization rates are in the range of 10^{-8}–10^{-6}). Only the electrons react to the rf-field and are heated to an average temperature of a few eV. Energetic

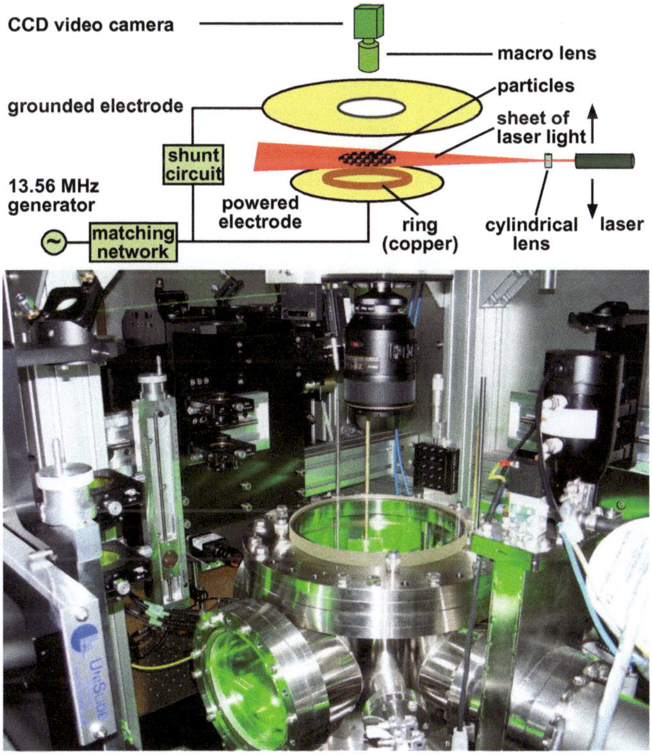

FIGURE 1.3
(See color insert following page 242). The sketch shows a schematic drawing of a typical rf electrode system with lower driven electrode and the grounded upper-ring electrode. The microparticles are injected by a dispenser (not shown) and levitate in the sheath electric field, additionally trapped horizontally by a parabolic potential formed by a ring positioned on the electrode. The microparticles are illuminated by a laser beam expanded into a sheet parallel to the electrode, and the reflected light is observed at 90° by a CCD camera. The image shows a typical assembly of the GEC-RF-Reference Cell.

electrons ionize neutral gas and thus are responsible for sustaining the plasma. In the central region of the discharge a homogeneous bulk plasma of densities between 10^8–10^{10} cm^{-3} is formed.

The high voltage sheath with a negative self bias of a few tens of Volts is very important for the complex plasma under gravity conditions. Microparticles are injected manually or by electric/electromagnetic dispensers. In the plasma they get charged in a fraction of a second to thousands of elementary charges (dependent on their size) and then they are trapped by the plasma generated electric potential.

The upward electrostatic force acting on the microparticles can balance gravity

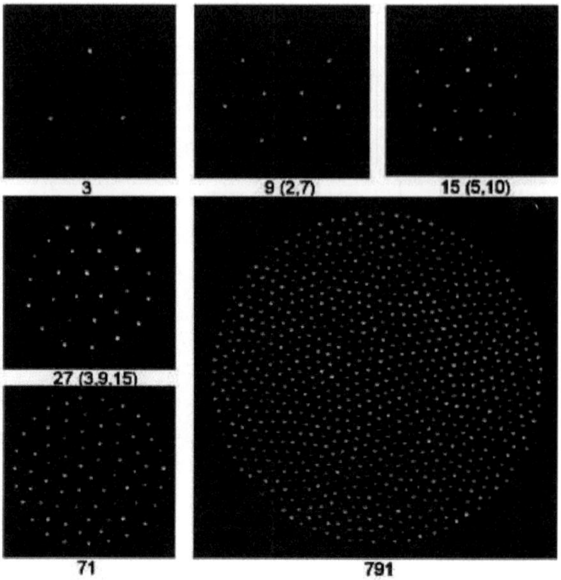

FIGURE 1.4
Original images of particles forming 2-dimensional clusters of different numbers of particles (Juan *et al.* 1998).

and allows them to levitate in a narrow region in the sheath. Every particle with a certain charge to mass ratio should levitate at the same height above the lower electrode; the smaller the ratio, the lower the leviation height. Since particles with monodispersive size distribution are most often used in the basic complex plasma experiments, they charge up to the same level in the plasma and form monolayer structures. Two-dimensional liquid and crystalline systems and their phase transitions are one of the interesting research topics in complex plasmas.

The particles are illuminated by a sheet of laser light parallel to the electrode and the reflected light is observed at 90° by a CCD camera (see Figure 1.3). This allows uss to study the full dynamics of the 2-dimensional system.

Single particles levitating in the sheath of the rf electrode can be used to probe the sheath electric field and potential distributions. Additionally, single particles are important for some basic complex plasma measurements, like the charge on the microparticle or the ion wake formation downstream from the particles. If the number of particles is increased step by step, one can investigate the formation of clusters (Juan *et al.* 1998), small crystalline structures dominated by surfaces, and observe the formation of a Mendeleev-like table, including stable configurations at magic numbers of particles as well as metastable states (see Figure 1.4 and Section 1.4.5). Adding more and more particles to the 2D cluster, finally a stable monolayer is reached, no longer dominated by surface effects – a transition to collective behavior takes place.

Large 2-dimensional systems are investigated extensively. The particles strongly interact via screened Coulomb forces and form, in most cases, 2-dimensional crystalline structures, so-called hexagonal lattices. Static analysis methods include the pair correlation and bond-orientational correlation function, the real space analogies to the structure factor measured in natural systems by Bragg reflexes. They show the state and structure of the complex plasma. Additionally, dynamical observations allow getting more details, e.g., on defect motion. Here, for example, supersonic motion of lattice defects through the crystal producing Mach cones was observed. Such lattice defects appear as 5-7-fold dislocations, as single events or in strings. They appear in regions of lower order through stress formation and overload. The information on the dynamics of the generation and motion of dislocations is of general interest in material sciences, the study of earthquakes, snow avalanches, colloidal crystals, 2D foams, and various types of shear cracks. Complex plasmas are virtually undamped and form macroscopic systems which allow this study on the most fundamental level – the kinetic level. In an experiment of Nosenko *et al.* (2008), a 2D finite plasma crystal was formed and observed with high time resolution for some period of time. The whole structure slightly rotated which led to shear stress inside the plasma crystal. The emerging shear strain and vorticity are discussed in Section 5.3.3.

The plasma chamber can be additionally modified by adding manipulation devices. Two kinds of such a manipulation are possible. With laser beams it is possible to manipulate the microparticles from outside, without a disturbance of the plasma. A static laser beam could be used to introduce shear forces, a movable beam, e.g., by scanning mirrors, local heating or propagating Mach cones. A plasma disturbing manipulation is a wire or a wire system put into the plasma and connected to an electrical signal. Through short voltage pulses or an alternating signal, the cloud, which is usually at the same height as the wire, can be excited. An interesting recent result of such an excitation is the melting and recrystallization which we discuss in some more detail.

The physics of 2-dimensional liquids and crystallites is rather rich and is therefore studied in many different systems. Of particular interest is the investigation of the transitions between different phases. Theoretically, these transitions are considered in the so-called KTHNY theory, developed and refined by Kosterlitz, Thouless, Halperin, Nelson and Young (Kosterlitz and Thouless 1973; Halperin and Nelson 1978; Nelson and Halperin 1979; Young 1979; Nelson 1983). In this theory the authors predict two continuous transitions between the two phases passing a so-called hexatic phase, characterized by a long range orientational but a short range translational order, instead of a first-order transition, which is usual for 3-dimensional systems.

An interesting crystallization experiment with a 2D complex plasma crystal was performed by Knapek *et al.* (2007). First, a crystalline monolayer with hexagonal lattice structure was formed (Figure 1.5a) by particles levitated above a horizontal rf electrode, similar to the above described GEC-RF-Reference setup, and confined from two sides by tungsten wires (at the same height). To induce melting, a short

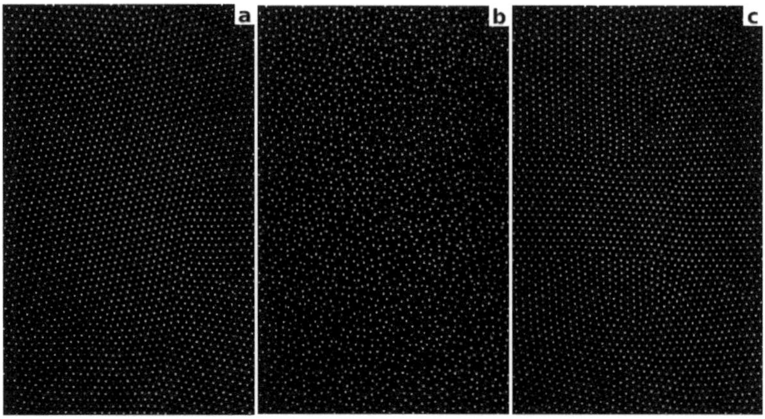

FIGURE 1.5
Melting and recrystallization in 2-dimensional complex plasmas. Original images show (a) initial crystalline state before the melting, (b) liquid state short after the melting (1.6 s after the pulse), and (c) metastable state at later stage of recrystallization (7.0 s after the pulse) (Knapek et al. 2007).

negative electric pulse was applied to the wires. The pulse caused a disturbance, which pushed the particles away from both wires to the center of the chamber leading to the melting of the plasma crystal (Figure 1.5b). After the pulse, the system starts to reorder, the particle velocities reduce significantly, and recrystallization takes place (Figure 1.5c). The kinetics of this re-crystallization was investigated in terms of the measured particle system temperature and structural properties, such as the defect fraction and translational and orientational order parameters (for detail see Section 5.3.2.

In the sheath and pre-sheath of a rf discharge, it is possible to levitate not only single layers of microparticles producing a 2-dimensional monolayer complex plasma but also a couple of layers in the vertical direction (the smaller the particle size, the larger the number of layers). This can finally end in an overall distribution of particles in the bulk of the plasma for sizes much below 1 μm, when gravity is no longer the dominant force. Depending on plasma and particle parameters complex plasma can form crystalline, liquid or gaseous (disordered) states. Usually, the crystal-like (highly ordered) state is observed at high pressures, when plasma instabilities and different mechanisms of the particle component heating are strongly suppressed.

First 3-dimensional observations of crystalline complex plasmas in a rf discharge, the so-called plasma crystals, were published by Pieper *et al.* (1996). The authors used an experimental set-up allowing a top view of the complex plasma in a plane parallel to the lower electrode. By scanning the laser and optics in the vertical direction a 3-dimensional tomographic view of the full system was possible. In Figure 1.6 scans from two different crystalline structures are shown, body center cubic (bcc) and

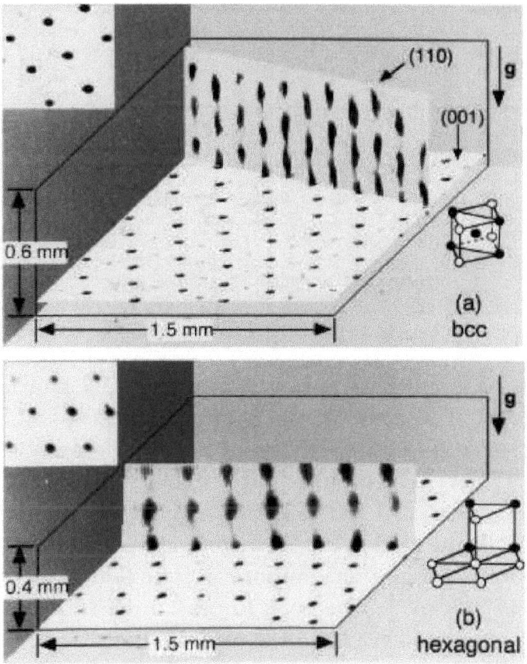

FIGURE 1.6
The two graphs show a 3-dimensional view of the crystalline structure of a multi-layer plasma crystal (Pieper *et al.* 1996). The structure is body centered cubic (bcc) in (a) and hexagonal in (b).

hexagonal aligned. Later, particle positions were determined from such scans (Zuzic *et al.* 2000). Using the 3D coordinates measured in this way, it was found that the horizontal interparticle distance Δ_{xy} decreases by $\simeq 40\%$ from top to bottom. This was attributed to the compression by the gravitational force. The particles are clearly forming layers. The distribution of particles in each layer is rather narrow, so that neighboring planes can be easily distinguished. A local order analysis shows that in most regions of the crystal, the metastable hcp structure dominates over the ground state fcc, which is a hint that the crystallization process is not fully finished at that time.

Usually, gravity is compensated by the electric field in the sheath above the lower electrode. Another way to compensate for gravity is to use the thermophoretic force (see Sec. 2.5.2). This force builds up when a temperature gradient is applied, and it acts in the direction of lower temperature. The thermophoretic force allows the formation of quasi-zero-*g* condition in a complex plasma, hence making possible the investigation the behavior of microparticles in the plasma bulk over a broad range of parameters. Figure 1.7 shows the effect of the thermophoretic force on a cloud of

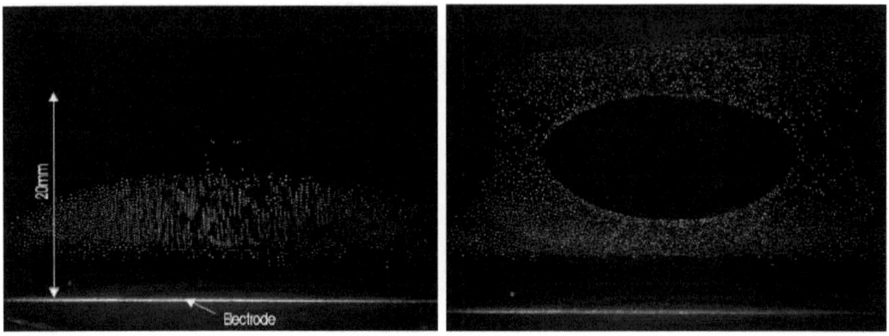

FIGURE 1.7
Left: Side view of a complex plasma formed by 3.4 μm microparticles suspended in a plasma, when no external temperature gradient is present. The gas is argon, the pressure is 48 Pa, the rf amplitude is 45 V peak to peak, and the discharge power is 17 mW. The number of particles is about 10^6, the field of view is 32×43 mm^2. The lowest five lattice planes of large interparticle distance consist of agglomerates. Right: Side view of a complex plasma at quasi-microgravity conditions accomplished by thermophoresis. The temperature gradient applied from outside is 1170 K m^{-1}. The peak to peak amplitude is 82 V, the discharge power is 57 mW, the pressure is 46 Pa, and the number of particles injected is on the order of 10^6. Both figures are from Rothermel *et al.* (2002).

particles studied by Rothermel *et al.* (2002). Left/right image shows the distribution of the cloud without/with thermophoretic force.

Another laboratory development is the use of the thermophoretic force in combination with a special confining system on the lower electrode to produce Coulomb or Yukawa balls – 3-dimensional clusters of a small number of microparticles forming spherical structures. By positioning a quadratic glass box of a certain hight on the lower rf electrode and adjusting the thermophoretic force acting on the microparticles levitating inside this box by heating the lower electrode, it is possible to form a potential trap for the particles which allows the formation of such a cluster (Arp et al. 2005). A schematic of the apparatus is shown in Figure 1.8. This sketch includes the forces acting on the particles and the original particle cloud distribution for different temperature gradients. With such a system many interesting static and dynamic analysis can be performed with a model system of a nano-cluster. For more details see Sec. 1.4.5.

1.1.2 Symmetrically driven rf discharge for microgravity experiments

Microparticles are strongly affected by gravity, contrary to the other plasma particles, the electrons and ions (Thomas et al. 1994). On the ground, a dc electric field is usually employed to compensate gravity in a small region of the plasma chamber, in the sheath. This allows measurements of the complex plasma under multi-layer

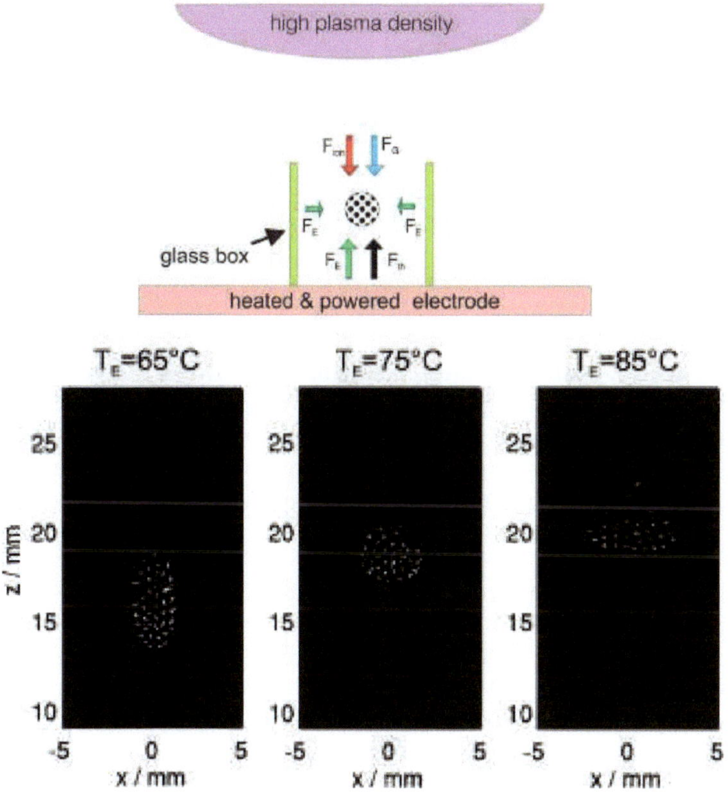

FIGURE 1.8
Top: The sketch shows the assembly for the formation of Coulomb/Yukawa clusters by using thermophoretic force and a confining glass box. The forces acting on the microparticles are shown, too. Bottom: Distribution of the microparticle cloud at different temperature gradients (Arp *et al.* 2005).

(or 2 1/2-dimensional) conditions. To perform certain precision measurements, especially of large 3-dimensional isotropic systems, microgravity conditions are necessary. Such experiments allow the study of systems and processes not attainable on the ground (Morfill *et al.* 1999; Ivlev *et al.* 2003b; Thomas *et al.* 2005).

Under microgravity conditions the complex plasma is forced out of the areas of strong electric fields, the sheath regions, and forms a homogeneous and isotropic system in the bulk of the discharge. As mentioned earlier, the symmetry of the electrode system has major influences on the symmetry of the complex plasma.

Figure 1.9 (left) shows that under gravity conditions only small complex plasma systems – of limited extent in the vertical direction – can be investigated (in a region

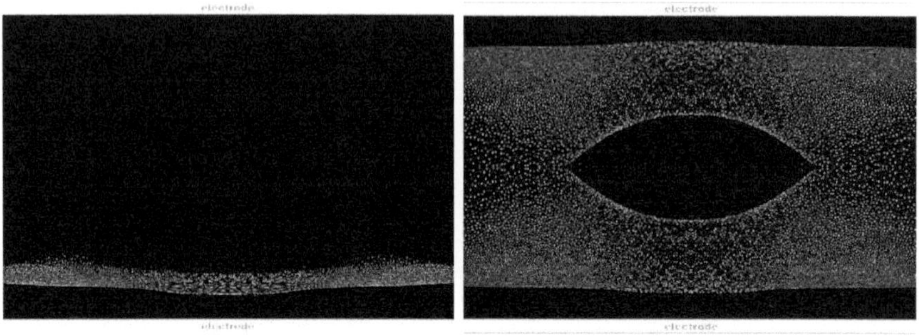

FIGURE 1.9
Microparticle (3.4 μm in diameter) distribution between the two electrodes under gravity (left) and microgravity conditions (right). Under gravity the charged particles sediment towards the lower electrode and can be levitated only by a strong electric field in the sheath. Under microgravity the particles are dispersed all over the experimental volume, forming large 3D complex plasmas.

where gravity is compensated by a strong electric field). Under microgravity we observe large complex plasma systems extended in all three space co-ordinates, as shown in Figure 1.9 (right). This is because stresses in the particle cloud under microgravity are relatively weak, which provides a much broader range of the complex plasma parameters available for investigations.

A well-balanced symmetrically driven rf electrode system provides a homogeneous distribution of the plasma with identical sheaths on both electrodes. This is a mandatory condition for a homogeneous distribution of the microparticles under microgravity conditions. A cross-sectional and perspective schematic of the newest microgravity setup, the PK-3 Plus chamber, is shown in Figure 1.10 (Thomas *et al.* 2008). The vacuum chamber consists of a glass cuvette of form of a cuboid with a quadratic cross section. Top and bottom flanges are metal plates. They include the rf electrodes, electrical feedthroughs and vacuum connections. The electrodes are circular plates of aluminum with a diameter of 6 cm. The distance between electrodes is 3 cm. The electrodes are surrounded by a 1.5 cm wide ground shield including three microparticle dispensers on each side. The dispensers are magnetically driven pistons with a storage volume at their ends. The storage volumes are filled with microparticles and covered with a sieve with an adapted mesh size. The microparticles are dispersed through the sieve into the plasma chamber by electromagnetically driven strokes of the piston.

The thermal concept of a microgravity setup is very important, because for micrometer particles a temperature gradient of only 1 $\mathrm{K\,cm^{-1}}$ would give rise to a thermophoretic force of $\approx 10^{-1}$ g, which obviously would destroy the microgravity conditions. Thus, the electrodes are thermally coupled to the metal flanges. An insulator ring of high thermal conductivity prevents a temperature gradient between the electrodes and the structure. The metallic parts of the chamber, the electrodes, flanges and connectors between the upper and lower flange are manufactured from

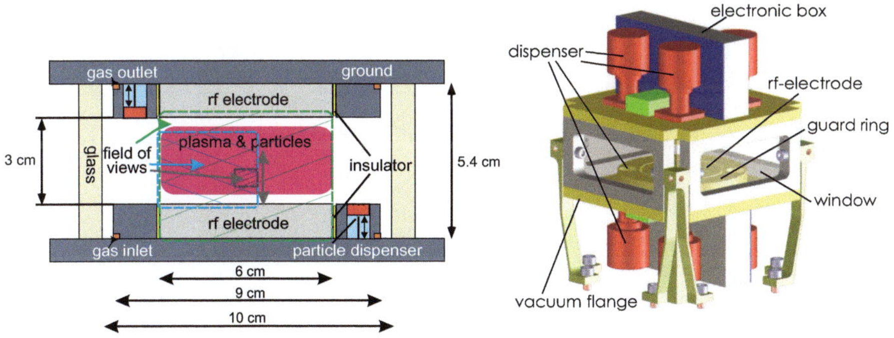

FIGURE 1.10
(See color insert following page 242). The sketches show the 2D (left) and 3D view of the plasma chamber (right) (Thomas *et al.* 2008).

highly conducting aluminum to additionally prevent a temperature drop between the upper and lower part of the chamber.

The optical particle detection system consists typically of laser diodes with cylindrical optics to produce a laser sheet perpendicular to the electrode surface and video cameras observing the reflected light at 90° with different resolutions [see Figure 1.10 (left) for typical field of views]. In the optical path interference filters at the laser wavelength are used to filter out the plasma glow. The cameras and lasers are mounted on a horizontal translation stage allowing a depth scan through the complex plasma.

With such a set-up microgravity experiments have been performed and have led to the observation of many interesting phenomena in liquid and crystalline complex plasmas. First, different kinds of crystalline structures (fcc, bcc and hcp) in a stable region of the complex plasma shown in Figure 1.11 (Nefedov *et al.* 2003a) were found. Additionally, shocks and waves (Samsonov *et al.* 2003a; Khrapak *et al.* 2003) and the boundary between the complex plasma and the microparticle-free plasma (void) (Annaratone *et al.* 2002) were studied in detail. An overview of the typical behaviour of complex plasmas under microgravity conditions is given below.

Figure 1.11 illustrates the typical static and dynamic behavior of complex plasmas under microgravity conditions. The main features clearly seen are

1. a microparticle-free "void" in the center of the system for most of the experimental parameters,
2. a sharp boundary between the void and the complex plasma,
3. demixing of complex plasma clouds formed by microparticles of different sizes,
4. vortices in different areas away from the central axis,
5. crystalline structures along the central axis.

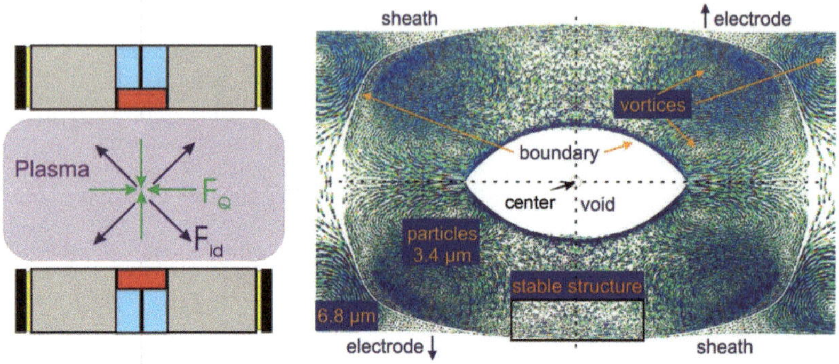

FIGURE 1.11
Left: Shown is the electrode system including the microparticle dispensers and the plasma area in between. The self-generated electric potential has a maximum in the center, decreasing radially and axially outwards. The radially symmetric forces on the particles are the electrostatic force F_e dragging the negative particles into the center and the ion-drag force F_{id} acting in the opposite direction. Right: Structure and dynamics of a complex plasma containing particles of two different sizes (3.4 μm and 6.8 μm diameter) under microgravity conditions. The trajectories of the microparticles are shown for an exposure time of 3 sec.

The void, the microparticle free region in the center of the discharge, prevents the formation of a homogeneous and isotropic distribution of the complex plasma. The origin of the void is the ion drag force which often overcomes the electric force in some vicinity of the discharge center and therefore pushes the particles out of the central region (see sketch in Figure 1.11, left). Under certain conditions, the void can be closed. This is very important for many dedicated experiments which are planned for the future. With the PK-3 Plus setup there exist three ways of void closure presently known: First by adjusting lowest rf power [as in the former ISS-laboratory PKE-Nefedov (Lipaev *et al.* 2007)]; second, by using a symmetrical gas flow; and third, by low frequency electric excitation (Thomas *et al.* 2008). The last can be used additionally to initiate a phase transition from an isotropic fluid into a so-called electrorheological string fluid, shown in Figure 1.12 (Ivlev *et al.* 2008).

The formation of such string fluids, or general electrorheological plasmas, is possible due to the manipulation of the interaction potential between the microparticles along the field line. It can be changed from an isotropic screened Coulomb to an asymmetric attractive potential through accelerating ions by the ac voltages at frequencies above the dust plasma frequency applied to the electrodes. Thus, the ions produce a wake region above and below the particles along the electric field axis, while the particles cannot respond. For future microgravity experiments, it is foreseen to manipulate the interaction potential in all three directions, to receive a fully isotropic attractive potential.

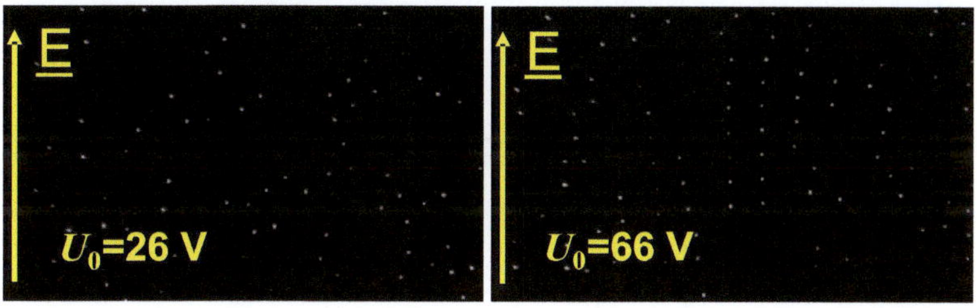

FIGURE 1.12
The figure shows the transition from an isotropic fluid (left) to a string fluid (right) observed in the PK-3 Plus laboratory onboard the ISS.

1.1.3 Complex plasmas in inductively coupled discharges

For completeness it should be mentioned that a plasma can be excited by an electric field generated from an rf current in a conductor, too. The changing magnetic field of this conductor induces an electric field in which the plasma electrons are accelerated. Two types of such inductively coupled discharges are used to study complex plasmas. The first one is a so-called circular inductor in the form of several rings around a glass tube similar to the dc discharge tubes (Fortov *et al.* 2000c) discussed below.

Compared to capacitively coupled discharges, it forms no constant electric field which allows the levitation of the microparticles. A potential trap for the particles is formed due to the ambipolar electric field. Such a trap exists at the lower end of the plasma, shown in Figure 1.13. The shape of the cloud which is trapped in this region depends on plasma conditions. Typical structures observed in experiments are shown in Figure 1.14.

The second inductively coupled device uses a flat inductor in addition to a usual rf electrode for generating the plasma. There exists a modification to the GEC-RF-Reference Cell, which employs such an inductor (Collins *et al.* 1996). The top electrode is replaced here with a stainless steel cylindrical fixture into which a 100-mm-diameter quartz window is installed. This window is parallel to the bottom electrode forming a 3.5 cm gap between the window and the surface of the bottom electrode. A copper coil is located on the atmospheric side of the window and is connected to a 13.56 MHz rf generator through a matching network. Another 13.56 MHz rf generator powers the bottom electrode.

Figure 1.15 shows different structures formed by the particles in the Inductively Coupled Plasma (ICP) discharge for different operating modes. Low power input to both electrodes provides a low plasma density and the particles levitate above the entire lower electrode. The shape of the cloud follows the contour of the disk placed on the lower electrode (Figure 1.15a). A high density mode can be reached by an increase of the power applied to the flat inductor. In this mode the trapping of the particles occurs in a narrow region just above the disk (Figure 1.15b).

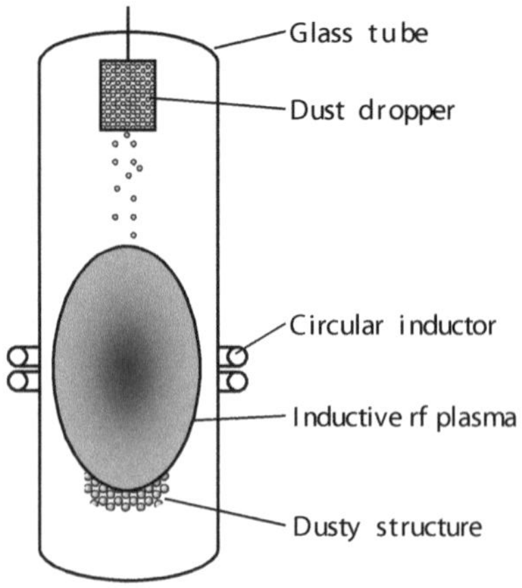

FIGURE 1.13
Schematic of an inductively coupled discharge consisting of a glass tube and a multi-ring inductor, which is mounted outside. The particle cloud can be levitated at the lower end of the plasma (Fortov *et al.* 2000c).

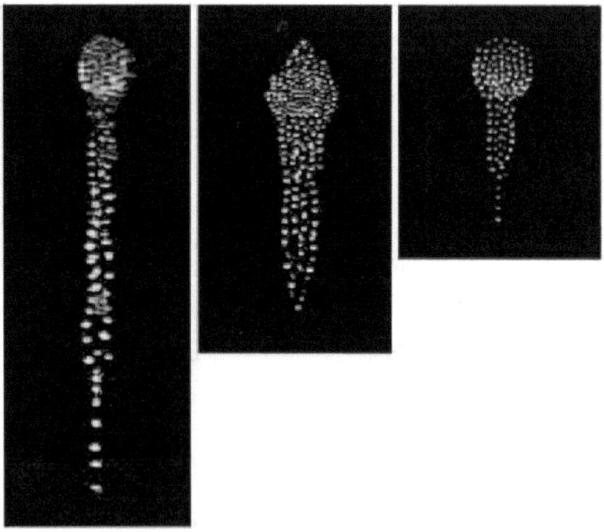

FIGURE 1.14
Typical shapes and structures of the particle cloud levitating in the inductively coupled discharge sketched in Figure 1.13 (Fortov *et al.* 2000c).

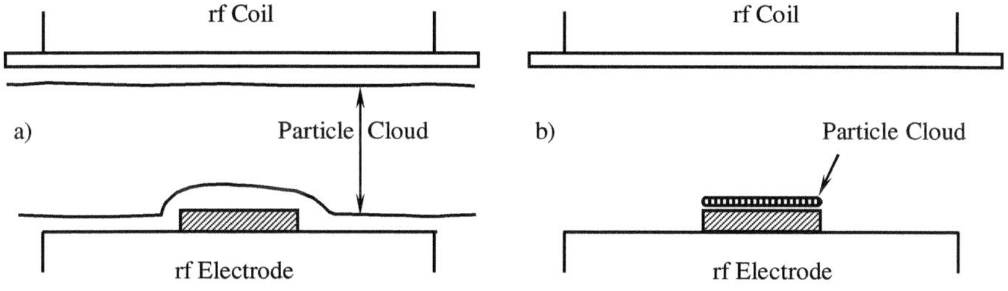

FIGURE 1.15
Sketch of the combined capacitively and inductively coupled GEC-RF-Reference Cell (Collins *et al.* 1996). The particle distribution is shown for low power (a) and high power (b).

1.2 Complex plasmas in dc discharges

1.2.1 Ground-based experiments

A dc gas discharge can also be used for the formation of ordered structures in complex plasmas (Fortov *et al.* 1996b, 1997b; Lipaev *et al.* 1997; Nefedov *et al.* 2000). The sketch of the typical experimental setup is shown in Figure 1.16. A discharge is usually generated in a vertically positioned cylindrical tube. The particles introduced into the plasma can levitate in the regions where the external forces are balanced. They are illuminated by a laser light and their positions are registered by a video camera. Typical conditions in the discharge are: A pressure in the range 0.1–5 Torr and a discharge current of $\sim$ 0.1–10 mA.

The ordered structures of particles are usually observed in standing striations of the positive column of the glow discharge, but can also be observed in an electric double layer formed in the transition region from the narrow cathode part of the positive column to the wide anode part, or in specially organized multi-electrode system having three or more electrodes at different potentials, etc. – that is, in the regions where the electric field can be strong enough to levitate the particles.

Most of the experiments were performed in standing striations of glow discharges. In the positive column of a low-pressure discharge, loss of electron energy in elastic collisions is small and the electron distribution function is formed under the action of the electric field and inelastic collisions. This can lead to the appearance of striations, that is, regions of spatial periodicity of the plasma parameters with the characteristic scale on an order of a few centimeters (Fortov *et al.* 2004a). The concentration of electrons, their energy distribution, and the electric field are highly nonuniform along the striation length. The electric field is relatively strong (around 10–15 V cm^{-1} at a maximum) at the head of striation – that is, a region occupying 25–30% of the total striation length – and relatively weak (around 1 V cm^{-1}) outside this region. The maximum value of the electron concentration is shifted relative to the maximum

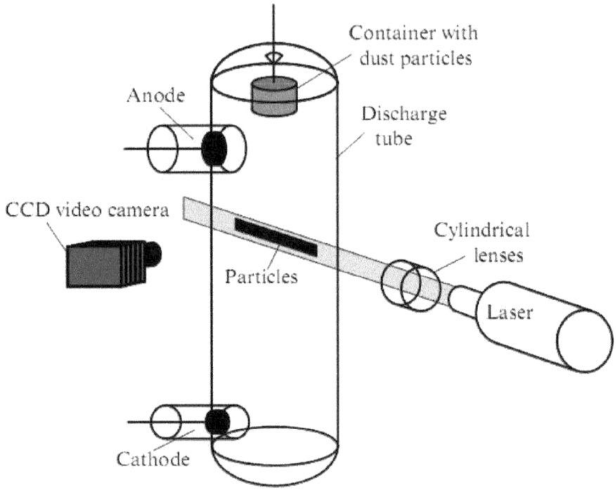

FIGURE 1.16
Schematic of the experimental setup for studying the formation of ordered structures in a dc gas discharge.

strength of the electric field in the direction of the anode. The electron energy distribution is substantially bimodal, with the head of the striation being dominated by the second maximum whose center lies near the excitation energy of neutral gas atoms. Due to the high floating potential of the discharge tube walls, the striations exhibit a substantially two-dimensional character: The center-wall potential difference at the head of the striation reaches 20–30 V. Thus, an electrostatic trap is formed at the head of each striation, which in the case of vertical orientation of the discharge tube is capable of confining particles with high enough charge and low enough mass from falling into the cathode positioned lower, while the strong radial field prevents particle sedimentation on the tube walls.

The process of structure formation proceeds routinely as follows: after being injected into the plasma of the positive column, the charged particles fall past their equilibrium position and then, over the course of several seconds, emerge and form a regular structure which is preserved sufficiently long (until the end of observation) provided that the discharge parameters are unchanged. The simultaneous existence of ordered structures in several neighboring striations can also be observed. Figure 1.17a demonstrates an image of two dust structures formed by hollow thin-walled microspheres made of borosilicate glass with a diameter of 50–63 μm in two neighboring striations. Figure 1.17b shows their coalescence into one rather extended formation, occurring due to varying the discharge parameters (Lipaev *et al.* 1997). This figure indicates the possibility of forming structures much more extended in the vertical direction than in rf discharges. In fact, the three-dimensional quasi-crystalline dusty structures were obtained for the first time by Lipaev *et al.* (1997).

Figure 1.18 shows a video image of a plasma crystal. This structure was obtained

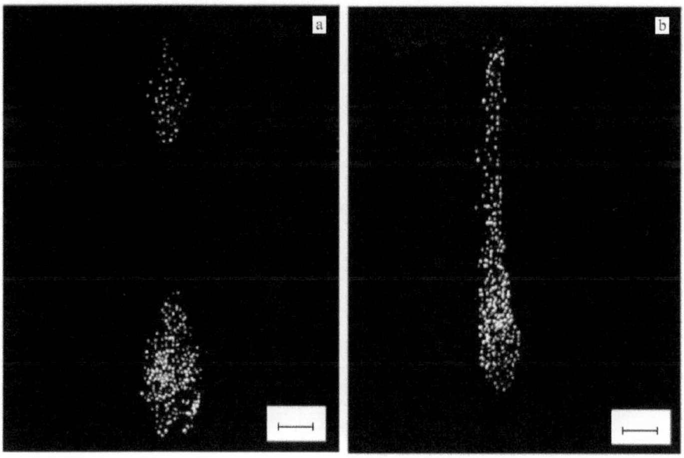

FIGURE 1.17
Video images of structures formed by charged microspheres of borosilicate glass: (a) in two neighboring striations (pressure around 0.5 Torr, discharge current of 0.5 mA); (b) after their coalescence (pressure around 0.4 Torr, discharge current of 0.4 mA). The scale length in the figures corresponds to 3 mm.

by Molotkov *et al.* (2004) in a dc discharge of neon-hydrogen mixture, where the striations are strongly flattened. The lattice constant of this crystal is of the order of 700 μm. Figure 1.19 is a video image of a plasma-dust structure of the liquid type (Fortov *et al.* 2000b). A convective motion of particles is observed in the lateral parts of the structure. The particles move upwards at the periphery and downwards at the center. Under some conditions in striations of the dc glow discharge, self-excitation of the dust acoustic waves occurs (see Figure 1.20).

In dc discharges, the transition of quasi-crystalline structures to fluid and gaseous states is also observed, similar to complex plasmas in rf discharges. This occurs either by lowering the pressure or increasing the discharge current. For example, for the structure comprising Al_2O_3 particles of diameter 3–5 μm at a pressure of 0.3 Torr and a current of 0.4 mA (estimated electron number density $n_e = 10^8$ cm^{-3} and temperature $T_e \sim 4$ eV), the pair correlation function reveals long-range order with four well-pronounced peaks (Lipaev *et al.* 1997). When the discharge current is increased by almost one order of magnitude to 3.9 mA ($n_e = 8 \times 10^8$ cm^{-3}), the structure "melts" and the pair correlation function reveals only short-range order. We note that during this "phase transition" the interparticle distance, equal to 250 μm, remains approximately constant but the ion Debye radius decreases considerably.

In order to obtain different values of the electric field and for a better stabilization of striations, Vasilyak *et al.* (2000) performed experiments in a specially designed, vertically arranged, conical discharge tube. The longitudinal electric field varied over the conical tube length exhibiting a maximum in the lower narrow part. Parti-

FIGURE 1.18
A horizontal cross-sectional view of a complex plasma crystal. The parameters are: Melamine formaldehyde particles of 1.87 μm in diameter, discharge in a neon-hydrogen mixture (1:1), pressure 0.8 Torr, current 1.1 mA, and mean interparticle spacing $\sim$ 700 μm.

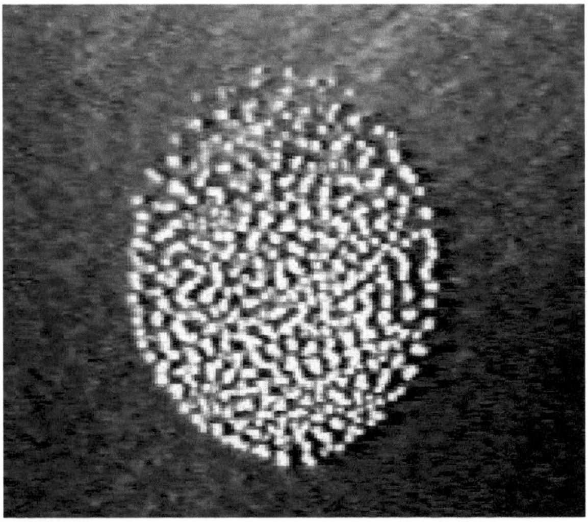

FIGURE 1.19
A vertical cross-sectional view of liquid-like structure. The parameters are: Melamine formaldehyde particles of 1.87 μm in diameter, neon pressure 0.3 torr, and current 0.6 mA.

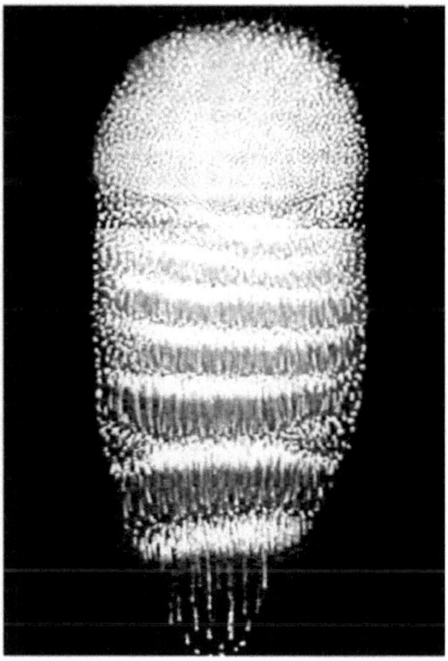

FIGURE 1.20
A self-excited dust acoustic wave in a dc gas discharge at p = 0.2 Torr neutral gas pressure (Fortov *et al.* 2000a). Wave characteristics are: Wave frequency $\omega \sim 60\ s^{-1}$, wave number $k \sim 60\ cm^{-1}$, phase velocity $v_{ph} \sim 1\ cm\ s^{-1}$.

cles of different mass were localized at different vertical positions in the discharge tube, which resulted in the separation of the grains by size, charge, and mass along the tube. It has been demonstrated by Vasilyak *et al.* (2000) that if a thin metal plate with several openings is placed in the conical tube, different complex plasma structures may be formed above the openings. Their dimensions depend on current and pressure. The structures may hang above an opening in the form of a disk, in the vicinity of an opening in the form of a ring, or even fill the entire surface above the plate except for the openings themselves (Figure 1.21). The discharge current increases from Figure 1.21a to Figure 1.21c.

Vasilyak *et al.* (2003) performed experiments in which a discharge tube 2 cm in diameter was cooled with the use of Peltier elements. The middle part of the discharge tube was cooled on both sides by two microrefrigerators 2.5 cm long, which tightly fitted the side surface of the tube. Two gaps 8 mm wide were situated between the microrefrigerators for observing dust structures. The particle structures (Figure 1.22) were concentrated in striations close to the tube axis at air pressures of 0.1–0.5 Torr and currents of 0.25–1 mA. Cooling the discharge tube walls by 20 K stretched the particle structure in the radial direction under the action of thermophoretic forces. First, the dust structure cross section transformed into an ellipse prolate toward cold walls. Next, the cloud of dust particles divided into two circles, which experienced

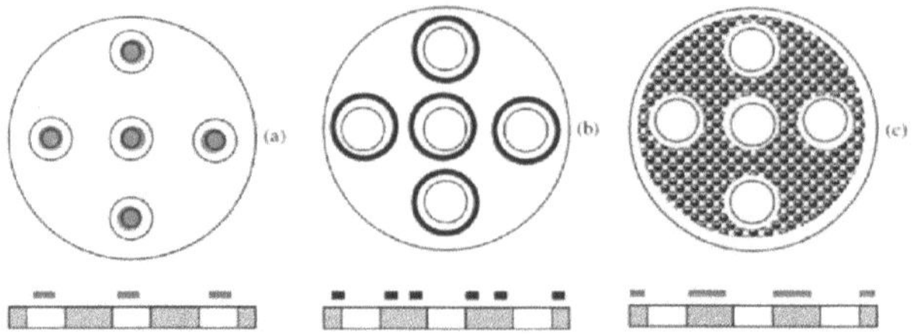

FIGURE 1.21
Dusty plasma structures above a thin plate with circular openings and evolution of these structures with an increase of a discharge current from (a) to (c) (Vasilyak *et al.* 2000).

deformations and were attracted to cold walls. Particles did not reach the walls under the action of the radial temperature gradient but were confined in a new equilibrium state, because the radial electric force directed toward the axis grew stronger as the radius increased. Dust clouds of various shapes can be created by varying the temperature field. If the walls of a discharge tube are locally cooled by microrefrigerators in the longitudinal direction, all charged dust particles are withdrawn by the longitudinal temperature gradient from the striation lying lower by 5–6 cm. They in part go to cold tube walls in the region of refrigerators. Particles remain in the striations that are situated above the refrigerators. Longitudinal thermophoretic forces are strong enough to draw dust particles upward from electrostatic traps, that is, from striations. As a consequence, a new trap arises. This trap is formed as a result of the superposition of thermophoretic and longitudinal electric field forces. Creating various thermal traps with the use of Peltier elements allows the shape of dust structures to be varied or even allows dust structures to be removed from the discharge region.

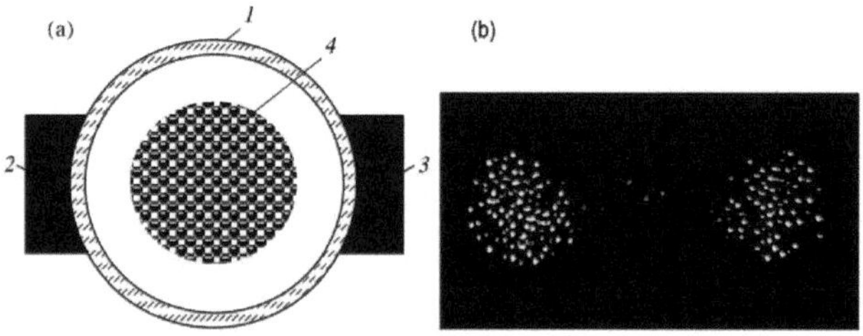

FIGURE 1.22
Particle structures obtained by cooling the two sides of the discharge tube: (a) Arrangement of microrefrigerators (the horizontal discharge tube cross section is shown; 1 – discharge tube, 2 and 3 – microrefrigerators, and 4 – dust structure); (b) Separation of the particle structure into two clouds (Vasilyak *et al.* 2003).

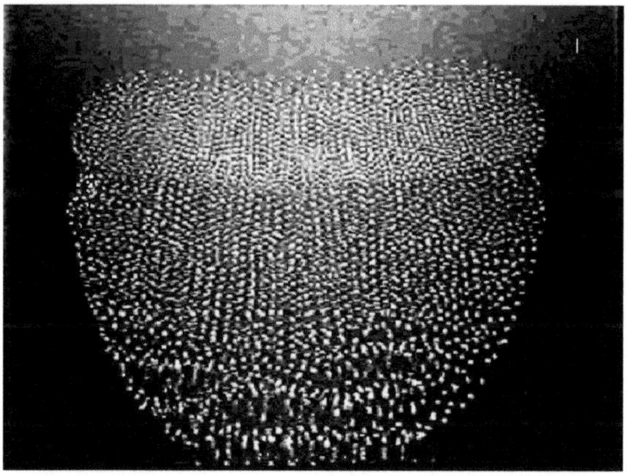

FIGURE 1.23
Vertical cross section of the ordered particle structure in a stratified dc discharge. In the lower region of the structure, the vertical oscillations of the particle density are present, in the central region high ordering appears, and in the periphery of the upper region the particles exhibit convective motion.

Termophoretic forces can be used for changing particle distribution in the horizontal cross section of relatively highly ordered structures in a dc glow discharge. Presuming that the boundary geometry can affect particle distribution, Karasev *et al.* (2008) used cylindrical coolers of diameter 2 cm and height 6 cm, held at 273 K and placed against a striation containing a particle structure, to change the geometry of its outer boundary. By varying the number of coolers, their positions, and their separations from the tube wall, azimuthally asymmetric thermophoretic forces can be used to form polygonal boundaries and vary the angles between their segments (in a horizontal cross section).

Under certain discharge conditions, an increase in the number of particles gives rise to complex structures where different regions coexist (see Figure 1.23): the high ordering region ("plasma crystals") and regions of convective and oscillatory motion of particles ("plasma liquids"). Usually, in the lower part of the structure, the particles oscillate in the vertical direction (dust density waves) at a frequency of 25–30 Hz and a wavelength of about 1 mm, the mean interparticle distance being 200 μm. Such self-excited oscillations can correspond to instabilities of the dust waves. Most of the central region of the structure is occupied by a crystal-like structure with a pronounced chain-like configuration. At the periphery of the upper part of the structure, the particles undergo convective motion whose intensity decreases towards the center of the structure. This complex picture is apparently associated with a peculiar distribution of plasma parameters and forces acting on the dust particles within a striation.

DC discharges give a possibility to observe different wave phenomena in complex plasmas. Using special excitation techniques one can excite waves. Interesting re-

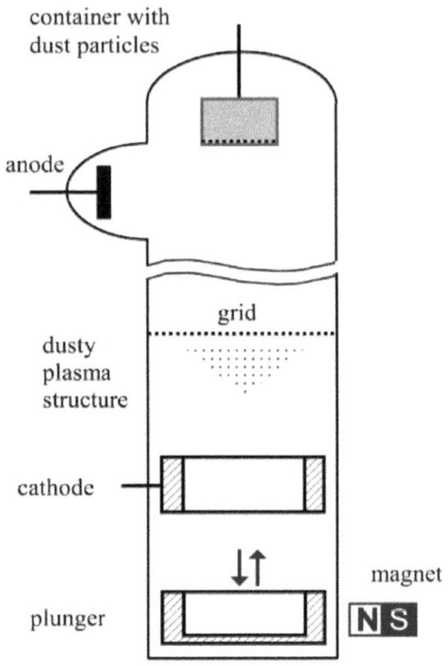

FIGURE 1.24
Scheme of the experimental setup with the plunger.

sults have been achieved with the help of a gas–dynamic impact (Fortov *et al.* 2004b; Molotkov *et al.* 2004; Pustyl'nik *et al.* 2004).

In the experimental setup similar to that presented in Figure 1.16, a plunger was set below the cathode to excite the waves (Figure 1.24). The plunger was a hollow thin-walled nickel cylinder of 26 mm diameter and height with the bottom made of a polymeric pellicle. It was put at the bottom of the tube (15 cm separated from the cathode) and moved with the help of a permanent magnet manually approached to the plunger from outside the tube. In this way the plunger could be moved upward and downward with a speed of about 30–40 cm s^{-1} and 4–5 cm space, creating a gas flow with the duration of about 0.1 s, which displaced the dust grains with respect to the striation. In some experiments a grid was inserted into the tube 7 cm above the upper cut of the cathode. The grid was kept under the floating potential.

In the experiments without the grid, when the plunger was moved downward, the dusty plasma structure was first slowly moving downward. During the time of 33 ms it was displaced for about 500 μm. The structure became unstable at this position. It moved rapidly back in the upward direction and several dust compressions appeared propagating downward. At the end of the process the waves were damped and the structure as a whole returnsed to its initial position. The characteristic wavelength and frequency were 1.3 mm and 14 Hz, respectively. These waves are very similar to the self-excited waves obtained by Fortov *et al.* (2000a). However they are different since the self-excited waves are usually observed in the lower parts of the structure in the area of higher electric field, whereas here they were formed in the upper part

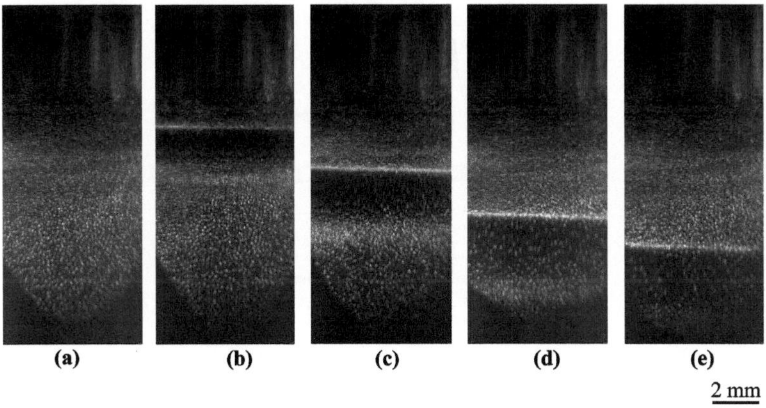

FIGURE 1.25
Sequence of videoframes presenting the evolution of the disturbance in the experiments with the grid, (a) the initial structure, time interval between each frame (b)–(e) is 83 ms.

of the structure.

It should be noted that the magnet which is used to transmit the motion to the plunger influences the cathode sheath of the discharge. Because of this when the plunger is moved downward, the striation performs an opposite motion. Consequently, the displacement of the dust grains with respect to the distributions of the plasma parameters is of the order of 2.5 mm. No changes in the shape of the particle structure are observed when the magnet is being moved. The estimated value of the magnetic field created by the magnet in the striation region is on the order of 0.1 G. This is a rather weak magnetic field, which cannot produce any significant disturbance in the plasma. It was experimentally verified that the influence of only the magnet is not enough to produce the waves.

In the experiments with the grid, a solitary wave was obtained. The pressure was again adjusted to 0.3 Torr and the discharge current was chosen in such a way that the lowest striation was formed exactly below the grid. This occurred at the current of 0.1 mA. The dusty plasma structure was very close to the grid (at a distance of 4 mm). After moving the plunger downward the structure was again for some time streaming downward, then it stopped and began moving towards its initial equilibrium position, and when it returned to the stable position, a disturbance propagating through it appeared (Figure 1.25). The disturbance observed is nearly plane, and therefore, it can be treated in terms of only one spatial variable. Figure 1.26 presents the shape of the compression factor ξ, which is the ratio of the distribution of brightness in the wave to the distribution of brightness in the initial structure, at different moments of time. The disturbance consists of three parts: A is the first compression, B is the rarefaction and C is the second compression. It is seen that the amplitudes for zones A, B and C reach the values of 2.0, 0.65, and 1.2, respectively.

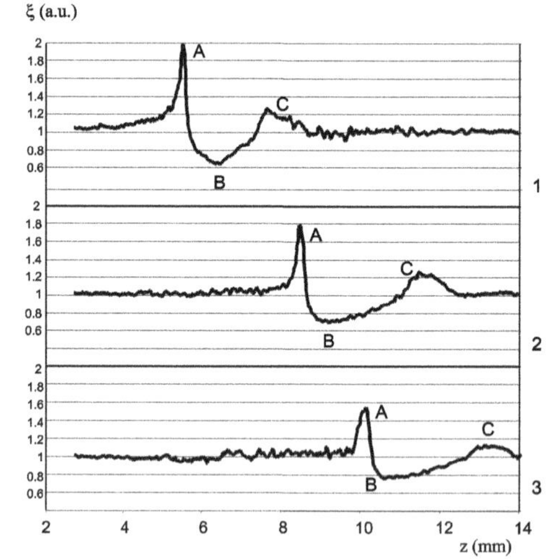

FIGURE 1.26
Shapes of the compressional factor in the wave at different moments of time. *A*, *B*, *C* denote three different structures of the wave: first compression, rarefaction, and the second compression, respectively. Time interval between curves 1 and 2 is 120 ms, between curves 2 and 3 is 60 ms.

Another method to produce waves in complex plasmas is to apply an impulse of axial magnetic field (Fortov *et al.* 2005c). To apply the electromagnetic impulse the scheme shown in Figure 1.27a was used. In this setup 16 loops of copper wire all in one horizontal plane were coiled around the discharge tube (impulse coil). A battery of high-voltage capacitors was charged up to 1.2 kV and then discharged onto the coil. The schematics of the impulse profile and its typical parameters are shown in Figure 1.27b. The current impulse amplitude is on the order of 90 A. The corresponding amplitude of the magnetic field inside the coil is estimated to be 150 G.

The impulse applied affects the striation only and produces no influence on the dust particles because of their high inertia. The striation could be observed by the videocamera as a bright background glow. The initial structure is shown in Figure 1.28a. When the impulse circuit is closed, the striation rapidly, i.e., for a time less than the frame duration, moves upward towards the anode (glow disappears in Figure 1.28b). As the striation moves away, the particles lose equilibrium and start falling down under the effect of gravity (Figure 1.28c). Then, as the current is decreasing the striation is moving backwards. The returning striation drags the particles upward (Figure 1.28d,e). Lower particles have more time to fall down than the upper ones. This leads to the "stretching" of the particle structure (compare Figure 1.28b and Figure 1.28f). In the next stage the pronounced division of the structure into two parts with different particle velocities and densities is observed (Figure 1.29). The

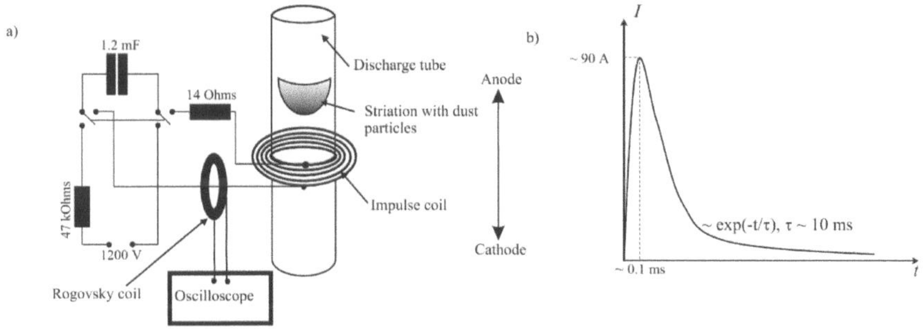

FIGURE 1.27
Scheme of the electromagnetic impulse experiments (a). The impulse shape, registered by the Rogovsky coil is given schematically in (b).

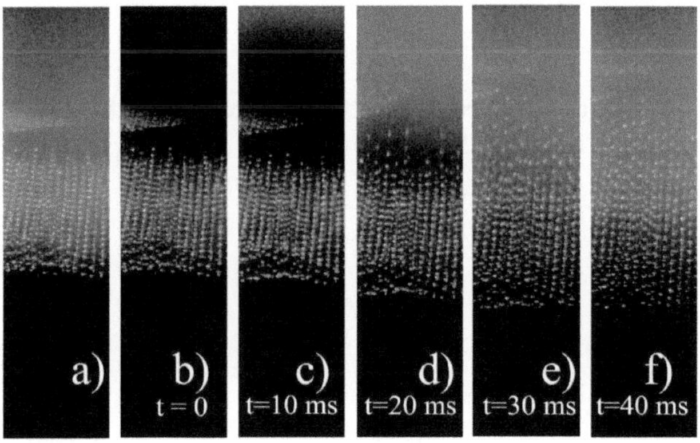

FIGURE 1.28
Sequence of videoimages, presenting the behavior of the particle cloud and the striation under the influence of the magnetic impulse. Here (a) corresponds to the initial structure at t = 0, the moment of the electromagnetic impulse launch. The images are adjusted so that the striation is seen as a bright background glow. Area shown in each image is 5.0×20.5 mm^2.

steepening of the perturbation is obvious in Figure 1.30, where the shapes of the dust density versus the vertical coordinate corresponding to the images in Figure 1.29, are presented.

The modified dc discharge facility (Thompson *et al.* 1997; Merlino *et al.* 1998) was used for experimental investigations of dust–acoustic waves. The facility is shown schematically in Figure 1.31a. A discharge was formed in nitrogen gas at a pressure $p \sim 100$ mTorr by applying a potential to an anode disk 3 cm in diameter located in the center of the discharge chamber. A longitudinal magnetic field of about 100 G provided radial plasma confinement. If the discharge current was

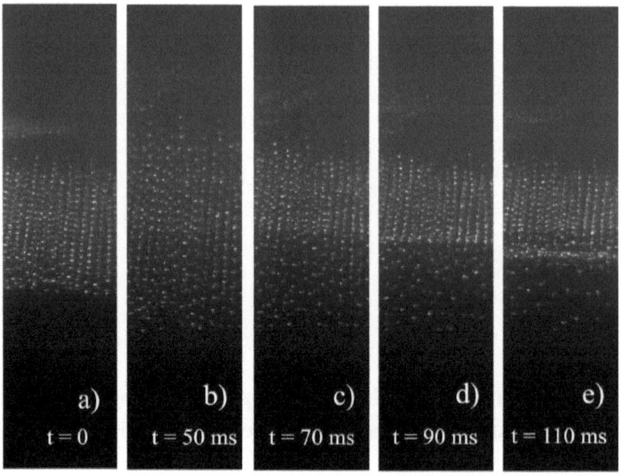

FIGURE 1.29
Sequence of the images, presenting the formation of the discontinuity front. Area shown in each image is 5.4×20.5 mm^2.

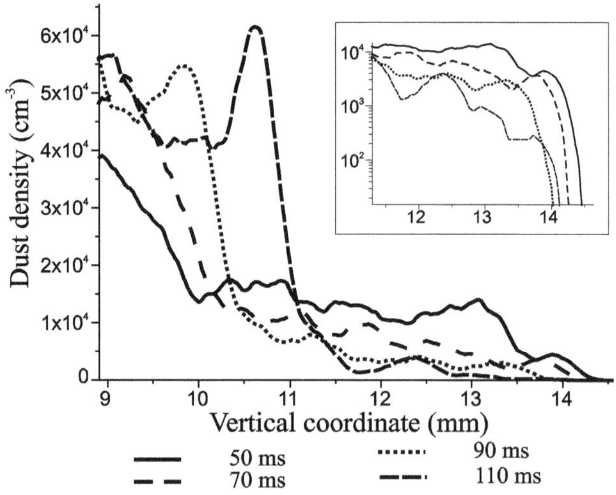

FIGURE 1.30
Spatial profiles of the particle density near the front of the perturbation at different moments of time corresponding to Figure 1.29. The curves indicate pronouced steepening of the perturbation. Low density parts of the perturbation are shown separately in the inset.

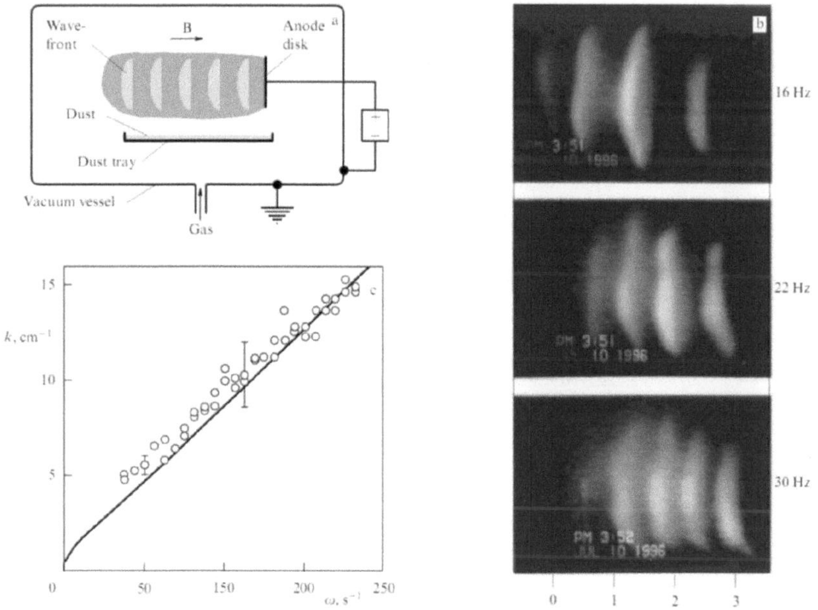

FIGURE 1.31
(a) Schematic diagram of the experimental facility used to investigate dust–acoustic waves in a dc gas discharge (Merlino *et al.* 1998). (b) Video images of typical wave structures in dusty plasma for three different values of the external excitation frequency (shown to the right of the images) (Thompson *et al.* 1997). (c) Experimentally obtained by Merlino *et al.* (1998) dust–acoustic wave dispersion relation (circles). The solid curve is computed from the theoretical dispersion relation which accounts for dust–neutral collisions.

sufficiently high (> 1 mA), dust–acoustic waves appeared spontaneously, similar to earlier experiments (Barkan *et al.* 1995). To investigate the properties of the waves in more detail, a low-frequency sinusoidal modulation with frequencies in the range of 6–30 Hz was applied to the anode. An example of the observed waves is shown in Figure 1.31b for three different excitation frequencies. Assuming that the wave frequency is determined by the external excitation frequency, the dispersion relation – the $k(\omega)$ dependence in this case – can be obtained by measuring the wavelength. The results are shown in Figure 1.31c. The observed waves exhibit linear dispersion $\omega \propto k$ and propagate with a velocity $v_{ph} \sim 12$ cm s^{-1}. From this it can be concluded that the observed waves correspond to the dust–acoustic waves.

Another way of producing complex plasma structures in a dc discharge has been developed by Uchida *et al.* (2000). A schematic diagram of the experimental apparatus is shown in Figure 1.32. A dc argon discharge plasma is produced at 220 mTorr by applying a negative dc potential of about 300 V to an upper electrode (cathode)

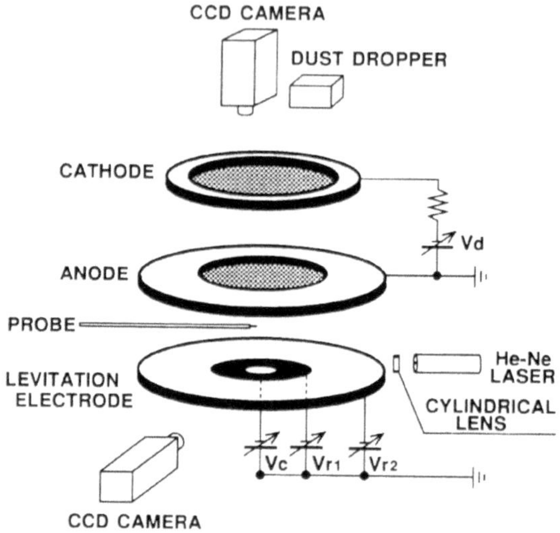

FIGURE 1.32
A sketch of the experimental apparatus used by Uchida *et al.* (2000).

with respect to a middle grounded mesh anode. The plasma diffuses downward through the mesh anode. The typical electron density and electron temperature of the diffused plasma are 10^8 cm^{-3} and 1 eV, respectively. Two centimeters below the mesh anode, a segmented particle levitation electrode is set up, consisting of three electrodes. At the center of the levitation electrode there is a disc of 0.5 cm in diameter, and two ring electrodes are set up around it. One is the ring electrode with inner and outer diameters of 0.5 cm and 1.5 cm, respectively, and the other with 1.5 and 19 cm, respectively. Different dc potentials can be applied to the three electrodes independently in order to control the potential profile in the particle levitation region. The dust particles are injected from a sieve into the glow region of the plasma through the mesh cathode and anode.

A quite interesting modification of the previous setup was suggested by Sato *et al.* (2000). It is shown in Figure 1.33 (left panel). The levitation (or confinement) electrode was made as a fine grid with 300 mesh/inch, below which an auxiliary dc discharge plasma was produced in addition to the main plasma. This additional plasma is used for supplying a low-energy electron beam (an electron shower) on the particles levitating above the levitation electrode which is provided by biasing the auxiliary plasma negatively for electrons in this plasma to pass through the levitation electrode. Increasing the auxiliary discharge current, it is possible to obtain a gradual phase transition of the particle structure from the solid (ordered lattice) to liquid state. Figure 1.33 (right panel) presents the top view of the solid-like structure obtained under the action of the low-energy electron beam leading to an increase of the negative charge on the grains. This double-plasma method is useful for levitation

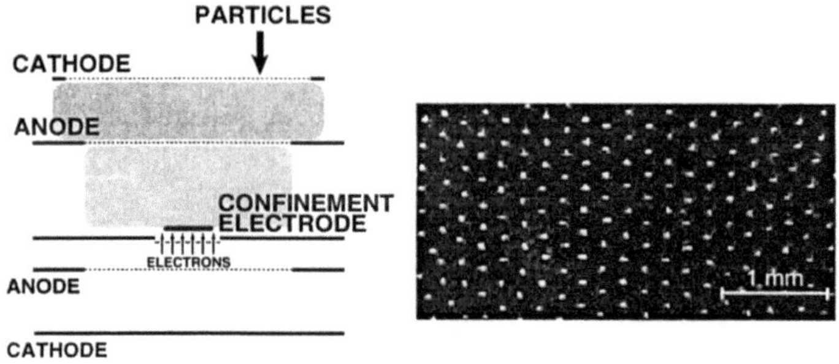

FIGURE 1.33
Left: Setup supplying electron shower on particles. Right: Top view of crystal-like structure of particles at an auxiliary dc discharge current equal to 1.2 mA (Sato *et al.* 2000).

of rather heavy dust particles (for example, Sato *et al.* (2000) observed a levitation of glass balloons of 50 μm in diameter).

1.2.2 Microgravity experiments

The results of investigations of complex plasma performed in a dc discharge in the laboratory ground-based experiments (Fortov *et al.* 1996b; Lipaev *et al.* 1997) were used as a base for the first experiments to study dynamics of microparticles under microgravity conditions. These experiments were performed onboard the Mir station (Nefedov *et al.* 2002, 2003b; Fortov *et al.* 2003). Note that in dc discharges some attention should be paid to compensate the action of the electric force in the dc discharge positive column and to eliminate a drift of the negatively charged microparticles in the anode direction. For this purpose the additional special electrode under the floating potential was used in the experiments.

The experimental setup is schematically shown in Figure 1.34. The main element of the working chamber was a gas discharge tube of radius $R = 1.6$ cm filled with neon to a pressure of $p = 1$ Torr. The distance between a plane anode and a cathode was 28 cm. An insulated electrode was mounted at 4.5 cm from the anode. The electrode was made as two steel grids (wire 60 μm in diameter) with 150 μm $\times$ 150 μm meshes, and the distance between the grids was 1 cm. During experiments, the electrode was under a floating potential and prevented negatively charged macroparticles from escaping to the anode. Bronze spherical particles (diameter of 70–180 μm, mean radius $a = 62.5$ μm, density of the material $\rho = 8.2$ g cm^{-3}) were placed between the grid electrode and the cathode. The diagnostics of macroparticles were performed with the use of a planar laser beam ("laser sheet" about 300 μm wide, wavelength 0.67 μm) and additional illumination of the dust cloud by an incandescent lamp. In the latter case, the number of detected particles was determined by the depth of view of the video system, which allowed tracking the particle positions for

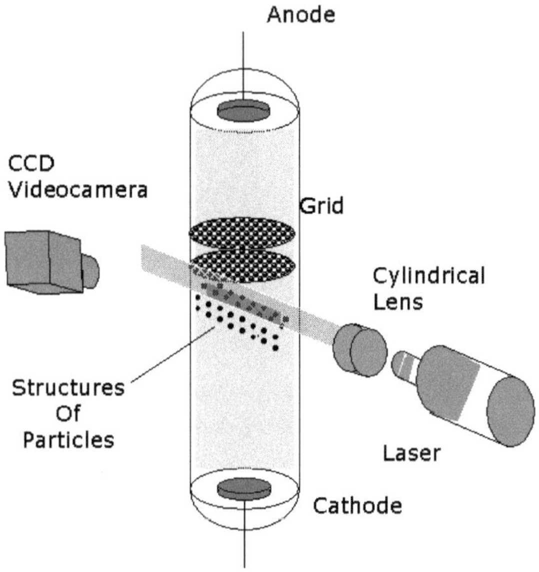

FIGURE 1.34
Sketch of the experimental setup in experiments onboard the Mir station.

long enough time to analyze their transport characteristics.

Experiments were performed at different discharge currents between 0.1 and 0.8 mA varied by the current source. Bronze particles were initially situated on the tube walls. For this reason, the system was subjected to a dynamic action (pushed) after switching on a discharge to shake off particles from the tube walls. After the dynamic action, bronze particles moved toward the grid electrodes, in the vicinity of which ordered structures were formed. The discharge was then "quenched", the particles relaxed to the initial state (returned to the tube walls), and the experiment was repeated at a new discharge current value.

The electric field of the grids provided axial confinement of the particle cloud, while the radial confinement was due to the radial electric field increasing sharply towards the tube walls. The basic plasma parameters were the following: Ne gas at a pressure $p \approx 130$ Pa, the electron temperature $T_e \approx 3$–7 eV, and the plasma number density in the range from 10^8 to 3×10^8 cm^{-3}. With increasing the discharge current from 0.1 to 0.8 mA, the interparticle separation Δ decreased from 1000 to 700 μm.

The measurements of the mean drift velocity of particles from the positive column to the grid electrode were used by Nefedov *et al.* (2003c) to estimate their electric charges. The average charge was found to be $|Z| \sim 1.5 \times 10^6$, practically independent of the discharge current.

The experiments performed in the dc glow discharge onboard the Mir space station confirmed that dc discharges could give new possibilities in studying complex plasmas under microgravity conditions. As a continuation of this direction of research

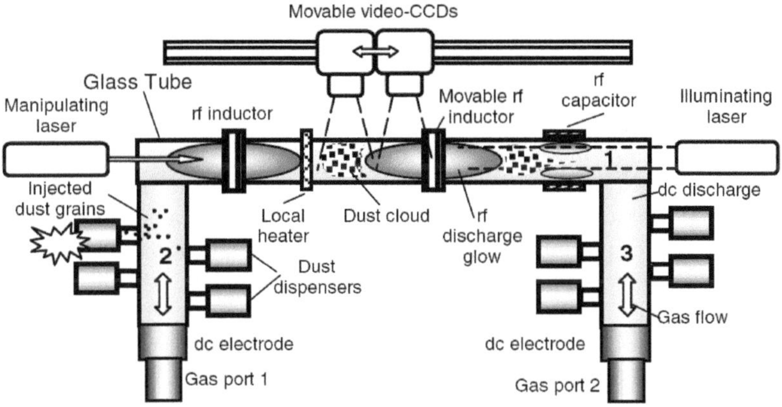

FIGURE 1.35
Scheme of the PK-4 setup: 1 – chamber area for complex plasma experiments; 2 and 3 – service parts.

activities, efforts have been made to develop quite a new experimental setup based on the combination of dc and rf discharges. The corresponding setup – called Plasma Crystal-4 facility (or PK-4 setup) – is planned for operation onboard ISS (Fortov *et al.* 2005b).

A scheme of the PK-4 experiment is presented in Figure 1.35. The gas-discharge chamber represents a 40 cm length Π-shape glass tube with an inner diameter of 3 cm. Such a shape of the chamber was determined by a number of technical and scientific requirements, in particular by the restricted size of space setups. The PK-4 chamber consists of an experimental part 1 and two identical service parts 2 and 3. All scientific experiments are performed in part 1. The service parts contain vacuum input/output ports to fill/pump a gas for the gas discharge. These ports contain the dc electrodes, which are isolated from the gas inlet/pump system via a Schottky locker. The service parts contain a set of dust dispensers (up to eight) to be used to inject different kinds of particles into the discharge chamber. In most cases, monodisperse spherical microparticles of size between 0.1 and 15 μm are used. The PK-4 setup is now equipped with four dust dispensers, and different combinations of dust particles can be used. In principle, nanoparticles synthesized directly in the discharge chamber as well as rod-like particles can also be used in future experiments. The injected particles are transported to the experimental part 1 with the help of the electric field of the dc discharge or by the gas flow. Here they are illuminated by a laser sheet (17 $\times$ 0.15 mm^2) and observed by CCD video cameras at 120 frames per second. The recorded frames are stored in a PC in digital form. The discharge chamber can be cleaned from dust particles by strong gas pulses.

The experimental part 1 is equipped with rf inductors and rf capacitors for excitation of local discharges. One of the rf inductors is movable. Basic discharge modes are: Pure dc discharge (dc mode), combinations of dc discharge with one rf inductive discharge [dc/rf(i) mode] or with two rf inductive discharges [dc/2rf(i) mode], two rf inductive discharges [2rf(i) mode], and combination of dc discharge with one rf capacitive discharge [dc/rf(c) mode]. The dc current and the power of the rf(i) discharge can be modulated with a low frequency. A local heater is intended to manipulate the dust particles with the help of temperature gradients in the discharge gas. A high power manipulating laser allows the performance of a precise manipulation of the complex plasma structures and flows. The application of a combination of the listed manipulators allows manipulating the topology of dusty plasma formations over a wide range. A photo of the PK-4 setup installed in a laboratory is presented in Figure 1.36a.

To measure the distribution of the basic parameters of the background discharge (the electron density n_e, electron temperature T_e, and space plasma potential φ), a movable Langmuir probe has been used. Typical results of measurements of the plasma parameters of the discharge in the dc mode are presented in Figure 1.36b. The dc mode plasma is characterized by a high uniformity, which is an essential condition for the creation of extended homogeneous particle structures and flows. Measurements also showed that with the help of the two rf(i) local discharges it is possible to control the space plasma potential distribution and to create a potential well for dust clouds.

To confirm the basic principles put in the PK-4 development, special tests have been performed under microgravity conditions of a short time duration. These tests were performed onboard the special airplane A-300 ZERO-G. The duration of the microgravity phase was about 22 s in each parabola. The main principles were confirmed. In particular, the transportation of charged particles from the service parts to the experimental one with the help of a constant electrical field of the dc discharge and the blocking of dust particles in the experimental part by the local rf(i) discharge were successfully tested. Some interesting results have been obtained: formation of extended 3D dust clouds in the positive column of the dc discharge, ordering of dust structures in the field of the local rf inductive discharge, availability of dust acoustic instabilities and nonlinear stationary dissipative waves. For the first time Usachev *et al.* (2009) observed formation of a boundary-free dust cluster in a low-pressure gas-discharge plasma.

In this experiment the dc discharge operated at $I_{dc} = 1.0$ mA and $U_{dc} = 884$ V, and its uniform positive column filled almost the entire chamber volume. In addition to the dc electrodes, the PK-4 discharge chamber in this experiment was equipped with an rf coil installed in the vicinity of the tube center. The coil was powered by an rf current at a frequency of 81.36 MHz and a power of 1.5 W. Under microgravity, monodisperse melamine formaldehyde dust particles with a radius of 1.28 μm were injected into the dc discharge plasma in the vicinity of the cathode. Being injected, the charged dust particles drifted to the anode due to the dc electric field of the discharge of about 2 V cm^{-1}. The camera field of view (FoV) was 12.6×9.5 mm^2 at the tube axis. During the experiments the injected small particles formed an elongated drifting uniform dust cloud with a diameter of 1.5 cm and a particle density of $n_d \sim 2\times10^{11}$ m^3, which was confined along the tube axis by the radial electric a)

b)

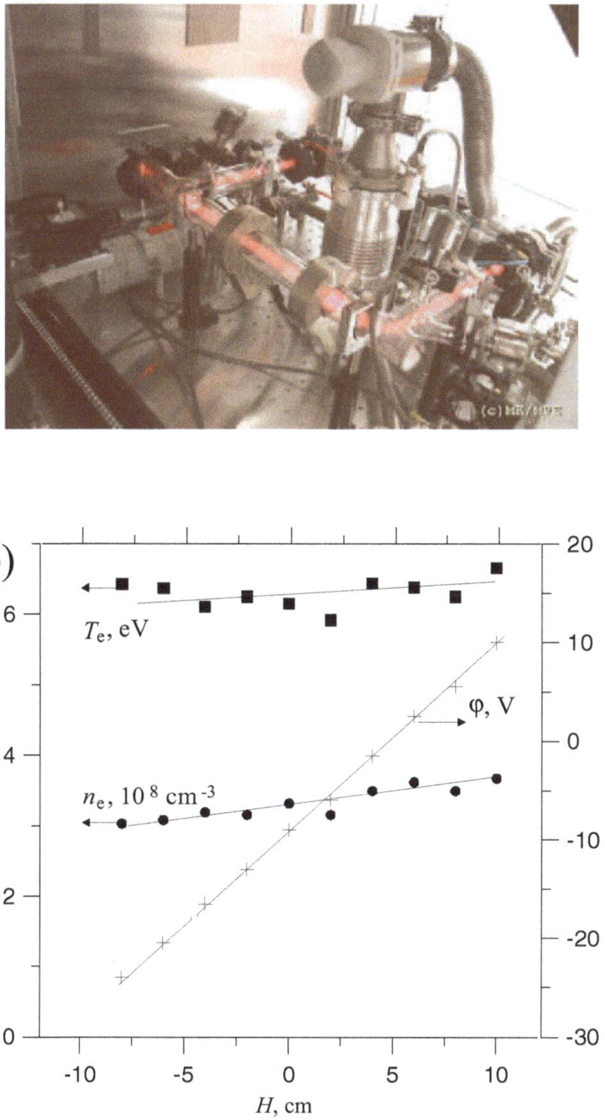

FIGURE 1.36

(a) Photo of PK-4 setup; (b) measured electron density n_e, electron temperature T_e, and plasma potential φ along the tube axis *H* for the dc discharge mode at gas pressure p = 50 Pa, discharge current I_{dc} = 1 mA, gas – Ne + 2%Xe. $H = 0$ corresponds to the center of the PK-4 chamber.

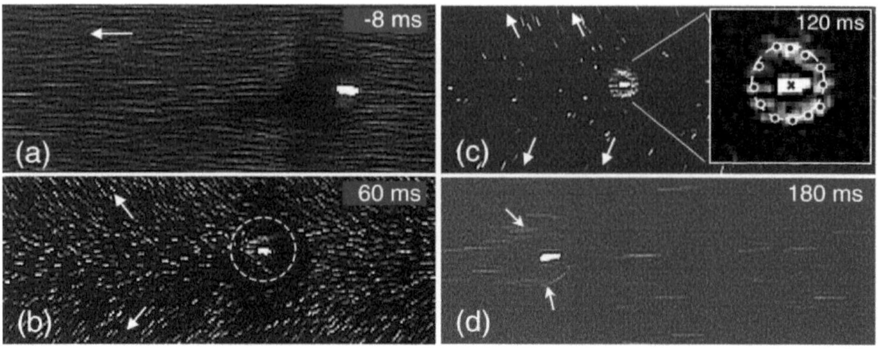

FIGURE 1.37
Cluster formation and destruction in the central part of the discharge tube. (a) Dust cloud in the uniform positive dc column just before the rf pulse: The arrow shows the direction of particle drift, the bright point is the big particle. (b) 60 ms after the rf pulse ignition: Beginning of the cluster formation, arrows show dispersing of small particles, dashed circle shows a sphere of attraction of small particles. (c) 120 ms after the rf pulse ignition: Formation of cluster is complete, arrows show drift of residual small particles; the insert shows an enlarged image of the cluster with 12 identified small particles. (d) Just after the ending of rf pulse: The cluster is disrupted instantaneously, arrows show small particles dispersed from the former cluster. The area shown is about 7 × 3 mm^2.

field in the cylindrical dc positive column. In addition to the injected small particles, heavy dust particle conglomerates randomly appeared in the dc discharge and FoV. As soon as the heavy particles appeared in the FoV, an experimentalist manually initiated an rf discharge pulse of a rectangular shape with a duration of 180 ms. In the process of these operations the formation of a boundary-free dust cluster, containing one big central particle with a radius of about 6 μm and about thirty 1 μm-sized particles situated on a sphere with a radius of 190 μm with the big particle in the center has been observed. Main stages of the cluster formation are shown in Figure 1.37. The reason for this boundary-free cluster formation is identified as the ion drag force acting on small particles in the ion flux directed to the big particle.

1.3 Thermal complex plasmas

Thermal complex plasma is a low temperature plasma with equal temperatures of all species that contains small sized liquid or solid particles (Sodha and Guha 1971; Yakubov and Khrapak 1989; Fortov *et al.* 2006a). The macroparticles interact efficiently with electrons and ions and often have major influence on plasma properties. Effects associated with the presence of particles were observed in early experiments (Sugden and Thrash 1951; Shuler and Weber 1954) with plasmas of hydrocarbon flames. Because of the high macroscopic charges that the particles can acquire (of the order of 10^2–10^4 e), strong interparticle interaction can lead to the formation of ordered structures of particles, analogous to structures in liquids or solids (Ichimaru 1982).

An experimental study of the formation of ordered structures in a classical quasineutal thermal plasma at atmospheric pressure and temperatures of 1700–2200 K was carried out (Fortov *et al.* 1996a, 1997a, 2004a, 2005a, and references therein). The rather large plasma dimensions (its volume is $\sim 10\ \text{cm}^3$, which corresponds to a particle number on the order of 10^8 at a particle density of $10^7\ \text{cm}^{-3}$), its uniformity, and the absence of external electric and magnetic fields made it possible to eliminate the effect of boundary conditions on phase transitions in the plasma.

1.3.1 Source of thermal plasma with macroparticles

The experimental setup includes a plasma generator and the diagnostic means for determining the parameters of the particles and gas (Kondrat'ev *et al.* 1994). The main part of the plasma source consists of a two-flare Mekker burner with propane and air fed into its inner and outer flares. The diameter of the inner flare is 25 mm and that of the outer, 50 mm. Particles are introduced into the inner flare of the burner. The burner design makes it possible to create a laminar flow of plasma with uniform parameter distributions (temperature, electron and ion densities). During operation, the velocity V_g of the plasma flow was varied over 2–3 m s^{-1} and the electron density in the flame, over 10^9–10^{11} cm^{-3}. The temperatures of the electrons and ions were equal and were varied over the range $T_i = T_e = T_g = 1700$–2200 K. Spectroscopic measurements of the temperature T_d of the particles (Nefedov *et al.* 1995) showed that it was close to the gas temperature ($T_d \cong T_g$). The combustion products were at atmospheric pressure.

In these experiments a thermal plasma with two types of chemically inert particles, Al_2O_3 and CeO_2 was studied. The particles of powder contain an admixture of alkali metal compounds. Spectral measurements in the plasma flow revealed the presence of alkali metal atoms (sodium and potassium) with low ionization potentials. Thus, the main components of the plasma in one case were charged CeO_2 particles, electrons, and singly charged Na^+ ions, and, in the other, charged Al_2O_3 particles, electrons, and Na^+ and K^+ ions.

In order to study ordered structures in the plasma, it is necessary to have data on the charge of the particles, as well as on the basic plasma parameters. The important feature of this plasma source is that it creates a large plasma volume. As a result, various diagnostic measurements of the plasma could be made. Various parameters of the gas and macroscopic particles were determined, such as the electron and alkali ion densities, the gas temperature, and the size and density of the macroparticles.

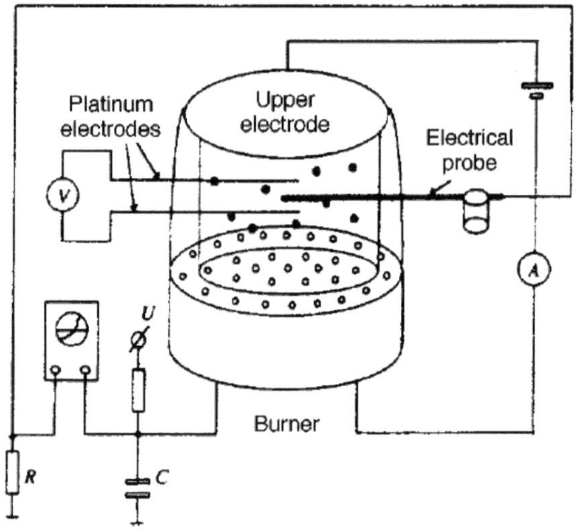

FIGURE 1.38
Sketch of the experimental setup and probe diagnostics in thermal complex plasmas.

1.3.2 Plasma diagnostics

The plasma was diagnosed by probe and optical techniques. The arrangement of the probe measurements is shown in Figure 1.38. The density n of the positive alkali metal ions was measured with an electrical probe (Benilov 1988; Kosov *et al.* 1991). The rms error in the density determination was 20%. The local electron density n_e was determined by a method based on measuring the current I and longitudinal electric field E in the plasma (Kosov *et al.* 1991). An electrode at constant voltage relative to the burner was placed in the plasma flow to determine the current I. Two platinum probes were introduced into the plasma to determine the tangential component E of the electric field. The electrical conductivity of the plasma, $\sigma = n_e e \mu_e$, was determined on the basis of Ohm's law $j = \sigma E$ (j is the current density and μ_e is the mobility of the electrons). The electron density was found from known μ_e. The error in n_e was less than 30%.

1.3.3 Particle diagnostics

A novel laser technique was used to determine the mean (Sauter) diameter D_d and density n_d of the macroparticles (Vaulina *et al.* 1996). This technique is based on measurements of the extinction of light in a dispersive medium at small scattering angles, and is intended for determining the characteristics of particles with sizes in the range 0.5–15 μm.

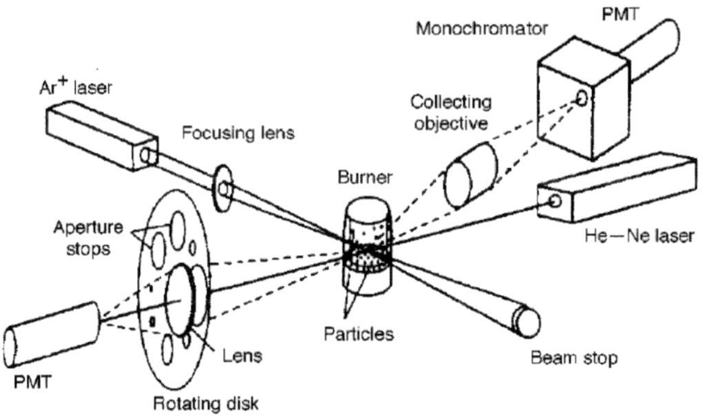

FIGURE 1.39
Sketch of the optical measurements of the sizes, density, and spatial structures of macroparticles.

The setup includes a rotating disk with aperture stops of various diameters located in front of a photodetector (Figure 1.39). An He–Ne laser ($\lambda = 0.633\ \mu$m) is used as the light source. For an error in the measurement of the extinction of about 2%, the errors in recovering the particle sizes and density were about 3% and 10%, respectively. The ordered structures were analyzed using the binary correlation function (Ichimaru 1982) obtained with a laser time-of-flight counter. The measurement volume is formed by focusing the beam of an Ar^+ laser ($\lambda = 0.488\ \mu$m) onto the axial region of the plasma stream. Radiation scattered by individual particles at an angle of 90° when they cross the laser beam spot is collected by a lens and directed onto the 15μm-wide monochromator entrance slit. The diameter of the measurement volume was less than 10 μm. The resulting pulsed signals were then processed to calculate the binary correlation function $g(r)$, which characterizes the probability of finding a particle a distance $r = V_d t$ from a given particle. Here t is time and V_d is the average particle velocity ($V_d \sim V_g$ for micron-sized particles). An analysis of $g(r)$ makes it possible to describe the spatial structure and interparticle correlations in the particle system.

Plasma diagnostics were performed in the temperature stabilized zone $h = 25$–40 mm above the top of the burner at various plasma temperatures and particle densities. The plasma temperature was changed by varying the propane/air ratio over 0.95–1.47. Thus, it was possible to change the Debye radius, the distance between particles, and the charge of particles in the plasma. Measurements of the spatial structures of the macroparticles were compared with data for an aerosol stream at room temperature. In the latter case, only air with particles of Al_2O_3 or CeO_2 was fed into the inner flare of the burner. This system simulates a plasma with a random spatial disposition of the macroparticles (a "gaseous" plasma).

In measurements with CeO_2 particles, the particle density n_d was varied over the range $(0.2\text{–}5.0)\times10^7$ cm^{-3} and the plasma temperature T_g over 1700–2200 K. As a consequence, the ion density n_i, varied from 0.4×10^{10} to 4.0×10^{10} cm^{-3} and the electron density, over $(2.5\text{–}7.2)\times10^{10}$ cm^{-3}. The CeO_2 particles are polydisperse, with a distribution half-width of less than 30% according to our measurements. The average Sauter diameter of the particles was about 0.8 μm.

Based on these data, the quasineutrality condition $Zn_d + n_i = n_e$ implies that the CeO_2 particles are positively charged to $\sim 10^3\,e$ with accuracy on the order of the factor 2. The observed magnitude and sign of the charge on the particles can be explained by thermal emission of electrons from the surface of heated CeO_2 particles (Sodha and Guha 1971; Fortov *et al.* 2006a), which are characterized by a low electron work function equal to ~ 2.75 eV (Fomenko 1981). In the following analysis of the data, we use the value $Z \approx 500$ for the particle charge.

1.3.4 Spatially ordered structures in thermal plasmas

Figure 1.40 shows typical binary correlation functions $g(r)$ for CeO_2 particles in an aerosol flow at room temperature ($T_g \simeq 300$ K) and in plasmas ($T_g \sim 2170$ and 1700 K) (Fortov *et al.* 1997b). It is quite evident that the correlation functions $g(r)$ for the plasma with temperature $T_g \sim 2170$ K and particle density $n_p \sim 2\times10^6$ cm^{-3} and for the aerosol stream are essentially the same. Thus, the particles in the plasma interact weakly and formation of ordered structures is impossible. This is also confirmed by the plasma diagnostic measurements. The optical and probe measurements showed that the average interparticle distance ($\Delta \sim 50$ μm) is roughly 3.5 times the Debye radius ($\lambda_D \sim 14$ μm), and thus, the interparticle coupling is weak.

For the lower plasma temperature $T_g \simeq 1700$ K and a particle density of $n_d \sim 5\times10^7$ cm^{-3}, Figure 1.40c shows that the binary correlation function $g(r)$ exhibits the short ordering characteristic for liquids. Under these conditions the interparticle coupling is much stronger and as a consequence the particles form an ordered liquid-like structure.

The formation of ordered structures was observed only at sufficiently high ($\sim 10^7$ cm^{-3}) particle densities. Lowering the concentration of CeO_2 particles increases the average distance between particles which lowers interaction energy. Then an ordered structure does not develop, as shown in Figure 1.40b ($n_d \sim 2\times10^6$ cm^{-3}). Plasmas with Al_2O_3 particles were studied at temperatures $T_g \sim$ 1900–2200 K. The higher concentration of Na^+ and K^+ ions (by ~ 10 times) leads to much stronger shielding of electric interactions between the particles in this case. As a result no spatially ordered structure was observed.

Another effect which can contribute to the formation of ordered structures in thermal complex plasmas is the electric attraction between positively charged particles in plasmas. The mechanism of this attraction is discussed in Section 2.3.1. Calculations of the interaction potential applicable for a highly collisional (atmospheric pressure) plasma (Khrapak *et al.* 2007) suggest that this mechanism could play some role in the experiments discussed above.

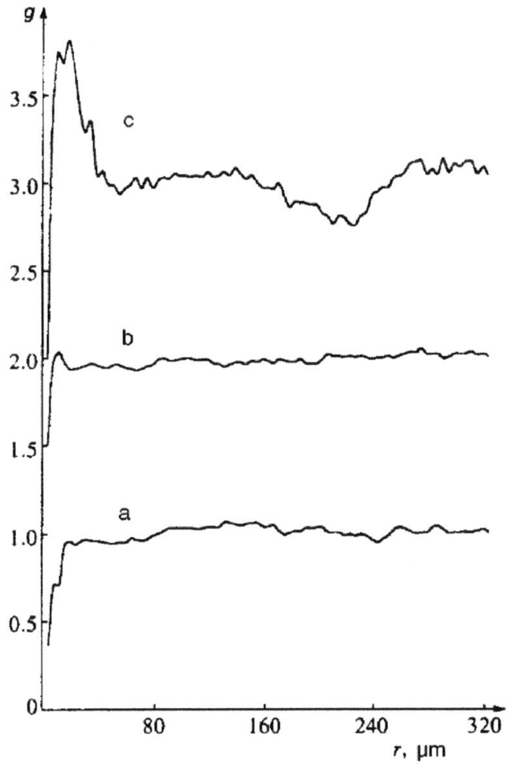

FIGURE 1.40
The binary correlation function $g(r)$ for CeO_2 particles in an air stream at room temperature $T_g \simeq 300$ K (a), in plasmas at a temperature $T_g \sim 2170$ K, corresponding to weak coupling (b), and at a temperature $T_g \sim 1700$ K, corresponding to stronger coupling (c).

1.4 Other types of complex plasmas

1.4.1 Complex plasmas at cryogenic temperatures

Structural and dynamical properties of ordered particle structures are greatly affected by thermal regime of the discharge which determines the temperature of the heavy plasma components – atoms, molecules, and ions. By cooling the discharge (e.g., to cryogenic temperature), one can reduce the plasma screening length. As a result, the mutual electric repulsion between charged microparticles significantly weakens; the particles may come closer to one another and form dense dust structures (Fortov *et al.* 2002; Asinovskii *et al.* 2006).

Traditionally, the technique of electrons and ions cooling down to temperatures in kelvin and millikelvin regimes is used to increase non-ideality of plasma systems. In this way it is possible to receive condensed crystalline ionic systems in cryogenic gas discharges (Fortov *et al.* 2006a), by laser cooling of atomic ions in non-neutral plasmas confined in Penning-type traps (Gilbert *et al.* 1988), in Paul rf traps, and in storage rings. Possibilities also exist for the observation of 2D crystalline structures of electrons on the surface of liquid helium. Complex plasmas at cryogenic temperatures were studied first by Fortov *et al.* (2002), where structures of polydisperse particles in rf and dc discharges cooled by liquid nitrogen (77 K) were experimentally observed. It was estimated that the dust density n_d in these experiments did not exceed 10^6 cm^{-3}. Moreover, the phenomenon of structure division into thin transversal layers with sharp boundaries was observed when gas pressure decreased. The layers were attributed to dust–acoustic instabilities. Super dense structures of polydisperse particles were observed by Asinovskii *et al.* (2006) in striation of dc discharge cooled by liquid helium (4.2 K). Although there was no possibility to determine the distances between the particles directly (in contrast to the work by Fortov *at al.* 2002), estimation made by the authors was $n_d \sim 10^9$ cm^{-3}.

The experiments designed to reveal the importance of the temperature factor in the formation of complex plasma structures were carried out in a cylindrically symmetric dc glow discharge generated in vertically oriented glass tube placed inside the cryostat – the cylindrical double glass Dewar system (Figure 1.41). The outer Dewar was used as a thermal guard and was filled with liquid nitrogen. The inner Dewar was filled with liquid nitrogen or liquid helium according to the temperature required. A glass discharge tube 1.2 cm in diameter and 42 cm in interelectrode distance was suspended from the cryostat lid. The upper electrode is the cylindrical hollow anode through which the particles were injected into the discharge. Particles were stored in the container with grid at bottom and positioned above the anode. When shaking the container the particles fell downwards through the grid. Monodisperse polystyrene particles with a diameter of 5.44 ± 0.09 μm were used. In order to illuminate the particles, a diode laser beam ($\lambda = 532$ nm) was introduced into the cryostat via optical fibre. The observations were performed through 1 cm wide Dewar's windows. For recording of scattered light from the particles, CCD video camera was used at a frame rate of 25 fps. The discharge was generated in He at a pressures $p = 2$–5 Torr and currents $I = 0.2$–1.3 mA.

The dust particles in the dc discharge acquire negative charge and are trapped in standing striations in the positive column. Each striation represents a parabolic potential well in the horizontal plane and is characterized by sharp increase of the electric field in the vertical direction, which is accompanied by only an insignificant decrease of the electron density. Thus, electrostatic force acting on a charged particle can balance gravity and lead to the levitation of particles in striations. In order to generate stable striations in a wide range of parameters, the discharge was locally constricted in its lower part. The discharge constriction was produced by means of a "capillary" – glass cylinder of 1 cm in diameter, which is narrowing to nozzle of about 0.1 cm in diameter in its upper part. A potential jump arises inside the "capillary" which can contribute to the generation of striations. Experiments demonstrated

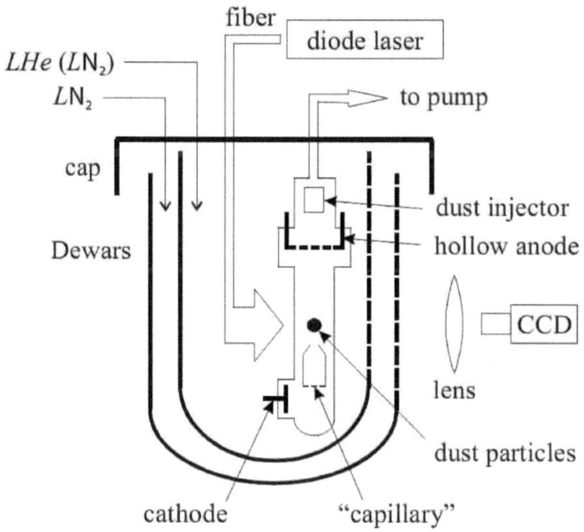

FIGURE 1.41
Scheme of experimental setup used to study complex plasma structures in a dc discharge at cryogenic temperatures.

that a standing striation was generated above the "capillary" at both room and cryogenic temperatures.

In the experiments at room temperature, levitation of the chains consisting of several dust particles aligned in the vertical direction was observed. The number of chains varied from several up to several dozens, so the overall number of particles did not exceed ~ 100. Depending on discharge parameters, particle chains could be found in the stationary state as well as moving around each other. It should be noted that the vertical chain-like ordering of dust structure is typical for dc low pressure glow discharges. Figure 1.42a represents the structure of particles obtained at discharge current $I = 0.5$ mA, discharge voltage $U = 4.2$ kV and gas pressure $p = 2$ Torr. With the assumption $T_i \simeq 300$ K, the ionic Debye radius (which determines the plasma screening length in this case) was estimated to be ~ 40 μm. The interparticle distance Δ was measured to be 500–750 μm ($n_d \sim 10^3$–10^4 cm^{-3}). Therefore, for the room temperature, we have $a \ll \lambda \ll \Delta$, where a is the particle radius.

Figure 1.42b shows side view of the particle structure in discharge cooled by liquid nitrogen at $U = 3.7$ kV and at the same discharge current and neutral gas density as we had in the experiments at room temperature. In this case the ionic Debye radius was estimated to be ~ 20 μm. At 77 K the structure consisted of long chains of approximately 15–20 particles with 200–250 μm interparticle separation. This structure is at least an order of magnitude denser ($n_d \sim 10^5$ cm^{-3}) than that at room temperature and consists of about 10^3 particles. Therefore, in the case of 77 K the

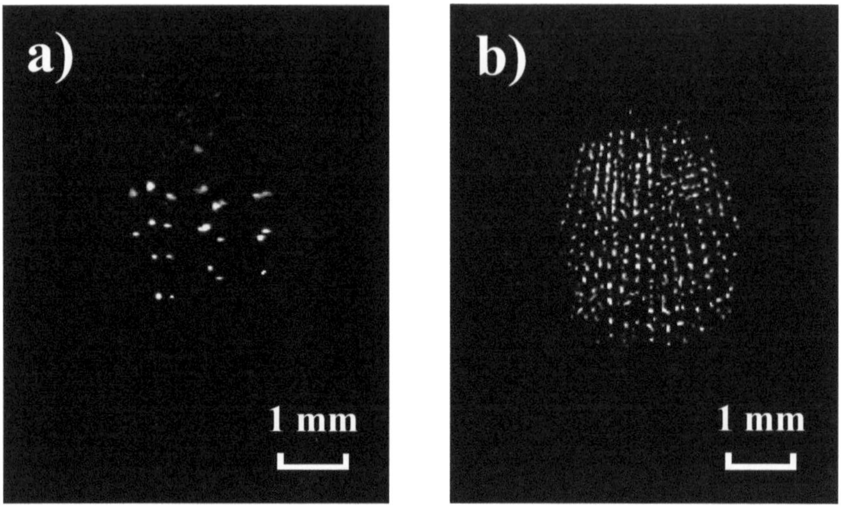

FIGURE 1.42
Typical side view of the particle structure in a striation of the dc glow discharge: a) at room temperature (300 K); b) at liquid nitrogen temperature (77 K). In both cases discharge current and neutral gas density are the same: I = 0.5 mA and $n = 6.4\times10^{16}$ cm^{-3} (corresponding to the pressure p = 2 Torr at 300 K).

same inequalities as for the room temperature case, $a \ll \lambda \ll \Delta$, are satisfied.

The experimental data were analyzed in order to compare the structural and dynamical properties of complex plasmas observed at temperatures near 300 and 77 K at the same discharge currents and neutral gas densities. Detailed analysis of video images did not reveal qualitative changes in the two particle structures. The measured pair correlation function $g(x)$, where $x = r/\Delta$ is the normalized distance, revealed close similarity, although the interparticle distances and the number of particles were considerably different in these two cases. Pair correlation functions also show that the particle structures were highly ordered in the preferred (vertical) direction.

Observations show that the particles at 77 K move with higher velocities than the particles at 300 K at the same discharge currents and neutral gas densities. The recorded particle velocity distribution functions over horizontal (x) and vertical (y) directions were found to be anisotropic (it can be related to a spatial inhomogeneity of the plasma-dust system) and close to the Maxwellian functions. Kinetic temperatures measured are $T_x \sim 1.0$ eV, $T_y \sim 0.2$ eV for particles in the discharge at 300 K and $T_x \sim 0.9$ eV, $T_y \sim 1.3$ eV at 77 K. Therefore, we can conclude that cooling of the discharge down to cryogenic temperatures leads to some "heating" of the particle system.

Glow discharge cooled by liquid helium has properties that are radically different from discharges at room and liquid nitrogen temperatures. Contrary to traditional dc discharges, investigations of stationary cryogenic discharge near 4.2 K are not numerous. At present, elementary processes and plasma kinetics of such discharges remain practically unexplored. Also, there are no investigations concerning discharge instabilities and, particularly, ionization waves (striations of positive column). Existing estimation indicate that the ionic Debye radius can be extremely small in this case, $\sim 4\ \mu$m, which is of the order of the particle size. It was observed by Asinovskii *et al.* (2006) that injection of monodisperse particles into the discharge could lead to the formation of sphere-like particle structure within the striation as in the case of polydisperse particles. The spherical structure vibrated (oscillated) in the horizontal direction at a frequency of about 10 Hz. Obtained estimates for the observed structure indicate that here complex plasma was in a state characterized by $a \sim \lambda \sim \Delta$.

1.4.2 Experiments with complex plasma induced by UV-radiation

One of the dust charging mechanisms in space is the photoemission. Under the influence of intensive fluxes of light, the dust particles can get positive electric charges $\sim (10^2–10^5)e$ and form crystal, liquid or gaseous dust structures (Rosenberg *et al.* 1996, 1999; Fortov *et al.* 1998; Vaulina *et al.* 2001). The phase state of these structures is closely connected with diffusion processes, which are one of the basic sources of the energy losses in complex plasmas.

The diffusion is a non-equilibrium mass transfer process caused by a thermal motion of particles, which leads to a steady state of the distribution of their concentrations and is one of basic sources of the energy losses in real physical systems. The diffusion occurs in various regimes, for example, Brownian diffusion of macroparticles suspended in a background gas, or self-diffusion of the particles. For the clouds consisting of positive ions and electrons, the joint diffusion transport of oppositely charged particles (the ambipolar diffusion) may appreciably influence the dynamical properties of the system. The case of ambipolar diffusion of low-ionized plasma in the absence of a magnetic field was surveyed by Schottky (1924). The first results of study of the polarization of oppositely charged particles for the two-component system consisting of macroparticles and emitted photoelectrons under the influence of sunlight in microgravity conditions were obtained by Vaulina *et al.* (2002) and Nefedov *et al.* (2003b). It should be noted that direct experimental observations of ambipolar diffusion for charged macroparticles are not feasible under usual laboratory conditions in the presence of gravity.

Detailed results of an experimental study of diffusion of dust particles, charged by photoemission, under microgravity conditions were presented in the works by Fortov *et al.* (1998), Vaulina *et al.* (2001, 2002), and Nefedov *et al.* (2003b). The data were obtained during an investigation of a complex plasma induced by solar radiation, onboard the Mir space station. These investigations have shown that under the action of intensive solar radiation the micron-size particles can acquire considerable positive electric charges (Fortov *et al.* 1998). The experimental study of the dust diffusion was performed for bronze particles with the mean radius $a \simeq 37.5\ \mu$m in the background gas (neon) at a pressure of $p = 40$ Torr. The particles were contained in a cylindrical glass tube, the bottom of which was the uviol window intended for

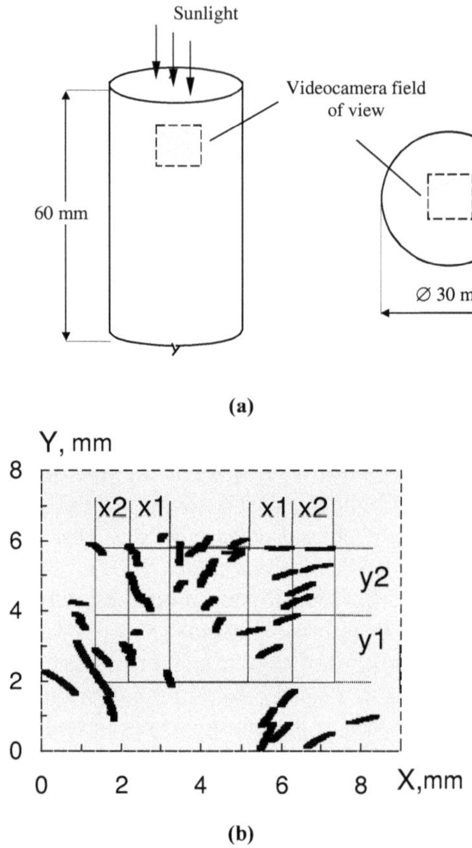

FIGURE 1.43
(a) Sketch of the experimental setup. (b) Trajectories of moving particles after the action of the solar radiation.

the solar illumination of the particle cloud. Extra illumination of particles by a laser beam was used for improved diagnostics. The image was registered by a video camera with the field of view $\sim 8 \times 9$ mm^2, the definition in depth was about 9 mm (see Figure 1.43a). Subsequently video-records were handled by a special program for the identification of the displacements of individual particles. Under solar radiation the number of recorded particles was determined by the definition in depth of the video-system, which allowed tracing positions of particles during times sufficient for the analysis of the particle dynamics. Trajectories of 40 particles (for 5 sec after the beginning of solar irradiation) are shown in Figure 1.43b.

The first step of the experiment was the observation of the particles without the action of the solar radiation. During the observations (~ 20 min) the number of par-

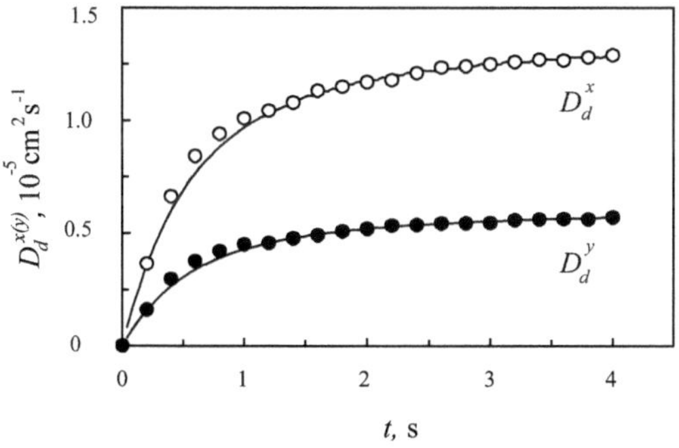

FIGURE 1.44
Experimental dependency of D_d^x (open circles) and D_d^y (solid circles) on time t, averaged for the different experiments, and their fit (solid lines) using the functional form of Equation (1.1) at $\nu_{fr} \approx 3$ s^{-1}.

ticles in the field of view of video system did not vary significantly. The second step was the observation of macroparticles while irradiating the cloud with the solar radiation. Since in the initial state, the bronze particles were deposited on the ampoule walls, the experiment was performed as follows: (1) dynamic effect (impact) on the system with the closed window; (2) exposure in darkness ~ 4 s $\gg \nu_{fr}^{-1}$ (ν_{fr} is the frequency of collision of dust with the gas molecules) to reduce the initial random dust velocities; (3) illumination of the system by solar radiation; (4) relaxation of the particles to the initial state (clinging to the ampoule walls) for the time $\sim$ 3–5 min. This interval is about three orders of magnitude shorter than the time for full diffusion loss of particles at room temperature due to Brownian motion.

The measured particle kinetic temperatures and coefficients of thermal diffusion are anisotropic: $T_x \sim 51$ eV, $T_y \sim 22$ eV; $D_0^x \sim 1.4 \times 10^{-5}$ cm^{2}s^{-1} and $D_0^y \sim 6.2 \times 10^{-6}$ cm^{2}s^{-1}. Coefficients of the thermal diffusion of particles were calculated from their measured velocities and displacements. The behavior of the measured functions $D_d^{x(y)}(t)$ (see Figure 1.44) at initial times of observation was close to that of non-interacting particles (Vaulina and Khrapak 2001):

$$D_d^{x(y)}(t) = D_0^{x(y)}(t)\{1 - [1 - \exp(-\nu_{fr}t)]/\nu_{fr}t\}. \qquad (1.1)$$

Note, that the function $D_d(t)$ for particles in strongly interacting systems is not monotonic, but has a pronounced maximum (Vaulina and Khrapak 2001), unlike the results presented in Figure 1.44.

The initial dust concentration n_0 was varied from ~ 200 to ~ 300 cm^{-3}. The dependencies of the relative dust concentration $n_d(t)/n_0$ on the time t are shown in Figure 1.45. The charge of the particles was obtained from the approximations of the curves $n_d(t)/n_d$ for $t > 40$ s by the method detailed by Fortov *et al.* (1998) and was close to $Z \sim 4 \times 10^4$ ($\pm 15\%$). Illustration of molecular dynamic simulations of

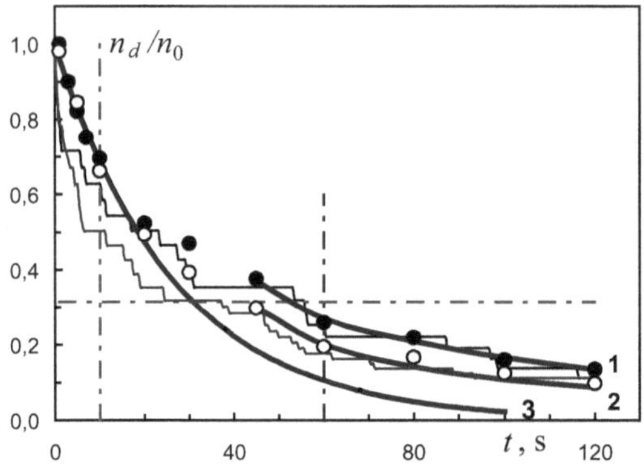

FIGURE 1.45
Dependencies of $n_d(t)/n_0$ on time t for the different initial concentration n_0: (filled circles) – 195 cm^{-3}; (open circles) – 300 cm^{-3}. Thin line is the calculation by the method of molecular dynamics; curves (1) and (2) are the approximation of Fortov *et al.* (1998); curve (3) is the function $n_d/n_0 = \exp(-\nu_d t)$.

the particle transport due to the electric repulsion between the particles along with the corresponding analytical approximations used to determine the particle charge are presented in Figure 1.45 for conditions close to experimental ones (Fortov *et al.* 1998).

Since the considered system consists of the positively charged macroparticles and the photoelectrons with the density $n_e \sim Zn_d$ emitted by them, it is possible to assume, that the transport properties of such a system depend on the ambipolar diffusion of the particles. Because of the large difference of the electron μ_e and dust μ_d mobilities, a negative surface charge appears on the tube walls. The incipient polarization electric field blocks further separation of the charged components. Therefore, the electrons and heavy dust particles can diffuse "together" with some effective coefficient D_a of the ambipolar diffusion (Raizer 1991):

$$D_a = (D_e\mu_d + D_d\mu_e)/(\mu_d + \mu_e). \tag{1.2}$$

Here D_e and D_d are the free diffusion coefficients for electrons and dust particles, respectively. In the case $\mu_e \gg \mu_d$, we have

$$D_a/D_d \approx 1 + ZT_e/T. \tag{1.3}$$

With the measured temperatures (T_x , T_y), the ratios of the diffusion coefficients can be estimated as $D_a/D_d^x \sim(0.8\text{–}1.6)\times 10^3$, and $D_a/D_d^y \sim(1.8\text{–}3.6)\times 10^3$ for $T_e \simeq 1\text{–}2$ eV.

Since polarizing effects are possible only at weak perturbation of the electroneutrality of a system ($\delta n/n < 0.1$) and the analyzed system could be close to electroneutral one only at the initial stages of the experiment (for times smaller than

~ 10 seconds), the approximation of the experimental data on the evolution of the particle concentration $n_d(t)/n_0$ by the function characterizing the velocity of particle diffusion losses at the ampoule walls (see Figure 1.45) has been used:

$$dn_d/dt = -n_d \nu_d. \tag{1.4}$$

Here $\nu_d = D_a/\Lambda^2$ is the frequency of diffusion losses, and Λ is the effective diffusion length (Vaulina *et al.* 2001). For a cylinder with radius $R = 15$ mm and height $\sim 4R$, the value of $R/2 = 0.75$ cm. The value of ν_d was obtained from the experimental curve $n_d(t)/n_0$ at $t < 10$ s, where the function $n_d(t)/n_0$ agrees well with the solution $n_d = n_0 \exp(-\nu_d t)$ of Equation (1.4) for $\nu_d \simeq 0.032$ s^{-1}. An estimation of the ambipolar diffusion coefficient yields $D_a \sim 2 \times 10^{-2}$ cm^{2}s^{-1}, which is in reasonable agreement with theoretical result $D_a \simeq ZT_eD_d/T \sim (1.2-2.4) \times 10^{-2}$ cm^{2}s^{-1} obtained for the measured values of $Z \simeq 4 \times 10^4$ and $D_d/T \simeq 3 \times 10^{-7}$ cm^{2}s^{-1}eV^{-1} for $T_e \simeq 1-2$ eV.

1.4.3 Nuclear-induced and track complex plasmas

Nuclear-induced plasma is produced by nuclear-reaction products which, passing through a medium, create ion–electron pairs, as well as excited atoms and molecules in their tracks. In terms of physical characteristics, the nuclear-induced plasma of inert gases differs significantly from thermal and gas discharge plasmas. At relatively low intensities of a radioactive source, typical for laboratory conditions, this plasma has a distinct track structure. The tracks are randomly distributed in space.

The experiments were performed by Fortov *et al.* (1999) and Vladimirov *et al.* (2001) in an ionization chamber which was placed in a hermetical transparent cell

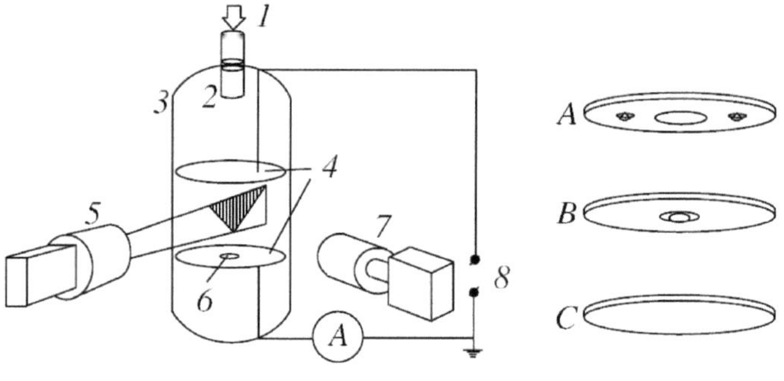

FIGURE 1.46
Experimental setup: Injection of gas and particles mixture from the evacuation and gas-filling system (1), container of dust particles (2), glass walls (3), metallic electrodes (4), laser with a cylindrical lens (5), 2D radioactive source (6), video camera (7), dc source (8), and various types of high-voltage electrode (A, B, C).

TABLE 1.1
Major characteristics of ionizing particles.

Ionizing particle	Average initial energy, keV	Total range in solid matter (CeO_2), μm	Total range in air (p = 1 atm), cm	Number of secondary electrons per particle
β-particle	138	58	5.6	≈ 5
Average fission fragment	9×10^4	5.5	2.3	≈ 250
α-particle	6×10^3	20	4.7	≈ 10

FIGURE 1.47
(a) Dust structures of micron-sized Zn particles in a nuclear-induced plasma in neon at a pressure of 0.75 atm and an electric field strength of 30 V cm^{-1}; (b) rotating dust structures of micron-sized Zn particles in neon. The agglomeration of particles is visible.

(Figure 1.46). Either β-particles (decay products of ^{141}Ce) or α-particles and fission fragments (decay products of ^{252}Cf) were used as ionizing particles. The energies of reaction products, their mean free paths in air and dust particle material, as well as the average number of secondary electrons emitted in collisions of the ionizing particle with a dust particle are given in Table 1.1.

In a plasma of atmospheric air at an external electric field strength of 20 V cm^{-1}, the levitating particles form liquid-like structures – the pair correlation function has one maximum. At stronger electric fields, the dust particles move in closed trajectories which form a torus with the axis aligned with that of the cylindrical chamber.

In an inert gas, sufficiently dense dust clouds with sharp boundaries exist, and the dust particles form a liquid-like structure inside these clouds (Figure 1.47a). In a nonuniform electric field, the particles form rotating structures in which agglomeration of small particles into coarse fragments proceeds in the course of time (Figure 1.47b).

In the experiments, particles of Zn and CeO_2 were used. The particle radius was estimated experimentally from the particle steady-state falling velocity after the removal of the electric field and was found to be 1.4 μm. The electric charge was inferred from the equilibrium condition for slowly moving levitating particles. The particle charge depended on the particle radius and lay in the range from 400 to 1000 elementary charges.

For the theoretical calculation of the charge on a spherical particle in the nuclear-induced plasma, nontraditional approaches are used because this plasma is spatially and temporarily inhomogeneous. The charging process is considered after the track plasma has separated into two clouds: one of electrons and the other of ions. When the electron or the ion cloud meets a dust particle on its drift to the corresponding electrode, then this particle acquires some charge from the cloud. A statistical treatment of these charging events constitutes the essence of the mathematical model developed for calculating the particle charge. The electron current to the particle is determined by the electron collection cross section in the collisionless limit. To obtain the ion current, the diffusion approximation is used. In the numerical simulation by Fortov *et al.* (2001b) and Rykov *et al.* (2003) the code first generates the event of creation, the direction of emission from the source, and the type of ionizing particle (alpha-particle or a fission fragment). The simulation shows that the particle charge fluctuates with time. On the one hand, the dust particle acquires a charge in the electron attachment process; on the other hand, its charge decreases substantially in the less frequent events of interaction with the ions. The contribution from alpha-particles leads to a complicated dependence of the particle charge on time. The characteristic timescale for changing the charge is 10^{-3}–10^{-2} s. This time is short enough, and hence the particle interaction with external fields can be expressed in terms of effective constant charge. A time-averaged charge calculated from this model is in most cases close to the values obtained for levitating particles from the balance between gravity and electrostatic forces. The charge fluctuations with amplitudes comparable to the charge itself can be one of the reasons preventing the formation of highly ordered particle structures.

When a particle gets into the track region, the cascade electrons having a mean energy of $\sim$ 100 eV could cause charges which are sufficient for dust system crystallization. To investigate the charging process, a numerical model based on a system of equations describing two-dimensional space-time track evolution has been developed. The system of equations includes the kinetic equation for electrons, the continuity equation for heavy components (ions, atoms, etc.), the Poisson equation and equations describing chains of plasma-chemical reactions. It follows from the calculations that a particle can collect no more than 10 electrons from one track. This means that the influence of many tracks is required in order to induce high charges on the particle.

In order for the charging rate in the track regions to dominate over that due to drift flows, the ionizing flux of a value of approximately 10^{13} cm^{-2} s^{-1} is required. Such a flux can be obtained in the beam of a charged particle accelerator. The experiments have been performed by Fortov *et al.* (2005a, 2006a,b) in different gases (He, Ne, Ar, Xe, Kr). Electrostatic proton accelerator has been used for the experiments. The experimental cell had the form of a rectangular parallelepiped with a basement of 16×16 cm^2 and a height of 12 cm (Figure 1.48). The cell's side faces

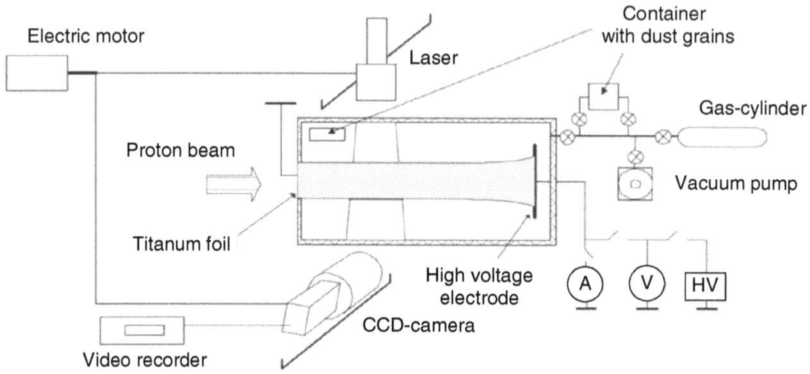

FIGURE 1.48
The experimental setup for investigating complex plasmas under the influence of a proton beam.

were made as glass windows, the particle structure behavior was observed by CCD camera registering the light scattered by the particles. Horizontal proton beam with an energy of 2 MeV was passed via titanium foil and a diaphragm 8 mm in diameter. Monodisperse melamine-phenol particles with radii a of 0.505, 0.875, 1.50, 2.41, and 2.75 μm were used in the experiments. The gas-particle mixture at a required pressure was created in the cell after pumping by high vacuum pump.

The particle behavior proved to depend on the gas pressure considerably, whereas the dependence on the gas type was weak. Main results of the experiment are as follows. After pumping the gas-dust mixture into the cell, a gas pressure of about 25 Torr was established. Near the high-voltage electrode, in the paraxial area of the proton beam, a dense particle structure was formed with an initial density of 5×10^6–10^7 cm^{-3}. The process of particle structure formation in the proton beam takes about 2–3 s. The particle structures have a cylindrical symmetry in equilibrium with the maximum diameter 8 mm approximately coinciding with the beam diameter. The structures can exist for 10 min or more at the same pressure. The structure is destroyed when the proton beam is closed or when the voltage applied to the high-voltage electrode is decreased or turned off.

By reducing the gas pressure in the experimental cell to less than several Torr, the particle component is crystallized at a distance of $\approx$ 1 cm from the electrode. The process of particle-ordered structure formation has a weak dependence on the gas type. The crystal observed (for an example, see Figure 1.49) has a simple cubical lattice with the mean distance between the particles of 90 μm in helium and 140 μm in krypton and xenon. Crystal structures have been obtained for all types of gases used in experiments and for particles of different diameters.

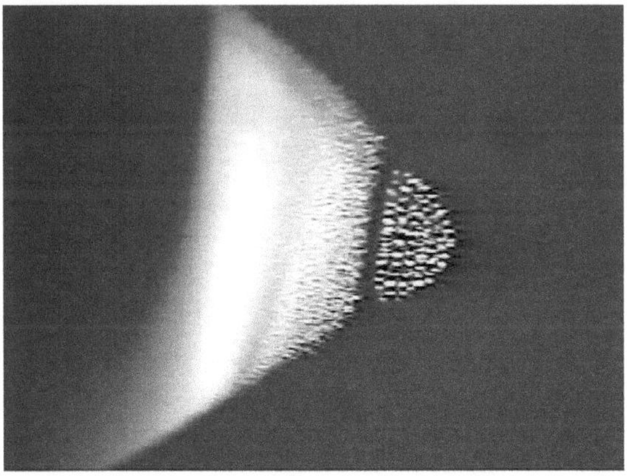

FIGURE 1.49
An example of the crystal-like particle structure. The experiment is performed in krypton gas with particles of 3 μm in diameter (voltage on a main electrode is 140 V, beam current is 3 μA).

1.4.4 Particle structures in a dc discharge in the presence of magnetic fields

The investigation of the response of complex plasmas to various external actions is of great interest. Such actions can be used to control the spatial position, ordering, and dynamics of complex plasma structures. Moreover, laboratory complex plasmas under external actions are a good experimental model for investigating the formation of dust and dusty plasma structures in space and various industrial and power facilities (Fortov *et al.* 2004a). For instance, the effect of the magnetic field is important for the analysis of the behavior of particles in the near-wall plasma in tokamaks. As shown by Konopka *et al.* (2000), Klindworth *et al.* (2000), Sato *et al.* (2001), Cheung *et al.* (2003), Samsonov *et al.* (2003b), Paeva *et al.* (2004), Hou *et al.* (2005), Dzlieva *et al.* (2005, 2006), and Karasev *et al.* (2006), the vertical magnetic field can lead to the rotation of particle structures in the horizontal plane due to the tangential component of the ion drag force. The majority of these experiments were performed with rf and dc discharges in magnetic fields up to 400 G. However, Sato *et al.* (2001) studied the effect of magnetic fields up to 4×10^4 G on dusty plasma clouds in an rf discharge and pointed out difficulties in obtaining a stable dc discharge in high magnetic fields of thousands of Gauss.

In the experiment described below the possibility of the formation of complex plasma structures and their dynamical characteristics in a dc glow discharge in axial magnetic fields up to 2500 G are discussed. The experimental setup schematically shown in Figure 1.50 (Vasil'ev *et al.* 2007) was used to investigate the action of the magnetic field on complex plasma structures. A stratified dc glow discharge was created in a vertically oriented cylindrical glass tube with cold electrodes. The inner

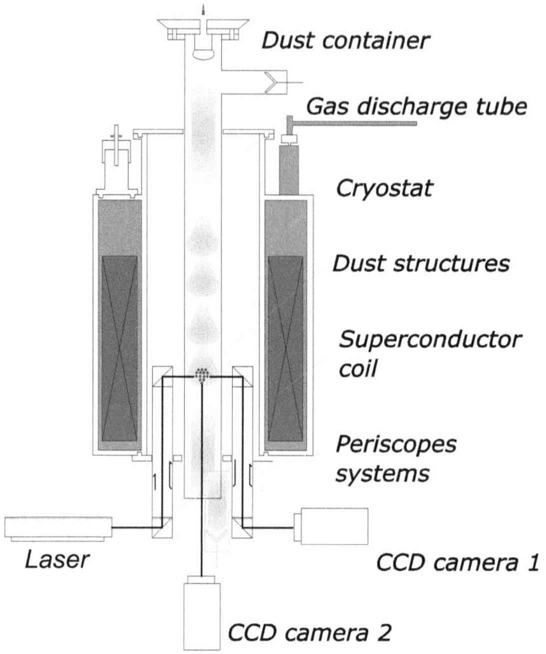

FIGURE 1.50
Sketch of the experimental setup consisting of a cryostat of a superconducting magnet, in which a gas discharge tube is placed, and a visualization system including an illuminating laser and two video cameras is shown.

diameter of the tube was 36 mm and the distance between the electrodes was 600 mm. Spherical melamine formaldehyde particles of 5.5 μm in diameter were used to form complex plasma structures in discharge striations. The particles were placed in a container that had a mesh bottom and was located in the upper part of the discharge tube. Particles were injected into the discharge with the use of a piezoelectric plate, whose vibration ensured particle injection in the region of the positive column of the discharge. Most of the experiments were carried out in neon gas at pressures of several tenths of Torr with discharge currents of several tenths of microampere.

A superconducting cylindrical solenoid located in a liquid helium cryostat was used as a generator of the magnetic field. The gas discharge tube was placed in a so-called "warm hole" 150 mm in diameter at the center of the cryostat. When the cooled solenoid was in operation, the temperature in this hole was no lower than 273 K. The direction of the magnetic field can be changed by changing the current direction in the solenoid. A system of two identical monocular optical periscopes was developed for the observation and diagnostics of complex plasma structures in the discharge inside the warm hole. Each periscope is a cylindrical tube with glass prisms in the upper and lower ends. For the observation of a certain region of the

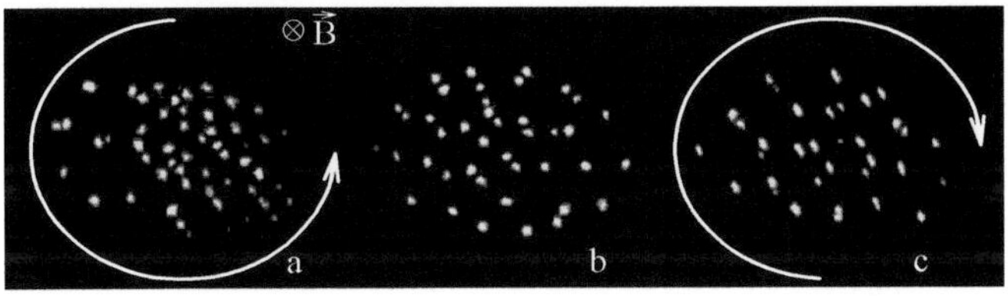

FIGURE 1.51
Horizontal section of the dusty plasma structure in the dc glow discharge with rotation in the magnetic field *B* equal to: (a) 75, (b) 500, and (c) 630 G. The rotation direction in (a) and (c) is shown by the arrows. Rotation is not observed in case (b).

gas discharge, each periscope can be vertically displaced by a distance equal to its working height (300–450) mm. The particles were illuminated with 532-nm laser radiation introduced inside the warm hole through the prisms of one of the periscopes. The complex plasma structures were detected by means of two CCD video cameras through the prisms of the second periscope and through the lower flat end of the gas discharge tube (see Figure 1.50). This system allowed us to obtain images of the structures in the horizontal and vertical planes.

The density of the electrons, their energy distribution, and the electric field are strongly nonuniform over the length of a striation in the stratified positive column of a dc glow discharge. The electric field is relatively high in the head of the striation, which covers 25–30% of the striation length, is equal to about 10–15 V cm^{-1} in the maximum, and is low outside this region (about 1 V cm^{-1}). Due to the high floating potential of the walls of the discharge tube, the striations are two-dimensional and the potential difference between the center and wall in the striation head reaches 20–30 V. Thus, an electrostatic trap, which can confine injected micron-size particles levitating in the axial region of the vertical discharge, exists in the head of each striation.

The experiments were performed with the discharge parameters such that the high magnetic field did not induce the contraction of the discharge. The maximum longitudinal magnetic field equal to 2500 G at which the vertical striations hold was obtained for the discharge in hydrogen at a pressure of several tenths of Torr. However, the particles injected into the discharge were not detected in the observation region for this field magnitude. The particle structures in the discharge in H_2 were detected only in fields up to 1000 G in the form of flat monolayers consisting of a small number of particles.

Small complex plasma structures were observed in the discharge striations in neon. In the axial magnetic field, they rotated in the horizontal plane about the vertical symmetry axis of the discharge. In a magnetic field of 75 G, the angular velocity of the dusty cloud was directed against the magnetic field (Figure 1.51a). However, with further increasing of the magnetic field, rotation was decelerated and terminated at 500 G (see Figure 1.51b). The rotation of the dusty structure in the opposite direction

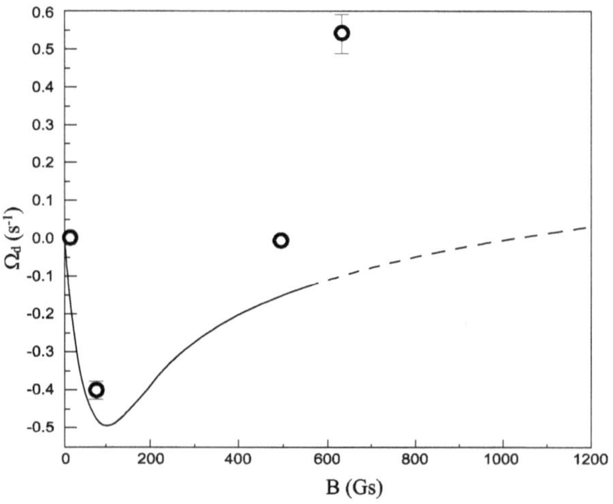

FIGURE 1.52
Measured angular velocity of the particle structure as a function of the magnetic field strength. Solid line corresponds to calculation using Equation (1.5).

was observed in a field of 630 G (see Figure 1.51c), and the angular velocity of the structure was aligned with the magnetic field. Figure 1.52 shows the experimental angular velocities of the particle cloud in the striation of the dc glow discharge as a function of the strength of the magnetic field. With further increase in the axial magnetic field up to 700 G, the particles forming the structure in the axial region of the discharge are displaced toward the periphery of the discharge, i.e., toward the walls of the discharge tube. In this case, the angular velocity of the particles does not change and is equal to 1–2 rad s^{-1}. Small oscillations of dust particles are also observed in this case. These oscillations are likely to be caused by instabilities of the glow discharge in the presence of the magnetic field.

The particle structures containing $\sim 10^3$ dust particles were obtained in the experiments with neon in fields up to 300 G. The rotation of the structures was not observed. In order to determine the structure and dynamical characteristics, the mass transfer curves were calculated from the particle displacements in successive video frames. These curves were then analyzed using a model developed by Vaulina *et al.* (2005). Figure 1.53 shows the resulting kinetic energy and diffusion coefficient of particles for various magnetic fields. It is seen that both quantities decrease with an increase in the axial magnetic field. Correspondingly, the nonideality parameter of the particle structure increases. As it is well known, the magnetization of the plasma with the conservation of the discharge current leads to a decrease in the axial electric field (see, e.g., Granovskii 1952; Golant *et al.* 1980). This decrease is likely to be the cause of the freezing of the particle system when increasing the magnetic field strength. However, the exact mechanism of crystallization requires further investigation.

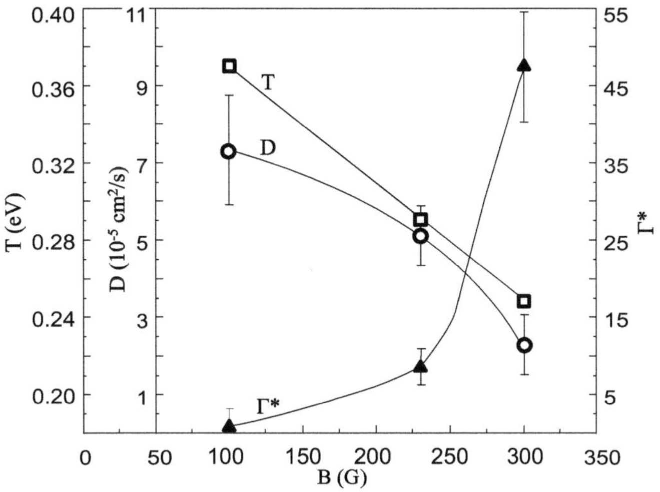

FIGURE 1.53
Magnetic field dependence of the kinetic temperature T, diffusion coefficient D, and nonideality parameter Γ^* of the dusty plasma structure.

The rotation of particles in the axial magnetic field occurs due to the ion drag force (Fortov *et al.* 2004a; Dzlieva *et al.* 2005, 2006; Karasev *et al.* 2006). The ions azimuthally drift in the crossed axial magnetic field and radial electric field. The ion drag force in the uniform rotation is balanced by the drag force with the neutral gas of atoms.

The inversion of the radial electric field and, therefore, the angular velocity of dust particles in the magnetic field can be caused by a change in the sign of the radial component of the ion density gradient dn_i/dr and magnetization of electrons to a degree such that their mobility becomes lower than the ion mobility.

Applying the expressions for the ion drag and neutral drag forces discussed in detail in Section 2.5, one can estimate the dependence of Ω_d on B. For the conditions of an experiment by Vasil'ev *et al.*, it was assumed that the radial distribution in the striation region near the discharge axis, where the particle structure is located, has the Bessel profile, which does not contradict the measurements reported by Golubovsky and Nisimov (1995). In this case, $dn_i/dr \approx 0.128r$ in the particles location ($r <$ 0.2 cm); i.e., the angular velocity should be independent of the radius in agreement with the observations. The substitution of the experimental parameters yields the following approximate dependence of Ω_d on B:

$$\Omega_d \approx -10^{-2}B\frac{1-10^{-6}B^2}{1+10^{-4}B^2}, \tag{1.5}$$

which is shown by the solid line in Figure 1.52. Since the effect of the magnetic field and the presence of the particles on the $n_i(r)$ is neglected in deriving Equation (1.5), this formula is valid only for small particle structures and low magnetic fields $B \lesssim 100$ G. Equation (1.5) provides a qualitatively correct magnetic field dependence of the angular velocity of the particle structure. Qualitative agreement is also reasonable up to $B \lesssim 100$ G. Note that the change in the sign of the angular velocity, which occurs at $B \approx 1000$ G, is attributed to the magnetization of electrons (their lower mobility as compared to ions). However, in this case, the destruction of the potential trap confining the dusty structure should be expected, but this destruction is observed at higher magnetic fields ($B \approx 700$ G) than $B \approx 500$ G at which the direction of rotation changes.

The reversal of the rotation direction is likely to be attributed to a change in the direction of the diffusion plasma flux: The derivative dn_i/dr near the structure becomes positive. In this case, dn_i/dr outside the dusty structure remains negative and the trap continues to exist. It is known that plasma recombination occurs on the surface of dust particles, i.e., plasma is absorbed by the particle structure. The radial diffusion flux from the axis to the walls in low magnetic fields $B \lesssim 100$ G prevails over the flux absorbed by the structure. As the magnetic field increases, the plasma is magnetized and the radial flux toward the wall decreases when the discharge current is unchanged. The absorption of plasma by the particles also weakens, but to a smaller degree, because magnetization does not affect the axial component of this flux. As a result, at a certain value of B (≈ 500 G in our experiment), the total plasma flux on the particles is larger than the plasma flux generated in the discharge near the particle structure. Therefore, the inversion of the radial plasma flux occurs in the central region of the discharge which leads to the change in the rotation direction of the particle structure. With further increase in B, the region of the inversed diffusion flux expands, the potential trap disappears, and the particle structure is destroyed at $B \approx 700$ G. According to the experimental results, the trap does not disappear completely, but it shifts to the peripheral region of the discharge.

1.4.5 "Small" dust structures: Coulomb or Yukawa clusters and balls

1.4.5.1 Cluster types

Coulomb clusters are the ordered systems which consist of a finite number of microparticles interacting via a repulsive Coulomb potential and confined by external forces (e.g., of electrostatic nature). In many cases, the interaction potential is believed to be of the Debye–Hückel (Yukawa) form, and therefore such systems are called "Yukawa clusters". The difference between the dust clusters and the dust crystals is rather conditional: both systems in fact consist of a finite number of particles. The term "clusters" is usually reserved for systems with the number of particles $N \leq 10^2$–10^3, while larger formations are refereed to as "crystals". A more precise definition of clusters would be the ratio of the number of particles in the outer shell to the total number of particles in the system. For crystals, this ratio should be small. Similar systems can be formed, for instance, in non-neutral plasmas in Penning or Paul traps (Gilbert *et al.* 1988; Dubin and O'Neil 1999), where the vacuum chamber is filled with the ions, as well as in colloidal solutions with macroscopic charged particles (Grier and Murray 1994). Examples of two-dimensional clusters are electrons on the surface of liquid He (Leiderer *et al.* 1987) and electrons in quantum dots (Ashoori 1996; Filinov *et al.* 2001). The distinctions between systems are mainly due to different types of interaction potential and different forms of confining potential.

Historically, clusters consisting of repulsive particles in an external confining potential were first investigated with the use of numerical modeling (mostly by Monte Carlo and molecular dynamic methods). Taking into account the possibility of applying the simulation results to dust particle clusters, we mention here works of Bedanov and Peeters (1994), Schweigert and Peeters (1995), Candido *et al.* (1998), Lai and I (1999), Astrakharchik *et al.* (1999a,b), Totsuji (2001), Totsuji *et al.* (2001, 2006), Drocco *et al.* (2003), Ichiki *et al.* (2004), and Kamimura *et al.* (2007). Most simulations were performed for two-dimensional clusters in an external harmonic (parabolic) potential. Such a configuration is usually realized in ground-based experiments with complex plasmas in gas discharges. The simulations show that for a relatively small number of particles in the cluster the "shell structure" is formed with the number of particles N_j in the jth shell ($\sum_j N_j = N$).

The first experimental investigation of dust clusters was reported by Juan *et al.* (1998). The experiment was performed in the sheath of an rf discharge. A hollow coaxial cylinder of 3 cm in diameter and 1.5 cm in height was put on the bottom electrode to confine the particles. Clusters with a number of particles from a few up to 791 were investigated. Figure 1.4 shows images of typical clusters with different numbers of particles, and Figure 1.54 demonstrates a series of the observed configurations. For large N, the inner particles arrange themselves into a quasi-uniform hexagonal structure, whereas near the outer boundary particles form several circular shells. The mean interparticle separation increases up to about 10% from the center to the boundary.

The rotation of dust clusters around the symmetry axis was studied by Klindworth *et al.* (2000), Ishihara *et al.* (2002), Cheung *et al.* (2004), Karasev *et al.* (2006), Vasil'ev *et al.* (2007), and Carstensen *et al.* (2009). In Ishihara *et al.* work, the cluster rotation was caused by the laser pressure. Both the rigid body and the differential (intershell) rotation had been observed. In other works, the cluster rotation was initiated by the presence of the magnetic field parallel to the cluster axis of symmetry. The rotational frequency was found to be linear with the weak magnetic field, although it saturates at moderate magnetic field strength (Cheung *et al.* 2004), and it is inversely proportional to the field in the strong field limit (Ishihara *et al.* 2002).

In an experiment by Cheung *et al.* (2004) performed in an inductively coupled argon discharge at a pressure of 100 mTorr, the applied voltage on the confinement electrode was set at +10 V. Melamine-formaldehyde particles were injected into the plasma sheath above the circular electrode. The 2D dust particle clusters with different numbers of particles in a horizontal plane were formed. In particular, planar-2 (two particles in the horizontal plane) to planar-16 (16 particles) dust clusters were formed using larger dust particles (radius $a = 3.105$ μm), and planar-2 to planar-12 dust clusters were formed using smaller dust particles (radius $a = 1.395$ μm).

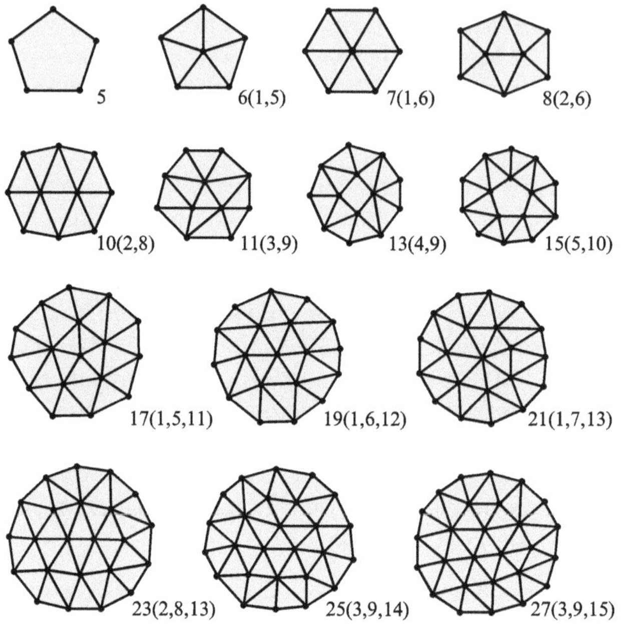

FIGURE 1.54
Typical shell configurations of several dust particle clusters consisting of different numbers of particles (Juan *et al.* 1998).

Similarly to the simulation results of Bedanov and Peeters (1994), dust particle clusters with a number (up to a few dozens) of particles observed in experiments by Cheung *et al.* (2004) are arranged in shell configurations in the form of concentric rings. The equilibrium configuration strongly depend on the exact number of particles in the cluster. On the other hand, large clusters, containing hundreds of particles, have a core region which is of hexagonal order, like in a 2D monolayer. If the horizontal confining potential is radially symmetric, then in order to match the boundary of the confinement well, the outermost shells contain dislocations and are therefore not perfectly hexagonal.

Also, we mention results of experiments by Annaratone *et al.* (2004), Arp *et al.* (2004), and Antonova *et al.* (2008), where the three-dimensional clusters were observed. The experiment by Annaratone *et al.* was performed in the adaptive rf electrode chamber (the rf electrode is an assembly of small pixels, each having an independent control of the rf voltage; Annaratone *et al.* 2003) filled with the argon gas at a pressure in the range 40–80 Pa, with plastic particles of 6.8 μm diameter. By a fine adjustment of the rf amplitude applied to a dc-grounded pixel it was possible to control the number of particles in the cluster and also its shape. The number of particles varied from 4 up to about 200. In the equilibrium positions the vertical confinement is provided by the electric field of the double layer/striation combined

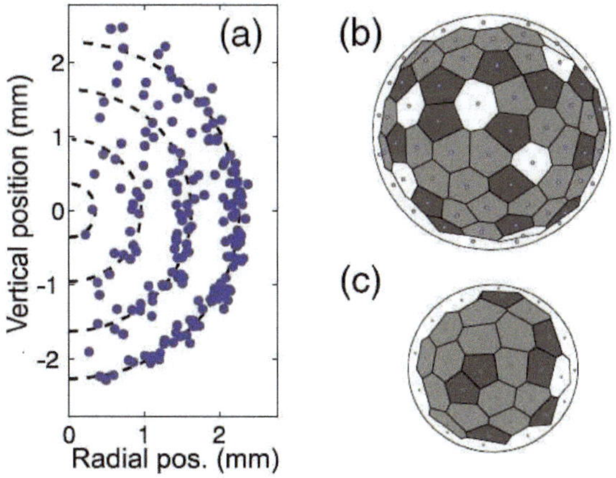

FIGURE 1.55

The cluster consisting of 190 particles. (a) The shell structure becomes evident by projecting all particles into the (ρ, z)-plane irrespective of their angular position with regard to the z axis; (b) and (c) are bottom views of Coulomb ball, showing the arrangement of the particles (small circles) in the shells superimposed by the Voronoi cell analysis. Hexagons are shaded light gray and pentagons dark gray (Arp *et al.* 2004).

with suitable conditions for the particle charging. It is unclear if the horizontal confinement is due to plasma pressure, internal forces among the cluster components, or the ion drag force. More work is certainly needed in order to explain the formation of such structures.

In the experiment by Arp *et al.* (2004, 2005), the so-called "Coulomb balls" – spherical particle clouds, in which hundreds or thousands of identical plastic spheres of 3.4 μm diameter are arranged in clearly separated crystalline shells – were observed in an rf discharge at pressures of 50–150 Pa. The particles were levitated by the thermophoretic force, which is excited by heating the lower plate, and the radial confinement was provided by a short upright glass tube. The highest order was observed in the outer shells, whereas in the center the particles were in a liquid (amorphous) state with no significant orientational order.

Figure 1.55 shows the analysis of the structure for a cluster containing 190 particles. According to MD simulations (see, for example, Dubin and O'Neill 1999), one can expect Coulomb crystals to form distinct "onion-like shells", in which the particles are evenly arranged in patterns with five or six neighbors. Such kind of shell structure becomes clearly visible when the particle positions are plotted in cylindrical coordinates $(x^2+y^2)^{1/2}$ and z (Figure 1.55a). The spherical shells are clearly separated in the lower half of the Coulomb ball. In the upper half a certain number of defects with particles at intershell positions is found. This is a general tendency seen in most of the experiments. Figure 1.55b shows the outer shell of the 190 particle cloud, where the particles are found evenly distributed. The number of neighbors for each particle is determined by a Voronoi analysis on the sphere. Besides the expected hexagons and pentagons, defect structures with seven neighbors were found. The previous shell (1.55c) has only hexagons and pentagons. This kind of surface structure was expected because of the incompatibility of a pure hexagonal lattice with a curved surface and the incommensurability of particle numbers in adjacent shells (Arp *et al.* 2004).

In dust clusters at zero temperature, the unique equilibrium configurations of particles in shells ($N_1, N_2, N_3, \ldots$) exist for a given particle number N. Such configurations are to some extent analogous to Mendeleev's Periodic Table of elements. The structure of these clusters depends on the shape of the interaction potential, confining potential and their relative strengths. At finite temperatures, metastable states with energies close to the ground state can also be realized (Block *et al.* 2008; Kählert *et al.* 2008). Further analysis is related to the stability of the cluster types. Assuming that dust particle–particle interactions are of the screened Coulomb (Debye–Hückel) type, the stability characteristics of a cluster can be analyzed by considering its total energy (Bedanov and Peeters 1994)

$$E = \frac{1}{2}\Omega_r \sum_{i=1}^{N} r_i^2 + Q^2 \sum_{i>j}^{N} \frac{1}{r_{ij}} \exp\left(-\frac{r_{ij}}{\lambda_D}\right), \qquad (1.6)$$

where Ω_r is the strength of the radial (horizontal) confining potential energy in the parabolic approximation, $r_i = \sqrt{x_i^2 + y_i^2}$ is the radial coordinate of the ith particle, and $r_{ij} = |r_i - r_j|$ is the distance between particles i and j. Note that instead of the parabolic confinement, some studies (Bedanov and Peeters 1994) employ the hard-wall confinement when the external radial potential energy is zero for radii less than some radius, and infinity for larger radii.

The ground states and metastable structures of 2D dust particle clusters, their dynamical properties and phase transitions have been also studied theoretically by Lozovik and Pomirchy (1990), Lozovik and Mandelshtam (1990, 1992) and, in applications to dusty plasmas by Astrakharchik *et al.* (1999a,b). A number of numerical simulations were devoted to studies of packing and defects of strongly coupled 2D clusters (Lai and I 1999) as well as to their structure and melting (Totsuji 2001; Drocco *et al.* 2003). Generally, the observed configurations of dust particle clusters agree well with theoretical predictions for a system of charged particles confined by an external field (Lozovik and Pomirchy 1990; Lozovik and Mandelshtam 1990, 1992; Bedanov and Peeters 1994). This can be proved by a comparison of the theoretically calculated numbers of particles in the shells and the experimentally observed clusters (Vladimirov *et al.* 2005). Detailed experimental and theoretical investigations of the metastable states of three-dimensional Yukawa clusters were performed by Block *et al.* (2008) and Kählert *et al.* (2008).

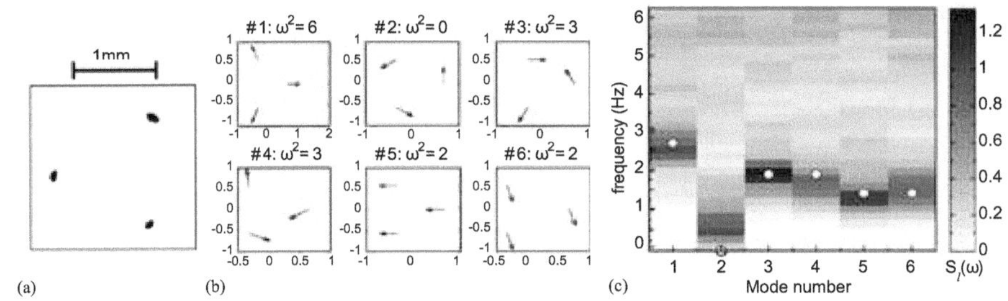

FIGURE 1.56
Normal mode spectrum of a three-particle cluster (Melzer 2003). (a) Particle trajectories over 1 min. (b) Six normal modes of the cluster, the mode frequencies ω_ℓ^2 are normalized to $\frac{1}{2}\omega_0^2$. (c) Measured mode spectrum of the modes; the spectral power density is shown in gray scale, the circles correspond to the calculated mode frequencies.

1.4.5.2 Oscillation spectra in 2D clusters

The oscillations in particle clusters were investigated by Schweigert and Peeters (1995), Melzer *et al.* (2001), Amiranashvili *et al.* (2001), Melzer (2003), Kong *et al.* (2004), Barkby *et al.* (2008) and Kedziora *et al.* (2008). To illustrate experimental investigations, we first discuss the work by Melzer, where the normal modes of 2D particle clusters of $N = 1$–145 particles trapped in the sheath above the lower rf electrode were studied. The normal modes were obtained from the analysis of the thermal Brownian motion of the particles around their equilibrium positions in the cluster. This method extends the thermal excitation technique by Nunomura *et al.* (2002) (developed for extended 2D lattices) to the case of finite particle clusters.

In Figure 1.56a the particle trajectories in an $N = 3$ cluster are shown. One can see that the thermal fluctuations of the microspheres around their equilibrium positions are small, but they are nevertheless sufficient to determine the mode spectrum. The six eigenmodes of this cluster calculated for the Yukawa potential are depicted in Figure 1.56b. There are the following modes: the breathing mode ($\ell = 1$), rotation of the entire cluster ($\ell = 2$), a twofold degenerate "kink" modes ($\ell = 3,4$), and the two sloshing modes ($\ell = 5,6$). The mode frequencies ω_ℓ^2 (in units of $\frac{1}{2}\omega_0^2$) are also indicated for $\kappa = 0$. For $\kappa > 0$ the oscillation pattern of the eigenmodes is unchanged. Their frequencies, however, depend on κ (see Figure 1.57). Obviously, the cluster rotation and the sloshing mode are independent of κ. The frequencies of the kink mode and the breathing mode increase with κ.

A typical analysis of the mode spectra uses forces that determine evolution and dynamics of particles in a cluster. The confinement potential is usually assumed to be parabolic structure. [Sometimes the square well potential – the "hard wall" potential (Kong *et al.* 2004) is also used.] The interparticle interactions are usually modelled by (unscreened or screened) Coulomb repulsion. A frictional force also can be present to simulate the particles slowing down as they move through the plasma.

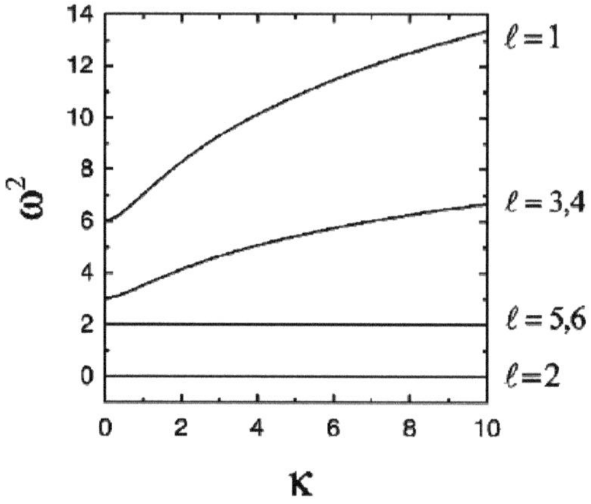

FIGURE 1.57
Mode frequencies of the three-particle cluster as a function of screening strength κ calculated by Melzer (2003). The modes are the breathing mode ($\ell = 1$), the cluster rotation ($\ell = 2$), the kink modes ($\ell = 3,4$), and the sloshing modes ($\ell = 5,6$).

Once the equilibrium is established the total energy of the cluster is calculated using Equation (1.6) (Schweigert and Peeters 1995; Melzer 2003).

For the general case of a cluster with N particles, there are $2N$ cluster modes. These modes describe different ways in which the dust particles in the cluster can oscillate. To determine these modes we can use the dynamical matrix calculated on the basis of the total energy from Equation (1.6):

$$E_{\alpha\beta,ij} = \frac{\partial^2 E}{\partial r_{\alpha,i} \partial r_{\beta,j}}, \tag{1.7}$$

where α and β can both be equal to x or y, and i and j denote the particle number. This results is a $2N \times 2N$ matrix. The eigenvalues of this matrix are then the squared frequencies of oscillation, where the corresponding eigenvectors describe the possible oscillation patterns.

The lowest frequency mode (LFM) was identified for different cluster sizes. For small particle numbers the intershell rotation was shown to be the LFM, whereas for larger clusters the formation of vortex–antivortex pairs was observed. This behavior is generally within the theoretical expectations for pure Coulomb systems (Schweigert and Peeters 1995). Analysis of the highest frequency modes (HFM) showed that for small clusters, the breathing mode with a coherent radial motion of all particles is the HFM. For larger clusters, modes with a relative three-particle oscillation localized in the center of the cloud provide the highest frequencies. The mode-integrated spectrum shows two broad maxima which are explained from "shear-like" or "compression-like" modes. The transition from finite number to crystal-like properties was observed to occur at around $N = 12$ particles.

In the analysis of cluster oscillation modes, most studies assume constant dust charges although the latter can vary in a plasma environment. The modes of clusters formed by two and three dust particles were analyzed taking into account spatial variations of the particle charge and shielding parameters by Kompaneets *et al.* (2006). Variations of dust charges include the random fluctuating component related to plasma fluctuations (always present in gas discharges) via the dust charging process. The fluctuations can in turn influence the kinetic energy of dust particles (Vaulina *et al.* 2006) as well as the collective modes in the particle structures.

The charges on dust particles fluctuate for several reasons (Kedziora *et al.* 2008). For example, thermal fluctuations are not correlated and one should expect the presence of an uncorrelated charge fluctuation component. However, the charge on a dust particle immersed in a plasma is not an independent variable and is related to fluctuations in a neighboring plasma. If the plasma fluctuations and, correspondingly, the associated charge fluctuations have the correlation length exceeding the size of the cluster, the fluctuations are correlated. Generally, a mixture of correlated and uncorrelated fluctuations should be expected in experiments with the particle clusters. In recent studies by Barkby *et al.* (2008) and Kedziora *et al.* (2008) the analysis of the oscillation normal modes in particle clusters with fluctuating charges was performed for correlated/uncorrelated dust charge fluctuations. This analysis demonstrated the normal mode splitting related to the variance of the fluctuations. It was reported that the fundamental pure rotational modes are mostly affected by the charge fluctuations while the least affected are the pure translational modes.

1.4.6 Complex plasmas with non-spherical particles

Most of the experimental and theoretical works dealing with the investigation of complex plasmas were performed with spherical particles. However, some experimental and theoretical investigations of complex plasmas with asymmetric particles have been performed, too (Mohideen *et al.* 1998; Molotkov *et al.* 2000; Annaratone *et al.* 2001; Fortov *et al.* 2001a; Vladimirov and Nambu 2001; Shukla and Mamun 2002; Ivlev *et al.* 2003a; Vladimirov and Ostrikov 2004; Maiorov 2004). Note that in work by Mohideen *et al.* (1998) the geometrical aspect ratio was $\alpha \sim 3$, and the first experiments with strongly asymmetric particles, $\alpha = (40-80) \gg 1$, were carried out by Molotkov *et al.* (2000) and Annaratone *et al.* (2001).

It is well known that colloidal solutions, which have much in common with complex (dusty) plasmas, show a much broader spectrum of possible states in the case of strongly asymmetric cylindrical or disk particles. In such solutions, liquid phase and several liquid-crystal and crystal phases with different degrees of orientational and positional ordering can be observed. It is also well known that the use of cylindrical probes (in addition to spherical) considerably broadens the possibilities of low–temperature plasma diagnostics. It is therefore obvious that the use of cylindrical particles can considerably broaden the frontiers of complex plasma research.

In an experiment by Molotkov *et al.* (2000), where the experimental setup analogous to that shown in Figure 1.16 was employed, nylon particles ($\rho = 1.1$ g cm^{-3}) of length 300 μm and diameters 7.5 and 15 μm, as well as particles of lengths 300 and 600 μm and diameter 10 μm, were introduced into the plasma of a dc discharge. The discharge was initiated in neon or a neon/hydrogen mixture at a pressure of 10–250 Pa. The discharge current was varied from 0.1 to 10 mA. In this parameter range, standing striations were formed in the discharge, which made particle levitation possible. A neon/hydrogen mixture was used to levitate heavier particles of larger diameter (15 μm) or larger length (600 μm). In this case, the particles formed structures consisting of 3–4 horizontal layers. Lighter particles levitated in pure neon and formed much more extended structures in the vertical direction. In Figure 1.58, a part of a horizontal cross section of an ordered structure levitating in a striation of a dc discharge excited in a neon/hydrogen mixture (1 : 1) at a pressure of 120 Pa and discharge current of 3.8 mA is shown.

The observed structures formed by microcylinders revealed clear ordering. All particles lay in the horizontal plane and were oriented in a certain direction. One could expect that their orientation should be determined by the cylindrical symmetry of the discharge tube. However, no correlation between the particle orientation and discharge tube symmetry was found. Nor could the preferential orientation of the particles be explained by the interparticle interaction, because individual particles were oriented in the same direction. Presumably, the preferential orientation was related to a weak asymmetry in the discharge. This was confirmed by the fact that the orientation could be changed by introducing an artificial perturbation into the discharge. In later experiments by Fortov *et al.* (2001a), nylon particles of lengths 300 and 600 μm and diameter 10 μm coated by a thin layer of conducting polymer were utilized. In a dc discharge they formed structures identical to those formed by uncoated particles of the same size.

Levitation of cylindrical particles was also observed near the sheath edge of a capacitively coupled rf discharge by Annaratone *et al.* (2001). In this experiment,

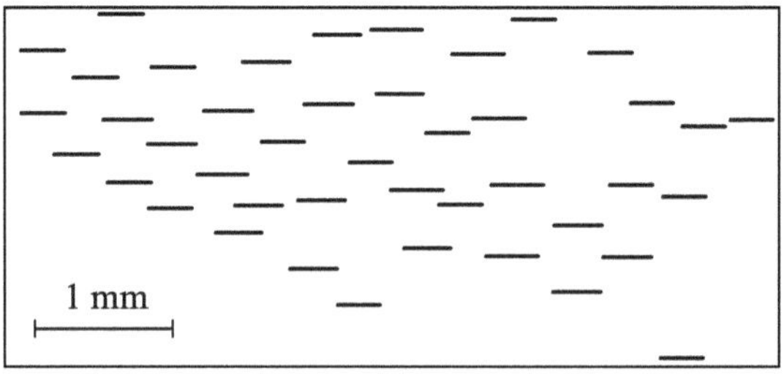

FIGURE 1.58
Digitized image of a part of a horizontal section of a structure formed by cylindrical macroparticles of length 300 μm and diameter 15 μm levitating in a striation of a dc discharge in a neon/hydrogen mixture at a pressure of 120 Pa and discharge current 3.8 mA.

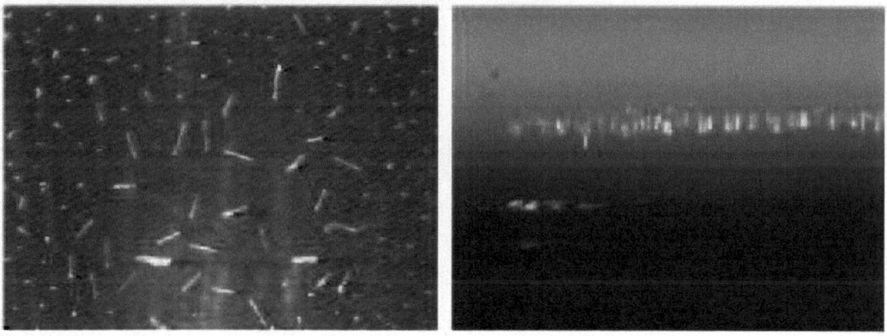

FIGURE 1.59
Typical video images of structures formed by cylindrical particles levitating near the sheath edge of an rf discharge. The discharge was initiated in krypton at a pressure of 52 Pa and discharge power of 80 W. Left panel shows a top view, dots correspond to vertically oriented particles; right panel gives a side view.

cylindrical particles of 300 μm in length and 7.5 and 15 μm in diameter were used, and a small fraction of very long particles (up to 800 μm in length and 7.5 μm in diameter) was also present. A typical picture of a structure formed by these particles is shown in Figure 1.59. Longer particles were oriented horizontally and mainly located in the central part limited by a ring placed on the electrode, while shorter particles were oriented vertically along the electric field. Levitation and ordering of the cylindrical particles occurred only for pressures higher than 5 Pa and for discharge power above 20 W. An increase in the discharge power did not significantly affect particle levitation. The average distance between vertically oriented particles varied from 1 to 0.3 mm. An increase in the particle density leaded to quasi-crystalline structure degradation and to an increase in the particle kinetic energies. The further increase in density was impossible because the particles started to fall down from the structure. Levitation of particles coated by a conducting polymer was not observed in an rf discharge for the conditions at which the dielectric particles of the same size and mass could levitate. Instead, the conducting particles stuck to the electrode, preserving vertical orientation, and some stuck to each other forming multi-particle fractal complexes with up to 10 particles.

The preferential orientation of cylindrical particles is determined by the interplay between the interaction of nonuniform electric fields in striations or sheaths with the particle charge and induced dipole and quadrupole moments. In a dc discharge, the particle charge is typically larger than in an rf discharge, allowing particle levitation in weaker electric fields. In this case, the dipole moment, which is proportional to the electric field strength squared, is much smaller than in an rf discharge. This can explain the different orientations of similar-sized particles: horizontal in a dc discharge and vertical in an rf discharge.

The effect of the rf plasma parameters on the orientation of the levitating cylin-

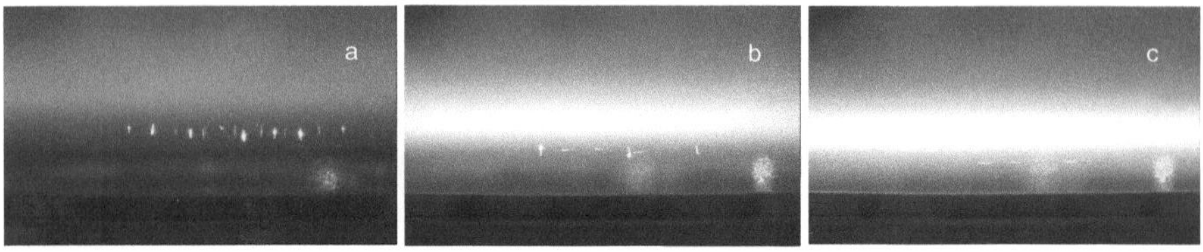

FIGURE 1.60
Levitation of the nylon particles L = 600 μm, a = 2.5 μm above an electrode driven at V_{pp} = 430 V: (a) at 32 Pa (dc bias = 124 V); (b) at 60 Pa (dc bias = 95 V); (c) at 98 Pa (dc bias = 82 V). Each picture corresponds to an area of 16.3 $\times$ 12.2 mm^2.

drical particles was investigated experimentally by Annaratone *et al.* (2009). The experiments were performed in an rf parallel plate reactor in argon plasma used previously by Annaratone *et al.* (2004). In this reactor the lower electrode, 40 mm in diameter, was rf driven at 13.56 MHz, the upper electrode and the chamber walls were grounded. The strong geometrical asymmetry of the reactor is confirmed by the building up of a high dc bias on the electrode, about two thirds of the rf voltage amplitude. This rf enhanced sheath has proved to be essential for the levitation of the nylon micro-rods 5 μm in diameter and 300–600 μm in length used in the experiment. The particles were visualized by illuminating the chamber with a laser, fanned out in a vertical plane, and recorded with a video camera from a side window. The plasma parameters were measured by a Langmuir probe. With a driving peak-to-peak voltage on the lower electrode $V_{pp} = 430$ V, the 600 μm micro-rods injected in plasma levitated vertically for pressures lower than 60 Pa (see Figure 1.60a). For pressures lower than 22 Pa, they formed two layers or even three layers at very low pressures before falling down (about 15 Pa). At 60 Pa there is a co-existence of vertical and horizontal orientations, Figure 1.60b, and for higher pressure the particles are normally horizontal, Figure 1.60c. Observed change of the microrods orientation is in good agreement with the theoretical model of Ivlev *et al.* (2003a).

When many particles are injected, some of them fall on the lower electrode. The falling behavior is still strongly dominated by electrostatic effects. These particles are shown, to give an example, in Figure 1.61 for $p = 22$ Pa and in Figure 1.62 for $p = 115$ Pa. At low pressures the particles tend to attach vertically on top of another already placed vertically (Figure 1.61). The attachment is more probable with particles coated by conducting polymer than with the equivalent non-coated nylon particles. A possible explanation of this effect is the dipole charge forming micro currents at the very moment of the particles touching. The head to tail attachment is typical of low pressure plasmas when the sheath is much larger than the particle dimension and it was observed in several experiments. However, when the size of the particle structure is comparable with the sheath dimension, some particles attach with a tilt or even horizontally. This is due to the sheath deformation by the structure itself. This effect is shown in Figure 1.63 where the micro-rods are inserted

FIGURE 1.61
Nylon, polypirol coated particles 300 μm in length and 2.5 μm in radius suspended in an argon plasma at a pressure of 22 Pa above the lower electrode driven at V_{pp} = 300 V. The area shown is 2.33 × 4.01 mm^2.

FIGURE 1.62
Nylon, polypirol coated particles (300 μm length and 2.5 μm radius) suspended in an argon plasma at pressure 115 Pa above an electrode driven at V_{pp} = 300 V. The area shown is 4.01 × 1.44 mm^2.

FIGURE 1.63
Nylon, polyaniline coated particles (300 μm length and 2.5 μm radius) in a complex plasma formed by 10 minutes etching of carbon in argon at a pressure 100 Pa. Symmetric rf discharge (PKE chamber) is driven at V_{pp} = 100 V. The area shown is 15.0 × 11.3 mm^2.

in an argon plasma in which carbon had been etched for about 10 minutes. Suspended nanoparticles are negatively charged and are repelled by the rods assembly that, being coated by conducting polymer, is at the potential of the electrode. The Mie-scattering of the laser light shows well the sheath profile. It is clear that the electric field in the upper part of the assembly is not vertical. A simple theoretical model of the effect of the multiparticle complexes near the bottom electrode is discussed by Annaratone *et al.* (2009).

Following Ivlev *et al.* (2003a) and Fortov *et al.* (2005a), let us consider levitation of a micro-rod of mass m and charge Q suspended in the electric field E of the rf discharge sheath. Parameters of the equilibrium state of the particle (the vertical coordinate of its center of mass h_0 and the orientation angle α_0 with respect to the vertical axis) can be found from minimization of the potential energy of the particle $U(h,\alpha)$. We now assume that among the different forces that act on the suspended micro-rods the electric force is dominant. Interaction between particles can be ignored in low particle density structures when the Debye length is shorter than all the other distances involved. Assuming for simplicity that the particle charge does not depend on h and α, the potential energy is determined by the following expansion:

$$U(h,\alpha) \simeq mgh + Q\phi(h) - \frac{dE(h)}{2}\cos^2\alpha - \frac{DE'(h)}{12}(3\cos^2\alpha - 1) + \dots, \quad (1.8)$$

with $E < 0$ and $E' > 0$. The magnitudes of the dipole and quadrupole moments are $d \simeq \frac{1}{24}EL^3/\Lambda < 0$ and $D \simeq \frac{1}{6}QL^2 < 0$, respectively [here $\Lambda = \ln(L/a)$ with the rod length L and radius a]. Equation (1.8) was obtained in the approximation of a "weakly inhomogeneous field" [$\ell_E = |E_0/E_0'| \gg L$, where $E(h) \simeq E(h_0) + E'(h_0)(h-h_0)$] for particles with high aspect ratio, $L/a \gg 1$ (Ivlev *et al.* 2003a). The equilibrium state of particles is determined by the absolute minimum of the potential energy. The conditions $\partial U/\partial h = 0$ (force balance) and $\partial U/\partial \alpha = 0$ (torque balance) result in

$$mg \simeq QE_0 + dE_0' \cos^2 \alpha, \quad \left(\frac{E_0^2 L}{2\Lambda} + QE_0'\right) \sin 2\alpha = 0. \tag{1.9}$$

The first equation shows that in addition to the gravity and monopole electric forces, the dipole force contributes to the balance condition in a vertical directions. However, this force does not affect the balance noticeably. From the second equation it follows that only two equilibrium orientation, vertical ($\alpha_0 = 0$) and horizontal ($\alpha_0 = \pi/2$), are possible. The condition for the stable angle one can get from the second derivative of the potential energy:

$$\partial^2 U/\partial \alpha^2 \sim (K-1)\cos 2\alpha_0 > 0, \tag{1.10}$$

where K is the "orientational parameter",

$$K = \frac{2d\,\ell_E}{D} \equiv \left(\frac{e|E_0|L}{\gamma_r T_e}\right)\frac{\ell_E}{L}, \tag{1.11}$$

and

$$\gamma_r = \frac{2\Lambda e|Q|}{LT_e} \tag{1.12}$$

is the absolute value of the dimensionless particle potential. This shows that the equilibrium is determined by the competition between the dipole and quadrupole terms in Equation (1.8). The dipole torque turns the rod along the electric field, whereas the quadrupole torque tends to make it horizontal. Hence, particles levitate horizontally ($\alpha_0 = \pi/2$) when $K < 1$, and vertically ($\alpha_0 = 0$) when $K > 1$. It follows from the obtained solutions that rotation of the rodlike particles in the vertical plane is impossible (it was never observed experimentally either).

For comparison of the obtained experimental results with the described theoretical model, it is necessary to estimate some plasma and particle parameters. The mass of the particle is $m = \pi a^2 L\rho \simeq 1.3 \times 10^{-11}$ kg ($\rho = 1.14 \times 10^3$ kg m^{-3}). The logarithm of the aspect ratio is $\Lambda = 5.48$. The electron energy is $T_e = 3$ eV. The particle charge is determined from Equation (1.9) using the measured value of the electric field strength, $Q \simeq mg/E_0$, and the dimensionless particles potential γ_r is determined by Equation (1.12). Then, using experimental values of ℓ_E, the parameter K is calculated. Results of these estimations are given in Table 1.2. One can see that in agreement with our theoretical model, the vertical orientation of the rodlike particles corresponds to $K = 4.3 \gg 1$, the horizontal one to $K = 0.10 \ll 1$, and transition between these two orientations takes place at $K \sim 1$. Thus, experimental results are in

TABLE 1.2
The rodlike particle charge Q, electric field strength E, spatial scale of the field variation ℓ_E, and orientation parameter K for different orientations.

	Vertical orientation	Mixed: Mainly vertical	Mixed: Mainly horizontal	Horizontal orientation
$-Q/e$	9.8×10^4	1.6×10^5	2.1×10^5	2.7×10^5
$-E$, kV/m	8.4	5.0	3.8	3.0
ℓ_E, mm	1.27	0.53	0.38	0.23
K	4.3	0.63	0.26	0.10

reasonable quantitative agreement with the predictions of the theoretical model by Ivlev *et al.* (2003a).

Technologies for manufacturing artificial flock materials most often utilize methods of pre-charging fibers in atmospheric-pressure corona discharges and their subsequent deposition at a given angle onto a glue substrate in an external electric field. Bulychev *et al.* (2003) and Dubinov and Sadovoy (2007) investigated experimentally the possibility of the orienting very large fibers (Kapron threads 100 μm in diameter and 3 mm in length) in glow discharges. Such fibers were about 1000 times heavier than those used by Molotkov *et al.* (1999) and Annaratone *et al.* (2001). Methods for extracting oriented fibers from the discharge region and for depositing them onto glue substrate, as well as methods for controlling their orientation angle with respect to the substrate surface were developed. The design of the camera and injector are described in detail by Bulychev *et al.* (2001) and Dubinov and Sadovoy (2007). The experimental results were as follows. When the fibers were injected into a unionized gas at a pressure from 0.1 Torr to an atmospheric one with no discharge initiated, they were observed not to be oriented and fell down on the substrate to lie there horizontally with random orientation. In contrast, when the fibers were injected into a steady uniform glow discharge, they fell down strictly vertically to orient themselves transverse to the discharge axis. After the fibers had reached the substrate, their lower ends were glued to it and the fibers themselves remained in the vertical position for an arbitrarily long time. The fibers glued to the substrate are shown in an enlarged fragment of the photograph in Figure 1.64a.

Transverse orientation of the fibers was explained by Dubinov and Sadovoy (2007) using comparison of the longitudinal and transverse components of the electric field in the discharge. The longitudinal component is known to be rather weak (Raizer 1991), while the transverse component, which is associated with the plasma density gradient near the dielectric wall of the chamber, can be as strong as a few hundred volts per centimeter. These experimental results yield a simple method of controlling the orientation angle of the fibers as they fall down on the substrate. Specifically, this can be done by changing the direction of the transverse plasma density gradient near the substrate surface with the help of a transverse limiter (see Figure 1.64b). Using

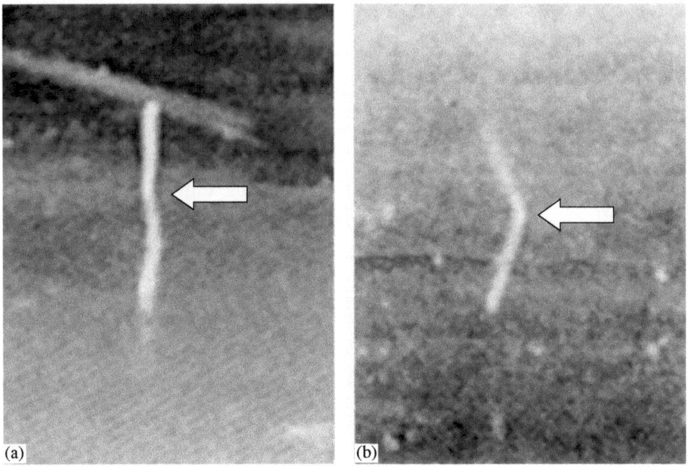

FIGURE 1.64
Photographs of fibers glued to a substrate: (a) in a longitudinally uniform glow discharge; (b) in a glow discharge with a limiter. The arrows show the sites where the fibers are glued (Dubinov and Sadovoy 2007)

.

this technique Bulychev *et al.* (2003) demonstrated experimentally that flock materials with a given inclination angle of fibers can be produced artificially by depositing fibers onto a glue substrate in a low-pressure glow discharge.

1.5 Formation and growth of dust particles

Generation of dust particles, ranging in size from a few nanometers to several tens of microns, is frequently reported for various plasma processing facilities. The particulate matter in the processing plasmas has numerous implications for the semiconductor micro-fabrication and materials processing. Here, the underlying physicochemical processes of the origin and growth of fine particles in reactive silane-, hydrocarbon- , and fluorocarbon-based plasmas, are discussed (Vladimirov and Ostrikov 2004; Vladimirov *et al.* 2005). Despite a difference in the process kinetics and the plasma chemistry involved, the growth scenario can be similar. The dust growth in chemically active plasmas starts with the formation of sub-nanometer/nano-meter-sized protoparticles nucleated as a result of homogeneous or heterogeneous processes. Thereafter, agglomeration/coagulation processes result in the generation of a few tens of nanometers particles, which quickly acquire negative electric charge as a result of collection of the plasma electrons and ions (Boufendi and Bouchoule

1994). The dust growth then usually proceeds to sub-micron and micron sizes via a relatively slow process of accretion of neutral or ionic monomers (e.g., deposition of SiH_x radicals on the grain surface in silane-based reactive plasmas).

Dust growth under the plasma conditions is not merely limited by the silane-, hydrocarbon-, and fluorocarbon-based chemistries. For example, carbon nitride particles, with diameters of a few hundred nanometers, can grow at room temperatures in rf capacitive discharges of N_2+CH_4 gas mixtures (Boufendi and Bouchoule 2002). Generally speaking, solid particle growth can certainly take place in many other reactive gas mixtures supporting polymerization and clustering processes in the ionized gas phase. Furthermore, chemical nucleation in the ionized gas phase is not the only possible mechanism of the dust growth in the processing plasmas. The particle growth can also be induced by physical and reactive sputtering of the wall or electrode material in the plasma-assisted dc/rf magnetron and other sputtering facilities.

1.5.0.1 Origin and mechanisms of growth of dust in silane plasmas

Plasmas of pure silane (SiH_4) and its mixtures are widely used for applications in the semiconductor industry (e.g., integrated circuitry and silicon-based microchips, flat panel displays, amorphous silicon solar cells). It is believed that understanding of the fine particle generation processes in silane-based plasmas is the most comprehensive as compared to other reactive plasma chemistries (Beck *et al.* 1994; Hollenstein 2000). The initial stages of the particle growth in pure silane discharges can adequately be described by the steady-state homogeneous nucleation model (Kortshagen and Bhandarkar 1999; Gallagher 2000). The basic assumption of the model is that the particle growth process is triggered by SiH_3^- anions and/or SiH_m neutral radicals, which polymerize into Si_nH_m radicals with larger numbers n of silicon atoms. With an increase of n, larger molecular clusters, and eventually subnano- or nano-sized particles of hydrogenated silicon are generated.

The apparent puzzle is to identify the precursor species and dominant gas-phase/surface reactions for the growth of larger ($> 10^4$ silicon atoms) particles and relate the dust growth to the discharge control parameters. There exist three major classes of possible catalyst candidates in the silicon hydride clustering process (Bhandarkar *et al.* 2000), namely, anions (negative ions), neutrals, and cations (positive ions). Overall, the underlying physics and chemistry of the dust origin and growth in chemically active plasmas critically depends on the prevailing experimental conditions.

For example, short lifetime neutral radicals SiH_2 can play the role at several stages of the dust growth (Watanabe *et al.* 2001). Neutral complexes are able to incorporate into larger saturated molecules and can thus be considered as viable nanoparticle growth precursors in reactive silane plasmas (Hollenstein 2000). Hence, in the short residence time situations one could expect that short-lifetime, highly reactive neutral radicals can efficiently support numerous homogeneous nucleation processes. In particular, neutral radicals SiH_m ($m = 0$–2) can be responsible for the nanopowder formation in dense helium or argon-diluted silane discharges (Watanabe 1997; Koga *et al.* 2000). Likewise, positive ions can also be regarded as potential powder precursors despite high activation barriers preventing the formation of higher-mass cations. In particular, cationic silicon clusters that contain up to ten silicon atoms have been detected in argon/hydrogen thermal plasmas by means of time-resolved mass-spectrometry by Leroux *et al.* (2000).

On the other hand, the anionic pathway is another viable route for the powder generation in silane-based plasmas. Invoking a simple argument that the formation of particulates does require critically large clusters, one can conclude that typical residence times of the neutrals are not sufficient to trigger the efficient dust growth process (Choi and Kushner 1993). However, the clustering process can involve negative ions trapped by the ambipolar potential in the plasma. Furthermore, the negatively charged intermediaries can increase the average residence time of the clusters and enable their growth to the critical size (Choi and Kushner 1993). Similar, hydrosilicon anions can be efficiently confined in the near-electrode areas and participate in the plasma-assisted clustering process. Thus, a large number of negative ions can accumulate and grow towards higher masses according to the homogeneous model (Gallagher 2000). Relevant time-resolved mass-spectrometry data have revealed that the anionic pathway is the most likely route for the nanoparticle generation in low-pressure rf silane plasmas (Hollenstein 2000). For example, the dust evolves from the molecular to the particulate form in low-pressure silane rf capacitively coupled plasmas (Howling *et al.* 1996). In this case the negative ions play a crucial role in the powder nucleation and growth process, and the entire range of negatively charged species, ranging from monosilicon anions through to nanometer-sized clusters, can be observed (Howling *et al.* 1996). Furthermore, the anion confinement correlates with the pronounced particle formation. Conversely, detrapping of the negative ions strongly inhibits the entire growth process.

Results of recent numerical simulation of dust particle formation mechanisms in silane discharges confirm that the anion SiH_3^- is the most dominant primary precursor of the particle formation. In fact, over 90% of the silicon hydride clustering proceeds through the silyl anion ($Si_nH_{2n+1}^-$) pathway, starting from SiH_3^-, whereas only $\sim$ 10% is through the siluene anion ($Si_nH_{2n}^-$) pathway, starting from SiH_2^- (De Bleecker *et al.* 2004). This conclusion is valid for negatively charged silicon hydride clusters $Si_nH_m^-$ containing up to 12 silicon and 25 hydrogen atoms.

The second phase of the dust particle growth can proceed via a rapid agglomeration of small clusters into larger (usually $\sim$40–50 nm-sized) particles (Kortshagen and Bhandarkar 1999). After the agglomeration phase is complete, the grain size increases with the relevant thin film growth rates. Note that the key dust nucleation and growth processes discussed above are most relevant to the plasmas of pure silane discharges. However, many real thin film fabrication processes require a substantial dilution of silane by hydrogen and/or argon. The offset and dynamics of the particulate growth appear quite different as compared to pure silane. In particular, silane dilution complicates the discharge chemistry and elongates the time scales required for the powder detection. Thus, the particulate size, bonding states, architecture, and surface morphology of the particles grown in the pure and buffer gas diluted silane plasmas can be quite different and critically depend on the reactive gas feedstock.

Physically, by varying the gas composition one can control the residence time t_{res}

of the precursor species in the discharge. Moreover, t_{res} appears to be a critical factor in the nanoparticle generation and growth. There is a direct correlation between the residence time of the precursor radicals and the size of fine particles detected (Bouchoule and Boufendi 1993). The selective trapping model of Fridman *et al.* (1996) assumes that the neutrals should reside in the reactor volume long enough to acquire a negative charge through the electron non-dissociative attachment and/or heavy particle charge exchange collisions. In this case, the nano-sized particles can be trapped in the near-electrode areas, building up the minimum number density for the coagulation onset.

In mixtures of reactive gases currently used for fabrication of various functional thin films and nanomaterials, the processes of fine powder generation are usually more complicated than in pure silane plasmas. For example, high-density plasmas of highly reactive SiH_4+O_2+Ar gas mixture (involving two electronegative gases – silane and oxygen) are used for the fabrication of silica nanoparticles. The challenge for the adequate understanding of the fine particle nucleation and growth processes is to incorporate the complex effects of the numerous chemical reactions involved, high reactivity of high-density (especially inductively coupled plasmas currently used as benchmark plasma reactors for semiconductor manufacturing) plasmas, and finite reactor size effects (most of the currently existing models deal with spatially averaged discharge models). From the plasma chemistry point of view, the polymerization of critical clusters can involve a combination of several clustering pathways. The clustering process in high-density SiH_4+O_2+Ar plasmas can proceed through various ion–neutral and neutral–neutral clustering channels (Suh *et al.* 2003). Ion–neutral clustering reactions can involve either positive or negative ionic precursors. On the other hand, the neutral–neutral clustering develops through the self-clustering reaction or by addition of SiO or SiO_2 radicals to the growing polymeric chain. An interesting peculiarity of dust-generating high-density plasmas of SiH_4+O_2+Ar mixtures is that the rates of positive–negative ion neutralization are very high due to comparable number densities of positively and negatively charged species. In this case, the neutralization of anions (e.g., SiH_3^-, currently believed to be one of the most important precursors for the dust growth in pure silane plasmas) occurs much faster than the clustering of neutral species. This explains why neutral clusters exhibit higher concentrations relative to anionic clusters (Suh *et al.* 2003).

Therefore, in each particular case the plasma chemistry behind the dust nucleation and growth can be quite different even in the presence of the same silicon-bearing precursor gas, which is an essential component for the fine particle nucleation. To this end, one cannot a priori state which particular precursor radical triggers the plasma polymerization of the critical clusters that nucleate into larger particles. However, it is quite possible to identify a few dominant clustering pathways and, relative role of the species in any particular charge state (positive, negative, or neutral). In this regard, one can perform the sensitivity analysis, which can give an answer as to which reactions dominate the production and steady-state number densities of higher silicon oxides [e.g., anions $(SiO_x)_{11}^-$ or neutrals $(SiO_x)_{11}$)] (Suh *et al.* 2003).

1.5.0.2 Growth of dust in hydrocarbon plasmas

The existing understanding of the nanoparticle growth in hydrocarbon (C_mH_n, e.g., methane, CH_4, or acetylene, C_2H_2) discharges is currently less developed as compared with the similar processes in the silane-based plasmas. However, the study of the plasma chemistry and growth of nano-sized particles in relevant ionized gas mixtures is gradually gaining momentum. For example, a numerical model of the nanoparticle clustering kinetics in the low-pressure rf discharge in acetylene by Stoykov *et al.* (2001) incorporates numerous gas-phase processes including the electron impact dissociation, electron attachment leading to the negative ion generation, ion–ion recombination, ion–neutral clustering, chemical reactions involving the hydrocarbon (chain and aromatic) neutrals, as well as diffusion losses of the plasma species to the discharge walls.

It is usually assumed that the carbon hydride clustering process is triggered by the electron-impact abstraction of hydrogen from the acetylene monomer followed by the efficient generation of C_mH_n radicals (with higher numbers of carbon and hydrogen atoms) via a chain of polymerization reactions (Stoykov *et al.* 2001). The model allows one to predict the most probable clustering pathways as well as the temporal evolution of the number densities of the major charged and neutral species. The most likely clustering process proceeds through the addition of the anion species $C_iH_j^-$ to the neutrals C_mH_n accompanied by the elimination of hydrogen and generation of the higher-mass anions. Eventually, the rapid chemical nucleation stage evolves into the equilibrium state, which can usually be reached when the particle loss to the walls is compensated by the production of the new species. The equilibrium state is strongly affected by the neutral gas temperature, rf power input, and working gas pressure. Similarly, depending on the external parameters, the particle nucleation process can be either enhanced or inhibited. Even though clustering occurs mainly through the formation of linear molecules, the proportion of aromatic hydrocarbons increases and becomes significant at higher working gas temperatures.

The results of numerical modeling of the clustering processes in acetylene plasmas by Stoykov *et al.* (2001) demonstrate that after the initial increase in the species concentrations, the production rates slow down and eventually a steady state is reached. This indicates that a balance between the species production and diffusion losses is achieved. Note that the rates of the diffusion losses are proportional to the species concentration and this loss channel plays only a minor role at the early stages of particulate development. This certainly favors a quick initial rise in the number densities of the reactive plasma species. Eventually, the diffusion losses are balanced by the gas-phase reactions that lead to the particle production and a steady state of the discharge can be established.

A comparison of the number densities of the structurally similar neutrals and anions reveals that the anion concentrations are much lower. Since the density of positive ions is an upper limit for the combined anion and electron densities, the above difference can be attributed to high growth rates of the neutral particles in acetylene plasmas. This does not necessarily mean that nano-sized particles are mostly neutral. In fact, one can note that most of the negative species are formed in the particle form, with the number densities of the same order of magnitude as the concentrations of the neutral particles. The ratio between the number densities of neutral and negative particles is important for the understanding of the details of the further growth processes, which are affected by the grain charge (Kortshagen and Bhandarkar 1999). To this end, the formation of neutral particles is favored at lower temperatures, higher degrees of ionization, and higher pressures (Stoykov *et al.* 2001). It is interesting that Fourier Transform Infra Red (FTIR) spectroscopy data collected in situ and from the powder samples confirm a predominant production of the acetylenic compounds in the ionized gas phase, although the presence of aromatic compounds appears to be non-negligible. The mass spectrometry studies of the rf plasma in acetylene by Deschenaux *et al.* (1999) show the presence of aromatic compounds such as benzene, substituted benzenes and toluenes.

The above results are in an agreement with the experiments of Hong *et al.* (2002) on the dust particle generation, size-controlled growth, diagnostics and deposition in 13.56 MHz rf plasmas of Ar/CH_4 and Ar/C_2H_2 gas mixtures. The most efficient dust particle generation is commonly observed for the elevated rf power levels, which indicates on the importance of the adequate amounts of the particle growth precursors (Hong *et al.* 2002). Another in situ FTIR spectroscopy and the plasma-ion mass spectrometry measurements of Kovacevic *et al.* (2003) evidence the highly-monodisperse size distributions of nanoparticles grown in rf plasmas of Ar+C_2H_2 gas mixtures. This conclusion is also cross-referenced by the scanning electron microscopy of the powder samples collected during different growth phases. Measurements of the intensity of the Rayleigh–Mie scattering of the infrared signal reveal that the process of the fine particle generation, growth and disappearance is periodic. The oscillation period of the infrared signal is approximately 35 min under prevailing experimental conditions (Kovacevic *et al.* 2003). The time scales when the electron-impact ionization is enhanced and the plasma parameters in Ar+C_2H_2 rf discharges noticeably change due to the dust growth $\tau_{C_2H_2}$ and appear to be consistently longer than the corresponding time scales $\tau_{\mathtt{SiH_4}}$ in silane-based plasmas.

The observed periodicity of the Reyleigh–Mie scattering signal can be attributed to the fact that the negatively charged particles are confined in the plasma potential as long as the different forces acting on the single particle are balanced (Kovacevic *et al.* 2003). Since the major forces acting on dust particles scale differently with the grain radius, the actual particle confinement critically depends on their size. As soon as the particles reach the critical size, they are either dragged out of the plasma bulk or fall down onto the lower electrode, which results in a decrease of the scattered signal. A quick drop in the intensity of the above signal evidences a highly monodispersive character of the powder growth process in Ar+CH_4 plasmas. Furthermore, the *ex situ* scanning electron microscopy suggests that the particles collected 10 min after the ignition of the discharge have a spheroidal shape with a particle diameter of about 150 nm and a fractal surface texture. It is thus likely that the accretion (uniform deposition of the neutral species onto the particle's surface) is probably a dominant particle growth mechanism.

An interesting observation related to the dust growth process is a high consumption of the acetylene monomer for the plasma polymerization as evidenced by the neutral mass spectrometry (Kovacevic *et al.* 2003). This is consistent with the findings of Deschenaux *et al.* (1999) and Hong *et al.* (2002) that acetylene as a monomer plays an important role in the fine powder formation in hydrocarbon plasmas. It also turns out that acetylenic compounds play a vital role in the dust nucleation and growth processes. There are two relevant experimental observations of Hong *et al.* (2003). First, in Ar+C_2H_2 plasmas the fine particles usually nucleate spontaneously at low discharge powers. On the other hand, the particle growth in Ar+CH_4 plasmas usually starts only after a transient elevation of the rf power or a quick inlet of the C_2H_2 monomer into the discharge volume. This can presumably be attributed to different nucleation scenarios in Ar+CH_4 and Ar+C_2H_2 discharges. Apparently, the procedure of adding more C_2H_2 or rf power to the discharge is required to trigger the nucleation of primary clusters and protoparticles, which is a quite slow process in Ar+CH_4 plasmas. Once the precursors are formed, the further growth process can proceed under normal discharge operation conditions.

Note that the elevated abundance of the C_2H_2 monomer species in the Ar+CH_4+H_2 inductively coupled plasmas for the PECVD (plasma enhanced chemical vapor deposition) of various carbon-based nanostructures (Tsakadze *et al.* 2004, 2005) can be achieved by operating the discharge at elevated rf powers. One can thus presume that the relevant nanostructure growth process can be strongly affected by the pronounced formation of fine powder particles in the ionized gas phase. On the other hand, the dynamics of the dust formation in Ar+C_2H_2 plasmas is periodic and results in the following scenario: nucleation followed by further growth followed by development of dust-free regions (dust voids, mostly due to the action of the ion drag force that pushes the dust grains from the plasma bulk) followed by new nucleation in the dust-free regions. A possible explanation for the differences in the dust growth dynamics in methane-based and acetylene-based reactive environments is that the nucleation process strongly depends on the concentration of C_2H^- negative ions efficiently generated in the Ar+C_2H_2 plasmas (Hong *et al.* 2003).

Another way to trigger dust generation in low-density methane-based plasmas is to use a pulsed Nd:YAG laser (Stoffels *et al.* 1999). If the photon energy fits the dissociation energy of the C-H bond, the absorption of UV photons results in the rapid dissociation of methane molecules and creation of active radicals, which is otherwise inefficient in the pristine methane plasma. In this way, it appears possible to synthesize submicron-sized dust particles that can subsequently arrange into larger agglomerates and structures levitating in the vicinity of the powered rf electrode (Stoffels *et al.* 1999). Some of the resulting particle arrangements can be further deposited and continue growing on the surface.

There are numerous indications that dust powder formation can also be induced by the surface and reactor contamination effects. For example, in pure methane discharges in a clean reactor chamber, the powder formation process takes at least a few hundred seconds. However, in a contaminated reactor, the dust particle appearance can be detected much faster. Thus, the dust formation might be affected by surface effects as is the case for SiN dusty plasmas. However, no high mass neutrals, cations or anions have been detected by the mass spectrometry, in contrast to the silane plasmas (Hollenstein 2000). Hence, it is very likely that large particles are formed via heterogeneous processes which are most common for the situations when the plasma species are non-reactive and direct gas-phase reactions leading to the formation of critical clusters are not efficient. In this case the dust particle growth can proceed via the electron-induced surface desorption of nano-sized clusters. The initially neutral clusters can migrate into the near-electrode/plasma sheath area where the probability of their excitation/ionization via collisions with high-energy electrons is quite high. Ion-molecular reactions can further contribute to the particulate growth. Finally, a pronounced coagulation process can lead to the formation of larger agglomerates (Hollenstein 2000).

1.5.0.3 Growth of dust in fluorocarbon plasmas

Fluorocarbon (C_xF_y) based plasmas have recently been widely used for ultrafine and highly selective etching of polysilicon and a number of PECVD processes including many common applications in thew microelectronic industry. Furthermore, many plasma etching processes of silicon and its components, as well as deposition of chemically resistant barriers, dry lubricants, etc., involve CF_4, C_2F_6, CHF_3, C_2F_4, aromatic fluorocarbons, etc. (Buss and Hareland 1994). A gas phase particulate formation is possible in capacitively coupled rf fluorocarbon plasmas. A sequence of monochromatic images of particulate suspension and growth obtained from a 13.56 MHz capacitively coupled vinilydene fluoride plasma at 27 mTorr sustained with 30 W rf powers reveals that the time of the initial particle detection usually varies in the $\sim$ 10 to 250 s range (Buss and Hareland 1994). In this case, the particles are usually non-agglomerated, have an almost spherical shape and can accumulate during the extended discharge operation. The grain diameter typically ranges from 110 to 270 nm.

Similar to silane- and hydrocarbon-based plasmas, the particles develop in size and evolve into a certain spatial pattern (e.g., dust cloud) usually suspended between the two electrodes (Buss and Hareland 1994). The time of the first appearance of particles is quite sensitive to the total gas pressure and the discharge chemistry. The addition of hydrogen or hydrogen-containing gas (e.g., CH_4) to a fluorocarbon discharge can result in an increase of the particle growth rate and the corresponding shortening of their first detection time. This effect can be attributed to the enhanced production of free radicals by hydrogen atom abstraction of fluorine. The appearance time for particles in a C_2F_4 plasma at 140 mTorr turns out to be approximately 110 s (Buss and Hareland 1994). Assuming a constant radial growth rate (and estimating a minimal diameter for the detection by the laser light scattering, one can obtain 0.5–1.4 $\mathrm{nm\,s^{-1}}$ for the particle growth rate. It is worth noting that the fluorocarbon film growth on the substrate placed on the lower electrode has a comparable rate of 2 $\mathrm{nm\,s^{-1}}$.

RF discharges of octafluorocyclobutane-based gas mixtures also generate large amounts of highly polymerized molecules, which correlates with the plasma polymerization processes in the gas phase (Takahashi and Tachibana 2001a). The chemistry behind the gas-phase nucleation processes can be quite similar in the silane- and fluorocarbon-based plasmas. Indeed, higher fluorocarbons polymerized in the ionized gas phase can act as efficient precursors for the generation of nano-sized particles and also take part in the thin film deposition processes.

Solid grains and agglomerates can also be abundant in fluorocarbon plasmas for ultra-fine selective etching of SiO_2 and PECVD of low-dielectric constant polymeric films. There is a remarkable correlation between the polymerization in the ionized gas phase and the relevant surface processes, which can shed some light on the prevailing powder formation mechanisms in fluorocarbon plasmas. For example, fluorinated carbon particles can be generated in a parallel plate 13.56 MHz plasma reactor, where a capacitively coupled plasma of c-C_4F_8 is sustained with the rf power density of 0.15 W cm^{-2} within the pressure range from 23 to 250 mTorr, which is typical of the PECVD of fluorinated amorphous carbon (a-C:F) thin films (Takahashi and Tachibana 2001b). Under such conditions, numerous nano- and micron-sized particles and agglomerates dispersed over the wafer surface can be observed. The diameter of the gas-phase grown particles typically ranges from 0.5 to 2.3 microns. In the intermediate pressure range ($>$50 mTorr), generation of the agglomerates with the size in the few tens of micrometer range and composed of the primary spherical particles takes place (Takahashi and Tachibana 2001b). The number of primary particles building up the agglomerates increases with pressure. A typical size of the fluorocarbon-based agglomerate at 250 mTorr pressure is about 30 μm.

The gas-phase particulate polymerization can be inferred through the dependence of the film deposition rate on the gas feedstock pressure. Specifically, the film deposition rate decreases when the gas pressure exceeds 50 mTorr (Takahashi and Tachibana 2001b). Presumably, this can be attributed to the enhanced loss of the gas-phase polymer precursors to the particle generation processes. It is also worthwhile to mention that no significant particulate formation in the gas phase is observed in CF_4 or C_2F_6-based discharges. Thus, it seems reasonable to conclude that stable CF_4 and C_2F_6 molecules cannot be as efficient as precursors for the particle growth. Even though the trigger catalyst plasma species are yet to be conclusively identified, one can presume the nanocluster route for the fine particle growth. Furthermore, there is a correlation between the clusterizing rates and the gas-phase concentrations of the source gas molecules and the main products of the first-order reactions (Takahashi and Tachibana 2001b).

We emphasize that similarly to SiH_4 and C_mH_n plasmas considered above, the negative ions also play an important role in the clustering reactions in fluorocarbon plasmas. Thus, elucidation of the dust generation pathways, including a detailed experimental investigation of the catalyst species and gas-phase reactions for polymerization is an apparent forthcoming challenge for the coming years. For example, solid C:F particles can be polymerized in C_4F_8 plasmas under conditions of pronounced generation of molecular species CF_4, C_2F_6, and C_2F_4 (Takahashi and Tachibana 2001b). Among them, the C_2F_4 molecule plays a leading role in the gas-phase synthesis of dust grains (Takahashi and Tachibana 2002). A possible reason is that this molecule can be activated in the plasma and subsequently transformed into highly reactive species $-CF{=}CF-$ and $-CF{=}CF_2$ involved in numerous branching and polymerization reactions. Thus, high molecular weight compounds appear and act as the particle nucleation precursors. It is interesting to mention that under similar experimental conditions the particle growth and film deposition is usually not observed in pure CF_4 or C_2F_6 plasmas (Takahashi and Tachibana 2002).

1.5.0.4 Dust growth in plasma-enhanced sputtering facilities

Nano and micron-sized particles of various materials (graphite, titanium, copper, silicon, etc.) can also be generated in plasma-enhanced sputtering facilities (Buss and Hareland 1994; Samsonov and Goree 1999). Contaminant particles appear in several kinds of plasma discharges with various sputtering targets. In particular, silicon/silica, carbon, copper, etc., particles can be synthesized in dc/rf plasmas of various gas feedstocks (Selwyn *et al.* 1990; Jellum *et al.* 1991; Ganguly *et al.* 1993).

For example, submicron- to micron-sized particles can be formed in the gas phase of sputtering capacitively coupled discharges by using a variety of target materials (Samsonov and Goree 1999). The dust clouds usually appear after a few seconds or minutes of the discharge operation and can be detected by a sensitive video camera. This period will further be referred to as the particle detection time $t_{\rm det}$. Initially, the particle cloud fills the entire volume between the electrodes except for the plasma sheaths, with the highest density near the upper (powered) rf electrode. Once the particles reach a critical diameter, which is approximately 120 nm in the experiments of Samsonov and Goree (1999), the discharge becomes unstable. At the end of the instability cycle one can observe an empty region (void) in the center of the particle cloud. The void expands as the particles grow in size until the void fills in nearly the whole inter-electrode region. This marks the end of the growth cycle $t_{\rm growth}$.

Typically, $t_{\rm det}$ varies from 15 s for copper to 10 min for aluminum, whereas $t_{\rm growth}$ varies from 3 min for Torr Seal epoxy to 3 hr for titanium sputtering targets (Samsonov and Goree 1999). After the end of the growth stage, the sizes of the graphite, titanium, stainless steel, and tungsten particles are usually in the submicron range (typically 300–400 nm in diameter), whereas aluminum and copper particles grow to micrometer sizes (typically 1–5 μm). Note that particles grown from different materials feature different shapes. Some particles, such as copper or aluminium, are filamentary fractals. In contrast, carbon particles usually have a bumpy spherical shape. Other materials, such as titanium and stainless steel, form compact coagulants of a few spheres. On the other hand, tungsten particles form compact agglomerates.

As compared with reactive (e.g., silane) plasmas, particle growth rates are usually lower in the sputtering discharges mostly because of the lower number densities of clustering/agglomerating species. However, the sputtering discharges have an obvious advantage that they can produce particles from almost any solid material that can be sputtered without decomposition. Similar to chemically active plasmas, the particle growth process in sputtering discharges also develops in several stages. However, in this case the particles originate as clusters released from the sputtering target or the walls. Afterwards, the clusters nucleate into primary particles, which can further agglomerate and form particulates of various shapes and architectures (e.g., spongy and filamentary or compact and spheroidal).

Note that the electric charge is a critical factor in determining the shape of the plasma-grown nanoparticles (Huang and Kushner 1997). Indeed, when particles have a small (typically negative) charge and a high velocity, they can easily overcome Coulomb repulsion and form compact or spheroidal agglomerates. On the other hand, when the charge is larger and the velocity is lower, the electrostatic repulsion is stronger and an incoming particle is more likely to strike the end of a particle chain than the middle (mostly because of the plasma shielding of the distant elements of the agglomerate), and this process tends to promote a filamentary or fractal shape (Huang and Kushner 1997).

The picture of the particulate growth is certainly more complex in magnetron sputtering discharges (Selwyn *et al.* 1997). The mechanisms for particle generation, transport and trapping during the magnetron sputter deposition are different from the mechanisms reported in etching processes in reactive plasmas, due to the inherent spatial non-uniformity of magnetically enhanced plasmas. Since the magnetron sputtering facilities are usually operated at low pressures, the contribution from the homogeneous mechanism (which is a dominant one in silane plasmas) is likely to be small. Hence, most contamination problems in magnetron sputtering processes can be attributed to heterogeneous contamination sources, such as wall flaking. Furthermore, highly non-uniform plasmas typical of magnetron sputtering processes are subject to simultaneous material removal and redeposition in different target regions. Thus, the formation of filament structures can be favored. Meanwhile, the filaments can be resistively heated by intense current flows, which can cause violent mechanical failures and the removal of the filament into the plasma bulk. Combined with the repulsion between the negatively charged filament and the sheath region, this process can result in an acceleration of the filaments away from the sputter target, which can be a source of hot and fast particles capable of damaging the substrate being processed (Selwyn *et al.* 1997).

Note that dc/rf sputtering belongs to a larger group of particle generation mechanisms from the surrounding solid surfaces, encompassing the reactive ion etching (Anderson *et al.* 1994), the filtered cathodic vacuum arc deposition (Beilis *et al.* 1999), the hollow cathode discharges, and some other processes. For example, in the anisotropic etching controlled by directed ion fluxes, small columnary etch residues are usually formed. As a result of a slight under-etching, the columns become thinner at their base and thus unstable. Since the structures are negatively charged, the Coulomb repulsion from the surface causes them to break off and to be ejected into the discharge. The split etch residues are finally trapped in the glow by the plasma force balance. The above mechanism is quite similar to the magnetically-enhanced sputtering systems.

References

Amiranashvili, S. G., Gusein-zade, N. G., and Tsytovich, V. N. (2001). Spectral properties of small dusty clusters. *Phys. Rev. E*, **64**, 016407/1–6.

Anderson, H. M., Radovanov, S., Mock, J. L., and Resnick, P. J. (1994). Particles in C_2F_6-CHF_3 and CF_4-CHF_3 etching plasmas. *Plasma Sources Sci. Technol.*, **3**, 302–309.

Annaratone, B. M., Khrapak, A. G., Ivlev, A. V., Söllner, G., Bryant, P., Sütterlin, R., Konopka, U., Yoshino, K., Zuzic, M., Thomas, H. M., and Morfill, G. E. (2001). Levitation of cylindrical particles in the sheath of rf plasma. *Phys. Rev. E*, **63**, 036406/1–6.

Annaratone, B. M., Khrapak, S. A., Bryant, P., Morfill, G. E., Rothermel, H., Thomas, H. M., Zuzic, M., Fortov, V. E., Molotkov, V. I., Nefedov, A. P., Krikalev, S., and Semenov, Y. P. (2002). Complex plasma boundaries. *Phys. Rev. E*, **66**, 056411/1–4.

Annaratone, B. M., Glier, M., Stuffler, T., Raif, M., Thomas, H. M., and Morfill, G. E. (2003). The plasma-sheath boundary near the adaptive electrode as traced by particles. *New J. Phys.*, **5**, 92/1–12.

Annaratone, B. M., Antonova, T., Goldbeck, D. D., Thomas, H. M., and Morfill, G. E. (2004). Complex-plasma manipulation by radiofrequency biasing. *Plasma Phys. Control. Fusion*, **46**, B495–B509.

Annaratone, B. M., Khrapak, A. G., and Morfill, G. E. (2009). Peculiar properties of rodlike particles levitating in the sheath of an rf plasma. *IEEE Plasma Phys.*, **37**, 1110–1115.

Antonova, T., Annaratone, B. M., Thomas, H. M., and Morfill, G. E . (2008). Energy relaxation and vibrations in small 3D plasma clusters. *New J. Phys.* **10**, 043028/1–13.

Arp, O., Block, D., and Piel, A. (2004). Dust Coulomb balls: Three-dimensional plasma crystals. *Phys. Rev. Lett.*, **93**, 165004/1–4.

Arp, O., Block, D., Klindworth, M., and Piel, A. (2005). Confinement of Coulomb balls. *Phys. Plasmas*, **12**, 122102/1–9.

Ashoori, R. C. (1996). Electrons in artificial atoms. *Nature*, **379**, 413–419.

Asinovskii, E. I., Kirillin, A. V., and Markovets, V. V. (2006). Plasma coagulation of microparticles on cooling of glow discharge by liquid helium. *Phys. Lett. A*, **350**, 126–128.

Astrakharchik, G. E., Belousov, A. I., and Lozovik, Y. E. (1999a). Properties of two-dimensional dusty plasma clusters. *Phys. Lett. A*, **258**, 123–130.

Astrakharchik, G. E., Belousov, A. I., and Lozovik, Y. E. (1999b). Two-dimensional mesoscopic dusty plasma clusters: Structure and phase transitions. *JETP*, **89**, 696–703.

Barkan, A., Merlino, R. L., and D'Angelo, N. (1995). Laboratory observation of the dust–acoustic wave mode. *Phys. Plasmas*, **2**, 3563–3565.

Barkby, S., Vladimirov, S. V., and Samarian, A. A. (2008). Oscillation modes in Coulomb clusters with variable charges. *Phys. Lett. A*, **372**, 1501–1507.

Beck, S. E., Collins, S. M., and O'Hanlon, J. F. (1994). A study of methods for moving particles in rf processing plasmas. *IEEE Trans. Plasma Sci.*, **22**, 128–135.

Bedanov, V. M. and Peeters, F. M. (1994). Ordering and phase transitions of charged particles in a classical finite two-dimensional system. *Phys. Rev. B*, **49**, 2667–2676.

Beilis, I. I., Keidar, M., Boxman, R. L., and Goldsmith, S. (1999). Macroparticle separation and plasma collimation in positively biased ducts in filtered vacuum arc deposition systems. *J. Appl. Phys.*, **85**, 1358–1365.

Benilov, M. S. (1988). Theory of electrical probes in flows of high-pressure weakly ionized plasma (a review). *High. Temp.*, **26**, 780–793.

Bhandarkar, U. V., Swihart, M. T., Girshik, S. L., and Kortshagen, U. (2000). Modeling of silicon hydride clustering in a low-pressure silane plasma. *J. Phys. D: Appl. Phys.*, **33**, 2731–2746.

Block, D., Kading, S., Melzer, A., Piel, A., Baumgartner, H., and Bonitz, M. (2008). Experiments on metastable states of three-dimensional trapped particle clusters. *Phys. Plasmas*, **15**, 040701/1–4.

Bouchoule, A. and Boufendi, L. (1993). Particulate formation and dusty plasma behaviour in argon-silane rf discharge. *Plasma Sources Sci. Technol.*, **2**, 204–213.

Boufendi, L. and Bouchoule, A. (1994). Particle nucleation and growth in a low-pressure argon-silane discharge. *Plasma Sources Sci. Technol.*, **3**, 262–267.

Boufendi, L., and Bouchoule, A. (2002). Industrial developments of scientific insights in dusty plasmas. *Plasma Sources Sci. Technol. A*, **11**, A211–A218.

Bulychev, S. V., Dubinov, A. E., Zhdanov, V. S., L'vov, I. L., Mikheev, K. E., Sadovoy, S. A., Saikov, S. K., and Selemir, V. D. (2001). Investigation of scattering of dust micro-particles by plasma oscillations. *Prikl. Mekh. Tekh. Fiz.*, **42**, 19–25.

Bulychev, S. V., Dubinov, A. E., Kudasov, Yu. B., L'vov, I. L., Mikheev, K. E., Sadovoy, S. A., Saikov, S. K., and Selemir, V. D. (2003). Controlled orientation of fibers in a glow discharge plasma. *Tech. Phys. Lett.*, **29**, 636–637.

Buss, R. J. and Hareland, W. A. (1994). Gas phase particulate formation in radiofrequency fluorocarbon plasmas. *Plasma Sources Sci. Technol.*, **3**, 268–272.

Candido, L., Rino, J.-P., Studart, N., and Peeters, F. M. (1998). The structure and spectrum of the anisotropically confined two-dimensional Yukawa system. *J. Phys.: Condens. Matter*, **10**, 11627–11644.

Carstensen, J., Greiner, F., Hou, L.-J., Maurer, H., and Piel, A. (2009). Effect of neutral gas motion on the rotation of dust clusters in an axial magnetic field. *Phys. Plasmas*, **16**, 013702/1–8.

Cheung, F., Samarian, A., and James, B. (2003). The rotation of planar-2 to planar-12 dust clusters in an axial magnetic field. *New J. Phys.*, **5**, 75/1–15.

Cheung, F., Samarian, A., and James, B. (2004). Angular velocity saturation in planar dust cluster rotation. *Phys. Scripta*, **107**, 229–232.

Choi, S. J. and Kushner, M. J. (1993). The role of negative ions in the formation of particles in low-pressure plasmas. *J. Appl. Phys.*, **74**, 853–861.

Chu, J. and I, L. (1994). Direct observation of Coulomb crystals and liquids in rf dusty plasmas. *Phys. Rev. Lett.*, **72**, 4009–4012.

Collins, S. M., Brown, D. A., O'Hanlon, J. F., and Carlile, R. N. (1996). Particle trapping, transport, and charge in capacitively and inductively coupled argon plasmas in a Gaseous Electronics Conference Reference Cell. *J. Vac. Sci. Technol. A*, **14**, 634–638.

De Bleecker, K., Bogaerts, A., Gijbels, R., and Goedheer, W. (2004). Numerical investigation of particle formation mechanisms in silane discharges. *Phys. Rev. E*, **69**, 056409/1–16.

Deschenaux, C., Affolter, A., Magni, D., Hollenstein, C., and Fayet, P. (1999). Investigations of CH_4, C_2H_2 and C_2H_4 dusty rf plasmas by means of FTIR absorption spectroscopy and mass spectrometry. *J. Phys. D: Appl. Phys.*, **32**, 1876–1886.

Drocco, J. A., Reichhardt, C. J. O., Reichhardt, C., and Janko, B. (2003). Structure and melting of two-species clusters in a parabolic trap. *Phys. Rev. E*, **68**, 060401/1–4.

Dubin, D. H. E. and O'Neil, T. M. (1999). Trapped nonneutral plasmas, liquids, and crystals (the thermal equilibrium states). *Rev. Mod. Phys.*, **71**, 87–172.

Dubinov, A. E. and Sadovoy, S. A. (2007). Mechanical effects in a rarified plasma. *Plasma Phys. Rep.*, **33**, 316–328.

Dzlieva, E. S., Karasev, V. Yu., and Eikhval'd, A. I. (2005). The effect of a longitudinal magnetic field on the plasmadust structures in strata in a glow discharge. *Opt. Spectrosc.*, **98**, 569–573.

Dzlieva, E. S., Karasev, V. Yu., and Eikhval'd, A. I. (2006). The onset of rotational motion of dusty plasma structures in strata of a glow discharge in a magnetic field. *Opt. Spectrosc.*, **100**, 456–462.

Filinov, A. V., Bonitz, M., and Lozovik, Yu. E. (2001). Wigner crystallization in mesoscopic 2D electron systems. *Phys. Rev. Lett.*, **86**, 3851–3854.

Fomenko, V. S. (1981). *Emission properties of materials*. Naukova Dumka, Kiev. [in Russian].

Fortov, V. E., Nefedov, A. P., Petrov, O. F., Samarian, A. A., Chernyschev, A. V., and Lipaev, A. M. (1996a). Experimental observation of Coulomb ordered structure in sprays of thermal dusty plasmas. *JETP Lett.*, **63**, 187–192.

Fortov, V. E., Nefedov, A. P., Torchinskii, V. M., Molotkov, V. I., Khrapak, A. G., Petrov, O. F., and Volykhin, K. F. (1996b). Crystallization of a dusty plasma in the positive column of a glow discharge. *JETP Lett.*, **64**, 92–98.

Fortov, V. E., Nefedov, A. P., Petrov, O. F., Samarian, A. A., and Chernyschev, A. V. (1997a). Highly nonideal classical thermal plasmas: Experimental study of ordered macroparticle structures. *JETP*, **84**, 256–261.

Fortov, V. E., Nefedov, A. P., Torchinsky, V. M., Molotkov, V. I., Petrov, O. F., Samarian, A. A., and Lipaev, A. M. (1997b). Crystalline structures of strongly coupled dusty plasmas in dc glow discharge strata. *Phys. Lett. A*, **229**, 317–322.

Fortov, V. E., Nefedov, A. P., Vaulina, O. S., Lipaev, A. M., Molotkov, V. I., Samaryan, A. A., Nikitski, V. P., Ivanov, A. I., Savin, S. F., Kalmykov, A. V., Solov'ev, A. Y., and Vinogradov, P. V. (1998). Dusty plasma induced by solar radiation under microgravity conditions: Experiments in the Russian space station "Mir". *JETP*, **87**, 1087–1097.

Fortov, V. E., Nefedov, A. P., Vladimirov, V. I., Deputatova, L. V., Molotkov, V. I., Rykov, V. A., and Khudyakov, A. V. (1999). Dust particles in a nuclear-induced plasma. *Phys. Lett. A*, **258**, 305–311.

Fortov, V. E., Khrapak, A. G., Khrapak, S. A., Molotkov, V. I., Nefedov, A. P., Petrov, O. F., and Torchinsky, V. M. (2000a). Mechanism of dust–acoustic instability in a direct current glow discharge plasma. *Phys. Plasmas*, **7**, 1374–1380.

Fortov, V. E., Molotkov, V. I., and Torchinsky, V. M. (2000b). Plasma crystals and liquids in dc glow discharge. In *Frontiers in dusty plasmas*, Nakamura, Y., Yokota, T., and Shukla, P. K. (eds.), pp. 445–448. Elsevier, Amsterdam.

Fortov, V. E., Nefedov, A. P., Sinel'shchikov, V. A., Usachev, A. D., and Zobnin, A. V. (2000c). Filamentary dusty structures in rf inductive discharge. *Phys. Lett. A*, **267**, 179–183.

Fortov, V. E., Khrapak, A. G., Molotkov, V. I., Nefedov, A. P., Poustylnik, M. Y., Torchinsky, V. M., and Yoshino, K. (2001a). Behavior of rod-like dust particles in striations. In *Proceedings of the XXV international conference on phenomena in ionized gases (ICPIG–2001)*, Nagoya, Japan. Contributed papers, Vol. 3, pp. 35–36.

Fortov, V. E., Nefedov, A. P., Vladimirov, V. I., Deputatova, L. V., Budnik, A. P., Khudyakov, A. V., and Rykov, V. A. (2001b). Dust grain charging in the nuclear-induced plasma. *Phys. Lett. A*, **284**, 118–123.

Fortov, V. E., Vasilyak, L. M., Vetchinin, S. P., Zimnukhov, V. S., Nefedov, A. P., and Polyakov, D. N. (2002). Plasma-dust structures at cryogenic temperatures. *Dokl. Phys.*, **47**, 21–24.

Fortov, V. E., Vaulina, O. S., Petrov, O. F., Molotkov, V. I., Lipaev, A. M., Torchinsky, V. M., Thomas, H. M., Morfill, G. E., Khrapak, S. A., Semenov, Yu. P.,

Ivanov, A. I., Krikalev, S. K., Kalery, A. Yu., Zaletin, S. V., and Gidzenko Yu. P. (2003). Transport of microparticles in weakly ionized gas-discharge plasmas under microgravity conditions. *Phys. Rev. Lett.*, **90**, 2450005/1–4.

Fortov, V. E., Khrapak, A. G., Khrapak, S. A., Molotkov, V. I. and Petrov, O. F. (2004a). Dusty plasmas. *Phys. Uspekhi*, **47**, 447–492.

Fortov, V. E., Petrov, O. F., Molotkov, V. I., Pustyl'nik, M. Yu., Torchinsky, V. E., Khrapak, A. G., and Chernyshov, A. V. (2004b). Large-amplitude dust waves excited by the gas–dynamic impact in a dc glow discharge plasma. *Phys. Rev. E*, **69**, 016402/1–5.

Fortov, V. E., Ivlev, A. V., Khrapak, S. A., Khrapak, A. G., and Morfill, G. E. (2005a). Complex (dusty) plasmas: Current status, open issues, perspectives. *Phys. Rep.*, **421**, 1–103.

Fortov, V., Morfill, G., Petrov, O., Thoma, M., Usachev, A., Hoefner, H., Zobnin, A., Kretschmer, M., Ratynskaia, S., Fink, M., Tarantik, K., Gerasimov, Yu., and Esenkov, V. (2005b). The project 'Plasmakristall-4' (PK-4) – a new stage in investigations of dusty plasmas under microgravity conditions: First results and future plans. *Plasma Phys. Control. Fusion*, **47**, B537–B549.

Fortov, V. E., Petrov, O. F., Molotkov, V. I., Pustyl'nik, M. Yu., Torchinsky, V. E., Naumkin, V, N., and Khrapak, A. G. (2005c). Shock wave formation in a dc glow discharge dusty plasma. *Phys. Rev. E*, **71**, 036413/1–5.

Fortov, V. E., Iakubov, I. T., and Khrapak, A. G. (2006a). *Physics of strongly coupled plasma*. Oxford University Press, Oxford.

Fortov, V. E., Rykov, V. A., Budnik, A. P., Filinov, V. S., Deputatova, L. V., Rykov, K. V., Vladimirov, V. I., Molotkov, V. I., Zrodnikov, A. V., and Dyachenko, P. P. (2006b). Dust crystals in plasma created by a proton beam. *J. Phys. A: Math. Gen.*, **39**, 4533–4537.

Fortov, V. E., Rykov, V. A., Filinov, V. S., Vladimirov, V. I., Deputatova, L. V., Petrov, O. F., Molotkov, V. I., Budnik, A. P., D'yachenko, P. P., Rykov, K. V., and Khudyakov, A. V. (2006c). Vortex dust structures in a track plasma of a proton beam. *Plasma Phys. Rep.*, **31**, 570–576.

Fridman, A. A., Boufendi, L., Hbid, T., Potapkin, B. N., and Bouchoule, A. (1996). Dusty plasma formation: Physics and critical phenomena. Theoretical approach. *J. Appl. Phys.*, **79**, 1303–1314.

Gallagher, A. (2000). Model of particle growth in silane discharges. *Phys. Rev. E*, **62**, 2690–2706.

Ganguly, B., Garscadden, A., Williams, J., and Haaland, P. (1993). Growth and morphology of carbon grains. *J. Vac. Sci. Technol. A*, **11**, 1119–1125.

Gilbert, S. L., Bollinger, J. J., and Wineland, D. J. (1988). Shell-structure phase of magnetically confined strongly coupled plasmas. *Phys. Rev. Lett.*, **60**, 2022–2025.

Goertz, C. K. (1989). Dusty palsmas in the solar system. *Rev. Geophys.*, **27**, 271–292.

Golant, V. E., Zhilinskioe, A. P., and Sakharov, I. E. (1980). *Fundamentals of plasma physics*. Wiley, New York.

Golubovsky, Yu. B. and Nisimov, S. U. (1995). Two-dimensional character of striations in a low-pressure discharge in inert gases. *J. Tech. Phys.*, **40**, 24–28.

Granovskii, V. L. (1952). *Electric current in a gas*. Gostekhizdat, Moscow, Vol. 1 [in Russian].

Grier, D. G. and Murray, C. A. (1994). The microscopic dynamics of freezing in supercooled colloidal fluids. *J. Chem. Phys.*, **100**, 9088–9095.

Grün, E., Morfill, G. E., and Mendis, D. A. (1984). Dust-magnetosphere interactions. In *Planetary rings*, Greenberg, R. and Brahic, A. (eds.), pp. 275–332. Univ. Arizona Press, Tucson.

Halperin, B. and Nelson, D. (1978). Theory of two-dimensional melting. *Phys. Rev. Lett.*, **41**, 121–124.

Hargis Jr., P. J., Greenberg, K. E., Miller, P. A., Gerardo, J. B., Torczynski, J. R., Riley, M. E., Hebner, G. A., Roberts, J. R., Olthoff, J. K., Whetstone, J. R., van Brunt, R. J., Sobolewski, M. A., Anderson, H. M., Splichal, M. P., Mock, J. L., Bletzinger, P., Garscadden, A., Gottscho, R. A., Selwyn, G., Dalvie, M., Heidenreich, J. E., Butterbaugh, J. W., Brake, M. L., Passow, M. L., Pender, J., Lujan, A., Elta, M. E., Graves, D. B., Sawin, H. H., Kushner, M. J., Verdeyen, J. T., Horwath, R., and Turner, T. R. (1994). The Gaseous Electronics Conference radio-frequency reference cell: A defined parallel-plate radio-frequency system for experimental and theoretical studies of plasma-processing discharge. *Rev. Sci. Instrum.*, **65**, 140–154.

Hartquist, T. W., Havnes, O., and Morfill, G. E. (1992). The effects of dust on the dynamics of astronomical and space plasmas. *Fund. Cosmic Phys.*, **15**, 107–142.

Hayashi, Y. and Tachibana, K. (1994). Observation of Coulomb-crystal formation from carbon particles grown in a methane plasma. *Jpn. J. Appl. Phys.*, **33**, L804–L806.

Hollenstein, C. (2000). The physics and chemistry of dusty plasma. *Plasma Phys. Control. Fusion*, **42**, R93–R104.

Hong, S., Berndt, J., and Winter, J. (2002). Study of dust particle formation and its applications in argon/methane and argon/acetylene mixtures, *Proc. Third Int. Conf. Dusty Plasmas (ICPDP-2002)*, Durban, South Africa, 20–24 May 2002, C14.

Hong, S., Berndt, J., and Winter, J. (2003). Growth precursors and dynamics of dust particle formation in the Ar/CH_4 and Ar/C_2H_2 plasmas. *Plasma Sources Sci. Technol.*, **12**, 46–52.

Hou, L. J., Wang, Y. N., and Miskovic, Z. L. (2005). Formation and rotation of two-dimensional Coulomb crystals in a magnetized complex plasma. *Phys. Plasmas*, **12**, 042104/1–9.

Howling, A. A., Courteille, C., Dorier, J. L., Sansonnens, L., and Hollenstein, C. (1996). From molecules to particles in silane plasmas. *Pure Appl. Chem.*, **68**, 1017–1022.

Huang, F. and Kushner, M. (1997). Shapes of agglomerates in plasma etching reactors. *J. Appl. Phys.*, **81**, 5960–5965.

Ichiki, R., Ivanov, Y., Wolter, M., Kawai, Y., and Melzer, A. (2004). Melting and heating of two-dimensional Coulomb clusters in dusty plasmas. *Phys. Rev. E*, **70**, 066404/1–4.

Ichimaru, S. (1982). Strongly coupled plasmas: High-density classical plasmas and degenerate electron liquids. *Rev. Mod. Phys.*, **54**, 1017–1059.

Ikezi, H. (1986). Coulomb solid of small particles in plasmas. *Phys. Fluids*, **29**, 1764-1766.

Ishihara, O., Kamimura, T., Hirose, K. I., and Sato, N. (2002). Rotation of a two-dimensional Coulomb cluster in a magnetic field. *Phys. Rev. E*, **66**, 046406/1–6.

Ivlev, A. V., Khrapak, A. G., Khrapak, S. A., Annaratone, B. M., Morfill, G., and Yoshino, K. (2003a). Rodlike particles in gas discharge plasmas: Theoretical model. *Phys. Rev. E*, **68**, 026403/1–10.

Ivlev, A. V., Kretschmer, M., Zuzic, M., Morfill, G. E., Rothermel, H., Thomas, H. M., Fortov, V. E., Molotkov, V. I., Nefedov, A. P., Lipaev, A. M., Petrov, O. F., Baturin, Y. M., Ivanov, A. I., and Goree, J. (2003b). Decharging of complex plasmas: First kinetic observations. *Phys. Rev. Lett.*, **90**, 055003/1–4.

Ivlev, A. V., Morfill, G. E., Thomas, H. M., Rath, C., Joyce, G., Huber, P., Kompaneets, R., Fortov, V. E., Lipaev, A. M., Molotkov, V. I., Reiter, T., Turin, M., and Vinogradov, P. (2008). First observation of electrorheological plasmas. *Phys. Rev. Lett.*, **100**, 095003/1–4.

Jellum, G. M., Daugherty, J. E., and Graves, D. B. (1991). Particle thermophoresis in low pressure glow discharges. *J. Appl. Phys.*, **69**, 6923–6934.

Juan, W. T., Huang, Z., Hsu, J., Lai, Y., and I, L. (1998). Observation of dust Coulomb clusters in a plasma trap. *Phys. Rev. E*, **58**, R6947–R6950.

Kählert, H., Ludwig, P., Baumgartner, H., Bonitz, M., Block, D., Kading, S., Melzer, A., and Piel, A. (2008). Probability of metastable configurations in spherical three-dimensional Yukawa crystals. *Phys. Rev. E*, **78**, 036408/1–13.

Kamimura, T., Suga, Y., and Ishihara, O. (2007). Configurations of Coulomb clusters in plasma. *Phys. Plasmas*, **14**,123706/1–11.

Karasev, V. Yu, Dzlieva, E. S., Ivanov, A. Yu., and Eikhvald, A. I. (2006) Rotational motion of dusty structures in glow discharge in longitudinal magnetic field. *Phys. Rev. E*, **74**, 066403/1–12.

Karasev, V. Yu., Ivanov, A. Yu., Dzlieva, E. S., and Eikhvald, A. I. (2008). Ordered dust structures in a glow discharge. *JETP*, **106**, 399–403.

Kedziora, D. J., Vladimirov, S. V., and Samarian, A. A. (2008). Mode-spectral analysis of 2D Coulomb clusters with fluctuating charges. *Evrophys. Lett.*, **84**, 55001/1–6.

Khrapak, S., Samsonov, D., Morfill, G., Thomas, H., Yaroshenko, V., Rothermel, H., Hagl, T., Fortov, V., Nefedov, A., Molotkov, V., Petrov, O., Lipaev, A., Ivanov, A., and Baturin, Y. (2003). Compressional waves in complex (dusty) plasmas under microgravity conditions. *Phys. Plasmas*, **10**, 1–4.

Khrapak, S. A., Morfill, G. E., Fortov, V. E., D'yachkov, L. G., Khrapak, A. G., and Petrov, O. F. (2007). Attraction of positively charged particles in highly collisional plasmas. *Phys. Rev. Lett.*, **99**, 055003/1–4.

Klindworth, M., Melzer, A. and Piel, A. (2000). Laser-excited intershell rotation of finite Coulomb clusters in a dusty plasma. *Phys. Rev. B*, **61**, 8404–8410.

Knapek, C. A., Samsonov, D., Zhdanov, S., Konopka, U., and Morfill, G. E. (2007). Recrystallization of a 2D plasma crystal. *Phys. Rev. Lett.*, **98**, 015004/1–4.

Koga, K., Matsuoka, Y., Tanaka, K., Shiratani, M., and Watanabe, Y. (2000). In situ observation of nucleation and subsequent growth of clusters in silane radio frequency discharge. *Appl. Phys. Lett.*, **77**, 196–198.

Kompaneets, R., Vladimirov, S. V., Ivlev, A. V., Tsytovich, V., and Morfill, G. (2006). Dust clusters with non-Hamiltonian particle dynamics. *Phys. Plasmas*, **13**, 072104/1–9.

Kondrat'ev, A. B., Nefedov, A. P., Petrov, O. F., and Samaryan, A. A. (1994). Optical diagnostics of conversion of coal particles in a flow of plasma combustion products. *High Temp.*, **32**, 425–431.

Kong, M., Partoens, B., Matulis, A., and Peeters, F. M. (2004). Structure and spectrum of two-dimensional clusters confined in a hard wall potential. *Phys. Rev. E*, **69**, 036412/1–10.

Konopka, U., Samsonov, D., Ivlev, A., Goree, J., Steinberg, V., and Morfill, G. E. (2000). Rigid and differential plasma crystal rotation induced by magnetic fields. *Phys. Rev. E*, **61**, 1890–1898.

Kortshagen, U. and Bhandarkar, U. (1999). Modeling of particulate coagulation in low pressure plasmas. *Phys. Rev. E*, **60**, 887–898.

Kosov, V. F., Molotkov, V. I., and Nefedov, A. P. (1991). Measurement of the charge-particle density in combustion-product plasmas by electric probes. *High Temp.*, **29**, 489–495.

Kosterlitz, J. and Thouless, D. (1973). Ordering, metastbility and phase transitions in two-dimensional systems. *J. Phys. C*, **6**, 1181–1203.

Kovacevic, E., Stefanovic, I., Berndt, J., and Winter, J. (2003). Infrared fingerprints and periodic formation of nanoparticles in Ar/C_2H_2 plasmas. *J. Appl. Phys.*, **93**, 2924–2930.

Lai, Y.-J. and I, L. (1999). Packings and defects of strongly coupled two-dimensional Coulomb clusters: Numerical simulation. *Phys. Rev. E*, **60**, 4743–4753.

Langmuir, I, Found, C. G., and Dittmer, A. F. (1924). A new type of electric discharge – The streamer discharge. *Science*, **60**, 392–394.

Leiderer, P., Ebner, W., and Shikin, V. B. (1987). Macroscopic electron dimples on the surface of liquid helium. *Surf. Sci.*, **113**, 405–411.

Leroux, A., Kessels, W. M. M., Schram, D. C., and van de Sanden, M. C. M. (2000). Modeling of the formation of cationic silicon clusters in a remote Ar/H_2/SiH_4 plasma. *J. Appl. Phys.*, **88**, 537–543.

Liberman, M. A. and Lichtenberg, A. J. (1994). *Principles of plasma discharges and materials processing*. Wiley, New York.

Lipaev, A. M., Molotkov, V. I., Nefedov, A. P., Petrov, O. F., Torchinskii, V. M., Fortov, V. E., Khrapak, A. G., and Khrapak, S. A. (1997). Ordered structures in a nonideal dusty glow-discharge plasma. *JETP*, **85**, 1110–1118.

Lipaev, A. M., Khrapak, S. A., Molotkov, V. I., Morfill, G. E., Fortov, V. E., Ivlev, A. V., Thomas, H. M., Khrapak, A. G., Naumkin, V. N., Ivanov, A. I., Tretschev, S. E., and Padalka, G. I. (2007). Void closure in complex plasmas under microgravity conditions. *Phys. Rev. Lett.*, **98**, 265006/1–4.

Lozovik, Yu. E. and Mandelshtam, V. A. (1990). Coulomb clusters in a trap. *Phys. Lett. A*, **145**, 269–272.

Lozovik, Yu. E. and Mandelshtam, V. A. (1992). Classical and quantum melting of a Coulomb cluster in a trap, *Phys. Lett. A*, **165**, 469–472.

Lozovik, Yu. E. and Pomirchy, L. M. (1990). Shell structure of two-dimensional electron clusters. *Phys. Stat. Sol. B*, **161**, K11–K13.

Maiorov, S. A. (2004). Charging of a rodlike grain in a plasma flow. *Plasma Phys. Rep.*, **30**, 766–771.

Melzer, A. (2003). Mode spectra of thermally excited two-dimensional dust Coulomb clusters. *Phys. Rev. E*, **67**, 016411/1–10.

Melzer, A., Klindworth, M., and Piel, A. (2001). Normal modes of 2D finite clusters in complex plasmas. *Phys. Rev. Lett.*, **87**, 115002/1–4.

Merlino, R. L., Barkan, A., Thompson, C., and D'Angelo, N. (1998). Laboratory studies of waves and instabilities in dusty plasmas. *Phys. Plasmas*, **5**, 1607–1614.

Mohideen, U., Rahman, H. U., Smith, M. A., Rosenberg, M., and Mendis, D. A. (1998). Intergrain coupling in dusty-plasma coulomb crystals. *Phys. Rev. Lett.*, **81**, 349–352.

Molotkov, V. I., Nefedov, A. P., Torchinskii, V. M., Fortov, V. E., and Khrapak, A. G. (1999). Dust acoustic waves in a dc glow-discharge plasma. *JETP*, **89**, 477–480.

Molotkov, V. I., Nefedov, A. P., Pustyl'nik, M. Y., Torchinsky, V. M., Fortov, V. E., Khrapak, A. G., and Yoshino, K. (2000). Liquid plasma crystal: Coulomb crystallization of cylindrical macroscopic grains in a gas-discharge plasma. *JETP Lett.*, **71**, 102–105.

Molotkov, V. I., Petrov, O. F., Pustyl'nik, M. Yu., Torchinskii, V. E., Fortov, V. E., and Khrapak, A. G. (2004). Dusty plasma of a dc glow discharge: Methods of investigation and characteristic features of behavior. *High Temp.*, **42**, 827–841.

Morfill, G. E., Thomas, H. M., Konopka, U., Rothermel, H., Zuzic, M., Ivlev, A., and Goree, J. (1999). Condensed plasmas under microgravity. *Phys. Rev. Lett.*, **83**, 1598–1601.

Nefedov, A. P., Petrov, O. F., and Vaulina, O. S. (1995). Analysis of radiant energy emission from high-temperature medium with scattering and absorbing particles. *J. Quant. Spectrosc. Radiat. Transfer*, **54**, 453–470.

Nefedov, A. P., Petrov, O. F., Molotkov, V. I., and Fortov, V. E. (2000). Formation of liquidlike and crystalline structures in dusty plasmas. *JETP Lett.*, **72**, 218–226.

Nefedov, A. P., Vaulina, O. S., Petrov, O. F., Molotkov, V. I., Torchinskii, V. M., Fortov, V. E., Chernyshev, A. V., Lipaev, A. M., Ivanov, A. I., Kaleri, A. Yu., Semenov, Yu. P., and Zaletin, S. V. (2002). The dynamics of macroparticles in a direct current glow discharge plasma under micro-gravity conditions. *JETP*, **95**, 673–681.

Nefedov, A. P., Morfill, G. E., Fortov, V. E., Thomas, H. M., Rothermel, H., Hagl, T., Ivlev, A. V., Zuzic, M., Klumov, B. A., Lipaev, A. M., Molotkov, V. I., Petrov, O. F., Gidzenko, Y. P., Krikalev, S. K, Shepherd, W., Ivanov, A. I., Roth, M., Binnenbruck, H., Goree, J. A., and Semenov, Y. P. (2003a). PKE-Nefedov: Plasma crystal experiments on the International Spacc Station. *New J. Phys.*, **5**, 33/1–10.

Nefedov, A. P., Vaulina, O. S., Petrov, O. F, Fortov, V. E., Dranzhevski, I. E., and Lipaev, A. V. (2003b). Dynamics of dust grains in two-copmonent dusty plasma initiated by solar radiation under microgravity conditions. *Plasma Phys. Rep.*, **29**, 31–41.

Nefedov, A. P., Vaulina, O. S., Petrov, O. F., Molotkov, V. I., Torchinskii, V. M., Fortov, V. E., Chernyshev, A. V., Lipaev, A. M., Ivanov, A. I., Kaleri, A. Yu., Semenov, Yu. P., and Zaletin, S. V. (2002). The dynamics of macroparticles in a direct current glow discharge plasma under micro-gravity conditions. *New J. Phys.*, **5**, 108/1–11.

Nelson, D. (1983). Defect-mediated phase transitions. In *Phase transitions and critical phenomena, Vol. 7.* Domb, C. and Leibowitz, J. (eds.), pp. 1–31. Academic Press, London.

Nelson, D. and Halperin, B. (1979). Dislocation-mediated melting in two dimensions. *Phys. Rev. B*, **19**, 2457–2484.

Nosenko, B. V., Zhdanov, S., and Morfill, G. (2008). Supersonic dislocations observed in a plasma crystal. *Phys. Rev. Lett.*, **99**, 025002/1–4.

Nunomura, S., Goree, J., Hu, S., Wang, X., Bhattacharjee, A., and Avinash K. (2002). Phonon spectrum in a plasma crystal. *Phys. Rev. Lett.*, **89**, 035001/1–4.

Paeva, G. V., Dahiya, R. P., Kroesen, G. W., and Stoffels, W.W. (2004). Rotation of particles trapped in the sheath of a radio-frequency capacitively coupled plasma. *IEEE Trans. Plasma Sci.*, **32**, 601–606.

Pieper, J. B., Goree, J., and Quinn, R. A. (1996). Three-dimensional structure in a crystallized dusty plasma. *Phys. Rev. E*, **54**, 5636–5640.

Pustyl'nik, M. Yu., Torchinskii, V. E., Molotkov, V. I., Khrapak, A. G., Chernyshov, A. V., Petrov, O. F., Fortov, V. E., Okubo, M., and Yoshino, K. (2004). Excitation of dust–acoustic waves in a dc glow-discharge dusty plasma by means of pulsed gasdynamic stimulation. *High Temp.*, **42**, 659–666.

Raizer, Y. P. (1991). *Gas discharge physics*. Springer–Verlag, Berlin.

Rosenberg, M., Mendis, D. A., and Sheehan, D. P. (1996). UV-induced Coulomb crystallization of dust grains in high-pressure gas. *IEEE Trans. Plasma Sci.*, **24**, 1422–1430.

Rosenberg, M., Mendis, D. A., and Sheehan, D. P. (1999). Positively charged dust crystals induced by radiative heating. *IEEE Trans. Plasma Sci.*, **27**, 239–242.

Rothermel, H., Hagl, T., Morfill, G. E., Thoma, M. H., and Thomas, H. M. (2002). Gravity compensation in complex plasmas by application of a temperature gradient. *Phys. Rev. Lett.*, **89**, 175001/1–4.

Rykov, V. A., Khudyakov, A. V., Filinov, V. S., Vladimirov, V. I., Deputatova, L. V., Krutov, D. V., and Fortov, V. E. (2003). Dynamic vortex dust structures in a nuclear-track plasma. *NJP*, **5**, 129/1–17.

Samsonov, D. and Goree, J. (1999). Particle growth in a sputtering discharge. *J. Vac. Sci. Technol. A*, **17**, 2835–2840.

Samsonov, D., Morfill, G., Thomas, H., Hagl, T., Rothermel, H., Fortov, V., Lipaev, A., Molotkov, V., Nefedov, A., Petrov, O., Ivanov, A., and Krikalev, S. (2003a). Kinetic measurements of shock wave propagation in a three-dimensional complex (dusty) plasma. *Phys. Rev. E*, **67**, 036404/1–5.

Samsonov, D., Zhdanov, S., Morfill, G., and Steinberg, V. (2003b). Levitation and agglomeration of magnetic grains in a complex (dusty) plasma with magnetic field. *New J. Phys.*, **5**, 24/1–10.

Sato, N., Uchida, G., Ozaki, R., Iizuka, S., and Kamimura, T. (2000). Structure controls of fine-particle clouds in dc discharge plasmas. In *Frontiers in dusty plasmas*, Nakamura, Y., Yokota, T., and Shukla, P. K. (eds.), pp. 329–336. Elsevier, Amsterdam.

Sato, N., Uchida, G., and Kaneko, T. (2001). Dynamics of fine particles in magnetized plasmas. *Phys. Plasmas*, **8**, 1786–1790.

Schottky, W. (1924). Diffusionstheorie der positiven saule. *Physik. Zeits.*, **25**, 635–643.

Schweigert, V. A. and Peeters, F. M. (1995). Spectral properties of classical two-dimensional clusters. *Phys. Rev. B*, **51**, 7700–7713.

Selwyn, G. S., Singh, J., and Bennet, R. S. (1989). In situ laser diagnostic studies of plasma-generated particulate contamination. *J. Vac. Sci. Technol. A*, **7**, 2758–2765.

Selwyn, G., McKillop, J., Haller, K., and Wu, J. (1990). In situ plasma contamination measurements by He–Ne laser light scattering: A case study. *J. Vac. Sci. Technol. A*, **8**, 1726–1731.

Selwyn, G. S., Weiss, C. A., Sequeda, F., and Huang, C. (1997). Particle contamination formation in magnetron sputtering processes. *J. Vac. Sci. Technol. A*, **15**, 2023–2028.

Shukla, P. K. and Mamun, A. A. (2002). *Introduction to dusty plasma physics*. IOP Publ., Bristol.

Shuler, K. E. and Weber, J. (1954). A microwave investigation of the ionisation of hydrogen-oxygen and acetylene-oxygen flames. *J. Chem. Phys.*, **22**, 491–502.

Sodha, M. S. and Guha, S. (1971). Physics of colloidal plasmas. *Adv. Plasma Phys.*, **4**, 219–309.

Stoffels, W. W., Stoffels, E., Ceccone, G., and Rossi, F. (1999). Laser-induced particle formation and coalescence in a methane discharge. *J. Vac. Sci. Technol. A*, **17**, 3385–3392.

Stoykov, S., Eggs, C., and Kortshagen, U. (2001). Plasma chemistry and growth of nanosized particles in a C_2H_2 rf discharge. *J. Phys. D: Appl. Phys.*, **34**, 2160–2173.

Sugden, T. M. and Thrash, B. A. (1951). A cavity resonator method for electron concentrationin flames. *Nature*, **168**, 703–704.

Suh, S. M., Girshick, S. L., Kortshagen, U. R., and Zachariah, M. R. (2003). Modeling gas-phase nucleation in inductively coupled silane-oxygen plasmas. *J. Vac. Sci. Technol. A*, **21**, 251–264.

Takahashi, K. and Tachibana, K. (2001a). Molecular composition of films and solid particles polymerized in fluorocarbon plasmas. *J. Appl. Phys.*, **89**, 893–899.

Takahashi, K. and Tachibana, K. (2001b). Solid particle production in fluorocarbon plasmas. I. Correlation with polymer film deposition. *J. Vac. Sci. Technol. A*, **19**, 2055–2060.

Takahashi, K. and Tachibana, K. (2002). Solid particle production in fluorocarbon plasmas. II. Gas phase reactions for polymerization. *J. Vac. Sci. Technol. A*, **20**, 305–312.

Thomas, H., Morfill, G. E., Demmel, V., Goree, J., Feuerbacher, B., and Möhlmann, D. (1994). Plasma crystal: Coulomb crystallization in a dusty plasma. *Phys. Rev. Lett.*, **73**, 652–655.

Thomas, H. M., Morfill, G. E, Ivlev, A. V., Nefedov, A. P., Fortov, V. E., Rothermel, H., Rubin-Zuzic, M., Lipaev, A. M., Molotkov, V. I., and Petrov, O. F. (2005). PKE-Nefedov – Complex plasma research on the International Space Station. *Micrograviry Sci. Technol.*, **16**, 317–321.

Thomas, H. M., Morfill, G. E., Fortov, V. E., Ivlev, A. V.,Molotkov, V. I., Lipaev, A. M., Hagl, T., Rothermel, H., Khrapak, S. A., Suetterlin, R. K., Rubin-Zuzic, M., Petrov, O. F., Tokarev, V. I., and Krikalev, S. K. (2008). Complex plasma laboratory PK-3 plus on the international space station. *New J. Phys.*, **10**, 033036/1–14.

Thompson, C., Barkan, A., D'Angelo, N., and Merlino, R. L. (1997). Dust acoustic waves in a direct current glow discharge. *Phys. Plasmas*, **4**, 2331–2335.

Totsuji, H. (2001). Structure and melting of two-dimensional dust crystals. *Phys. Plasmas*, **8**, 1856–1862.

Totsuji, H., Totsuji, C., and Tsuruta, K. (2001). Structure of finite two-dimensional Yukawa lattices: Dust crystals. *Phys. Rev. E*, **64**, 066402/1–7.

Totsuji, H, Ogawa, T., Totsuji, C., and Tsuruta, K. (2006). Structure of spherical Yukawa clusters. *J. Phys. A*, **39**, 4545–4548.

Tsakadze, Z. L., Ostrikov, K., Long, J. D., and Xu, S. (2004). Self-assembly of uniform carbon nanotip structures in chemically active inductively coupled plasmas. *Diam. Relat. Mater.*, **13**, 1923–1929.

Tsakadze, Z. L., Ostrikov, K., and Xu, S. (2005). Low-temperature assembly of ordered carbon nanotip arrays in low-frequency, high-density inductively coupled plasmas. *Surf. Coat. Technol.*, **191**, 49–53.

Uchida, G., Ozaki, R., Iizuka, S., and Sato, N. (2000). Formation of ring-shaped fine-particle clouds in a dc plasma. In *Frontiers in dusty plasmas*, Nakamura, Y., Yokota, T., and Shukla, P. K. (eds.), pp. 449–452. Elsevier, Amsterdam.

Usachev, A. D., Zobnin, A.V., Petrov, O. F., Fortov, V. E., Annaratone, B. M., Thoma, M. H., Höfner, H., Kretschmer, M., Fink, M., and Morfill, G. E. (2009). Formation of a boundary-free dust cluster in a low-pressure gas-discharge plasma. *Phys. Rev. Lett.*, **102**, 045001/1–4.

Vasil'ev, M. M., D'yachkov, L. G., Antipov, S. N., Petrov, O. F., and Fortov, V. E. (2007). Dusty plasma structures in magnetic fields in a dc discharge. *JETP Lett.*, **86**, 358–363.

Vasilyak, L. M., Vetchinin, S. P., Nefedov, A. P., and Polyakov, D. N. (2000). Ordered structures of microparticles in a glow discharge. *High Temp.*, **38**, 675–679.

Vasilyak, L. M., Vetchinin, S. P., Zimnukhov, V. S., Polyakov, D. N., and Fortov, V. E. (2003). Dust particles in a thermophoretic trap in a plasma. *JETP*, **96**, 436–439.

Vaulina, O. S. and Khrapak, S. A. (2001). Simulation of the dynamics of strongly interacting macroparticles in a weakly ionized plasma. *JETP*, **92**, 228–234.

Vaulina, O. S., Nefedov, A. P., Petrov, O. F., and Chernyshev, A. V. (1996). Determination of the size and refractive index of weakly absorbing particles. *J. Appl. Spectroscopy*, **63**, 253–259.

Vaulina, O. S., Nefedov, A. P., Petrov, O. F., and Fortov, V. E. (2001). Transport properties of macroparticles in dust plasma induced by solar radiation. *JETP*, **92**, 979–985.

Vaulina, O. S., Nefedov, A. P., Fortov, V. E., and Petrov, O. F. (2002). Diffusion in microgravity of macroparticles in dusty plasma induced solar radiation. *Phys. Rev. Lett.*, **88**, 035001/1–4.

Vaulina, O. S., Petrov, O. F., and Fortov, V. E. (2005). Simulations of mass-transport processes on short observation time scales in nonideal dissipative systems. *JETP*, **100**, 1018–1028.

Vaulina, O. S., Vladimirov, S. V., Repin, A. Yu., and Goree, J. (2006). Effect of electrostatic plasma oscillations on the kinetic energy of a charged macroparticle. *Phys. Plasmas*, 13, 012111/1–5.

Vladimirov, S. V. and Nambu, M. (2001). Interaction of a rod-like charged macroparticle with a flowing plasma. *Phys. Rev. E*, **64**, 026403/1–7.

Vladimirov, S. V. and Ostrikov, K. (2004). Dynamic self-organization phenomena in complex ionized gas systems: New paradigms and technological aspects. *Phys. Rep.*, **393**, 175–380.

Vladimirov, S. V., Ostrikov, K., and Samarian, A. A. (2005). *Physics and applications of complex plasmas*. Imperial College, London.

Vladimirov, V. I., Deputatova, L. V., Nefedov, A. P., Fortov, V. E., Rykov, V. A., and Khudyakov, A. V. (2001). Dust vortices, clouds, and jets in nuclear-excited plasmas. *JETP*, **93**, 313–323.

Watanabe, Y. (1997). Dust phenomena in processing plasmas. *Plasma Phys. Control. Fusion*, **39**, A59–A72.

Watanabe, Y., Shiratani, M., and Koga, K. (2001). Formation kinetics and control of dust particles in capacitively coupled reactive plasmas. *Phys. Scripta*, **89**, 29–32.

Yakubov, I. T. and Khrapak, A. G. (1989). Thermophysical and electrophysical properties of low temperature plasma with condensed disperse phase. *Sov. Tech. Rev. B. Therm. Phys.*, **2**, 269–337.

Young A. (1979). Melting and the vector Coulomb gas in two dimensions. *Phys. Rev. B*, **19**, 1855–1866.

Zuzic, M., Ivlev, A. V., Goree, J., Morfill, G. E., Thomas, H. M., Rothermel, H., Konopka, U., Sütterlin, R., and Goldbeck, D. D. (2000). Three-dimensional strongly coupled plasma crystals under gravity conditions. *Phys. Rev. Lett.*, **85**, 4064–4067.

2

Basic plasma–particle interactions

Sergey A. Khrapak and Alexey V. Ivlev

The study of interactions between an object and surrounding plasma is a basic physical problem with many applications ranging from astrophysical topics (Goertz 1989; Mendis 2002) to operation of gas discharges (Dimoff and Smy 1970), technological plasma applications (Bouchoule 1999; Kersten *et al.* 2001) and fusion related research (Winter 2000; Krasheninnikov *et al.* 2004, de Angelis 2006). In complex plasmas the plasma–particle interactions are especially important. They determine evolution of essentially any characteristic phenomenon occurring in complex plasmas, such as self-organization, formation of ordered (crystal-like) structures and other phase transitions, propagation and instabilities of low-frequency waves, transport.

To understand and explain these phenomena, a detailed knowledge of the properties of plasma–particle interactions in a wide range of plasma parameters is required. Of particular importance are basic processes such as particle charging; electric potential distribution around a charged particle in plasmas; interparticle interactions; momentum exchange between different plasma components and the dust particles giving rise to the ion-, electron-, and neutral-drag forces; processes on the particle surface. In this chapter we give a comprehensive overview of the current state understanding of these basic physical processes.

2.1 Charging of particles in complex plasmas

The particle charge is one of the most important parameters of complex plasmas. It determines the particle interactions with plasma electrons and ions, with electromagnetic fields, between the particles themselves, affect properties of low-frequency (dust acoustic waves), etc. Hence all studies of complex plasmas necessarily begin with a model for the particle charge.

A non-emitting particle (grain) immersed in a plasma collects electrons and ions and becomes charged. Its surface (floating) potential is determined from the balance of collected ion and electron fluxes. Since electrons are much more mobile than ions, the surface potential is negative and roughly equal to the electron temperature $\phi_s = -zT_e/e$, where z is a (positive) coefficient of the order of unity. This ensures that most of the electrons are unable to overcome the potential barrier between the particle surface and surrounding plasma, and hence, ion and electron fluxes can balance each other. The numerical coefficient z is not fixed, but in general depends on a number of plasma and particle parameters such as plasma composition (ion-to-electron mass ratio), electron and ion temperatures, plasma screening length, ion and electron mean free paths with respect to collisions with neutrals, ion and electron drift velocities (in the presence of the drifts), particle size and shape, particle material. Considerable efforts have been made in order to accurately determine the particle floating potentials (coefficient z) under different plasma conditions. This includes theory, experiments, and numerical simulations. In this section an overview of these activities is given.

The traditional and most frequently used approach to calculate the particle floating potential in complex plasmas is the orbital motion limited (OML) theory (see, e.g., Allen 1992). This approach is based on the analysis of ballistic (collisionless) ion and electron trajectories in the central field of an individual charged particle and allows the determination of the ion and electron collection cross sections from the conservation of energy and angular momentum. Ion and electron fluxes are obtained by integrating the collection cross sections over the corresponding velocity distribution functions. The application of the OML theory to the charging of particles in isotropic and anisotropic plasma conditions is considered in Section 2.1.1.2.

Orbital motion limited theory operates with ballistic ion and electron trajectories thus neglecting any effect associated with ion and electron collisions with neutrals. At the same time, there have been a number of investigations, including theories, experiments, and simulations, which clearly demonstrated that the largest variations in z (about one order of magnitude in some cases) can be associated with the plasma collisionality level (Zobnin *et al.* 2000, 2008; Lampe *et al.* 2001a, 2003; Bryant 2003; Fortov *et al.* 2004; Ratynskaia *et al.* 2004a; Khrapak *et al.* 2005b, 2006b; Maiorov 2005; Vaulina *et al.* 2006; D'yachkov *et al.* 2007; Hutchinson and Patacchini 2007; Rovagnati *et al.* 2007; Gatti and Kortshagen 2008; Khrapak and Morfill 2008b). Thus, plasma collisionality is a very important factor affecting the magnitude of the particle floating potential and charge. This issue will be discussed in detail in Section 2.1.2.

If emission processes are involved (e.g., thermo-, photo-, and secondary electron emission from the particle surface), the charge can be reduced in absolute magnitude or even assume positive values (e.g., Goree 1994; Walch *et al.* 1995; Rosenberg *et al.* 1999; Sickafoose *et al.* 2000). Effect of different emission processes on particle charging is briefly discussed in Section 2.1.4.

In this chapter we also briefly discuss the effect of the presence of particles on charge balance in plasmas (Section 2.1.5) and the properties of particle charge fluctuations (Section 2.1.6).

2.1.1 Charging in collisionless plasmas

We start with the description of particle charging in the collisionless regime. In order to clarify some of the assumptions underlying the collisionless OML approach, we first briefly discuss the properties of particle motion in the central field.

2.1.1.1 Motion of a particle in the central field

For a particle moving ballistically in the central field $U(r)$, the total energy, $\mathscr{E} = \frac{1}{2}m(v_r^2 + v_\theta^2) + U(r)$, and the angular momentum, $m\rho v$, are conserved. Here m is the particle mass, ρ is the impact parameter, v_r and v_θ are the radial and tangential velocities, respectively, and v is the velocity at $r \to \infty$. It follows from the conservation laws that the radial motion of a particle is fully determined by the *effective* potential energy,

$$U_{\text{eff}}(r,\rho) = \frac{\rho^2}{r^2} + \frac{2U(r)}{mv^2}, \tag{2.1}$$

where U_{eff} is normalized to the initial kinetic energy, $\mathscr{E}_0 = \frac{1}{2}mv^2$. The particle motion is restricted to the region where $U_{\text{eff}} \leq 1$. This means that if for a given ρ an equation

$$U_{\text{eff}}(r,\rho) = 1 \tag{2.2}$$

has root(s), then the largest of them determines the distance of the closest approach to the center, r_0. If a sphere of radius a is placed in the center, then the particle is collected by this sphere when $r_0 \leq a$. For $r_0 > a$, it experiences elastic scattering by the center potential, but does not reach the sphere surface.

For a *repulsive* potential, $U(r) > 0$, the effective potential is a positive, monotonically decreasing function of r. Therefore, Equation (2.2) has one solution. The maximum impact parameter corresponding to particle collection is

$$\rho_c^-(v) = \begin{cases} a\sqrt{1 - \frac{2U(a)}{mv^2}}, & \frac{2U(a)}{mv^2} < 1, \\ 0, & \frac{2U(a)}{mv^2} \geq 1. \end{cases} \tag{2.3}$$

For an *attractive* potential, $U(r) < 0$, there are several possibilities, depending on the particular behavior of $U(r)$ (Al'pert *et al.* 1965). The extremum values of $U_{\text{eff}}(r)$ are determined from the condition

$$r^3(dU/dr) = mv^2\rho^2. \tag{2.4}$$

If $|U(r)|$ decreases everywhere more slowly than $1/r^2$, then the left-hand side of Equation (2.4) grows monotonically, and there is only one solution to this equation. This solution corresponds to a minimum in U_{eff}. Hence, Equation (2.2) has only one solution, and similarly to Equation (2.3), we obtain for the maximum impact parameter corresponding to collection

$$\rho_c^+(v) = a\sqrt{1 - \frac{2U(a)}{mv^2}}. \tag{2.5}$$

On the other hand if $|U(r)|$ decreases more slowly than $1/r^2$ at small r but decreases faster than $1/r^2$ at large r, then Equation (2.4) can have two solutions, and, hence, U_{eff} has both maximum and minimum. Maximum is determined by the conditions $U'_{\text{eff}}(r_M) = 0$ and $U''_{\text{eff}}(r_M) < 0$, where primes denote derivatives with respect to r. The maximum in U_{eff} always occurs at distances larger than the minimum.

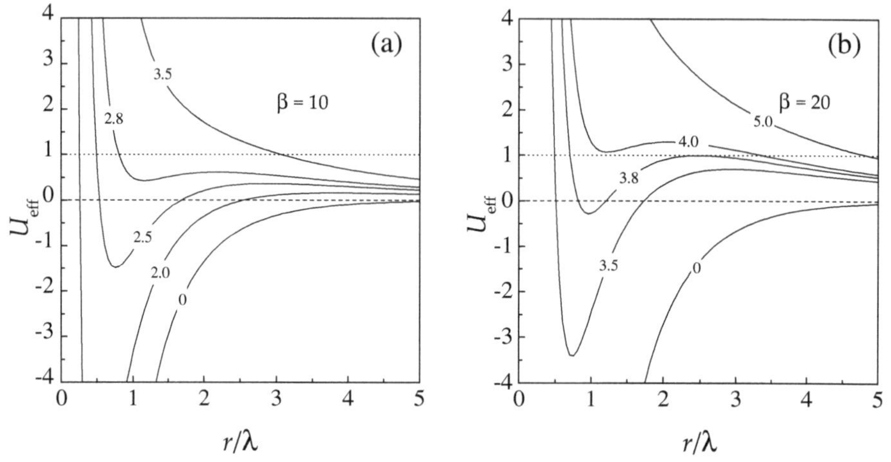

FIGURE 2.1
Curves of the effective potential for the radial particle motion in the central field (for the Debye–Hückel potential) for two values of the scattering parameter β (see text) and different dimensionless impact parameters ρ (indicated in the figures). The potential barrier is absent at $\beta = 10$ and present at $\beta = 20$.

If $U_{\mathrm{eff}}(r_{\mathrm{M}}) \geq 1$, then Equation (2.2) can have multiple solutions: Physically, this means that the potential barrier emerges, which reflects the particle. This is illustrated in Figures 2.1 and 2.2. From Equation (2.1), we see that there is a *transitional impact parameter*

$$\rho_* = r_{\mathrm{M}}\sqrt{1 - \frac{2U(r_{\mathrm{M}})}{mv^2}},$$

which separates particle trajectories into two groups: no barrier for $\rho < \rho_*$, but for $\rho > \rho_*$, the potential barrier emerges and the particle is reflected at $r \geq r_{\mathrm{M}}$. This causes a discontinuity in the dependence of the distance of closest approach on ρ (see Figure 2.2). In the case $r_{\mathrm{M}} > a$, the particles with $\rho > \rho_*$ cannot be collected by the center. Thus, Equation (2.5) should be modified:

$$\rho_{\mathrm{c}}^{+}(v) = \begin{cases} a\sqrt{1 - \frac{2U(a)}{mv^2}}, & a\sqrt{1 - \frac{2U(a)}{mv^2}} < \rho_*(v), \\ \rho_*(v), & a\sqrt{1 - \frac{2U(a)}{mv^2}} \geq \rho_*(v). \end{cases} \tag{2.6}$$

As a useful example let us consider the attractive Debye–Hückel (Yukawa) interaction potential, $U(r) = -(U_0/r)\exp(-r/\lambda)$, where λ is the effective plasma screening length. Usually (but not always), the plasma screening length coincides with the linearized Debye radius, $\lambda \equiv \lambda_{\mathrm{D}} = \lambda_{\mathrm{D}i}/\sqrt{1 + (\lambda_{\mathrm{D}i}/\lambda_{\mathrm{D}e})^2}$, where $\lambda_{\mathrm{D}i(e)} = \sqrt{T_{i(e)}/4\pi e^2 n_{i(e)}}$. In this case $\lambda \simeq \lambda_{\mathrm{D}i}$, provided $T_e \gg T_i$. Some exceptions, including screening of the particle charge in the regime of highly non-linear ion–particle

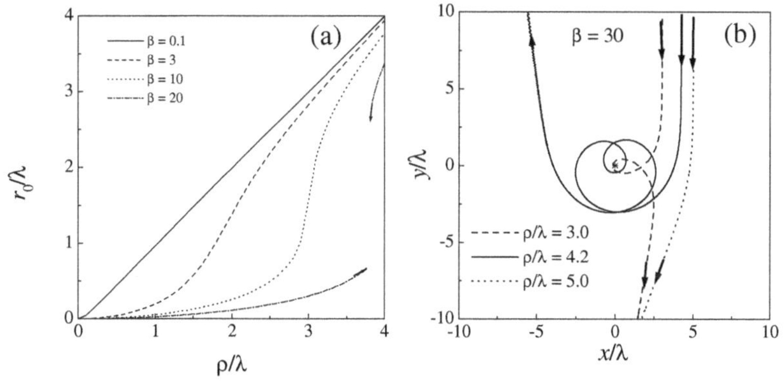

FIGURE 2.2
(a) Normalized distance of the closest approach to the center r_0 vs. normalized impact parameter ρ. The curves are calculated for the Debye–Hückel potential for different values of the scattering parameter β. When $\beta \geq \beta_{cr} \simeq 13.2$ a discontinuity appears due to a barrier in the effective potential. (b) Particle trajectories during collisions for different impact parameters for the Debye–Hückel interaction potential and scattering parameter $\beta = 30$. Impact parameters are chosen to be below, about, and above the transitional impact parameter $\rho_* \simeq 4.24\lambda$.

interaction and the notion of effective screening length in anisotropic plasmas, will be considered later in this chapter.

Let us normalize all the distances to λ. Then the behavior of the effective potential U_{eff} is governed by two dimensionless parameters, the so-called *scattering parameter*

$$\beta = \frac{U_0}{mv^2\lambda}, \tag{2.7}$$

and the normalized impact parameter, viz., $U_{\text{eff}}(r) = \rho^2/r^2 - 2(\beta/r)\exp(-r)$. Curves of the effective potential for two values of β and different values of ρ are displayed in Figure 2.1. The potential barrier is absent at $\beta = 10$, while at $\beta = 20$, the existence of the barrier leads to an abrupt jump in the distance of the closest approach, from $r_0 \simeq 0.7$ to $r_0 \simeq 2.6$ at $\rho \simeq 3.8$. An inflection point (the point where maximum and minimum in the curve of the effective potential coincide) is determined from the condition $U''_{\text{eff}}(r_{\text{M}}) = 0$. For the Debye–Hückel potential, it occurs at $r = (1+\sqrt{5})/2 \simeq 1.62$ (Al'pert *et al.* 1965). The position of the emerging potential barrier is the solution of the transcendental equation $r_{\text{M}}\exp(r_{\text{M}}) = \beta(r_{\text{M}} - 1)$. A solution exists only if $\beta \geq \beta_{\text{cr}} \simeq 13.2$ and grows monotonically with β. The transitional impact parameter is

$$\rho_* = r_{\text{M}}\sqrt{\frac{r_{\text{M}}+1}{r_{\text{M}}-1}}$$

and also increases with β starting from $\rho_*(\beta_{cr}) \simeq 3.33$. For large β, Khrapak *et al.* (2003b) obtained the following asymptotic solutions

$$r_M \simeq \ln\beta - \ln^{-1}\beta, \qquad \rho_* \simeq \ln\beta + 1 - \tfrac{1}{2}\ln^{-1}\beta. \tag{2.8}$$

The trajectories of scattered particles and the dependence of the distance of closest approach on the impact parameter calculated for the attractive Debye–Hückel potential are shown in Figure 2.2.

2.1.1.2 Orbital motion limited (OML) approximation

In the OML approach three major assumptions are employed: (i) The particle is isolated in the sense that other particles do not affect the motion of electrons and ions in its vicinity; (ii) electrons and ions do not experience collisions during their approach to the particle; (iii) the barriers in the effective potential are absent. Then the cross sections for electron and ion collection are determined from the laws of conservation of energy and angular momentum.

The collection cross section is $\pi\rho_c^2$, where ρ_c is the maximum impact parameter for the collection. It is given by Equation (2.3) for the electrons and by Equation (2.5) for the ions. Thus, the (velocity-dependent) collection cross sections are

$$\sigma_e(v) = \begin{cases} \pi a^2\left(1 + \frac{2e\phi_s}{m_e v^2}\right), & \frac{2e\phi_s}{m_e v^2} > -1, \\ 0, & \frac{2e\phi_s}{m_e v^2} \le -1, \end{cases} \tag{2.9}$$

and

$$\sigma_i(v) = \pi a^2\left(1 - \frac{2e\phi_s}{m_i v^2}\right), \tag{2.10}$$

where $m_{e(i)}$ is the electron (ion) mass, v is the velocity of the electrons and ions relative to the particle, the particle surface potential ϕ_s is negative, and the ions are singly charged. An obvious advantage of the OML approximation is that the cross sections are independent of the plasma potential distribution around the particle. This is, however, only true when the potential satisfies certain conditions so that the barrier in the effective potential is absent.

Electron and ion fluxes to the particle surface are determined by the integral of the corresponding cross sections with the velocity distribution functions $f_{e(i)}(v)$:

$$J_{e(i)} = n_{e(i)} \int v\sigma_{e(i)}(v) f_{e(i)}(v) d^3v,$$

where $n_{e(i)}$ is the electron (ion) number density. Using the Maxwellian velocity distribution of plasma particles $f_{e(i)}(v) = (2\pi v_{T_{e(i)}}^2)^{-3/2}\exp(-v^2/2v_{T_{e(i)}}^2)$, where $v_{T_{e(i)}} = \sqrt{T_{e(i)}/m_{e(i)}}$ is the electron (ion) thermal velocity, we get after the integration

$$J_e = \sqrt{8\pi} a^2 n_e v_{T_e} \exp\left(\frac{e\phi_s}{T_e}\right), \tag{2.11}$$

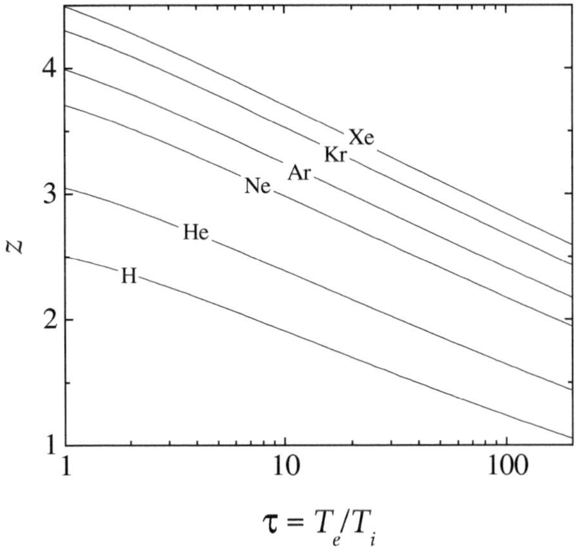

FIGURE 2.3
Dimensionless floating potential $z = -e\phi_s/T_e$ of an isolated spherical particle as a function of electron-to-ion temperature ratio $\tau = T_e/T_i$ for isotropic plasmas of different gases.

$$J_i = \sqrt{8\pi}a^2 n_i v_{T_i}\left(1 - \frac{e\phi_s}{T_i}\right). \tag{2.12}$$

The evolution of the particle charge is governed by the equation

$$\frac{dZ}{dt} = J_i - J_e, \tag{2.13}$$

and the stationary charge is determined from the flux balance,

$$J_e = J_i. \tag{2.14}$$

In dimensionless form Equation (2.14) can be rewritten as

$$v_{T_e}\exp(-z) = v_{T_i}(1 + z\tau), \tag{2.15}$$

where $\tau = T_e/T_i$ is the electron-to-ion temperature ratio and plasma is quasineutral, $n_e \simeq n_i$. Thus, within OML approximation the dimensionless floating potential z is a function of only two parameters – the electron-to-ion temperature and mass ratios. In Figure 2.3, values of z are presented for different gases (H, He, Ne, Ar, Kr, Xe) as functions of τ. The particle floating potential decreases with $\tau = T_e/T_i$ and increases with the gas atomic mass. For typical values of $\tau \sim 10-100$ in gas discharges, the dimensionless floating potential is in the range $z \sim 2$–4. For a particle with $a \sim 1\ \mu$m

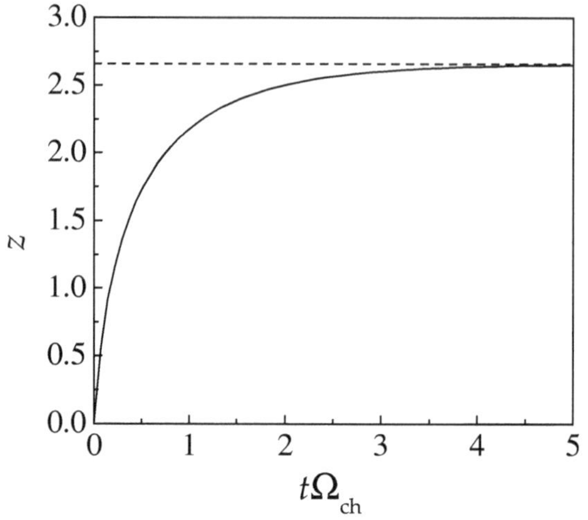

FIGURE 2.4
Dimensionless floating potential $z = -e\phi_s/T_e$ of an isolated spherical particle as a function of dimensionless time $t\Omega_{ch}$ for an argon plasma with $\tau = 50$. The particle is initially uncharged. The horizontal dashed line corresponds to the stationary value of the potential.

and $T_e \sim 1$ eV, the characteristic charge number is $|Z| \sim (1-3)\times 10^3$, if we assume "vacuum" relation between the particle surface potential and charge, $Q = a\phi_s$. This is usually a good approximation in complex plasmas as long as the particles are small ($a \ll \lambda$), while, in principle, there may be some deviations due to strongly nonlinear screening and/or non-equilibrium distribution of the electrons and ions around the particle.

To characterize the dynamics of particle charging let us introduce the characteristic *charging frequency* Ω_{ch}. It is defined as the relaxation frequency for small deviations of the charge from the stationary value Z_0: $\Omega_{ch} = d(J_i - J_e)/dZ|_{Z_0}$. Using Equations (2.11) and (2.12), we obtain

$$\Omega_{ch} = \frac{1+z}{\sqrt{2\pi}}\frac{a}{\lambda_{Di}}\omega_{pi}, \tag{2.16}$$

where $\lambda_{Di} = \sqrt{T_i/4\pi e^2 n_i}$ is the ionic Debye radius, and $\omega_{pi} = v_{T_i}/\lambda_{Di}$ is the ion plasma frequency. In deriving Equation (2.16) it is also assumed $\tau \gg 1$. The solution of nonlinear Equation (2.13) with fluxes from Equations (2.11) and (2.12) and the initial condition $z|_{t=0} = 0$ is shown in Figure 2.4. One can see that the overall time scale of the nonlinear charging is of the order of Ω_{ch}^{-1}.

Complex plasmas are often subject to electric fields. For instance, in ground-based experiments with rf discharges the particles can levitate in the (pre)sheath above the lower electrode, while in dc discharges the particles are often trapped in striations. Both these regions are characterized by high degree of plasma anisotropy and strong electric fields. The larger the particle size is, the stronger the electric field is required in order to compensate for gravity. The presence of the electric fields causes plasma drifts relative to the particle component. This in turn can affect particle charging by, for example, changing the collection cross sections and velocity distribution functions of ions and electrons. The situation becomes in general more complicated than in the case of an isotropic plasma, and we are not aware of any self-consistent analytic solutions existing for this case. Therefore, let us only briefly discuss the simplest approach to describe charging in anisotropic regime.

To get an idea how the plasma drifts can affect particle charging, we use the following simplifications. We take the OML collection cross sections, but instead of isotropic Maxwellian distribution function use the shifted Maxwellian distributions, viz.,

$$f_{i(e)}(\mathbf{v}) = \left(2\pi v_{T_{i(e)}}^2\right)^{-3/2} \exp\left[\frac{(\mathbf{v}-\mathbf{u}_{i(e)})^2}{2v_{T_{i(e)}}^2}\right], \tag{2.17}$$

where $\mathbf{u}_{i(e)}$ is the average drift velocity of ions (electrons). Integration of the cross section Equation (2.10), with the shifted Maxwellian function Equation (2.17), yields the following expression for the ion flux (Whipple 1981; Uglov and Gnedovets 1991; Kilgore *et al.* 1994):

$$J_i = \sqrt{\pi}\frac{a^2 n_i v_{T_i}^2}{u_i}\left[\sqrt{\pi}(1+2\xi^2+2z\tau)\mathrm{erf}(\xi)+2\xi\exp(-\xi^2)\right], \tag{2.18}$$

where $\xi = u_i/\sqrt{2}v_{T_i}$. Similarly, integrating the cross section Equation (2.9), with shifted Maxwellian distribution Equation (2.17), the electron flux can be written as (Uglov and Gnedovets 1991):

$$J_e = \sqrt{\pi}\frac{a^2 n_e v_{T_e}^2}{u_e}\left\{\sqrt{\pi}(1/2-\xi_+\xi_-)\left[\mathrm{erf}(\xi_+)-\mathrm{erf}(\xi_-)\right] + \xi_+ e^{-\xi_-^2} - \xi_- e^{-\xi_+^2}\right\}, \tag{2.19}$$

where $\xi_\pm = \sqrt{z} \pm u_e/\sqrt{2}v_{T_e}$. In the limit $u_i \ll v_{T_i}$ and $u_e \ll v_{T_e}$ Equations (2.18) and (2.19) reduce to Equations (2.12) and (2.11), respectively. In the opposite limit we have

$$J_i = \pi a^2 n_i u_i\left[1+(1+2z\tau)(v_{T_i}/u_i)^2\right] \tag{2.20}$$

and

$$J_e = \pi a^2 n_e u_e\left[1+(1-2z)(v_{T_e}/u_e)^2\right]. \tag{2.21}$$

Often, the drift of electrons is negligible while the ion drift is large. For example, this occurs in the regime of ambipolar plasma, in the (pre)sheath regions, i.e., where the electron distribution is close to Boltzmann, implying that the electric force acting on the electrons is compensated by the electron pressure. In this case the electron flux to the particle surface is given by Equation (2.11) while for the ions one should use Equation (2.18). The resulting dimensionless floating potential of the particle as

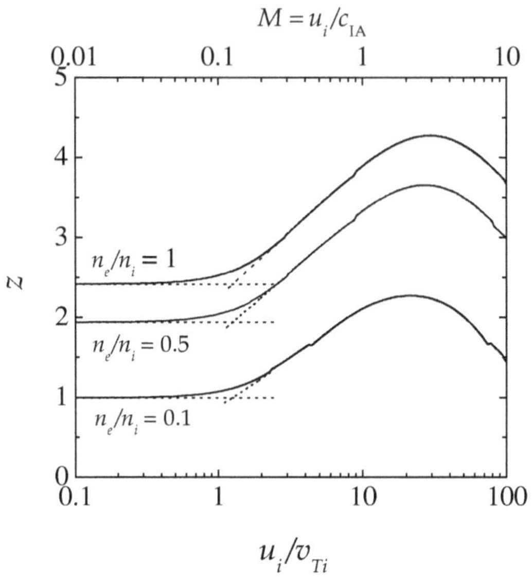

FIGURE 2.5
Dimensionless floating potential $z = -e\phi_s/T_e$ of an isolated spherical grain as a function of the ion drift to ion thermal velocity ratio, u_i/v_{T_i} (or Mach number $M = u_i/C_{IA}$), for a plasma with drifting ions. The calculations are for three different electron-to-ion density ratios and correspond to an argon plasma with $\tau = 100$.

a function of the ion drift velocity is shown in Figure 2.5 for three values of n_e/n_i. The charge is practically constant for $u_i \leq v_{T_i}$, then it increases with u_i, attains a maximum at $u_i \sim (2\text{–}3)C_{\text{IA}}$ (where $C_{\text{IA}} = \sqrt{T_e/m_i}$ is the ion–acoustic velocity), and starts decreasing. Comparison of results calculated with exact flux Equation (2.18), and with asymptotic Equations (2.12) and (2.20) (the latter are indicated by dotted lines) shows almost no discrepancy, except for a narrow region near $u_i \sim v_{T_i}$.

Figure 2.5 illustrates the behavior of the particle charge in a sheath above the electrode in rf/dc discharges. The averaged electric field here increases almost linearly towards the electrode (e.g., Tomme *et al.* 2000 and references therein). The ions are accelerated by the electric field, which leads to an increase of z until u_i becomes several times larger than C_{IA}. A positive space charge is developed in the sheath so that $n_i/n_e > 1$ as the electrode is approached. This causes z to decrease compared to the quasineutral region. Thus, when approaching the electrode from the unperturbed bulk plasma, the dimensionless charge z first somewhat increases, reaches a maximum, and then decreases. The charge can even reach positive values sufficiently close to the electrode. Examples of the dependence of the particle surface potential on the distance from the electrode in collisionless and collisional sheaths of rf and dc discharges were calculated by Nitter (1996) for a set of plasma parameters.

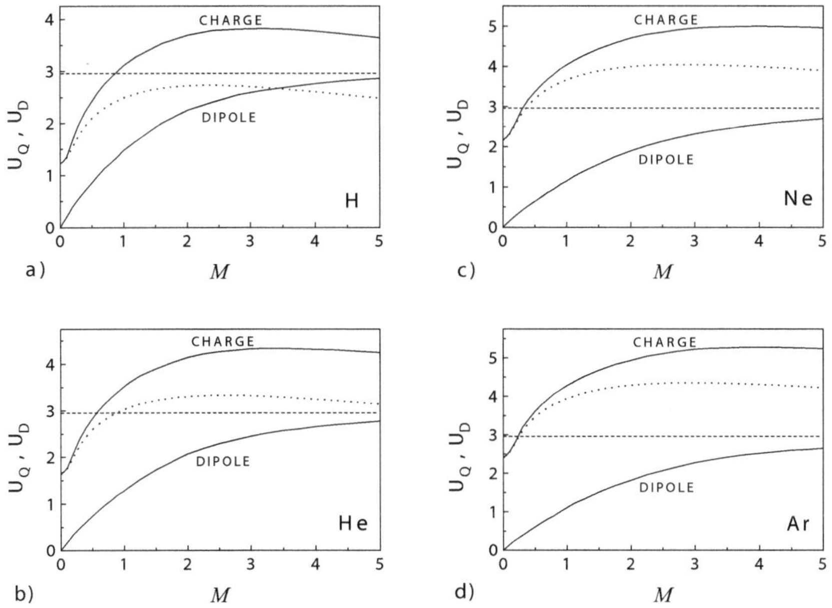

FIGURE 2.6
The dimensionless charge, $U_Q = e|Q|/aT_e$ (in the text z is adopted), and the dimensionless dipole moment, $U_D = ed/a^2T_e$, of the dielectric particle (solid lines) versus Mach number of the plasma flow, M (Ivlev *et al.* 1999). The dotted line corresponds to a metal particle. The dashed line is the asymptote for the dipole curve. The figures correspond to (a) H, (b) He, (c) Ne, and (d) Ar.

The accuracy of neglecting the potential anisotropy caused by the ion flow (i.e, the assumption of the OML collection cross section) was checked by Lapenta (1999) and Hutchinson (2003, 2005) using particle-in-cell codes. It was shown that the potential asymmetry is virtually negligible with respect to the total ion flux for a conducting particle. However, this is generally not true for a dielectric particle. The latter acquires a significant dipole moment and the absolute magnitude of the charge of dielectric particle is larger than that of conductive particle. Such trend was predicted analytically by Ivlev *et al.* (1999). Figure 2.6 shows that the effect can be quite significant. Another important circumstance which can affect the accuracy of expression (2.18) is the deviation of the ion velocity distribution function from shifted Maxwellian (Schweigert 2001; Ivlev *et al.* 2005) discussed below. This important issue has been addressed by Alexandrov *et al.* (2008), but needs to be investigated in more detail.

To conclude this section, let us discuss the major reasons that can make the OML approximation inapplicable. Apart from central symmetry breaking in anisotropic plasmas, briefly discussed above, these include:

(i) *Effect of "closely packed" particles.* The effect is two-fold. The particle component contributes to the quasineutrality condition, making the ion density larger than the electron density (Havnes *et al.* 1984). This increases the ratio of the ion-to-electron flux and hence reduces the absolute magnitude of the particle charge compared to the case of an individual particle. The strength of the effect can be characterized by the dimensionless parameter $P = |Z|n_d/n_e \equiv z(aT_e/e^2)(n_d/n_e)$ (often called "Havnes parameter"), which is the ratio of the charge residing on the dust particle component to that on the electron component (n_d is the particle number density). If we simply use expressions (2.11) and (2.12) for the electron and ion fluxes together with the quasineutrality condition $n_e = n_i + Zn_d$, we get in dimensionless units

$$\sqrt{\tau}\exp(-z) = \sqrt{\frac{m_e}{m_i}}(1+z\tau)(1+P).$$

The charge tends to that of an individual particle when $P \ll 1$, while it is reduced considerably for $P \gg 1$. In addition, when the interparticle separation Δ is smaller than the characteristic length of interaction between ions (electrons) and the particle, the ion (electron) trajectories are affected by the presence of neighboring particles. The analysis based on the consideration of particle motion in the central field is no more applicable and the OML approximation fails. Barkan *et al.* (1994) demonstrated experimentally that the effect of "closely packed" particles can lead to substantial charge reduction.

(ii) *Potential barriers.* The OML theory presumes the absence of a barrier in the effective potential energy. As discussed in Section 2.1.1.1 the barrier is absent for repulsive interaction (i.e., for the electrons), but it can emerge for attractive interaction (i.e., for the ions). In this case some (low energy) ions will be reflected from this barrier, thus reducing the net ion flux to the particle and, therefore, increasing the absolute magnitude of the floating potential. Allen *et al.* (2000) showed that for a Maxwellian velocity distribution, there are always sufficiently slow ions, which are reflected from the barrier. However, if the fraction of the ions which are not collected because of the barrier is small then the corrections to OML are also small. Using Equation (2.6) this requirement can be formulated in terms of the ion thermal velocity $a\sqrt{1+2U(a)/T_i} < \rho_*(v_{T_i})$. For the Debye–Hückel interaction potential with the large ion thermal scattering parameter $\beta(v_{T_i}) \simeq z\tau(a/\lambda) > \beta_{\rm cr}$, Khrapak *et al.* (2004a) derived the following condition of the applicability of the OML approximation:

$$\sqrt{2z\tau}(a/\lambda) \lesssim \ln[z\tau(a/\lambda)].$$

For typical complex plasma parameters $z \sim 1$ and $\tau \sim 100$, we get that OML is applicable for $a/\lambda \lesssim 0.2$. Based on similar arguments Lampe (2001) showed that the OML approximation is well justified when $(a/\lambda)\tau \ll 100$. Thus, the OML theory works well for sufficiently small grains. Essentially the same conclusion is drawn by Kennedy and Allen (2003) from a consistent numerical solution for the surface potential and potential distribution around the probe in a collisionless plasma.

(iii) *Ion and electron collisions with neutrals.* As mentioned above, plasma collisionality is one of the most important factors determining the magnitude of the particle floating potential in complex plasmas. The OML approximation deals with ballistic ion trajectories and is, therefore, applicable only in the limit of very weak collisionality. At the same time in practical situations, ion mean free path can often be comparable to the plasma screening length, i.e., plasma is in the moderately collisional regime. Moreover, in high pressure plasmas, ion and electron motion to the particle can be completely collision dominated. Detailed discussion of the dependence of the particle surface potential on the plasma collisionality is the subject of the next section.

2.1.2 Effect of plasma collisionality on the particle charging

The qualitative picture of the dependence of the normalized floating potential z on plasma collisionality is sketched in Figure 2.7. The plasma collisionality, measured in units of the ratio of the plasma screening length to the ion mean free path λ/ℓ_i, can be conventionally divided into three regimes: weakly collisional (WC), $\ell_i > \lambda$; strongly collisional (SC), $\ell_i < \lambda$; and fully collisional plasma (FCP), $\ell_i, \ell_e \ll \lambda$. These regimes are indicated in Figure 2.7. In the collisionless limit ($\ell_i, \ell_e \gg \lambda$), OML approximation yields very accurate values for the floating potential, provided the particle radius is considerably smaller than the plasma screening length (see above). When ion collisionality increases (although the ratio ℓ_i/λ can still be relatively large, of the order of ten), the collisions start to affect the ion flux to the particle. As explained below, in this (weakly collisional) regime, collisions increase the ion flux which leads to a decrease in z. The minimum is reached when the ion mean free path is on the order of the plasma screening length. As the ion collisionality increases further (ion mean free path decreases), a transition to the mobility-limited (hydrodynamic) regime of ion collection occurs. In this regime the ion mobility is suppressed by collisions, ion flux decreases, and z increases with increasing ion collisionality. However, the magnitude of z does not exhibit constant growth. When the FCP regime is reached, electron transport to the particle becomes collisional too. Then both electron and ion fluxes are equally suppressed by collisions, and the floating potential saturates at some value which is independent of plasma collisionality.

Figure 2.7 demonstrates that the calculation of the particle floating potential in gas discharges operating under different conditions requires an expression applicable in a wide range of ion collisionality. Several such expressions have been proposed in the literature. Below we review some of the proposed approximations, compare the results obtained using these approximations, and briefly discuss their reliability.

A simple interpolation formula for the dependence of z on ion collisionality has been proposed by Khrapak and Morfill (2008b). This interpolation (hereinafter referred to as KM interpolation) is based on properly combining the expressions applicable in the cases of weak and strong ion collisionality. Let us, therefore, discuss the corresponding limits.

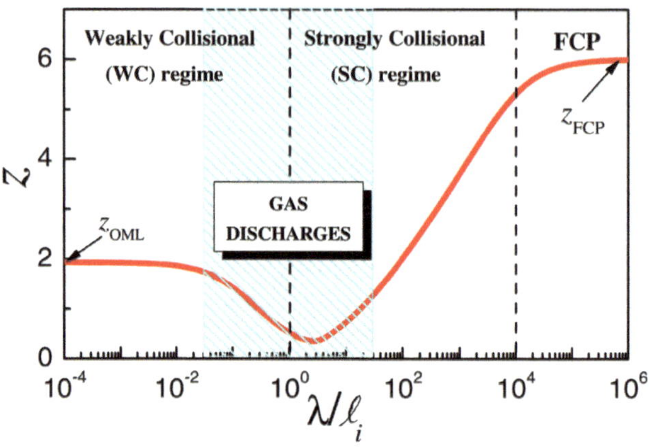

FIGURE 2.7
Sketch of the dependence of the particle normalized floating potential $z = -\phi_s e/T_e$ on the ion collisionality index λ/ℓ_i. The entire range of ion collisionality is conventionally divided into three regimes: weakly collisional (WC), strongly collisional (SC), and fully collisional plasma (FCP); for details see text. A shaded region in the figure corresponds approximately to the plasma collisionality pertinent to gas discharges operating in a wide range of neutral gas pressures. The curve shows the dependence of z on λ/ℓ_i calculated using the model of D'yachkov *et al.* (2007) for neon plasma with $\tau = 200$, $\lambda/a \simeq 35$, and $\ell_e/\ell_i = 30$. The corresponding OML and FCP values of z are $z_{OML} \simeq 1.9$ and $z_{FCP} \simeq 6.0$.

An approximation for the collected ion flux in the weakly collisional regime can be derived using the ideas discussed by Zakrzewski and Kopiczinski (1974) in the context of plasma probe theory and more recently by Zobnin *et al.* (2000) and Lampe *et al.* (2003) in the context of complex plasmas. Following the arguments of Lampe *et al.* let us consider a sphere of radius R_0 surrounding a small particle ($a \ll R_0$). The distance R_0 determines the region of strong ion-particle coupling. It can be roughly estimated from the condition $|U(R_0)| \simeq T_i$, where $U(r)$ denotes the potential energy of the ion–particle interaction. When an ion orbiting around a particle makes a collision with an atom within $r \lesssim R_0$ a fast ion (which probably would not reach the particle surface) loses its energy and angular momentum and has a low probability to escape from the potential well. Such an ion has a high probability to be collected by the particle, either directly after collision or through subsequent collisions. Especially effective are charge exchange collisions which lead to a substitution of the original ion by a low-energy ion created from a neutral atom. Essentially every charge-exchange collision within $r \lesssim R_0$ results in an ion collection by the particle. Thus, ion–neutral collisions enhance the ion flux to the particle in the weakly collisional regime. A rough estimate of the collision-induced contribution to the collected ion flux is given by the product of the random (thermal) ion influx through the spherical surface of radius R_0 [for $r > R_0$ the ion density and (Maxwellian) velocity distribution are weakly perturbed by the presence of the particle] and the probability to experience a collision within this sphere. For simplicity let us normalize all the fluxes to the flux of Maxwellian ions collected by an uncharged sphere of radius a, so that $J_\alpha = \sqrt{8\pi}a^2 n_i v_{T_i} j_\alpha$. The normalized ion flux into the interaction sphere is $j_{R_0} \simeq (R_0/a)^2$; the probability to experience a collision is roughly $\sim R_0/\ell_i$, provided $\ell_i > R_0$, i.e., ions are weakly collisional. There are also ions which do not experience collisions; their flux can be estimated as $\simeq j_{\rm OML}(1 - R_0/\ell_i)$, where $j_{\rm OML} = 1 + z\tau$ is the normalized OML flux. The net ion flux to the particle in this weakly collisional regime is therefore $j_i^{\rm WC} \simeq 1 + z\tau + (R_0^3/a^2\ell_i)$, which basically coincides with Equation (31) from Lampe *et al.* (2003). Assuming that the electric potential distribution around the particle follows the Debye–Hückel form, the expression for the ion flux can be further simplified (Khrapak *et al.* 2005b) as

$$j_i^{\rm WC} \simeq 1 + z\tau + 0.1(z\tau)^2(\lambda/\ell_i). \tag{2.22}$$

This expression is known as the collision enhanced collection (CEC) approximation (Khrapak and Morfill 2006).

In the FCP limit, a fluid description of the ion and electron components can be used. The system of equations to be solved includes conservation of the ion and electron fluxes (Su and Lam 1963),

$$4\pi r^2 D_\alpha \left[\nabla n_\alpha \pm (n_\alpha e/T_\alpha)\nabla\phi\right] = J_\alpha \tag{2.23}$$

and the Poisson equation for the electric potential around the particle

$$\Delta\phi = -4\pi e[n_i - n_e]. \tag{2.24}$$

Here $\alpha = i(e)$ corresponds to ions (electrons), $D_\alpha = \ell_\alpha v_{T_\alpha}$ is the diffusion coefficient of the corresponding species and the positive (negative) sign on the left-hand side of Equation (2.23) corresponds to ions (electrons). In this simplest formulation all plasma production and loss mechanisms in the vicinity of the particle are neglected. The boundary conditions for the potential are $\phi(a) = \phi_s$ and $\phi(\infty) = 0$. For an absorbing particle surface in the continuum regime ($\ell_i, \ell_e \ll a$), the boundary conditions are $n_i(a) = n_e(a) = 0$ and $n_i(\infty) = n_e(\infty) = n_0$, where n_0 is the unperturbed plasma density. Analytic solution of this system of equations exists in the case of sufficiently small grain, $a \ll \lambda_{\rm De}$, where $\lambda_{\rm De} = \sqrt{T_e/4\pi e^2 n_0}$ is the electron Debye radius. The corresponding expressions for the ion and electron fluxes are (Chang and Laframboise 1976; Khrapak *et al.* 2006b)

$$j_i \simeq \sqrt{2\pi}\frac{\ell_i}{a}\frac{z\tau}{1 - \exp(-z\tau)} \tag{2.25}$$

and

$$j_e \simeq \sqrt{2\pi}\frac{v_{T_e}}{v_{T_i}}\frac{\ell_e}{a}\frac{z\exp(-z)}{1 - \exp(-z)}, \tag{2.26}$$

respectively. Chang and Laframboise found that in a one-temperature plasma ($\tau = 1$) deviations from these expressions exceed 10% for $a/\lambda_{De} \gtrsim 1$ for the electron flux and $a/\lambda_{De} \gtrsim 0.1$ for the ion flux, correspondingly. Equating expressions (2.25) and (2.26), we get the FCP limiting value of the floating potential. For a two-temperature plasma with $\tau \gg 1$, we have $z_{FCP} \simeq \ln[1 + (\ell_e/\ell_i)\sqrt{T_i m_i / T_e m_e}]$. For a one-temperature plasma, the result is particularly simple, $z_{FCP} = \ln(D_e/D_i)$. Very good agreement between this expression and floating potentials obtained from the exact numerical solution in the limit $a/\lambda_{De} \to 0$ was documented by Baum and Chapkis (1970). We will not further discuss the FCP regime here, since it is beyond the typical range of ion collisionality in gas discharges, see Figure 2.7. We only mention a few papers which addressed different aspects of charging in the FCP regime in the context of complex plasmas (Pal' *et al.* 2001, 2002; Bystrenko and Zagorodny 2003; Filippov *et al.* 2003).

Much more important from a practical point of view is a wide pressure range where the electron transport to the particle is collisionless while the ion transport is collision dominated. Such partially collisional plasma (PCP) can, in particular, be realized in noble gases, where the characteristic cross section for ion–neutral collisions σ_{in} is typically between one and two orders of magnitude larger than that for electron–neutral collisions σ_{en}. In addition, in argon, krypton, and xenon, σ_{en} has a pronounced minimum for electron energies of about 1 eV – the so-called Ramsauer–Townsend effect (Massey *et al.* 1969). In this partially collisional plasma regime the electron flux to the particle can be calculated using the OML approximation. The ion flux can be well approximated by the expression (2.25). This is because the electron density distribution around the negatively charged particle follows closely the Boltzmann relation, independently of whether electrons are collisionless (Al'pert *et al.* 1965) or highly collisional (Su and Lam 1963). The boundary conditions for electron density at the particle surface are also virtually identical: In both cases electron density close to the grain surface is much smaller than the plasma density far from the particle, provided $z \gtrsim 1$. Hence, ion transport to the particle is governed by essentially the same equations with identical boundary conditions both in FCP and in PCP cases (Khrapak *et al.* 2006b). Taking also into account that usually in complex plasmas $z\tau \gg 1$, the normalized ion flux that the particle collects in the regime of strongly collisional ions can be further simplified,

$$j_i^{SC} \simeq \sqrt{2\pi}(\ell_i/a)z\tau. \tag{2.27}$$

In the considered regime the ion flux decreases with increasing ion collisionality (decreasing ℓ_i) while the electron flux is independent of plasma collisionality. Therefore, the partile floating potential increases in the absolute magnitude (see Figure 2.7).

Expressions (2.22) and (2.27) are then combined into a simple form

$$j_i^{eff} = \left(1/j_i^{WC} + 1/j_i^{SC}\right)^{-1}. \tag{2.28}$$

This formula accurately describes the ion fluxes in the limits of weak and strong ion collisionality and provides an interpolation between these two limits. It has been postulated by Khrapak and Morfill (2008b) that the applicability conditions of this interpolation formula are $a/\lambda \ll 1$ and $\beta_T \lesssim 10$, where $\beta_T = (a/\lambda)z\tau$ is the ion thermal scattering parameter which measures the strength of the ion-particle coupling in isotropic plasmas. In a wide pressure range (excluding only the FCP regime) the floating potential can be calculated by equating $j_i^{\rm eff}$ from Equation (2.28) to j_e from the OML approximation. This yields

$$z = \ln(v_{T_e}/v_{T_i} j_i^{\rm eff}). \tag{2.29}$$

In this approximation z is a function of four dimensionless parameters: electron-to-ion mass and temperature ratios (m_e/m_i and τ), normalized particle radius a/λ, and ion collisionality index λ/ℓ_i.

Another approximation for the ion flux that the particle collects in a collisional stationary plasma has been proposed by Hutchinson and Patacchini (2007). This is a fit (hereinafter referred to as Hutchinson's fit) based on the computation of the ion flux to a floating sphere using the Specialized-Coordinate Electrostatic Particle and Thermals in Cell (SCEPTIC) code (Hutchinson 2003). The authors propose an expression for the ion flux in the regime of weak collisionality that accounts for the collision-induced flux enhancement. This expression can be written in the present notation as

$$j_{i1} = (1 + \tau z_{\rm OML})\left[1 + \ln(1 + 17\nu_s + 5\nu_s^2)\right], \tag{2.30}$$

where $\nu_s = (\lambda/\sqrt{2}\ell_i)/[0.9 + 0.1(100/\tau)^{1.5}]$ is the scaled ion collision frequency ($\nu = v_{T_i}/\ell_i$) and $z_{\rm OML}$ is the OML value of the particle surface potential. In the opposite strongly collisional (continuum) limit, the authors use an expression which can be written in the present notation as

$$j_{i2} = 1.30(\ell_i/a)\tau\left\{\ln\left[1.1 + 30(a/\ell_i)\sqrt{m_i T_i/m_{\rm H} T_e}\right]\right\}^{1.15}, \tag{2.31}$$

where $m_{\rm H}$ is the proton mass. Combination of the two fluxes in the form

$$j_i^{\rm eff} = (1/j_{i1}^w + 1/j_{i2}^w)^{-1/w} \tag{2.32}$$

then agrees with the individual expressions in the corresponding limiting cases of weak and strong collisionality and provides an interpolation between these limits. Here w is an adjustable parameter which controls the maximum value of the ion flux in the intermediate region. An empirical expression for w based on SCEPTIC results is

$$w = (0.37 + 0.067\ln\tau)\ln(\lambda_{\rm De}/a) - 1. \tag{2.33}$$

Equations (2.30)–(2.33) describe the dependence of the collected ion flux on ion collisionality. Note that in the considered approximation the ion flux is independent of the actual value of the particle surface potential, but depends only on its OML limiting value $z_{\rm OML}$. The floating potential is obtained from Equation (2.29). In this approximation z is also a function of m_e/m_i, τ, a/λ, and λ/ℓ_i.

Another analytic fit for the ion flux in the intermediate regime of ion collisionality has been proposed by Zobnin *et al.* (2008). This fit (hereinafter referred to as

Zobnin's fit) is based on the results of direct numerical solution of the ion kinetic equation with the Bhatnagar–Gross–Krook (BGK) collision integral. The expression for the ion flux in the present notation is

$$j_i = (1+z\tau) \times \left\{1+\frac{(a/\ell_i)z\tau}{0.07+2(a/\lambda)+2.5(a/\ell_i)+[0.27(a/\lambda)^{1.5}+0.8(a^2/\lambda\ell_i)]z\tau+0.4(a/\ell_i)^2 z\tau/[1-0.4(a/\ell_i)]}\right\}. \quad (2.34)$$

This expression reduces to the OML result in the limit of low collisionality ($\ell_i \to \infty$) and tends to the continuum limit result in the opposite limit of high collisionality, unless collisionality is too high, as discussed below. The floating potential is determined from Equation (2.29), which is now transcendent, since j_i depends on the actual value of the floating potential z in the considered approximation. Similarly to the previous approximations, z depends on m_e/m_i, τ, a/λ, and λ/ℓ_i.

An analytic expression for the particle potential in collisional plasmas has been derived by D'yachkov *et al.* (2007). In this model (hereinafter referred to as D'yachkov's model) the influence of a collisionless layer (between the particle surface and surrounding plasma) on the charging currents is investigated under the assumptions $\ell_{i(e)} \ll \lambda$ and $a \ll \lambda$. The motion of ions (electrons) in the collisionless region is determined from the energy and momentum conservation laws. Beyond the collisionless region the drift-diffusion approximation is used. A proper matching procedure at the boundary between collisionless and collisional regions allows the determination of ion and electron fluxes to the particle surface. The floating potential is then obtained from the balance of these fluxes. The corresponding transcendental equation for z can be written in the present notation as

$$\frac{v_{T_i}}{v_{T_e}}\exp z = \exp\left[-\frac{z\tau}{1+\ell_i/a}\right]\left\{(1+\tfrac{\ell_i}{a})^2-\tfrac{\ell_i}{a}(2+\tfrac{\ell_i}{a})\exp\left[-\frac{z\tau}{(1+\ell_i/a)(2+\ell_i/a)}\right]\right\}^{-1} + \frac{a}{\sqrt{2\pi}z\tau\ell_i}\left\{1-\exp\left[-\frac{z\tau}{1+\ell_i/a}\right]-\sqrt{\frac{T_e m_e}{T_i m_i}}\frac{\ell_i}{\ell_e}\left[\exp\left(\frac{z}{1+\ell_e/a}\right)-1\right]\right\}. \quad (2.35)$$

It can be shown that in the highly collisional limit ($\ell_i, \ell_e \to 0$), the solution of Equation (2.35) yields $z_{\rm FCP}$ value, while in the collisionless limit ($\ell_i, \ell_e \to \infty$), the OML result $z_{\rm OML}$ is recovered. In this model z is a function of m_e/m_i, τ, ℓ_i/a, and ℓ_e/ℓ_i (in all calculations reported below $\ell_e/\ell_i = 100$ is fixed); i.e., apart from the other approximations, z is independent of the normalized particle radius a/λ. Note, however, that the explicit use of the condition $a \ll \lambda$ in the derivations limits the applicability of the D'yachkov's model to the case of very small point-like grains.

Comparison between the results obtained from different approximations considered above is presented in Figures 2.8 and 2.9. Figure 2.8 shows the calculations for an argon plasma with $\tau = 100$ and two different particle sizes: $\lambda_{\rm De}/a = 666.7$ ($\lambda/a = 66.3$) in Figure 2.8(a) and $\lambda_{\rm De}/a = 66.7$ ($\lambda/a = 6.6$) in Figure 2.8(b). These parameters are the same as in Figures 1 and 3 of the paper by Hutchinson and Patacchini (2007). In Figure 2.9 are shown calculations for a neon plasma. The plasma parameters in Figure 2.9 (a), $\lambda_{\rm De}/a = 1490$ ($\lambda/a \simeq 100$) and $\tau = 220$, are relevant to experimental results of Ratynskaia *et al.* (2004a) and Khrapak *et al.* (2005b) for $a = 0.6$ μm particles, which are shown by open circles in this figure. (Note that in

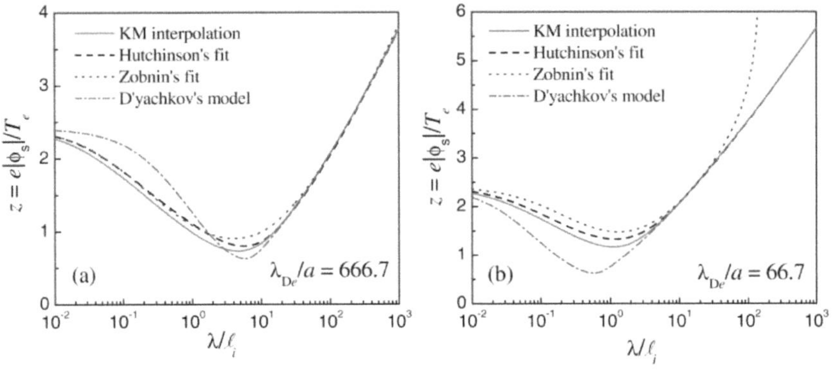

FIGURE 2.8
Dimensionless particle surface (floating) potential $z = e|\phi_s|/T_e$ as a function of ion collisionality index λ/ℓ_i for argon plasma. Calculations are for electron-to-ion temperature ratio $\tau = 100$ and two different particle sizes: $\lambda_{De}/a = 666.7$ (a) and $\lambda_{De}/a = 66.7$ (b). Solid curves correspond to the KM interpolation formula (Khrapak and Morfill 2008b), dashed curves to the fit of Hutchinson and Patacchini (2007), dotted curves to the fit by Zobnin *et al.* (2008), and dash-dotted curves to the model by D'yachkov *et al.* (2007).

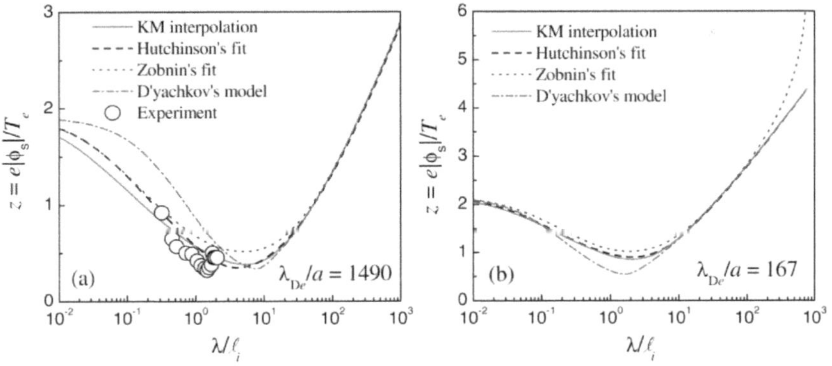

FIGURE 2.9
Dimensionless particle surface (floating) potential $z = e|\phi_s|/T_e$ as a function of ion collisionality index λ/ℓ_i for neon plasma. Calculations are for electron-to-ion temperature ratio $\tau = 220$ and normalized particle size $\lambda_{De}/a = 1490$ (a) and for $\tau = 100$ and $\lambda_{De}/a = 167$ (b). Solid curves correspond to the KM interpolation formula (Khrapak and Morfill 2008b), dashed curves to the fit of Hutchinson and Patacchini (2007), dotted curves to the fit by Zobnin *et al.* (2008), and dash-dotted curves to the model by D'yachkov *et al.* (2007). Open circles in Figure 2.9(a) correspond to experimental results by Ratynskaia *et al.* (2004a).

reality each experimental point corresponds to particular values of τ and λ/a, since in the experiment discharge pressure was varied, which inevitably led to some variations in plasma parameters. Therefore, "average" values of τ and λ/a have been chosen for comparison.) The plasma parameters in Figure 2.9 (b), $\lambda_{\rm De}/a = 167$ ($\lambda/a \simeq 16.6$) and $\tau = 100$, are identical to those in Figure 9 of the paper by Hutchinson and Patacchini (2007).

The results presented in Figures 2.8 and 2.9 demonstrate that all the approximations considered exhibit the same qualitative dependence of the normalized floating potential on ion collisionality, which has been discussed above. Note the rather close agreement between Hutchinson's and Zobnin's fits and the KM interpolation formula. Usually, the curve corresponding to Hutchinson's fit lies in between the curves corresponding to Zobnin's fit and the KM interpolation. The curves calculated from D'yachkov's model tend to the accurate result in the limit of high collisionality, but deviate from other approximations in the regime of weak and moderate collisionality. A typical trend is that this model yields larger values of z as compared to the other approaches for smaller particles and smaller values of z for larger particles. Thus, the quantitative results of this model are apparently less reliable in predicting the particle charge in weakly and moderately collisional regimes. This is not very surprising since in the D'yachkov's model, the condition $\lambda \gg \ell_i$ is assumed to be satisfied, i.e., disagreement occurs outside the range of direct applicability of this model. Note also that the curves based on Zobnin's fit diverge from the other curves at sufficiently high ion collisionality. The larger the particle, the smaller is the collisionality level at which Zobnin's fit becomes invalid.

A large number of calculations corresponding to the wide range of particle sizes $50 \leq \lambda_{\rm De}/a \leq 10^3$ and electron-to-ion temperature ratios $10 \leq \tau \leq 200$ in argon and neon gases performed by Khrapak and Morfill (2008b) demonstrates that the maximum relative difference between Zobnin's and Hutchinson's fits and between the KM interpolation and Hutchinson's fit usually does not exceed 10–15%. Thus, except perhaps for some extreme regimes, not common for complex plasmas, Hutchinson's and Zobnin's fits along with the KM interpolation formula demonstrate rather good agreement. The present accuracy of experimental charge measurements (see below) does not allow us to make a definite choice between the considered approximations. Hence, they can be equally recommended for use in estimating particle floating potentials and charges in practical situations.

2.1.3 Experimental determination of the particle charge

2.1.3.1 Quasi-isotropic plasmas

A unique experimental determination of the particle charges in a wide range of complex plasma parameters (including wide range of neutral gas pressures) has been reported by Ratynskaia *et al.* (2004a) and Khrapak *et al.* (2005b). This experiment was performed with the PK-4 facility (Usachev *et al.* 2004) in ground-based conditions. It uses horizontally oriented dc-discharge tube filled with neon gas at pressures 20–150 Pa and particles of radii $a \simeq 0.6$ μm, $a \simeq 1.0$ μm, and $a \simeq 1.3$ μm. For these particle sizes the weak ambipolar radial electric field in the bulk plasma is sufficient to compensate against gravity so that the particles can levitate in an isotropic plasma near the tube axis.

The dynamics of particles of different sizes is studied varying the neutral gas pressure p and the number of injected particles (controlled by settings of the particle dispenser), which allows the change of the particle number density n_d inside the tube. For a sufficiently low number of injected particles the flow is stable for all pressures investigated. The flow is recorded for a number of different pressures. The charge is then estimated from the *force balance condition* using the measured particle velocities. The most important forces are the electric force, the neutral drag force, and the ion drag force. For the ion drag force the model of Khrapak *et al.* (2002) is used. For a larger number of injected particles (and larger n_d), an easily identifiable transition to unstable flow (with a clear wave behavior) occurs at a certain threshold pressure p_*, which can be found experimentally with an accuracy of about 1 Pa (Ratynskaia 2004a,b). This transition is a manifestation of the ion streaming instability caused by the relative drift between the particle and the ion components. The value of p_* depends on n_d (shifting towards higher pressures when n_d is increased). In this case the charge can be estimated from a *linear dispersion relation* describing the transition of the particle flow to the unstable regime at p_*. The applicability of the linear dispersion relation method is, however, limited due to non-negligible effect of particles on plasma parameters at large particle density (Khrapak *et al.* 2005b). In practice this method was used only for smallest grains and low particle densities. The charges determined from this experiment are shown in Figure 2.10 by open symbols.

Another experimental technique to determine the particle charge is based on excitation of low-frequency (dust acoustic) waves and analysis of their dispersion relations. Two experiments reporting estimates for the particle charge using this technique have been performed under microgravity conditions with the use of the PKE–Nefedov facility (Nefedov *et al.* 2003). The waves in the particle cloud were excited by applying a low-frequency modulation voltage to the electrodes. The charge was then estimated by comparing the measured dispersion relations with the theoretical ones. In this way the dimensionless charge was found to be $z \sim 0.4$ (at $p \simeq 25$ Pa argon gas pressure) by Khrapak *et al.* (2003c) and $z \sim 0.8$ (at $p \simeq 12$ Pa argon gas pressure) by Yaroshenko *et al.* (2004). The results of these experiments are shown in Figure 2.10 by solid circle and square, respectively.

One more experimental method to determine the particle charge is based on gravity-driven heavy "test" particle collisions with smaller particles levitating in the quasi-isotropic region of an inductively coupled rf discharge plasma (Fortov *et al.* 2004). A heavy particle falls down in a vertical glass tube and interacts with the cloud of small particles suspended in the diffuse edge of the discharge. The interaction process is recorded with a high-speed video camera and individual elastic "collisions" are analyzed. Assuming the Debye–Hückel potential around each particle both the particle charge and effective screening length can be estimated. Fortov *et al.* (2004) estimated the particle charge for three different pressures (20, 30, and 50 Pa) of neon gas. The results are shown in Figure 2.10 by solid triangles.

Figure 2.10 summarizes the results of different experiments performed to measure

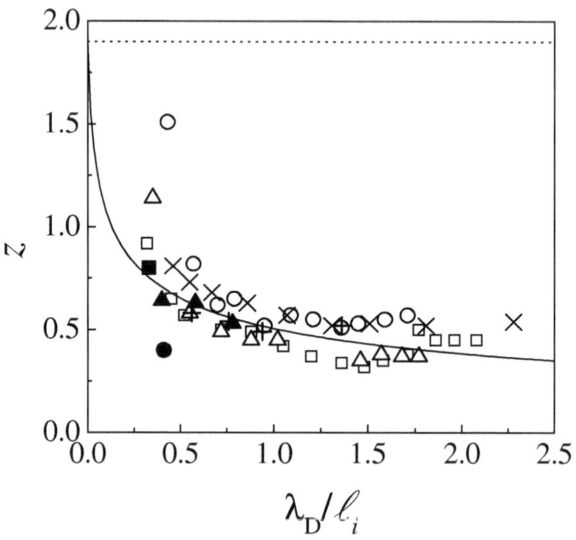

FIGURE 2.10
The dimensionless particle floating potential $z = |Q|e/aT_e$ as a function of the ion collisionality index, λ/ℓ_i (assuming $\lambda \simeq \lambda_D$). Open symbols correspond to the charges measured by Khrapak *et al.* (2005b) for particle radii $a \simeq 0.6\ \mu$m (squares), $a \simeq 1.0\ \mu$m (triangles), and $a \simeq 1.3\ \mu$m (circles). The symbols × and + are for the charges found from MD simulation for plasma conditions similar to those in Khrapak *et al.* (2005b) for $a = 0.6\ \mu$m and $a = 1.25\ \mu$m. Solid circle and square are the charges estimated from wave excitation technique in PKE–Nefedov facility by Khrapak *et al.* (2003c) and Yaroshenko *et al.* (2004), respectively. Solid triangles correspond to the charges obtained in the experiment by Fortov *et al.* (2004). The dotted line is calculated using the collisionless OML approach for an individual particle. The solid curve shows a calculation using the analytic CEC approximation of Equation (2.22). These analytic calculations are performed for plasma conditions relevant to the experiment by Khrapak *et al.* (2005b): neon plasma with $T_e = 7$ eV and $T_i = 0.03$ eV.

the particle charge. In analyzing data it is assumed here that the effective plasma screening length is $\lambda \simeq \lambda_D \simeq \lambda_{Di}$ for the quasi-isotropic plasma conditions investigated. Figure 2.10 demonstrates reasonable agreement between the results of different experiments, although they were performed under different plasma conditions (e.g., different types of discharges, different gases, different particle sizes, and different plasma parameters). This indicates that the ion collisionality index λ/ℓ_i is one of the most important parameters that controls the particle charge in isotropic complex plasmas. Also shown in the figure are the results of molecular dynamics (MD) simulations performed using the code developed by Zobnin *et al.* (2000) and modified for plasma conditions relevant to those in experiments by Ratynskaia *et al.* (2004a) and

Khrapak *et al.* (2005b). The solid curve corresponds to the floating potential calculated using CEC approximation of Equation (2.22). The agreement between theory, simulations and experiments is rather good, especially taking into account that the experimental accuracy is not expected to be much better than $\sim 30\%$ for both z and λ/ℓ_i. Note also that the obtained floating potentials are considerably smaller than those calculated from collisionless OML theory (dotted line) for all the conditions investigated.

Thus, ion–neutral collisions are experimentally identified as a main process affecting and regulating particle charging in the bulk of gas discharges. All the considered evidence clearly indicates that for a typical quasi-isotropic complex plasma, $z \lesssim 1$, which is considerably smaller than the result of the collisionless OML theory.

2.1.3.2 Anisotropic plasmas

Experimental determination of the particle charge is especially important in cases where the plasma parameters are unknown or cannot be determined with sufficient accuracy. This is usually relevant to anisotropic plasma conditions (e.g., sheaths near the plasma chamber walls or electrodes) where in addition to the charging model (already complicated by plasma anisotropy, non-neutrality, presence of "superthermal" ions and electrons, etc.), one needs to choose an appropriate model for the sheath, which is not trivial itself.

Several experimental methods were proposed to measure particle charge in sheath or striation regions of electrical discharges. Some of them are discussed below.

Vertical resonance method. Characteristics of the vertical oscillations of a single particle in the sheath are determined by the particle charge. Typical values of the vertical resonance frequency Ω_v are in the range 1–100 Hz and, hence, one can use low-frequency excitations to estimate the charge. As a simplest example, we refer to linear (harmonic) oscillations. By measuring the amplitude of the particle oscillations at different frequencies and then fitting the obtained frequency response curve with the well-known theoretical expression (Landau and Lifshitz 1976), it is possible to determine the resonance frequency Ω_v and the neutral damping rate ν_{dn} simultaneously. This technique was first employed by Melzer *et al.* (1994), and later it was used to determine the particle charges in different investigations (Trottenberg *et al.* 1995; Zuzic *et al.* 1996; Homann *et al.* 1999; Piel and Melzer 2002; Piel *et al.* 2003). The excitation is usually performed by applying a low-frequency signal to the rf electrode (in a modified variant, to a Langmuir probe or a wire inserted into a plasma in a vicinity of the levitated particle) or by using a focused laser beam.

The main difficulty in estimating the particle charge from experimental results is to establish the relation between Ω_v and Q. It is often supposed that the dependence of Q on the height is much weaker than that of the electric field; i.e., in the first approximation one can assume $Q \simeq const$, and therefore $\Omega_v^2 \simeq -QE_0'/m_d$, where E_0' denotes the derivative of the electric field evaluated at the particle equilibrium position. At sufficiently high pressures, the value of the derivative E' is practically constant over the sheath and can be estimated using a certain theoretical model. (Probe measurements in the sheath are not reliable due to uncertainty in their interpretation.) This introduces some inaccuracy in the measurement results. Nevertheless, this method is often used in experiments because of its particular simplicity.

Vertical equilibrium. The method is based on the fact that when the potential distribution in the sheath is suitably determined, the charge of an isolated particle levitated in the sheath can be estimated from their equilibrium heights by using balance between gravity and electric force. The method was used by Tomme *et al.* (2000) and later by Samarian and Vladimirov (2003). In this method care must be taken in choosing an appropriate model for the sheath. Another possible source of uncertainties is the effect of forces that are not taken into account (e.g., ion drag, thermophoresis, etc.)

Dust lattice waves. Excitation of dust waves in strongly ordered particle structures (e.g., particle chains or two-dimensional lattices) is often employed as a diagnostic tool in studying dusty plasmas. The dispersion relations of the dust lattice waves can be derived assuming a certain form of the interparticle interaction potential. Usually the Debye–Hückel potential is used. In this case the main parameters entering into the dispersion relation are the particle charge Q and the ratio of the interparticle distance to the effective screening length – the lattice parameter $\kappa = \Delta/\lambda$. Other parameters are either known in advance (e.g., particle mass) or can be easily determined in the experiment (e.g., interparticle distance). Hence, Q and κ can be estimated by comparing an experimental dispersion relation with a relevant theoretical model. The measured dispersion relations were used to estimate the particle charge in experiments by Peters *et al.* (1996), Homann *et al.* (1997, 1998), and Nunomura *et al.* (2000, 2002).

Other methods of charge determination were also used. In the *collision method* proposed by Konopka *et al.* (2000a) two-particle collisions are produced in a sheath region of an rf discharge using a horizontal electric probe, which allows the manipulation of the particles. The form of the interaction potential is reconstructed from the analysis of particle trajectories. It was found that for low discharge powers and pressures, the interaction potential can be fitted with the Debye–Hückel (Yukawa) potential within experimental uncertainties. From the fit the effective particle charge and plasma screening length can be estimated.

Measurements of the particle charge in a stratified dc discharge plasma were performed by Fortov *et al.* (2001). In this work, aperiodic oscillations of an isolated particle were excited by a focused laser beam. Analysis of particle trajectory yielded the charge.

A nonlinear dependence of the particle charge on its size was evidenced in experiments in anisotropic plasmas (see, e.g., Tomme *et al.* 2000; Fortov *et al.* 2001; Samarian and Vladimirov 2003). According to Samarian and Vladimirov (2002) this nonlinear dependence can be attributed to the dependence of surrounding plasma parameters on particle size: The particles with different radii levitate in different regions of the sheath or striation characterized by different plasma parameters, i.e., different ion and electron densities, ion drift velocity, electron temperature, etc. This makes the particle surface potential dependent on particle size and thus causes a nonlinear dependence of the particle charge on size, in contrast to the charging in the bulk of gas discharges, where the floating potential only weakly depends on the particle size.

The effect of ion–neutral collisions on particle charging, discussed in detail in connection with charging in isotropic plasmas, can apparently also be important in anisotropic plasmas. In this case the effect is two-fold: In addition to destroying collisionless ion trajectories in the vicinity of the particle, collisions change the structure of the sheath or striation. Thus the charge can depend on the ion collisionality in a quite complicated way, and we are not aware of any consistent theoretical study of this effect. We note in this context an experiment by Fortov *et al.* (2001) where some increase of the particle surface potential with pressure was reported.

2.1.4 Emission processes

The collection of ions and electrons from the plasma is not the only possible charging mechanism. Electrons can also be emitted from the particle surface due to thermionic, photoelectron, and secondary electron emission processes. These processes are important for dust charging in the working body of solid-fuel magnetohydrodynamic (MHD) generators and rocket engines (Sodha and Guha 1971; Yakubov and Khrapak 1989; Soo 1990), in the upper atmosphere, in space (Whipple 1981; Bliokh *et al.* 1995; Mendis 2002), and in some laboratory experiments, for instance, in thermal plasmas (Fortov *et al.* 1996a,b), or in plasmas induced by UV irradiation (Fortov *et al.* 1998), with photoelectric charging of particles (Sickafoose *et al.* 2000), charging by electron beams (Walch *et al.* 1995), etc. Emission of electrons increases the particle charge and, under certain conditions, the particles can reach a positive charge, in contrast to the situation discussed so far. Due to the emission processes two-component systems consisting of dust particles and the electrons emitted by them can in principle exist. In this case, the equilibrium potential (charge) of the particle is determined by the balance of the fluxes that are collected by the particle surface and emitted from it, so that the quasi-neutrality condition is $Zn_d \simeq n_e$. Such a system serves as the simplest model for investigating different processes associated with emission charging of particles (Yakubov and Khrapak 1989; Khrapak *et al.* 1999). Let us briefly consider each of the emission processes listed above.

Thermionic emission. For an equilibrium plasma characterized by a temperature T, it is common to use the following expressions for the flux of thermoelectrons (Sodha and Guha 1971):

$$J_{\rm th} = \frac{(4\pi a T)^2 m_e}{h^3} \exp\left(-\frac{W}{T}\right) \times \begin{cases} 1, & \phi_{\rm s} < 0, \\ \left(1+\frac{e\phi_{\rm s}}{T}\right)\exp\left(-\frac{e\phi_{\rm s}}{T}\right), & \phi_{\rm s} > 0. \end{cases}$$

Values of the work function W of thermoelectrons for different metals and semiconductors lie typically within the range from 2 to 5 eV. In the case of dielectric particles, where free electrons appear due to ionization, thermionic emission cannot play a significant role because the particles usually melt before thermionic emission makes a substantial contribution to the electron flux. For negatively charged particles its electric field accelerates the electrons from the particle surface and some increase in emission current can be expected due to reduction of the work function by the Schottky effect. Thermionic emission was identified as the dominant charging mechanism in thermal plasmas (Fortov *et al.* 1996a,b).

Photoelectron emission. The electron emission can be caused by an incident flux of photons with energies exceeding the work function of photoelectrons from the particle surface (Rosenberg and Mendis 1995; Rosenberg *et al.* 1996). The characteristic value of the work functions for most of the materials does not exceed 6 eV, and hence photons with energies ≤ 12 eV can charge dust particles without ionizing a buffer gas. The flux of emitted electrons can be written as (Goree 1994; Rosenberg *et al.* 1999):

$$J_{\rm pe} = 4\pi a^2 YJ \begin{cases} 1, & \phi_{\rm s} < 0, \\ \exp\left(-\dfrac{e\phi_{\rm s}}{T_{\rm pe}}\right), & \phi_{\rm s} > 0, \end{cases}$$

where J is the photon flux density and Y is the quantum yield for the particle material. It is also assumed that the radiation is isotropic; the efficiency of radiation absorption is close to unity, which occurs when the particle size is larger than the radiation wavelength; and the photoelectrons possess a Maxwellian velocity distribution with the temperature $T_{\rm pe}$. This temperature lies in most cases within the ranges from 1 to 2 eV. The quantum yield is very low just above the threshold, but for the most interesting regime of a vacuum ultraviolet, it can reach a value of one photoelectron per several photons. Therefore, the photoelectric emission mechanism of particle charging can be quite important in space.

Sickafoose *et al.* (2000) studied experimentally photoelectric emission charging of particles with diameters of $\sim 100\ \mu$m. Conducting particles acquired a positive floating potential and charge both increasing linearly with the decreasing work function of photoelectrons. Behavior of particles charged by solar radiation in microgravity conditions was investigated by Fortov *et al.* (1998). An analysis of particle dynamics after UV irradiation revealed that the particles with mean radius 37.5 μm were charged to approximately $10^4\ e$.

Secondary electron emission. The flux J_{se} of secondary electrons is connected to that of primary electrons, J_e, through the secondary emission coefficient δ, viz., $J_{se} = \delta J_e$. The coefficient δ depends both on the energy $\mathscr{E}$ of primary electrons and on the particle material. The dependence $\delta(\mathscr{E})$ turns out to be practically universal for different materials, if δ is normalized to the maximum yield $\delta_{\rm m}$ of electrons, and $\mathscr{E}$ is normalized to the value $\mathscr{E}_{\rm m}$ of energy at which this maximum is reached. The corresponding expressions for the case of monoenergetic electrons were given by Whipple (1981) and Goree (1994). The values of the parameters $\delta_{\rm m}$ and $\mathscr{E}_{\rm m}$ for some materials given by Whipple lie within the ranges $\delta_{\rm m} \sim (1-4)$ and $\mathscr{E}_{\rm m} \sim (0.2-0.4)$ keV. For the case of Maxwellian-distributed electrons, the expression for δ was given, for instance, by Goree.

Walch *et al.* (1995) experimentally investigated the charging of particles of various materials and diameters from 30 to 120 μm by thermal and monoenergetic superthermal electrons. When the charging was dominated by superthermal electrons, the particles were charged to the potential proportional to the electron energy and the charge proportional to the particle radius. However, when the electron energy reached a threshold value (different for various materials), from which the secondary electron emission became important, a sharp decrease in the absolute magnitude of the charge was found.

2.1.5 Quasineutrality of complex plasmas

The dust particles immersed in a plasma act as ionization and recombination centers. Particles that emit electrons may increase the electron concentration in the plasmas. Conversely, when the particles absorb electrons from the plasma, they become negatively charged and reduce the electron density compared to the ion density. From the quasi-neutrality condition, it is clear that the presence of particles influences the plasma charge composition when $|Z|n_d/n_e \equiv P \gtrsim 1$.

In the absence of emission processes, electrons and ions recombine on the particles. Plasma loss rates are determined by the expression $Q_{\mathrm{L}e(i)} = J_{e(i)}n_d$, where $J_{e(i)}$ is the flux of electrons (ions) absorbed by the particle surface. For large particle concentrations, the losses of electrons and ions on the particles can exceed the recombination losses in the particle-free plasma (volume recombination and/or plasma losses to the walls of a discharge camera). In self-sustained plasmas an increase in the recombination frequency has to be compensated for by a corresponding increase in the ionization frequency (Boeuf 1992; Lipaev *et al.* 1997). This can, for instance, lead to an increase in the electron temperature and the discharge electric field.

When particles emit electrons, they serve as ionization sources as well. The particle contribution to the ionization is characterized by the flux of emitted electrons (see the previous section). In the limiting case, emission from particles embedded into a neutral gas completely determines the charge composition of the plasma, playing the role of sources and sinks for the electrons.

2.1.6 Fluctuations of the particle charge

So far the particle charge was treated as a continuous *regular* variable. However, the charging currents represent in reality sequences of events bound to electron and ion absorption or emission by the particle surface. These sequences and time intervals between the successive acts of absorption and emission are random. As a result, the particle charge can fluctuate around its average value. The importance of charge fluctuations was recognized rather early: Morfill *et al.* (1980) suggested that charge fluctuations can have a major influence on dust transport in astrophysical plasmas. Several studies addressed the problem of charge fluctuations that arise from the random nature of the charging process (Cui and Goree 1994; Matsoukas and Russel 1995, 1997; Matsoukas *et al.* 1996). In particular, gas discharge plasmas, where particles are charged by collecting electrons and ions, were considered within the framework of the OML approach. Several different charging mechanisms, including thermionic and photoelectronic emission processes, were also considered by Khrapak *et al.* (1999).

Charge fluctuations due to the discrete nature of charging can be described as a stationary, Gaussian and Markovian process (or the Ornstein–Uhlenbec process after Uhlenbeck and Ornstein 1930). This process was originally adopted to describe the stochastic behavior of the velocity of a Brownian particle. In the above case, it describes the behavior of the deviation of a particle charge number from its average value: $Z_1(t) = Z(t) - Z_0$, where $Z_0 = \langle Z(t) \rangle$ is the average charge number ($Q_0 = eZ_0$). Let us derive the main properties of charge fluctuations. For simplicity, we limit consideration to the particle charging by electron and ion collection in the OML approximation. Generalization to other charging mechanisms is trivial. The Langevin equation for $Z_1(t)$ is

$$\frac{dZ_1}{dt} + \Omega_{\rm ch} Z_1 = f(t), \tag{2.36}$$

where $f(t)$ is the stochastic term, associated with random acts of electron/ion collection. Function $f(t)$ satisfies the following properties: $\langle f(t) \rangle = 0$ and $\langle f(t) f(t') \rangle = 2J_0 \delta(t - t')$, where $J_0 = J_e = J_i$ is the average flux of electrons and ions to the particle in the stationary state. Applying these properties to the solution of Equation (2.36),

$$Z_1(t) = Z_1(0) \exp(-\Omega_{\rm ch} t) + \exp(-\Omega_{\rm ch} t) \int_0^t f(t') \exp(\Omega_{\rm ch} t') dt',$$

we obtain the following properties of charge fluctuations:

(1) The charge fluctuation amplitude has zero average:

$$\langle Z_1 \rangle = 0.$$

(2) The charge autocorrelation function decays exponentially,

$$\langle Z_1(t) Z_1(t') \rangle = \langle Z_1^2 \rangle \exp\left(-\Omega_{\rm ch} |t - t'|\right), \tag{2.37}$$

where the relative charge dispersion (squared fluctuation amplitude) is

$$\sigma_Z^2 \equiv \frac{\langle Z_1^2 \rangle}{Z_0^2} = \frac{\gamma_Z}{|Z_0|}. \tag{2.38}$$

Using OML theory we get

$$\gamma_Z = \frac{1 + z\tau}{z(1 + \tau + z\tau)} \simeq \frac{1}{1 + z},$$

assuming that $\tau \gg 1$.

(3) The process $Y(t) = \int_0^t Z_1(t') dt'$ is Gaussian but neither stationary nor Markovian. With the help of Equation (2.37) we obtain

$$\langle Y(t)^2 \rangle = \frac{2\langle Z_1^2 \rangle}{\Omega_{\rm ch}^2} \left[\Omega_{\rm ch} t + \exp(-\Omega_{\rm ch} t) - 1\right].$$

Usually, it is enough to use these properties for investigating the influence of charge fluctuations on dynamic processes in complex (dusty) plasmas. In particular, the following investigations can be mentioned: dust particle "heating" (in terms of the kinetic energy) in an external electric field due to charge fluctuations was studied by Vaulina *et al.* (1999a,b, 2000) and by Quinn and Goree (2000); instabilities of particle oscillations due to charge fluctuations were considered by Morfill *et al.* (1999a) and Ivlev *et al.* (2000); dust diffusion across a magnetic field due to random charge fluctuations was investigated by Khrapak and Morfill (2002) with application to astrophysical plasma. Dynamics of particles with fluctuating charges is considered in details in Section 3.2.1.

We note that the discreteness of the charging process in not the only reason for particle charge fluctuation. Spatial and temporal variations in plasma parameters and collective effects in complex plasmas constitute other sources of charge fluctuations. These issues, however, are much less investigated.

2.2 Electric potential distribution around a particle

2.2.1 Isotropic plasmas

The distribution of the electric potential $\phi(r)$ around an individual spherical particle of charge $Q = Ze$ satisfies the Poisson equation (2.24) with the boundary conditions $\phi(\infty) = 0$ and $\phi(a) = \phi_s$. The particle potential and the charge are related through $d\phi/dr|_{r=a} = -Q/a^2$. It is often assumed that the electric potential around a small particle in isotropic plasmas can be described by the Debye–Hückel (Yukawa) form,

$$\phi(r) \simeq \phi_s(a/r)\exp[-(r-a)/\lambda] \simeq (Q/r)\exp(-r/\lambda). \tag{2.39}$$

This result can be obtained by assuming Boltzmann distributions for ions and electrons and solving the linearized Poisson equation. In the linear regime $\lambda = \lambda_D$. Linearization is often invalid in complex plasmas since the particle floating potential is $\phi_s \sim -T_e/e$, and therefore, ion-particle coupling is very strong close to the particle, provided $T_e \gg T_i$. Nevertheless, numerical solution of the nonlinear Poisson-Boltzmann equation shows that the functional form of Equation (2.39) still persists, but the actual value of the particle charge should be replaced by an effective charge which is somewhat smaller in the absolute magnitude (Nefedov *et al.* 1998; Bystrenko and Zagorodny 1999).

A greater influence on the potential structure is effected by the plasma absorption on the particle surface. The continuous ion and electron fluxes from the bulk plasma to the particle make their distributions non-Boltzmann. Although the deviations are only marginal for repelled electrons (Al'pert *et al.* 1965), for attracted ions they are quite substantial. In the absence of plasma production and loss in the vicinity of the particle, conservation of the plasma flux completely determines the far asymptote of the potential. As a result, at large distances the potential is not screened exponentially but exhibits a power law decay. In collisionless plasmas the far asymptote scales as $\phi(r) \propto r^{-2}$ (Al'pert *et al.* 1965; Tsytovich 1997; Fortov *et al.* 2005). Close to the particle (up to a distance of a few Debye radii from its surface), the Debye–Hückel

(DH) form works reasonably well, but with $\lambda = \lambda_{\rm eff}$, where the effective screening length $\lambda_{\rm eff}$ can deviate considerably from $\lambda_{\rm D}$. For example, numerical calculations by Daugherty *et al.* (1992) have demonstrated that for a particle smaller than the ion Debye radius, $\lambda_{\rm eff} \simeq \lambda_{\rm D}$. As the particle radius increases, $\lambda_{\rm eff}$ also increases and can reach values comparable to $\lambda_{{\rm D}e}$ when $a \sim \lambda_{{\rm D}i}$. Ratynskaia *et al.* (2006) found that the structure of the normalized electric potential computed numerically is rather insensitive to the separate values of the dimensionless parameters $a/\lambda_{{\rm D}i}$ and $z\tau$, but depends universally on their product, the ion thermal scattering parameter $\beta_{Ti} = z\tau(a/\lambda_{{\rm D}i}) \simeq \beta_T$ (in the following we will not distinguish between β_{Ti} and β_T). It is instructive to compare the dependencies of $\lambda_{\rm eff}/\lambda_{{\rm D}i}$ on β_T obtained by Daugherty *et al.* and Ratynskaia *et al.* It is worth noting that these calculations involve quite different approximations: the ion energy distribution far from the particle is assumed monoenergetic in Daugherty *et al.* and Maxwellian in Ratynskaia *et al.*; corrections to the electron density distribution due to electron absorption on the particle surface are taken into account in Daugherty *et al.*, but neglected in Ratynskaia *et al.*; barriers in the effective potential for the radial ion motion are accounted for in Daugherty *et al.*, but neglected in Ratynskaia *et al.* Nevertheless, the agreement between the results by Daugherty *et al.* and Ratynskaia *et al.* is rather close, as can be seen from Figure 2.11. A reasonable fit to the results by Daugherty *et al.* in the entire range of β_T investigated is given by

$$(\lambda_{\rm eff}/\lambda_{{\rm D}i}) \simeq 1 + 0.105\sqrt{\beta_T} + 0.013\beta_T. \tag{2.40}$$

Note that $\lambda_{\rm eff} \simeq \lambda_{{\rm D}i}$ for weak ion–particle coupling (linear screening), as expected.

Another important factor which influences the structure of the electric potential around an absorbing particle in plasmas is ion–neutral collisions. For example, it is well known that in strongly collisional plasmas, the far asymptote of the potential exhibits a Coulomb-like decay (Su and Lam 1963; Bystrenko and Zagorodny 2003; Khrapak *et al.* 2006b; Filippov *et al.* 2007a), instead of the $\propto r^{-2}$ decay in the collisionless limit. Moreover, the long-range asymptote of the electric potential is approximately proportional to the collected ion flux (in the absence of plasma production and loss in the vicinity of the particle). As has been shown in Section 2.1.2, the ion flux is considerably affected by collisions even in the weakly collisional regime. Therefore, the collisional contribution to the structure of the electric potential around an absorbing particle can be significant even when $\ell_i > \lambda$. Let us therefore discuss this issue in more detail.

A simple linear kinetic model which accounts for the combined effect of ion absorption on the particle and ion–neutral collisions has been proposed independently by Filippov *et al.* (2007b) and Khrapak *et al.* (2008). In this model a small (point-like) individual grain of negative charge Q immersed in a stationary isotropic weakly ionized plasma is considered. Plasma production and loss in the vicinity of the particle are neglected, except plasma absorption on the particle surface. This implies that the characteristic ionization/recombination length is considerably larger than the length scale under consideration. Electron density can be approximated with

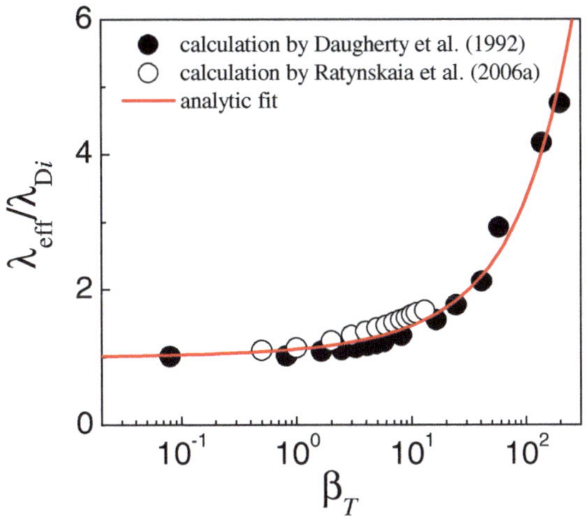

FIGURE 2.11
Ratio of the effective screening length to the ion Debye radius, $\lambda_{\rm eff}/\lambda_{{\rm D}i}$ as a function of the ion thermal scattering parameter $\beta_T = z\tau(a/\lambda_{{\rm D}i})$. Solid circles correspond to numerical calculations by Daugherty *et al.* (1992). Open circles correspond to numerical calculations by Ratynskaia *et al.* (2006). In both cases $\lambda_{\rm eff}$ is obtained from the best fit of the numerically calculated potential with the Debye–Hückel expression. To obtain the dependence of $\lambda_{\rm eff}/\lambda_{{\rm D}i}$ on β_T from the data presented in Daugherty *et al.*, it is assumed that $\lambda_{{\rm D}i} = \sqrt{2E_i/4\pi e^2 n_0}$, where E_i is the energy of monoenergetic ions, which yields the correct length scale for screening in the linear regime, and it is assumed that $\tau = T_e/2E_i$ in evaluating β_T. The solid curve corresponds to the analytic fit of Equation (2.40).

high accuracy by the Boltzmann distribution

$$n_e \approx n_0 \exp(e\phi/T_e), \tag{2.41}$$

where n_0 is the unperturbed plasma density. The kinetic equation for the ions is

$$\frac{\partial f}{\partial t} + \mathbf{v}\frac{\partial f}{\partial \mathbf{r}} + \frac{e\mathbf{E}}{m}\frac{\partial f}{\partial \mathbf{v}} = -\nu(f - n_i f_M) - \delta(\mathbf{r})v\sigma(v)f, \tag{2.42}$$

where f is the ion velocity distribution function, $f_M = (2\pi v_{T_i}^2)^{-3/2}\exp(-v^2/2v_{T_i}^2)$ is the Maxwellian distribution function normalized to unity and $n_i = \int f d^3v$ is the ion density (we assume that ions are in equilibrium with neutrals, so that $v_{T_i} = v_{T_n}$). The first term on the right-hand-side is the model collision integral in the Bhatnagar–Gross–Krook form (Bhatnagar *et al.* 1954) with a *constant* effective ion–neutral collision frequency ν. The second term represents the ion loss on a small particle and is expressed through the effective (velocity dependent) collection cross section

$\sigma(v)$. Such a model kinetic description of ion absorption on a point-like grain has been proposed by Filippov *et al.* (2007a) for collisionless plasmas. In this case $\sigma(v)$ is given by the OML model. In the collisional case $\sigma(v)$ has a less transparent physical meaning and apparently cannot be determined from first principles. However, Khrapak *et al.* (2008) proposed to use its relation to a "measurable" quantity – the ion flux that the particle collects, $J_i = n_0 \int v\sigma(v) f_M(v) d^3v$. Assuming a certain functional form for $\sigma(v)$, one can then easily express the potential through the collected ion flux J_i.

The equations for the electrons and ions are supplemented by the Poisson equation

$$\Delta\phi = -4\pi e(n_i - n_e) - 4\pi Q\delta(\mathbf{r}). \tag{2.43}$$

Standard linearization procedure then yields

$$\phi(r) = \frac{Q}{r}\exp(-k_{\rm D}r) - \frac{e}{r}\int_0^\infty \frac{k_{\rm D}\sin(kr)f(\theta)dk}{k^2+k_{\rm D}^2} \equiv \phi_{\rm I} + \phi_{\rm II}, \tag{2.44}$$

where

$$f(\theta) = \frac{8n_0}{\pi^{3/2}k_{\rm D}}\frac{\int_0^\infty \sigma(\xi)\xi^2 \arctan(\xi/\theta)\exp(-\xi^2)d\xi}{1-\sqrt{\pi}\theta\exp(\theta^2)[1-\mathrm{erf}(\theta)]}.$$

Here $k_{\rm D} = \sqrt{k_{\rm De}^2 + k_{\rm Di}^2}$ is the inverse linearized Debye radius, $k_{{\rm D}i(e)} = \lambda_{{\rm D}i(e)}^{-1}$, $\theta = (\nu/\sqrt{2}kv_{T_i})$, and $\xi^2 = v^2/2v_{T_i}^2$. The first term $\phi_{\rm I}$ in Equation (2.44) is the familiar Debye–Hückel potential. The second term $\phi_{\rm II}$ appears due to ion absorption by the particle and accounts for ion–neutral collisions. For a non-absorbing particle $[\sigma(v) \equiv 0]$ only the conventional DH form survives, as expected. In this case ion–neutral collisions do not affect the potential distribution.

In the collisionless (CL) limit we have $\nu = 0$, $\theta = 0$, and $\arctan(\xi/\theta) = \pi/2$. Using the OML collection cross section $\sigma(\xi) = \pi a^2\left[1+(z\tau)\xi^{-2}\right]$, one easily gets

$$\phi_{\rm II}(r) = -\frac{e}{r}\frac{\pi a^2 n_0(1+2z\tau)}{2k_{\rm D}}\mathscr{F}(k_{\rm D}r), \tag{2.45}$$

where $\mathscr{F}(x) = [{\rm e}^{-x}{\rm Ei}(x) - {\rm e}^{x}{\rm Ei}(-x)]$ and ${\rm Ei}(x)$ is the exponential integral. This expression has been derived by Filippov *et al.* (2007a). For sufficiently large distances $x \gg 1$, $\mathscr{F}(x) \approx 2/x$ and the corresponding potential is

$$\phi_{\rm II}(r) \simeq -\frac{T_e}{e}\left(\frac{a}{r}\right)^2\frac{1+2z\tau}{4(1+\tau)}, \tag{2.46}$$

which coincides with the well-known result of probe theory (Al'pert *et al.* 1965; Tsytovich 1997; Allen *et al.* 2000; Fortov *et al.* 2005). The transition from the DH form to the long-range asymptote (2.46) occurs at about the location where these two forms are equal (Lampe 2001), i.e., at

$$r \simeq \lambda_{\rm D}\left[\ln(2\lambda_{\rm D}/a) + \ln\ln(2\lambda_{\rm D}/a)\right].$$

In the opposite strongly collisional (SC) regime, we have $\theta \gg 1$, $\arctan(\xi/\theta) \approx \xi/\theta$, and $\mathrm{erf}(\theta) \approx 1 - \pi^{-1/2}\mathrm{e}^{-\theta^2}(\theta^{-1} - \frac{1}{2}\theta^{-2})$. The actual form of $\sigma(\xi)$ is not important since the integral in $f(\theta)$ is directly expressed through the ion flux J_i in this case. The potential is (Khrapak *et al.* 2008)

$$\phi_{\mathrm{II}}(r) \simeq -\frac{e}{r}\frac{J_i}{D_i k_{\mathrm{D}}^2}\left[1 - \exp(-k_{\mathrm{D}} r)\right], \tag{2.47}$$

where $D_i = v_{T_i}^2/\nu$ is the diffusion coefficient of the ions. This expression coincides with the results obtained using the hydrodynamic approximation (Filippov *et al.* 2007a; Khrapak *et al* 2007b,c).

The most interesting regime relevant to many complex plasma experiments in gas discharges is the weakly collisional (WC) regime, $\ell_i \gtrsim \lambda_{\mathrm{D}}$. In this case, assuming $\nu \to 0$ yields $\theta \ll 1$, $\arctan(\xi/\theta) \approx \pi/2 - \theta/x$, and $\mathrm{erf}(\theta) \to 0$. To calculate the potential, the functional form $\sigma(\xi)$ is required. Khrapak *et al.* (2008) have made a simple assumption of a constant cross section, which allowed the avoidance of divergence of the integrals in calculating $f(\theta)$. They assumed $\sigma(\xi) = \sigma_0 = \sqrt{\pi/8}(J_i/n_0 v_{T_i})$. Integration then yields

$$\phi_{\mathrm{II}}(r) \simeq -\frac{e}{r}\frac{\sqrt{\pi}}{4\sqrt{2}}\frac{J_i}{k_{\mathrm{D}} v_{T_i}}\left\{\mathscr{F}(k_{\mathrm{D}} r) + \frac{\varepsilon\pi^{3/2}}{\ell_i k_{\mathrm{D}}}\left[1 - \exp(-k_{\mathrm{D}} r)\right]\right\}, \tag{2.48}$$

where $\varepsilon = \sqrt{\pi} - 4\pi^{-3/2} \approx 0.60$ is a numerical factor. The two terms in the curly brackets of Equation (2.48) correspond to absorption induced "collisionless" and "collisional" contributions, respectively. The collisional contribution to the potential dominates over the collisionless one for $r \gtrsim (2/\varepsilon\pi^{3/2})\ell_i \approx 0.6\ell_i$. Equation (2.48) is approximate because of the assumption of a constant collection cross section. For instance, substituting the OML expression for the ion flux in Equation (2.48) yields the CL potential $\phi_{\mathrm{II}}(r) = -(e/r)(\pi a^2 n_0/2k_{\mathrm{D}})(1 + z\tau)\mathscr{F}(k_{\mathrm{D}} r)$, which is different from the exact expression (2.45) by a factor $\frac{1+z\tau}{1+2z\tau}$ ($\approx \frac{1}{2}$ since usually $z\tau \gg 1$). Note also that although Equation (2.48) is derived under the assumption $k_{\mathrm{D}}\ell_i \gg 1$, it yields a correct result (to an accuracy of a numerical factor very close to unity) also in the opposite limit, $k_{\mathrm{D}}\ell_i \ll 1$, as can be immediately seen by comparing Equations (2.47) and (2.48).

To determine the ion flux J_i entering into Equation (2.48) one can use the approximations discussed in Section 2.1.2. Khrapak *et al.* (2008) used the CEC approximation [Equation (2.22)] in the WC regime and continuum limit expression [Equation (2.27)] in the SC regime. An example of the electric potential distribution calculated by Khrapak *et al.* (2008) is shown in Figure 2.12. The plasma parameters used are representative for complex plasma experiments in gas discharges: argon gas, $\tau = 100$, and $k_{\mathrm{D}} a = 0.01$. The solid curves correspond to numerical integration of Equation (2.44) for three different ion collisionality indexes $k_{\mathrm{D}}\ell_i$, the dashed curve corresponds to the SC (in fact, FCP) limit and the dotted curve to the CL limit. For reference, the dash-dotted curve shows the DH potential. Note that the particle surface potentials are different for different curves. This reflects the fact that not only

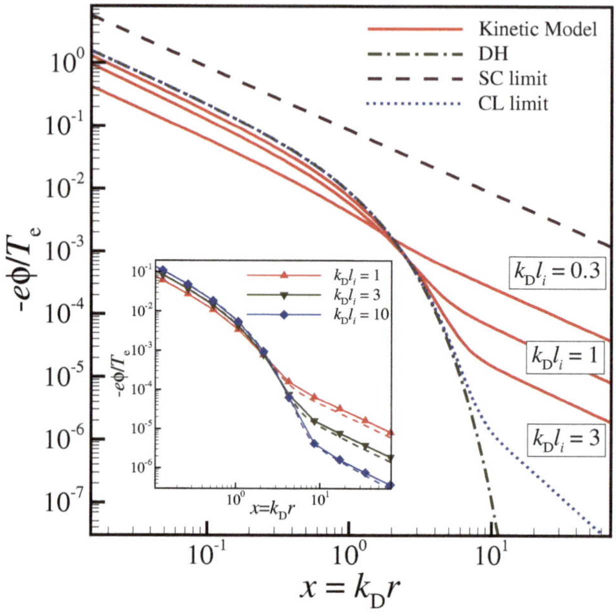

FIGURE 2.12

Normalized electric potential around a small particle in an isotropic weakly ionized plasma versus normalized distance. The solid curves are obtained using numerical integration in Equation (2.44). Dashed (dotted) curve corresponds to the analytic approximation in the strongly collisional (collisionless) limit. The dash-dotted curve shows the Debye–Hückel potential with the surface potential calculated from the (collisionless) OML model. The inset shows a comparison between direct numerical integration of Equation (2.44) (solid lines with symbols) and the approximate expression for the weakly collisional regime [Equations (2.44) and (2.48), marked by dashed curves].

the functional form but also the initial value of the potential at the particle surface depend on the ion collisionality, as discussed in Section 2.1.2. The inset shows the comparison between direct numerical integration of Equation (2.44) and the approximate expression for the WC regime [Equations (2.44) and (2.48)]. The agreement is rather good and improves with increasing $\ell_i k_D$, as expected.

Figure 2.12 demonstrates that the long-range asymptote of the potential is dominated by the combined effect of collisions and absorption. It exhibits Coulomb-like decay $\phi(r) \sim Q_{\rm eff}/r$, where the effective charge $Q_{\rm eff}$ is determined from the plasma parameters and *increases monotonically* in absolute magnitude with ion collisionality. At short distances the potential follows the DH form (2.39), but the surface potential shows a *non-monotonic* dependence on $k_D\ell_i$. In the WC regime z decreases with increasing collisionality, while in the SC regime z increases until it reaches a maximum value in the FCP regime (see Figure 2.7). In the WC regime the transition from short-range DH to the long-range Coulomb-like asymptote occurs through an intermediate $\propto r^{-2}$ decay. In the SC regime the potential is Coulomb-like starting practically from the particle surface.

Thus, the combined effect of continuous ion absorption and ion–neutral collisions determines both the amplitude and the functional form of the electric potential distribution around a small absorbing particle in plasmas. The results discussed here have been obtained using a simple linear kinetic model which neglects any plasma production and loss mechanisms in the vicinity of the particle (Filippov *et al.* 2007b; Khrapak *et al.* 2008). It should be noted, that direct application of linear models to complex plasmas is usually limited to the cases of small particles and/or low electron-to-ion temperature ratios when the range of "nonlinear" ion–particle interaction is considerably shorter than the plasma screening length. However, linear theory seems to be the only tool available at the moment to obtain analytic results and to demonstrate the important effect of ion–neutral collisions on the structure of the electric potential. A fully self-consistent nonlinear model of the potential distribution around an absorbing particle in collisional plasma is a challenge for future research. The last point to be mentioned here is that effects of plasma production and loss in the vicinity of the particle (which have been neglected so far) can also influence the structure of the electric potential. This topic has been investigated by Filippov *et al.* (2007a) and Chaudhuri *et al.* (2008) using the hydrodynamic approximation and assuming that plasma production is due to electron impact ionization, while plasma loss is due to either electron–ion volume recombination or ambipolar diffusion towards the discharge chamber walls and electrodes. Conditions under which plasma production and loss mechanisms are of minor importance have been identified by Chaudhuri *et al.*

2.2.2 Anisotropic plasmas

Strong electric fields are often present in laboratory conditions (e.g., in rf sheaths or dc striations). This induces an ion drift and, hence, creates a perturbed region of plasma density around the particle, caused by downstream focusing of ions – the so-called "plasma wake". One can apply the linear dielectric response formalism (see e.g., Alexandrov *et al.* 1984) to calculate the potential distribution in the wake. This approach is applicable provided ions are weakly coupled to the particle (i.e., the region of nonlinear ion-particle electric interaction is small compared to the plasma screening length). Note that higher ion drift velocities imply better applicability of the linear theory. The electrostatic potential created by a point-like charge at rest is defined in this approximation as

$$\phi(\mathbf{r}) = \frac{Q}{2\pi^2}\int \frac{e^{i\mathbf{kr}}d\mathbf{k}}{k^2\varepsilon(0,\mathbf{k})}, \tag{2.49}$$

where $\varepsilon(\omega,\mathbf{k})$ is the plasma permittivity and $\mathbf{u}$ is the ion flow velocity. Using a certain model for the permittivity, one can calculate the anisotropic potential distribution

(Nambu *et al.* 1995; Vladimirov and Nambu 1995; Vladimirov and Ishihara 1996; Ishihara and Vladimirov 1997; Xie *et al.* 1999; Lapenta 2000; Lemons *et al.* 2000). The potential profile can also be obtained from numerical modeling (Melandsøand Goree 1995; Lampe *et al.* 2000; Maiorov *et al.* 2001; Winske 2001; Lapenta 2002; Vladimirov *et al.* 2003).

Physically, the generation of plasma wakes in anisotropic dusty plasmas is similar to the generation of electromagnetic waves by a particle which is placed in a moving medium (Bolotovskii and Stolyarov 1992; Ginzburg 1996), and the analogy with the Vavilov-Cherenkov effect can be useful. The potential is no longer monotonic within a certain solid angle downstream from the particle, but has a well pronounced extremum (maximum for a negatively charged particle). Numerical modeling shows that the shape of the wake potential is sensitive to the ion–neutral collisions (Hou *et al.* 2003) and the electron-to-ion temperature ratio which governs Landau damping (Lampe *et al.* 2001b). In typical situations, these mechanisms can effectively "smear out" the oscillatory wake structure, leaving a single maximum.

Let us illustrate how the wake potential depends on the plasma flow. The ion drift velocity is conveniently characterized by the value of the "thermal" Mach number, $M_T = u_i/v_{T_i}$. The pronounced anisotropic wake structure appears in both subthermal and superthermal regimes of the drift (both regimes are ubiquitous for typical experimental conditions). In this context, one can mention the work of Lampe *et al.* (2000) where some examples of the wake structures, calculated numerically for different plasma conditions, are presented (for example, see Figure 2.13).

First we consider subthermal ion drift, $M_T \lesssim 1$. The potential profile in this case can be calculated from Equation (2.49) analytically within the BGK approach for the ion–neutral collision integral (Schweigert 2001; Ivlev *et al.* 2005). The far-field potential has a well-known $\propto r^{-3}$ asymptote (Montgomery *et al.* 1968). By combining this with the near-field Yukawa core, in the case of small collisionality (viz., small ratio of the ion–neutral collision frequency to the ion plasma frequency), we can approximate the potential by the following expression (Kompaneets 2007):

$$\phi(r,\theta) = Q\left[\frac{e^{-r/\lambda_{\rm D}}}{r} - 2\sqrt{\frac{2}{\pi}}\frac{M_T\lambda_{\rm D}^2}{r^3}\cos\theta - \left(2-\frac{\pi}{2}\right)\frac{M_T^2\lambda_{\rm D}^2}{r^3}(3\cos^2\theta - 1)\right]. \tag{2.50}$$

Equation (2.50) is written in spherical coordinates, where θ is the angle between $\mathbf{r}$ and $\mathbf{u}_i$ and is accurate to $o(M_T^2/r^3)$. It shows that microparticles attract each other in a certain solid angle along the flow, and repel in the transverse direction. Such behavior is usually observed in ground-based experiments – particles levitating in, e.g., (pre)sheaths of rf discharges (Melzer *et al.* 1996b) form stable vertical "strings". This result highlights the importance of the self-consistent consideration of the ion kinetics where the ion–neutral collisions are properly taken into account. Indeed, the (somewhat arbitrary) use of, e.g., a shifted Maxwellian distribution (Wang *et al.* 1981) to model a flowing plasma yields attraction between particles in the transverse direction, which contradicts the experimental observations.

In some ground-based experiments particles levitate in the regions where the electric field is so strong that the thermal Mach number can be significantly larger than

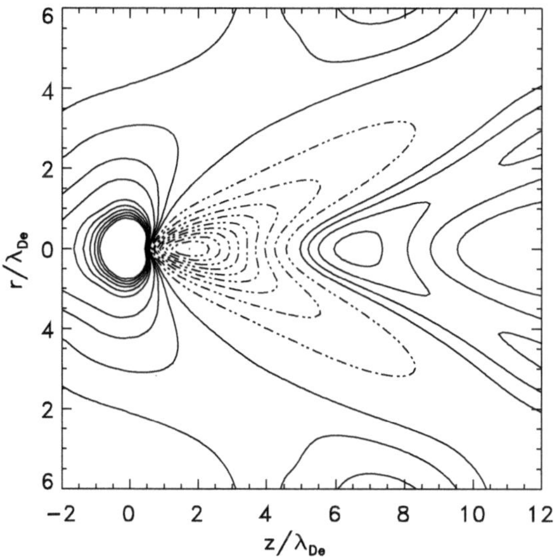

FIGURE 2.13
Example of plasma wake (Lampe *et al.* 2000). The figure illustrates the complex structure of the wake potential $\phi(\mathbf{r})$ (plasma flows to the right). Calculations are for collisionless ions with a shifted Maxwellian distribution ($M_T = 7.5$) and for Boltzmann electrons ($\tau = 25$). The (negatively charged) grain is at the center of the left-most node, solid and dashed curves indicate contour lines for negative and positive potentials, respectively, distance is in units of λ_{De}.

unity; also, the collisionality can be rather high. The BGK collision integral is no longer applicable in this case, because in highly suprathermal regimes the ion *mean free path* rather than the collision frequency should be considered constant. Solution of the kinetic equation with a constant mean free path yields the following asymptotic form of the ion velocity distribution $f(\mathbf{v})$ in the formal limit $M_T \to \infty$:

$$f(\mathbf{v}) = n_i \sqrt{\frac{2m}{\pi T_{\|}}} \exp\left(-\frac{mv_z^2}{2T_{\|}}\right) \delta(\mathbf{v}_{\perp}), \quad v_z > 0, \tag{2.51}$$

whereas for $v_z < 0$ we have $f = 0$ (electric field $\mathbf{E}$ is directed along the z axis). Here $T_{\|} = eE\ell_i$ is the field-induced "temperature" characterizing such *half-Maxwellian* distribution. [Note that similar distributions were directly measured in experiments (Zeuner and Meichsner 1995; Rao *et al.* 1996) and obtained in simulations (Maiorov 2006)]. The distribution function given by Equation (2.51) significantly deviates from the velocity distribution obtained in the framework of the BGK approach (see Section 2.5.1.2).

In such highly suprathermal collisional regime, $M_T \gg 1$, the wake potential given by Equation (2.50) is no longer applicable as well. The calculations based on the constant-mean-free-path model performed by Kompaneets *et al.* (2007) provide the potential distribution for this case (in cylindrical coordinates):

$$\phi(r_\perp, z) = \frac{2Q}{\pi \ell_i} \mathrm{Re} \int_0^\infty dt \frac{\exp[it(z/\ell_i)]}{1+(\ell_i/\lambda)^2 Y(t)} K_0\left(\frac{r_\perp}{\ell_i}\sqrt{\frac{t^2+(\ell_i/\lambda)^2 X(t)}{1+(\ell_i/\lambda)^2 Y(t)}}\right). \quad (2.52)$$

Here $\lambda = [eE\ell_i/(4\pi n_i e^2)]^{1/2}$ is the "field-induced" Debye length, and ℓ_i is the ion–neutral mean free path. Further, K_0 is the zero-order modified Bessel function of the second kind. Equation (2.52) is expressed in terms of two functions,

$$X(t) = 1 - \sqrt{1+it},$$

$$Y(t) = \frac{2\sqrt{1+it}}{it} \int_0^1 \frac{d\alpha}{[1+it(1-\alpha^2)]^2} - \frac{1}{it(1+it)},$$

where the square root should be taken with positive real part. The asymptotic behavior of Equation (2.52) at large distances is

$$\phi(r,\theta) = -\frac{Q\lambda^2}{\ell_i}\frac{\cos\theta}{r^2}\left(\frac{2}{1+\cos^2\theta}\right)^{3/2} + O\left(\frac{1}{r^3}\right). \quad (2.53)$$

Thus, at large distances the test charge produces a dipole-like field, with the dipole moment $Q\lambda^2/\ell_i$. For a negatively charged grain ($Q < 0$), this dipole moment is directed along the ion drift. Note the difference from the pure dipole field, due to the additional anisotropic factor $[2/(1+\cos^2\theta)]^{3/2}$.

Some other effects can affect the distribution of electric potential around a charged particle in flowing plasmas. For example, the effects of finite particle size and asymmetry of the charge distribution over its surface were considered by Ishihara *et al.* (2000) and Hou *et al.* (2001). The effect of ion absorption on the particle surface has been so far investigated only for subthermal ion flows in highly collisional plasmas (Chaudhuri *et al.* 2007). In this regime the absorption-induced ion rarefication behind the particle can dominate over the effect of ion focusing and a negative space charge region develops downstream from the particle.

2.3 Interparticle interactions

2.3.1 Isotropic plasmas

Let us first consider the electric interaction between a pair of particles. Assuming for simplicity that the particles have equal charges which are independent of their separation r, the interaction energy is

$$U(r) = Q\phi(r). \quad (2.54)$$

As discussed above, depending on plasma parameters and interparticle separation the interparticle electric interaction can exhibit properties of exponentially screened Coulomb (DH) or inverse power-law ($\propto r^{-2}$ or $\propto r^{-1}$) potentials. As shown by Khrapak *et al.* (2008), the strength of interaction can depend non-monotonically on ion collisionality (neutral gas pressure).

Besides electrical effects, there exist other mechanisms that can contribute to interparticle interactions in complex plasmas. These are associated with the specific property of complex plasmas – their thermodynamic openness caused by the continuous exchange of matter and energy between the particles and surrounding plasma. For instance, constant plasma absorption on the particle surfaces gives rise to a so-called "ion shadowing" interaction (Ignatov 1996; Tsytovich *et al.* 1996; Lampe *et al.* 2000; Khrapak *et al.* 2001), which is basically the ion drag force that one particle experiences as a consequence of the ion flux directed to another neighboring particle and vice versa. The ion shadowing force is always attractive. An approximate expression for the ion shadowing potential taking into account the effect of ion–neutral collisions in the weakly collisional regime has been derived by Khrapak *et al.* (2008) and Khrapak and Morfill (2008a). It can be written as

$$U_{\rm sh} \simeq -\frac{1}{3}\sqrt{\frac{2}{\pi}}\frac{Qe}{r}\frac{J_i}{v_{T_i}}\frac{Qe}{T_i}\Lambda, \tag{2.55}$$

where Λ is the modified Coulomb logarithm derived by Khrapak *et al.* (2002). For small particles $\Lambda \simeq \ln(1+\beta_T^{-1})$. Although the ion shadowing interaction is not pairwise, because it depends on the mutual orientation of the particles (when more than two particles are involved), for sufficiently rarefied systems (when interaction occurs mostly through binary collisions) Equation (2.55) is expected to be a good approximation.

Analysis of Equations (2.48) and (2.55) reveals that at large distances both electric and ion shadow interaction potentials are proportional to J_i, which is not surprising since both interaction mechanisms stem from the conservation of the ion flux collected by the particle. Both interactions have r^{-1} long-range asymptote. Therefore, depending on their relative magnitudes either attraction or repulsion occurs. An approximate condition for attraction has been derived by Khrapak and Morfill (2008a). It is $k_{\rm D}\ell_i \gtrsim 4$, i.e., rather low plasma collisionality is required in order to make ion shadowing attraction operational in complex plasmas.

A mechanism of interaction similar to ion shadowing can be associated with the neutral component, provided the particle surface temperature is different from the temperature of the surrounding neutral gas (Tsytovich *et al.* 1998). If the particle surface is hotter there is a net momentum flux from the particle into the plasma which results in the repulsion between a pair of particles. If the particle surface is colder, the momentum flux from neutral gas to the particle generates attraction between the particles. In the free molecular (kinetic) regime an expression for this "neutral shadowing" interaction potential is (Tsytovich *et al.* 1998)

$$U_n(r) = \frac{3\pi}{8}\frac{a^2 p}{r}\frac{\delta T}{T_n}, \tag{2.56}$$

where p is the neutral gas pressure and $\delta T = T_s - T_n$ is the difference between the temperatures of the particle surface and surrounding neutral gas. Experiments and theoretical estimations (see Section 2.6) demonstrated that in low pressure gas discharges the particle surface temperature is somewhat hotter that the neutral gas temperature (Daugherty *et al.* 1993; Swinkels *et al.* 2000; Khrapak and Morfill 2006). Estimation by Khrapak and Morfill has shown that under typical discharge conditions the magnitude of neutral shadowing is usually weaker compared to ion shadowing and electric interactions.

Another interesting feature related to the interparticle interactions in complex plasmas is the possibility of long-range electric attraction between the particles charged positively by electron emission processes. This electric attraction has been discussed by Delzanno *et al.* (2004, 2005) for collisionless plasma conditions and by Khrapak *et al.* (2007b) and D'yachkov (2008) for strongly collisional plasmas. The physics behind this attraction is the following. In the absence of plasma production and loss in the vicinity of the particles, the net ion and electron fluxes directed to the particle should balance each other. This implies that sufficiently far from the particle a weak (ambipolar) electric field should exist, which accelerates less mobile ions and decelerates more mobile electrons. This long-range electric field is directed to the particle center, independently of the sign of its charge. Another particle placed in this weak electric field is attracted to the test particle if its charge is positive and is repelled in the opposite case. Khrapak *et al.* (2007b) suggested that the considered mechanism can be responsible for the formation of ordered particle structures in the "dusty combustion" (thermal plasma) experiment by Fortov *et al.* (1996b,c), discussed in Section 1.3.

2.3.2 Anisotropic plasmas

In anisotropic plasmas the wake effect is important (see Section 2.2.2). This effect is usually invoked for explaining the vertical ordering of the particles (chain formation) often observed in ground-based experiments. In fact, the interparticle interactions become *nonreciprocal* due to the presence of wakes, so that the classical Hamiltonian approach to the particle dynamics is no longer valid in this case (see Section 3.2).

Note that another effect was pointed out, which might play some role in the vertical ordering of particles along the ion flow. This effect is connected with a distortion of the ion velocity field by the upstream particle and the appearance of a horizontal component of the force, caused by the ion momentum transfer to the downstream particle (Lapenta 2002). This force – the ion drag force – pushes the downstream particle back to the axis with the origin at the upstream particle position and parallel to the ion flow. Numerical modeling of the ion velocity field in the wake showed that for certain conditions the ion drag mechanism may prevail over the electrostatic one (Lapenta 2002).

2.3.3 Experiments

Determination of the interaction potential constitutes a delicate experimental problem. Only a few such experiments have been performed so far (Konopka *et al.* 1997, 2000a; Takahashi *et al.* 1998; Melzer *et al.* 1999; Hebner and Riley 2003; Hebner *et al.* 2003). Let us briefly discuss these experiments.

An elegant method based on an analysis of elastic collisions between the two particles was proposed by Konopka *et al.* (1997, 2000a). In this experiment the particles are introduced into an rf discharge through a small hole in the glass window built into the upper electrode and are levitated above the lower electrode, where the electric field compensates for gravitational force. To confine the particles horizontally, a ring is placed on the lower electrode, which introduces a horizontal parabolic confining potential. The manipulation of the particles and activation of elastic collisions between them is performed with the use of a horizontal electric probe introduced into the discharge chamber. During the collision, the particle trajectories are determined by the confining potential and the interparticle interaction potential which is a function of the interparticle distance. An analysis of recorded particle trajectories during collisions yields the coordinates and velocities of both the particles during collision. Then, the form of the interaction potential can be reconstructed from the equation of motion. Application of this method (Konopka *et al.* 2000a) showed that for low discharge powers and pressures the interaction potential can be well described by the Debye–Hückel (Yukawa) form (2.39) within experimental uncertainties. This is illustrated in Figure 2.14. Note that the measured potential is also in remarkably good agreement with the analytic model by Kompaneets *et al.* (2007) [see Equation (2.52)]. Unfortunately, the range of interparticle distances investigated in the experiment was not broad enough to discriminate the DH model and that of Kompaneets *et al.* More experiments are necessary to resolve this issue.

A method based on the laser manipulation of the particles was proposed to study interaction between the particles in anisotropic plasmas by Takahashi *et al.* (1998). Melzer *et al.* (1999) employed this method for two particles suspended simultaneously in the rf sheath: A single particle of radius $a \simeq 1.7$ μm, and a cluster of two particles (of the same size) stuck together. The particles were introduced into a plasma of an rf discharge in helium at a pressure of $p \sim 50-200$ Pa. Because of the different charge-to-mass ratios, the double particle levitated closer to the lower electrode. Both particles, meanwhile, were almost free to move in the horizontal plane. The first observation was the following: For sufficiently low pressures, the particles tend to form a bound state, in which the lower particle is vertically aligned to the upper one. With increasing pressure, the bound state can be destroyed, and then the particle separation in the horizontal plane is limited only by a very weak horizontal confinement produced by a specially concave electrode. The backward decrease of pressure brings the system back into the bound state. It was found that the effect exhibits hysteresis.

In order to prove that the observed bound state is not due to external confinement, the particles were manipulated by laser radiation. The laser beam was focused on either the upper or the lower particle, causing their motion. It was found that when

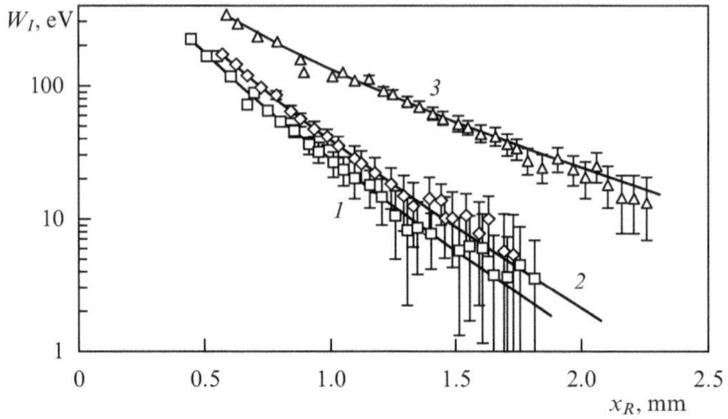

FIGURE 2.14
Potential energy W_I of interaction between two particles versus the distance x_R between them measured by Konopka *et al.* (2000a). Measurements were taken with the particles of radius $a \simeq 4.5$ μm at $p = 2.7$ Pa in argon and different rf peak-to-peak voltages U_p. Symbols correspond to experimental results, solid lines show their fit to the DH (Yukawa) potential $W_I(x_R) = (Q^2/x_R)\exp(-x_R/\lambda)$ leading to the following effective particle charge number $Z = Q/e$ and screening length λ (T_e is the measured electron temperature): 1: $|Z| = 13900$, $\lambda = 0.34$ mm, $T_e = 2.0$ eV, $U_p = 233$ V; 2: $|Z| = 16500$, $\lambda = 0.40$ mm, $T_e = 2.2$ eV, $U_p = 145$ V; 3: $|Z| = 16800$, $\lambda = 0.90$ mm, $T_e = 2.8$ eV, $U_p = 64$ V. Note that the screening length determined from the experiment is of the order of the electron Debye radius.

the upper particle is pushed, the lower particle follows its motion. This behavior proves that the lower particle is subject to an attractive horizontal force mediated by the upper particle. In contrast, when the lower particle is pushed, the response of the upper particle is much weaker and the bound state can be easily destroyed. Hence, the interaction between the particles is clearly non-reciprocal.

Another set of experiments to measure forces produced by the ion wake field from the upper (lighter) "target" particle by colliding the latter with the (heavier) "probe" particle levitated at a lower height in a sheath of an rf discharge was reported by Hebner and Riley (2003) and Hebner *et al.* (2003). In these experiments attractive and repulsive interactions between charged particles were calculated using Newton's equations of motion for various experimental conditions (using different particles sizes and neutral gas pressures). It was shown that the magnitude of attractive potential increases with lowering the gas pressure. It was also found that the attractive forces decay fairly rapidly as the vertical distance between the particles increases. Fits to repulsive and attractive forces were proposed.

It is obvious that the experimental results reported by Melzer *et al.* (1999), Hebner and Riley (2003) and Hebner *et al.* (2003) can be explained by the the wake effect. However, the question of whether the attractive force has an electrostatic nature or is associated with the ion drag mechanism (Lapenta 2002) still needs to be investigated.

2.4 Momentum exchange

The momentum exchange between different species plays an exceptionally important role in complex plasmas. For example, the momentum transfers in collisions with the neutral gas "cool down" the system, in particular grains and ions, introducing some damping. The forces associated with the momentum transfer from electrons and ions to the charged particles – i.e., the electron and ion drag forces – often determine static and dynamical properties of the particle component, affect wave phenomena, etc. The momentum exchange in grain–grain collisions and its competition with the momentum transfer in grain–neutral gas collisions governs grain transport properties, scalings in fluid flows, etc. While various aspects of electron–ion interaction (collisions) as well as electron, ion, and grain collisions with neutrals have been well studied, comparatively little work has been done on grain–electron, grain-ion and grain–grain collisions.

In this section, we assume the Debye–Hückel (Yukawa) potential around the particle and perform a detailed analysis of the binary collisions involving the particles. First, the momentum transfer cross section for different types of collisions is calculated and analytic approximations for some limiting cases are derived. These approximations are used to estimate the characteristic momentum exchange rates in complex plasmas. This provides us with a unified theory of momentum exchange in complex plasmas in the *binary collision approximation*. Some direct applications of the obtained results are also considered, e.g., calculations of the electron and ion drag forces in complex plasmas, classification of possible complex plasma states in terms of momentum exchange, the hierarchy of the momentum exchange in grain–grain and grain–neutral collisions and corresponding dynamical states of complex plasmas.

2.4.1 Momentum transfer cross section

We consider binary collision between two particles of masses m_1 and m_2 interacting via an isotropic potential of the form

$$U(r) = -(U_0/r)\exp(-r/\lambda),$$

where λ is the *effective* screening length, $U_0 > 0$ for attraction and $U_0 < 0$ for repulsion. The particle trajectories during collision are ballistic, i.e., any types of multiple collisions are neglected. The problem is equivalent to the scattering of a single particle of reduced mass, $\mu = m_1m_2/(m_1+m_2)$, in the central field $U(r)$ (whose center is at the center of masses of the colliding particles). The analysis of motion in the central field was given in Section 2.1.1.1 and below we employ the results of this analysis. First, we study the case of point-like particles. The role of the finite grain size will be addressed later.

The momentum transfer (scattering) cross section in this approximation is given by the integral over impact parameters

$$\sigma_s = 2\pi \int_0^\infty [1 - \cos\chi(\rho)]\rho d\rho, \tag{2.57}$$

where χ is the deflection (scattering) angle. The latter depends on the impact parameter in the following way: $\chi(\rho) = |\pi - 2\varphi(\rho)|$, where $\varphi(\rho) = \rho \int_{r_0}^\infty dr r^{-2}[1 - U_{\rm eff}(r,\rho)]^{-1/2}$ and $U_{\rm eff}(r,\rho) = \rho^2/r^2 + 2U(r)/\mu v^2$ is the normalized effective potential energy. The distance of closest approach, $r_0(\rho)$, in the integral above is the largest root of the equation $U_{\rm eff}(r,\rho) = 1$.

The scattering parameter, $\beta(v) = |U_0|/\mu v^2 \lambda$, introduced in Section 2.1.1.1 is the ratio of the Coulomb radius, $R_C = |U_0|/\mu v^2$, to the effective screening length λ. It characterizes the "coupling" between colliding particles: The coupling is weak when the characteristic distance of interaction $R_0 \sim R_C$, introduced through $|U(R_0)| = \frac{1}{2}\mu v^2$, is shorter than the screening length, i.e., when $\beta(v) \ll 1$. In the opposite limit, $\beta(v) \gg 1$, when $R_0 \gg \lambda$, the coupling is strong. In addition, the normalized momentum transfer cross section, σ_s/λ^2, depends only on β (Lane and Everhart 1960; Khrapak *et al.* 2003b, 2004b), which makes $\beta(v)$ a *unique parameter* which describes momentum exchange for Yukawa-type interactions.

Note that the theory of Coulomb scattering, which uses an unscreened (Coulomb) potential and a cutoff at $\rho_{\rm max} = \lambda$ in the integral (2.57), is widely used to describe momentum exchange in collisions between electrons and ions in conventional plasmas. It holds for $R_C \sim R_0 \ll \lambda$ or $\beta \ll 1$, i.e., in the limit of weak coupling. However, for $\beta \geq 1$ the theory of Coulomb scattering is not applicable: In this case the interaction range R_0 is larger than the screening length and a considerable fraction of the interaction occurs outside the Debye sphere providing substantial contribution to the momentum transfer. The use of a cutoff at $\rho_{\rm max} = \lambda$ considerably underestimates the momentum transfer in this case (Khrapak *et al* 2002).

Now let us estimate the characteristic values of the scattering parameter for different types of collisions involving grains. Taking into account that $|U_0| \sim |Z|e^2$ for electron–grain and ion–grain collisions, and $|U_0| \sim Z^2e^2$ for grain–grain collisions, we get the following hierarchy of characteristic scattering parameters: (i) *Grain–electron* collisions, $\beta_T^{de} \sim z(a/\lambda) \sim 0.01 - 0.3$; (ii) *Grain–ion* collisions, $\beta_T^{di} \sim z\tau(a/\lambda) \sim 1 - 30$; *Grain–grain* collisions, $\beta_T^{dd} \sim z_d(a/\lambda) \sim 10^3 - 3\times 10^4$, where $z_d = Z^2e^2/aT_d \equiv z|Z|(T_e/T_d)$ is the normalized potential energy of two grains which are just touching. We also assumed $z \sim 1$, $\tau \sim 10^2$, $a/\lambda \sim 0.01 - 0.3$, $|Z| \sim 10^3$, and $z_d = z|Z|\tau = 10^5$ (for $T_d = T_i$), which is typical for complex plasmas. These estimates show that the coupling is weak only for grain–electron collisions. At the same time, coupling for grain–ion and grain–grain collisions is usually strong, and the theory of Coulomb scattering fails to describe such collisions. In connection with grain–ion collisions, this issue has been discussed by Hahn *et al.* (1971), Khrapak *et al.* (2002, 2003b, 2004b), and Fortov *et al.* (2005).

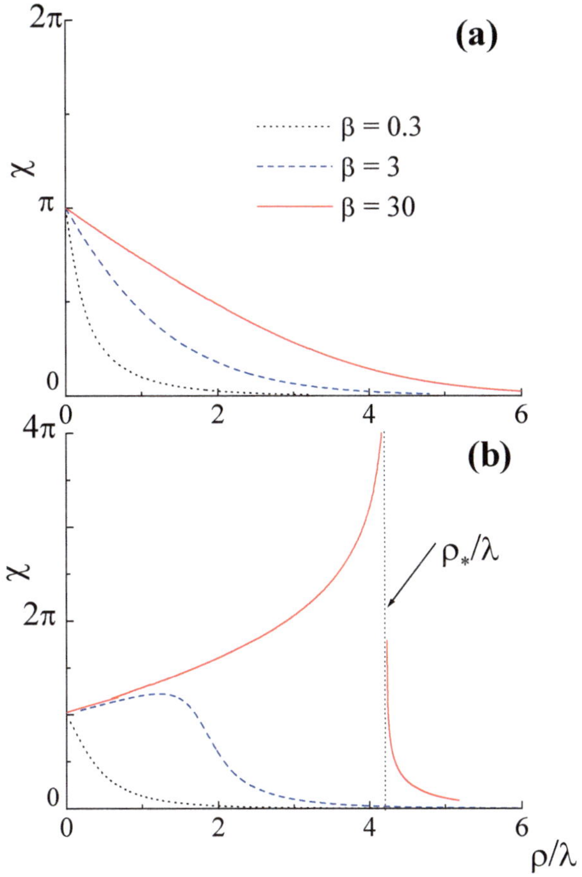

FIGURE 2.15
Scattering angle χ versus the normalized impact parameter ρ/λ, where λ is the effective screening length. The numerical calculations for a repulsive (a) and attractive (b) Yukawa interaction potential are plotted for three different scattering parameters $\beta = 0.3, 3$ and 30. The vertical dotted line at $\rho \simeq 4.2\lambda$ in (b) indicates the transitional impact parameter ρ_* at which χ diverges.

The numerical calculation of the momentum transfer cross sections for a wide range of β ($0.1 < \beta < 10^3$) for both attractive and repulsive Yukawa potential was reported by Khrapak *et al.* (2004a). First, the dependence of the scattering angle on the impact parameter, $\chi(\rho)$, was obtained. Then, Equation (2.57) was numerically integrated yielding the momentum transfer cross sections. The obtained results are presented in Figures 2.15 and 2.16.

The scattering angle $\chi(\rho)$ decreases monotonically for repulsive interactions for all β. In contrast, for attractive interactions a monotone decrease of the scattering

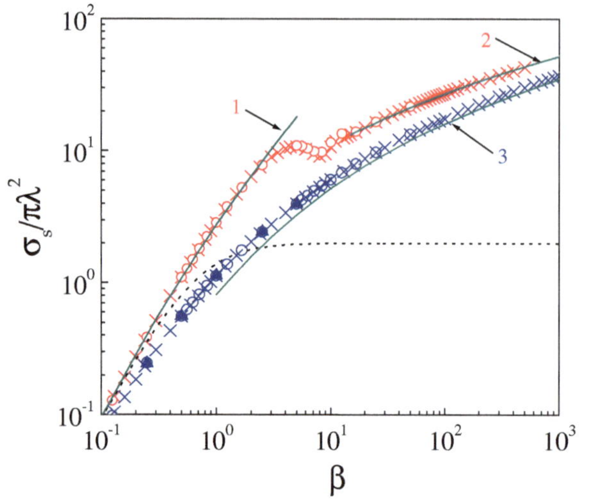

FIGURE 2.16
Momentum transfer cross section, σ_s, normalized to $\pi\lambda^2$ (where λ is the effective screening length), versus the scattering parameter β. The upper data are for attractive and the bottom data are for repulsive Yukawa potentials. Crosses correspond to the numerical results by Khrapak *et al.* (2004a), triangles are numerical results by Lane and Everhart (1960), and circles are numerical results by Hahn *et al.* (1971). Solid curves correspond to the following analytic expressions: 1: Equation (2.60); 2: Equation (2.61); 3: Equation (2.59). The dotted line corresponds to the Coulomb scattering theory [Equation (2.58)]. All the results are for pointlike particles.

angle is observed only for $\beta \lesssim 1$, while for $1 \lesssim \beta \lesssim \beta_{cr}$ it becomes a non-monotone function of ρ, and at $\beta > \beta_{cr} \simeq 13.2$ the scattering angle diverges at the "transitional" impact parameter $\rho_* \simeq \lambda(\ln\beta + 1 - \frac{1}{2}\ln^{-1}\beta)$, see Equation (2.8). The divergence of the scattering angle for attractive interactions arises from the barrier in the effective potential U_{eff}. Note also that when $\beta \ll 1$, the trajectories are mainly deflected within the plasma screening length (at $\rho/\lambda \lesssim 1$). In the opposite case $\beta \gg 1$, the scattering angle can be substantial even for $\rho \gg \lambda$, both for repulsive and attractive interaction. (This is another demonstration of the fact that the Coulomb scattering theory is nonapplicable for $\beta \gtrsim 1$, as discussed above.)

The results obtained for the momentum transfer cross section (Figure 2.16) show the following features: The cross section for the attractive potential is always larger than that for the repulsive potential (they converge in the limit of weak coupling $\beta \ll 1$). The cross section for the repulsive potential grows monotonically, while for the attractive potential a local maximum and minimum appear near $\beta = \beta_{cr}$. This non-monotonic behavior is a consequence of the bifurcation which the scattering angle $\chi(\rho)$ experiences in the range $1 \lesssim \beta \lesssim \beta_{cr}$. It is also evident from Figure 2.16 that the

Coulomb scattering theory (shown by the dotted line) considerably underestimates the cross section for both repulsion and attraction when $\beta \gtrsim 1$.

Now let us consider different limiting cases when an analytic description for the momentum transfer cross section is possible.

Repulsive potential. In the limit of weak coupling the Coulomb scattering theory is applicable as discussed above. The well-known expression for the *Coulomb* scattering cross section

$$\sigma_s^C/\pi\lambda^2 = 2\beta^2 \ln(1+1/\beta^2) \tag{2.58}$$

is shown by the dotted line in Figure 2.16. For $\beta \gtrsim 1$, Equation (2.58) is no longer applicable; however, an asymptotic analytic approximation for the case $\beta \gg 1$ can be obtained as follows. The relevant characteristic of the steepness of the potential is the parameter $\gamma_0 = |d\ln U(r)/d\ln r|_{r=R_0}$. The case $\gamma_0 \gg 1$ corresponds to a rapidly decreasing steep potential so that the momentum is mostly transferred in a spherical "shell" of radius R_0 and thickness $\sim R_0/\gamma_0$. Hence, the scattering resembles that of a hard sphere potential (Baroody 1962; Smirnov 1982) and with increasing γ_0 the momentum transfer cross section tends to

$$\sigma_s^{HS}/\pi\lambda^2 \simeq (R_0/\lambda)^2. \tag{2.59}$$

For the Yukawa potential $\gamma_0 = 1 + R_0/\lambda \gg 1$, provided $\beta \gg 1$. A rapidly converging analytic solution for $R_0(\beta)$ derived by Khrapak *et al.* (2004a) is $R_0/\lambda \simeq \ln 2\beta - \ln\ln 2\beta$.

Attractive potential. For weak coupling ($\beta \ll 1$) the theory of Coulomb scattering is applicable. The momentum transfer cross section is the same as for the repulsive potential and is given by Equation (2.58). It was shown by Khrapak *et al.* (2002) that even for moderate β, the extension of the standard Coulomb scattering theory is possible by taking into account all the trajectories with a distance of closest approach shorter than λ. The definition of the maximum impact parameter (cutoff) then becomes $r_0(\rho_{max}) = \lambda$ instead of $\rho_{max} = \lambda$ and leads to a modification of the Coulomb logarithm. The *modified Coulomb* momentum transfer cross section is

$$\sigma_s^{MC}/\pi\lambda^2 \simeq 4\beta^2 \ln(1+1/\beta). \tag{2.60}$$

Although this approach is not rigorous, Equation (2.60) shows very good agreement with numerical results by Hahn *et al.* (1971), Kilgore *et al.* (1993) and Khrapak *et al.* (2004a) up to $\beta \sim 5$ (see Figure 2.16) and agrees exactly, of course, with Coulomb scattering theory for $\beta \ll 1$.

The case of strong coupling ($\beta \gg 1$) requires a new physical approach. Such an approach was formulated by Khrapak *et al.* (2003b). The existence of the potential barrier in U_{eff} at $\beta > \beta_{cr}$ and the discontinuity in $\chi(\rho)$ it causes play a crucial role for the analysis of collisions. As shown in Figure 2.15 the dependence of the scattering angle on the impact parameter in the limit of long range interactions ($\beta = 30$) has the following features: For "close" ($\rho < \rho_*$) collisions we have $\chi \to \pi$ at $\rho \to 0$, and $\chi(\rho)$ grows monotonically until $\rho = \rho_*$, where it diverges; for "distant" collisions

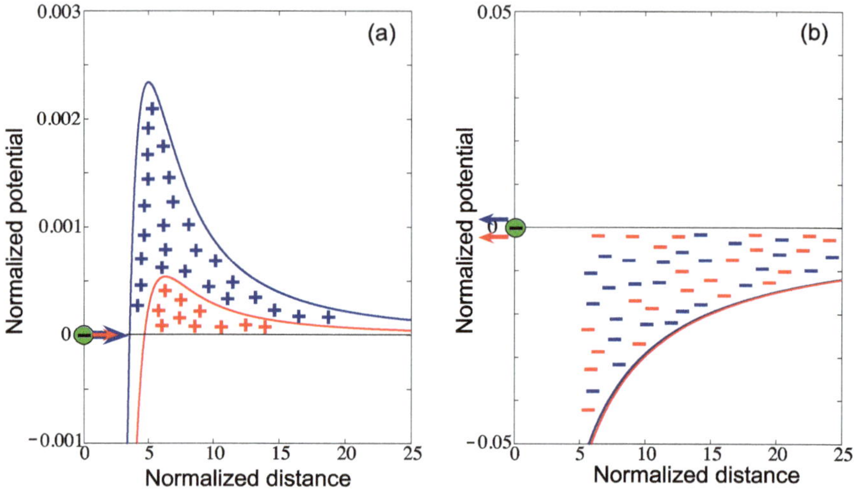

FIGURE 2.21
Electric potential behind a small negatively charged particle in a highly collisional plasma with slowly drifting ions. Plot (a) corresponds to a non-absorbing particle, while plot (b) corresponds to an absorbing particle. The ions are moving to the right. The direction of the ion drag force associated with ion focusing (a) and ion depletion (b) behind the particle is shown by arrows. The calculations are performed by Chaudhuri *et al.* (2007) using the linear plasma response technique for the following set of plasma parameters: $T_e = T_i$, $|Q|e/aT_e = 3$, $a/\lambda_D = 0.2$, $u/v_{T_i} = 0.003$, $\ell_i/\lambda_D = 0.03(0.01)$ for the upper (lower) curves. Signs "+" and "–" correspond to the positive and negative space charge regions, respectively.

of the ions causes a rarefication of the ion density downstream from the particle (see Figure 2.21b). The two effects are added in a simple superposition in the linear model and compete with each other. It turns out that in the limit of very small particle and very high ion collisionality, rarefication dominates over focusing. The ion drag force reverses its direction.

The sign reversal of the ion drag force in highly collisional plasmas has also been found by Filippov *et al.* (2008) using a model similar to that described here. Negative values of the ion drag force were reported from a simple model considering an absorbing sphere in a highly collisional plasma with flowing ions under the assumption of a central Coulomb-like interaction potential between the ions and the sphere (Khrapak *et al.* 2007a). Numerical simulation by Patacchini and Hutchinson (2008) demonstrates the following dependence of the ion drag force on collisionality: It slightly increases with collisionality in the weakly collisional regime, shows a local maximum when $\ell_i \sim \lambda_D$, and then decreases reaching negative values in the limit of high collisionality.

($\rho > \rho_*$) the scattering angle decreases rapidly, due to the exponential decay of the interaction potential.

It is convenient to consider the contributions from close and distant collisions into the momentum transfer separately. As shown by Khrapak *et al* (2003b), the behavior of χ as a function of the normalized impact parameter ρ/ρ_* is *practically independent* of β for $\rho < \rho_*$. This self-similarity allows us to present this contribution to the cross section (normalized to $\pi\lambda^2$) as $\simeq \mathscr{A}(\rho_*/\lambda)^2$, where $\mathscr{A} = 2\int_0^1[1-\cos\chi(\xi)]\xi d\xi$ and $\xi = \rho/\rho_*$. The value of the factor $\mathscr{A}$ can be determined by direct numerical integration. It was found that $\mathscr{A} = 0.81 \pm 0.01$ for all β in the range $\beta_{\rm cr} \leq \beta \leq 500$. For distant collisions the scattering angle decreases rapidly in the vicinity of ρ_*. This makes it possible to apply the small angle approximation to estimate their contribution to the cross section (normalized to $\pi\lambda^2$) as $\simeq 2.0 + 4.0\ln^{-1}\beta$. Combining these contributions and keeping terms up to $O(1)$, we can write the momentum transfer cross section in the limit of *strong coupling* attractive interaction as (Khrapak *et al* 2003b)

$$\sigma_{\rm s}^{\rm SC}/\pi\lambda^2 \simeq 0.81(\rho_*/\lambda)^2 + 2.0, \tag{2.61}$$

where $(\rho_*/\lambda)^2 \simeq \ln^2\beta + 2\ln\beta$. Expression (2.61) is valid for $\beta \geq \beta_{\rm cr}$ and pointlike particles. Figure 2.16 shows the very good agreement between Equation (2.61) and numerical calculations. A sufficiently accurate and even simpler approximation is $\sigma_{\rm s}^{\rm SC} \simeq \pi\rho_*^2$, which can be further justified when the finite size of the particle is taken into account.

Finally, let us briefly discuss the role of finite particle size. In this case a new length scale enters the problem. In contrast to the case of point-like particles, where the scattering is described by the single parameter β, we now have a second parameter, a/λ. If the distance of the closest approach, r_0, is smaller than a (or $2a$ for particle–particle collisions), then the direct (touching) collision takes place. In this case we will assume absorption for grain–electron and grain–ion collision, and specular reflection for grain–grains collisions.

A detailed discussion of the effect of finite particle size on the momentum transfer is given by Khrapak *et al.* (2004a). It is shown that for the repulsive interaction (grain–electron and grain–grain collisions) the effect of finite size considerably affects the momentum transfer only when coupling is very weak,

$$\beta \lesssim (a/\lambda)\Lambda^{-1/2}, \tag{2.62}$$

where $\Lambda \simeq \ln(1/\beta) \gg 1$ is the Coulomb logarithm. Recalling that $\beta_T^{de} \sim z(a/\lambda)$ and $\beta_T^{dd} \sim z_d(a/\lambda)$ and since $z \sim 1$, $z_d \gg 1$, we conclude that the effect of finite size can usually be neglected for grain–electron and grain–grain collisions.

The effect of finite size is more important for attractive (grain-ion) interactions. For example, for sufficiently large β the maximum impact parameter corresponding to ion collection is $\rho_{\rm c}^+ = \rho_*$, as follows from Equation (2.6). At the same time the contribution to the momentum transfer from ions with $\rho > \rho_*$ vanishes at large β. Hence in this case the momentum transfer is associated mostly with ions collected

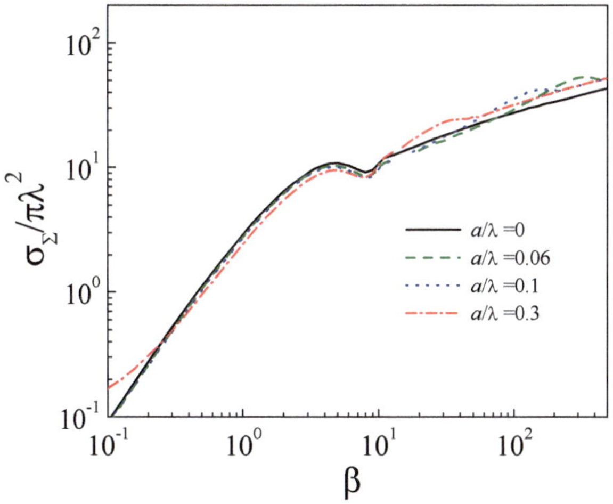

FIGURE 2.17
The total momentum transfer cross section, σ_Σ, normalized to $\pi\lambda^2$ (λ is the effective screening length), versus the scattering parameter β for the attractive Yukawa potential. The numerical results for different values of a/λ are shown to illustrate the role of finite particle radius a.

by the particle and the total momentum transfer cross section σ_Σ (sum of the contributions due to scattering and absorption) tends to $\pi\rho_*^2$. However, this does not imply much difference compared to the case of pointlike particles, because scattering with large angles at $\rho < \rho_*$ and absorption (which formally corresponds to the scattering at $\pi/2$) produce comparable effects. The dependence $\sigma_\Sigma(\beta)$ for an attractive Yukawa potential is shown in Figure 2.17 for different values of a/λ. One can see that the momentum transfer can decrease or increase (in comparison with pointlike particles), depending on the values of a/λ and β. At the same time, the momentum transfer cross section is not very sensitive to the particle size – the deviation of σ_Σ from σ_s for a pointlike particle does not exceed $\sim 50\%$.

2.4.2 Momentum exchange rates

Let us consider a *test* particle (grain) moving through a gas of *field* particles (electrons, ions, or grains) having an isotropic Maxwellian velocity distribution function. The test particle velocity u_d is assumed to be smaller than the field particle thermal velocity $v_{T\alpha}$. Introducing the momentum exchange rate $\nu_{d\alpha}$ through $du_d/dt = -\nu_{d\alpha}u_d$, we get (Khrapak *et al.* 2004a)

$$\nu_{d\alpha} = \frac{1}{3}\sqrt{\frac{2}{\pi}}\frac{n_\alpha \mu_{d\alpha}}{m_d v_{T\alpha}^5}\int_0^\infty v^5 \sigma_\Sigma(v)\exp(-v^2/2v_{T\alpha}^2)dv,$$

where $\sigma_\Sigma(v)$ is the corresponding total momentum transfer cross section (function of the relative velocity), $\mu_{d\alpha}$ is the reduced mass, and $\alpha = e,i,d$. Some results following from this expression are given below.

2.4.2.1 Grain–electron collisions

For grain–electron interactions usually $\beta_T^{de} \ll 1$ and the standard Coulomb scattering approach is applicable. This yields

$$\nu_{de} \simeq (2\sqrt{2\pi}/3)(m_e/m_d)n_e v_{T_e} a^2 z^2 \Lambda_{de}, \tag{2.63}$$

where n_e, m_e, and v_{T_e} are the density, mass, and thermal velocity of electrons, and

$$\Lambda_{de} = z\int_0^\infty \mathrm{e}^{-zx}\ln[1+4(\lambda/a)^2x^2]dx - 2z\int_1^\infty \mathrm{e}^{-zx}\ln(2x-1)dx$$

is the Coulomb logarithm for grain–electron collisions integrated over the Maxwellian distribution derived by Khrapak and Morfill (2004). In the typical case $(2/z)(\lambda/a) \gg 1$, we obtain $\Lambda_{de} \simeq 2\ln[(2/z)(\lambda/a)]$ with logarithmic accuracy.

2.4.2.2 Grain–ion collisions

For grain–ion interaction β_T^{di} often exceeds unity and then the Coulomb scattering approach is not applicable. In the case $\beta_T^{di} \lesssim 5$, Equation (2.60) can be used. This yields

$$\nu_{di} \simeq (2\sqrt{2\pi}/3)(m_i/m_d)n_i v_{T_i} a^2 z^2 \tau^2 \Lambda_{di}, \tag{2.64}$$

where n_i, m_i, and v_{T_i} are the density, mass, and thermal velocity of ions, and

$$\Lambda_{di} \simeq 2z\int_0^\infty \mathrm{e}^{-zx}\ln[1+2\tau^{-1}(\lambda/a)x]dx \tag{2.65}$$

is the *modified* Coulomb logarithm for grain–ion scattering (Khrapak *et al.* 2002) integrated over the Maxwellian distribution function [in Equation (2.65) we took into account that $\tau \gg 1$]. In the limit of small β_T^{di} or $(1/z\tau)(\lambda/a) \gg 1$, the result reduces to that of the Coulomb scattering theory and we have $\Lambda_{di} \simeq 2\ln[(2/z\tau)(\lambda/a)]$. In the opposite limit of very large scattering parameters, $\beta_T^{di} > \beta_{\mathrm{cr}} \simeq 13.2$, the total momentum transfer cross section is to good accuracy $\sigma_\Sigma \simeq \pi\rho_*^2$, where $\rho_* \sim \lambda\ln\beta_T^{di}$. This yields

$$\nu_{di} \simeq (8\sqrt{2\pi}/3)(m_i/m_d)n_i v_{T_i}\rho_*^2. \tag{2.66}$$

2.4.2.3 Grain–grain collisions

For grain–grain interactions the standard Coulomb scattering approach can be employed for only extremely small grain charges and/or extremely high grain energies, so that $\beta_T^{dd} = z_d(a/\lambda) \ll 1$. In this situation we have

$$\nu_{dd} \simeq (4\sqrt{2\pi}/3)n_d v_{T_d} a^2 z_d^2 \Lambda_{dd}, \tag{2.67}$$

where n_d, and v_{T_d} are the density, and thermal velocity of the grains, and

$$\Lambda_{dd} = z_d \int_0^\infty \mathrm{e}^{-z_d x} \ln[1 + (\lambda/a)^2 x^2] dx - 2z_d \int_1^\infty \mathrm{e}^{-z_d x} \ln(2x-1) dx,$$

is the Coulomb logarithm for the grain–grain collisions integrated over the Maxwellian distribution. If $(1/z_d)(\lambda/a) \gg 1$, the Coulomb scattering approach is applicable and we have $\Lambda_{dd} \simeq 2\ln[(1/z_d)(\lambda/a)]$ with logarithmic accuracy. In the regime $\beta_T^{dd} \gg 1$, which is more typical for complex plasmas, the analogy with hard sphere collisions can be used. The result is (Khrapak *et al* 2004a)

$$v_{dd} \simeq (4\sqrt{2\pi}/3) n_d v_{T_d} R_0^2. \tag{2.68}$$

The obtained results for the momentum exchange in grain–grain collisions will be used below to investigate the possible states of complex plasmas.

2.4.3 Momentum exchange diagram

The grain charges in complex plasmas, as well as the plasma screening length are not constant. This is why the strength of the electrostatic coupling between the grains can be easily changed experimentally over a fairly wide range (by varying, e.g., the discharge conditions) (Morfill *et al.* 2002). This is a major distinguishing feature of complex plasmas compared to conventional plasmas, where the ion charges are normally constant (single). In complex plasmas, one can observe the transitions from the disordered, weakly coupled to strongly coupled states and the formation of ordered structures of grains – plasma crystals (Chu and I 1994; Hayashi and Tachibana 1994; Melzer *et al.* 1994, 1996a; Thomas *et al.* 1994; Fortov *et al.* 1996a,b, 1997, 2005; Thomas and Morfill 1996; Lipaev *et al.* 1997; Tsytovich 1997; Morfill *et al.* 2002, 2004; Piel and Melzer 2002).

Another major distinguishing feature of complex plasmas is that the overall dynamical time scales associated with the particle component are relatively long (dust plasma frequency ~10–100 Hz) (Tsytovich 1997; Piel and Melzer 2002; Khrapak *et al.* 2003c). Furthermore, the grains themselves are large enough to be easily visualized individually. All together this makes it possible to investigate phenomena occurring in different phases at the most fundamental kinetic level (Morfill *et al.* 2002, 2004; Fortov *et al.* 2005). Although there is always some damping introduced into the complex plasma systems due to neutral gas friction (Piel and Melzer 2002), the resulting damping rate is many orders of magnitude smaller than that in colloidal suspensions, and it can easily be made much smaller than the major eigenfrequencies of the particle component dynamics. Hence the most interesting dynamical phenomena have usually enough time to evolve (Morfill *et al.* 2004).

Let us dwell upon these features of complex plasmas in detail.

Figure 2.18 represents different "phase states" of complex plasmas as functions of the electrostatic (Coulomb) coupling parameter Γ_S and the mean grain separation Δ, normalized either to the grain size a or to the screening length λ ("finiteness

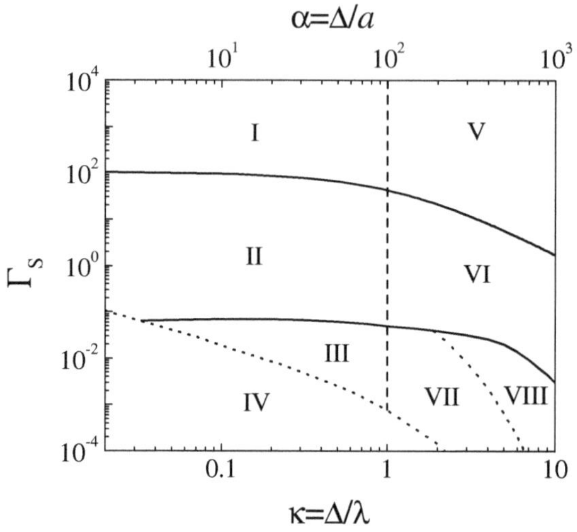

FIGURE 2.18
"Momentum exchange" diagram of complex plasmas in (Γ_S, κ) parameter space (Khrapak *et al.* 2004a). The vertical dashed line at $\kappa = 1$ conditionally divides the system into "Coulomb" and "Yukawa" parts. Different states are marked in the figure. Regions I (V) represent Coulomb (Yukawa) crystals; Regions II (VI) are for Coulomb (Yukawa) non-ideal plasmas; Regions III (VII and VIII) correspond to Coulomb (Yukawa) ideal plasmas. Note that in the Region VIII the electric Yukawa interaction asymptotically reduces to the hard sphere limit. In region IV the electrostatic interaction is not important and the system is like conventional system of hard spheres. For further explanations see text.

parameter" $\alpha = \Delta/a$ and "lattice parameter" $\kappa = \Delta/\lambda$, respectively). The parameter $\Gamma_S = \Gamma \exp(-\kappa)$, which characterizes the "actual" coupling ratio (potential energy/kinetic energy) for the Debye–Hückel interaction potential at the average intergrain distance, is expressed in terms of the (Coulomb) coupling scale $\Gamma = Q^2/\Delta T_d$ (note that in terms of Γ and κ the thermal scattering parameter is $\beta_T^{dd} = 2\Gamma\kappa$). The use of Γ_S implies that the calculations should be representative to some extent for other types of "similar" interaction potentials, too (viz., with "similar" long- and short-range asymptotes). The vertical line $\kappa = 1$ conditionally divides the diagram into weakly screened (Coulomb) and strongly screened (Yukawa) parts. In Figure 2.18 we have set $\lambda/a \equiv \alpha/\kappa = 100$, which is typical of complex plasmas studied so far, but there is in principle a wide range of variation, depending on grain size and plasma conditions chosen. The "melting line" which indicates the liquid-solid phase transition is shown by the upper solid line in Figure 2.18 is discussed in Section 5.1.

Further insight into the possible phase states shown in Figure 2.18 is obtained from the results on the momentum transfer cross section for grain–grain collisions. This approach allows us to obtain a clear physical classification of complex plasmas. The lower solid curve indicates the "transition" between "ideal" and "non-ideal" plasmas. We determine this transition from the condition $\sqrt{\sigma_\Sigma/\pi} = (4\pi/3)^{-1/3}\Delta$, which implies that the characteristic range of grain–grain interaction (in terms of the momentum exchange) is comparable to the intergrain distance (in terms of the Wigner-Seitz radius). Above this line the interaction is essentially multiparticle, whereas below the line only pair collisions are important. This refines the standard condition used to define a "boundary" between ideal and non-ideal plasmas, $\Gamma_S \sim 1$. From the thermodynamical point of view, this line determines the limit of employing expansions of the thermodynamical functions (e.g., virial expansion) over the (small) coupling parameter.

It is important to note that for a one-component system of particles interacting via Debye–Hückel (Yukawa) repulsive potential (as well as for any monotonic repulsive potential) no liquid–gas phase transition is possible (formally, the critical point occurs at $T_d = 0$). This is different, if the pair potential is not monotonic, e.g., a long-range attractive component (for instance, ion shadow interaction) added to a short-range repulsive electrostatic potential exists. This issue will be discussed in more detail in Section 5.1.

The regions where the system is similar to that of hard spheres are also shown in Figure 2.18: Below the left dotted curve the electrostatic interaction is too weak and the momentum exchange occurs due to direct interparticle collisions; i.e., we have a usual system of hard spheres where charges do not play any noticeable role. This line corresponds to $\beta_T^{dd} = (a/\lambda)\Lambda_{dd}^{-1/2}$ [see Equation (2.62)]. The right dotted curve marks the transition boundary for hard sphere-like interaction. Here the "mean" scattering parameter for grain–grain collisions exceeds unity ($\beta_T^{dd} > 1$), and hence, the strongly screened electrostatic interaction reduces asymptotically to the hard sphere limit with radius $R_0 \simeq \lambda \ln(2\beta_T^{dd})$.

Next we investigate complex plasma properties in terms of the competition between the momentum exchange in mutual grain–grain collisions and the interaction with the surrounding medium.

Complex plasmas can be "engineered" as essentially a "one-phase fluid" (when the interactions between the grains dominate) or as a "particle laden two-phase flow" (when the interactions with the background medium are of similar or greater importance). We have illustrated this by plotting contours of constant ratios of the grain–grain/grain–background momentum exchange rates, ν_{dd}/ν_{dn}, in the (Γ_S,κ) diagram in Figure 2.19.

In complex plasmas the exchange of momentum with the background medium is mostly through grain–neutral gas collisions,

$$\nu_{dn} = \delta(8\sqrt{2\pi}/3)(m_n/m_d)a^2 n_n v_{T_n}, \tag{2.69}$$

where m_n, n_n, and v_{T_n} are the mass, density, and thermal velocity of neutrals, respectively (Epstein 1924). The value of the numerical factor δ depends on the exact process of neutral scattering from the particle surface. For example, $\delta = 1$ for the cases of complete absorption and specular reflection, while $\delta = 1 + \pi/8 \simeq 1.4$ for

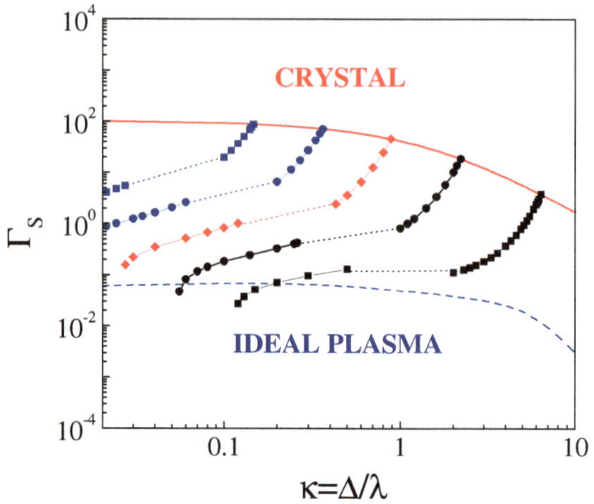

FIGURE 2.19
Typical contours are shown of constant ratios of the momentum exchange rates in grain–grain collisions relative to grain–background (neutral gas) collisions (Khrapak *et al.* 2004a). The values $\nu_{dd}/\nu_{dn} = 10^2, 10, 1, 10^{-1}$, and 10^{-2} are depicted in a phase diagram for complex plasmas in (Γ_S, κ) parameter space (from left to right). Also shown in the figure are the lines corresponding to crystal melting (solid line) and the boundary between ideal and non-ideal plasmas (dashed line). The following complex plasma parameters were used in these calculations: Grains of radius $a = 1$ μm and material mass density of 1 g/cm^3 in argon plasma at a neutral gas pressure of 100 Pa; room temperature ions and neutrals $T_i \sim T_n \sim 0.03$ eV, and $a/\lambda = 10^{-2}$.

diffuse scattering with full accommodation. According to Liu *et al.* (2003), the latter value is more consistent with the experimental results.

For the momentum exchange rate in grain–grain collisions, we use Equation (2.68) at $\beta_T^{dd} \gg 1$ (upper symbols in the figure) and Equation (2.67) at $\beta_T^{dd} \ll 1$ (lower symbols). In the transition regime $\beta_T^{dd} \sim 1$, none of these approximations is applicable and we have therefore simply linked the two regimes by dotted lines.

Figure 2.19 shows that there is a broad range of parameters where complex plasmas have the properties of one-phase fluids ($\nu_{dd}/\nu_{dn} \gg 1$), and those of two-phase fluids $\nu_{dd}/\nu_{dn} \sim 1$. In the extreme limit of very small ν_{dd}/ν_{dn}, we can also, of course, have "tracer particles" in the background medium, which provide practically no disturbance to the background flow. Taking into account that a number of plasma parameters (e.g., the neutral gas pressure, plasma screening length, the ratio a/λ) can be varied relatively easily within approximately one order of magnitude, most of the possible states can be investigated.

In concluding this section we note that the considered model uses a number of simplifying assumptions (e.g., fixed form of the potential, ballistic trajectories during collisions, fixed particle charges) which are not necessarily satisfied in real complex plasmas. Nevertheless, in many cases this simple model does provide reasonable predictions, and hence, it can be considered as the basis for more sophisticated models. The described results can be important for "engineering" experiments which aim to make use of special properties of complex plasmas.

2.5 Forces on particles

Knowledge of the major forces acting on microparticles in complex plasmas is essential for understanding dynamic phenomena and equilibrium configurations of complex plasmas observed in experiments. The forces can be naturally divided into two groups: The first one includes the forces which have electric nature: electron drag, ion drag, and electrostatic forces, whereas the second one includes the charge-independent forces: gravity, neutral drag, and thermophoretic forces. The calculation of the ion drag force is rather complicated in some cases. At the same time, this force is particularly important in complex plasmas and therefore it is a subject of a separate section.

2.5.1 Ion drag force

An exceptionally important example of plasma–particle interactions is the ion drag (wind) force which is caused by the momentum transfer from the flowing ions to charged particles. Knowledge of the ion drag force in a wide regime of plasma parameters is necessary for understanding a variety of phenomena occurring in space and in the laboratory. To illustrate the importance of the ion drag force for complex plasmas, let us mention that this force is responsible for the formation of the so-called voids – regions free of particles – in experiments under microgravity conditions (Goree *et al.* 1999; Morfill *et al.* 1999b; Kretschmer *et al.* 2005; Lipaev *et al.* 2007), it causes rotation of particle structures (e.g., clusters) in the presence of a magnetic field (Konopka *et al.* 2000b; Ishihara *et al.* 2002; Kaw *et al.* 2002), affects the properties of low-frequency waves (D'Angelo 1998; Khrapak and Yaroshenko 2003), contributes to the interparticle interactions (Khrapak *et al.* 2006a; Khrapak and Morfill 2008a), etc.

Despite the high importance of the ion drag force, a complete self-consistent model for this force, describing all cases of interest, has not yet been constructed. Rather, there exist several approximations which can be utilized in certain parameter regimes. Let us briefly mention the approaches used so far.

The traditional way to derive the ion drag force is based on the "binary collision formalism". Here the mechanical problem of the ion motion in the (central) field of the charged particle is solved. Analysis of ion trajectories yields the velocity dependent momentum transfer cross section. The force is then obtained by integrating the cross section with an appropriate ion velocity distribution function. This approach can be applied for any form and strength of the ion–particle interaction, but since one deals with ballistic ion trajectories, the effect of ion–neutral collisions, which is often important in complex plasmas, cannot be consistently accounted for.

An alternative way to calculate the ion drag force is based on the so-called "linear plasma response formalism" (see, e.g., Thompson and Hubbard 1960; Montgomery *et al.* 1968). Instead of calculating single ion trajectories and then the momentum transfer cross section, one can solve the Poisson equation coupled to the kinetic equation for the ions and obtain the self-consistent anisotropic component of the electric field induced by the ion flow at the position of the particle, which produces the (drag) force. This approach consistently accounts for ion–neutral collisions and potential anisotropy caused by the ion flow, but is applicable only for the weak ion–particle coupling since linearizations are involved.

In the limit of very high plasma collisionality fluid (hydrodynamic) description of plasma is possible. This basically corresponds to a highly collisional limit of the kinetic approach. However, using the hydrodynamic formulation, one can easily take into account ion absorbtion on the particle which inevitably occurs in plasmas. As will be demonstrated below, in the highly collisional regime absorption represents a very important factor which not only affect the magnitude of the force, but also its direction.

In this section we present the results of these approaches and discuss unresolved issues.

2.5.1.1 Binary collision approach

In collisionless plasmas it is natural to employ the binary collision (BC) approach to calculate the ion drag force. Typical assumptions used in the BC approach are an isotropic attractive Debye–Hückel (Yukawa) interaction potential between the ion and the particle and a shifted Maxwellian velocity distribution function for the ions. Initially, the ion drag force was calculated using the standard theory of Coulomb scattering (Uglov and Gnedovets 1991; Barnes *et al.* 1992). Basically this is the linear approximation assuming that the characteristic length of ion–particle interaction is much shorter than the plasma screening length and most of the ions are scattered with small angles within the Debye sphere. This theory is extensively used to describe collisions in conventional electron–ion plasmas (e.g. Spitzer 1962). However, it turns out that the Coulomb scattering theory underestimates considerably the momentum transfer cross section in ion–particle collisions and the ion drag force in typical complex plasmas. The reason for that has been discussed in Section 2.4.1. Let us review it here.

The point is that the negative (floating) surface potential of a non-emitting particle in plasmas is rather high, on the order of the electron temperature (see Section 2.1). Since the electron-to-ion temperature ratio is usually on the order of one (few) hundred(s), ions are strongly coupled to the particle in a wide surrounding region. This implies that the characteristic length of the ion–particle interaction can be comparable or even exceed the plasma screening length, which is of the order of the ion

Debye radius in an isotropic plasma. For this reason, the standard Coulomb scattering theory which neglects all collision events with impact parameters larger than the plasma screening length becomes inadequate. There were attempts to improve the situation by postulating the plasma screening length to be equal to the *electron* Debye radius (Kilgore *et al.* 1993; Zafiu *et al.* 2003a). This assumption causes an artificial enhancement of the ion drag force; however, it is arbitrary and physically unjustified (Khrapak *et al.* 2003a; Zafiu *et al.* 2003b), at least for small particles in quasi-isotropic plasmas.

The very fact that the standard Coulomb scattering theory fails to describe momentum transfer in ion–particle collisions was recognized by Khrapak *et al.* (2002) who proposed an extension of the Coulomb scattering theory to the regime of moderate ion–particle coupling. The modification merely affects the Coulomb logarithm, as discussed in Section 2.4.1. Later on, the limit of strong ion–particle coupling was investigated in detail and an analytic approach to calculate the ion drag force in this regime was developed (Khrapak *et al.* 2003b, 2004a,b). This approach has been discussed in Section 2.4.1.

Let us now present the quantitative results of the BC approach obtained for different strengths of ion–particle coupling.

In the framework of the BC approach, the ion drag force $F_{\rm id}$ is completely determined by the (velocity-dependent) total momentum transfer cross section for the ion–particle collisions, σ_Σ, which has been derived in Section 2.4.1. The force is $F_{\rm id} = m_d \nu_{di} u$, where u is the ion flow velocity and ν_{di} is the momentum exchange rate (the latter is given by averaging the cross section over the ion velocity distribution, see Section 2.4.2). The force depends on the magnitude of the thermal scattering parameter, $\beta_T = |Q|e/\lambda T_i$, where λ is the effective screening length (which does not necessarily coincide with the Debye radius, see Sections 2.2.1 and 2.5.1.4).

For subthermal flows (when the *thermal* Mach number is small, $M_T \equiv u/v_{T_i} \ll 1$), we can directly employ results of Section 2.4.2.2: At moderate coupling strength ($\beta_T \lesssim 5$), Equation (2.64) yields

$$F_{\rm id} = \frac{1}{3\sqrt{2\pi}} \left(\frac{T_i}{e}\right)^2 \Lambda_{di} \beta_T^2 M_T, \tag{2.70}$$

where Λ_{di} is given by expression (2.65). Here we also assume $\lambda \simeq \lambda_{Di}$ for subthermal ion flows (see Section 2.5.1.4). Equation (2.70) yields the scaling $F_{\rm id} \propto (Q/\lambda)^2 M_T$. In the linear regime $\beta_T \ll 1$ the logarithm is reduced to $\Lambda_{di} \simeq 2\ln\beta_T^{-1}$, which is identical to the results of the standard Coulomb scattering theory. In the opposite regime of strongly nonlinear scattering, $\beta_T \gg \beta_{\rm cr} \simeq 13$, we obtain from Equation (2.66)

$$F_{\rm id} \simeq \frac{2}{3}\sqrt{\frac{2}{\pi}} \left(\frac{T_i}{e}\right)^2 \ln^2 \beta_T M_T. \tag{2.71}$$

In this case the force depends logarithmically on the scattering parameter and, hence, on Q and λ.

For the case of arbitrary ion flow velocity the expressions for the ion drag force are more complicated. They were derived by Khrapak *et al.* (2005a) for the regime of moderate ion–particle coupling and by Nosenko *et al.* (2007) for the regime of strong ion–particle coupling, respectively. Note, that for highly superthermal ion flows with $M_T \gg 1$, the drift velocity rather than the thermal velocity should be used to evaluate the scattering parameter β. Also, the screening is determined by the electrons rather than by ions in this case, $\lambda \simeq \lambda_{De}$ (see Section 2.5.1.4). Therefore, we conclude from Equation (2.7) that the scattering parameter decreases rapidly with the Mach number, and we can expect the linear scattering (weak coupling, $\beta \sim \beta_T(\lambda_{Di}/\lambda_{De})M_T^{-2} \leq 1$) to be typical for $M_T \gg 1$. Then the momentum transfer cross section is given by Equation (2.58) and after the integration over the shifted Maxwellian distribution the force is

$$F_{id} \simeq \left(\frac{T_i}{e}\right)^2 \ln\left(\frac{\lambda_{De}}{\lambda_{Di}}\frac{M_T^2}{\beta_T}\right)\frac{\beta_T^2}{M_T^2}. \tag{2.72}$$

The ion drag decreases as $\propto M_T^{-2}$ at large Mach numbers (neglecting a weak logarithmic dependence). In the limit of a very high flow velocity, the momentum flux onto the grain (collection) dominates over the scattering part and then the force tends to the "geometrical asymptote", $F_{id} \simeq (T_i/e)^2(a/2\lambda)^2 M_T^2$, which does not depend on the particle charge.

2.5.1.2 Kinetic approach

The binary collision approach discussed above is applicable for any strength of ion–particle coupling. However, it is not intrinsically consistent. These reasons are the following: (i) Ion–neutral collisions are completely neglected; (ii) the approach presumes a certain potential distribution around the test charge (usually isotropic Yukawa-type potential), although in reality the potential is a self-consistent function of the plasma environment (e.g., ion collisionality, ion flow velocity); (iii) the approach presumes a certain velocity distribution function for the ions (usually the shifted Maxwellian distribution). These issues can be resolved by employing the linear kinetic approach.

A comprehensive kinetic model accounting for the effect of ion–neutral collisions on the ion drag force that a *non-absorbing* particle experiences in plasmas was developed by Ivlev *et al.* (2004a,b, 2005). In this model electrons are assumed to have Boltzmann distribution. The ion component is described by the kinetic equation with the collisional integral in the model Bhatnagar–Gross–Krook (BGK) form with constant ion–neutral collision frequency ν (see Section 2.2.1). The functional form of the BGK collision integral is particularly suitable for the description of the charge–exchange collisions. In reality, the ion–neutral collision cross section is a complicated (monotonically decreasing) function of the ion velocity which cannot be generally approximated by any simple scaling (Raizer 1991; Lieberman and Lichtenberg 1994). It is reasonable, therefore, to choose the approximation $\nu = const$ which allows us to represent the model collision operator in the convenient algebraic form.

The ion and electron density perturbations are coupled to the Poisson equation. Linearization of this system of equations allows one to obtain the magnitude of the polarization field that the test charge embedded into a flowing plasma induces at its origin. The product of this polarization field and the particle charge yields the ion drag force acting on a non-absorbing particle, $\mathbf{F}_{\rm id} = -Q\nabla\phi|_{\mathbf{r}=0}$. The direction of the ion drag force is obviously parallel to the ion flow.

The electric potential distribution around a non-absorbing test particle is given by Equation (2.49). It is expressed via the plasma permittivity $\varepsilon(0,\mathbf{k})$. The latter is determined by the electron and ion responses, $\varepsilon(\omega,\mathbf{k}) = 1+\chi_e+\chi_i$. For Boltzmann electrons we have $\chi_e \simeq (k\lambda_{\rm De})^{-2}$. The ion contribution χ_i should be in general obtained from the self-consistent solution of the linearized kinetic equation coupled to the Poisson equation. Ivlev *et al.* (2004a) used the conventional expression for Maxwellian collisional ions (see, e.g., Alexandrov *et al.* 1984). They obtained that in the limit $M_T \ll 1$ the force is

$$F_{\rm id} \simeq \frac{1}{3}\sqrt{\frac{2}{\pi}}\left(\frac{T_i}{e}\right)^2\left[\ln\beta_T^{-1} + \frac{1}{\sqrt{2\pi}}\mathscr{K}(\lambda_{\rm D}/\ell_i)\right]\beta_T^2 M_T + O(M_T^3), \qquad (2.73)$$

where $\mathscr{K}(x) = x\arctan x + (\sqrt{\frac{\pi}{2}}-1)\frac{x^2}{1+x^2} - \sqrt{\frac{\pi}{2}}\ln(1+x^2)$ is the "collision function" and $\lambda_{\rm D}$ is the linearized Debye length. For $\ell_i \geq \lambda_{\rm D}$ the function $\mathscr{K}$ is negligibly small compared to the Coulomb logarithm and Equation (2.73) yields the standard collisionless expression for the ion drag force [Equation (2.70) for $\beta_T \ll 1$] derived from the BC approach. In terms of the ion kinetics, the origin of this force is the Landau damping. In the opposite limit $\ell_i \ll \lambda_{\rm D}$, the hydrodynamic effects become more important, and the expression in the brackets in Equation (2.73) changes from $\ln\beta_T^{-1}$ to $\ln[(\ell_i/\lambda_{\rm D})\beta_T^{-1}] + \sqrt{\frac{\pi}{8}}(\lambda_{\rm D}/\ell_i)$. If collisions become "very frequent", $\ell_i \leq \beta_T\lambda_{\rm D}$, the kinetic effects disappear completely and the resulting ion drag force is

$$F_{\rm id} \simeq (1/6)(T_i/e)^2(\lambda_{\rm D}/\ell_i)\beta_T^2 M_T. \qquad (2.74)$$

Note that to avoid logarithmic divergence of the integral in deriving the ion drag force at large k, the upper limit of integration was set to $k_{\rm max} \sim (\lambda_{\rm D}\beta_T)^{-1}$. For larger k linearization procedure is not justified since ion–particle interaction in this regime is essentially nonlinear.

An important result obtained independently by Schweigert (2001) and Ivlev *et al.* (2005) is that when the ion drift is generated by an external electric field $\mathbf{E}$, the assumption of Maxwellian ions is a good approximation only for subthermal flow velocities. The point is that the kinetic equation with the BGK collision integral yields the following solution for the steady-state ion distribution function in the electric field (Schweigert 2001; Ivlev *et al.* 2005):

$$f(v_\parallel, v_\perp) = n_0\Phi_\perp(v_\perp)\int_0^\infty \Phi_\parallel(v_\parallel - ux)e^{-x}dx. \qquad (2.75)$$

Here $\Phi_\parallel = (2\pi v_{T_n}^2)^{-1/2}\exp(-v_\parallel^2/2v_{T_n}^2)$ and $\Phi_\perp = (2\pi v_{T_n}^2)^{-1}\exp(-v_\perp^2/2v_{T_n}^2)$ are the longitudinal and transverse factors of the Maxwellian neutral velocity distribution, respectively, so that $\Phi \equiv \Phi_\parallel\Phi_\perp$. As usual we assume that ion and neutral temperature are equal, and hence, $v_{T_i} = v_{T_n}$. The ion drift velocity in the mobility limit is $u = eE/m_i\nu$. When $u \to 0$ we have $f \to n_0\Phi$, where n_0 is the ambient (constant) ion density. For subthermal ion drift, distribution (2.75) is close to the shifted Maxwellian function, $f \simeq n_0\Phi(v)(1+\mathbf{uv}/v_{T_i}^2)$. However, for $u \geq v_{T_i}$ the deviation from the Maxwellian form is significant.

Thus, for superthermal ion drifts, Maxwellian response for the ions is no longer applicable. Using Equation (2.75) one can derive the self-consistent ion response in a collisional plasma with the external electric field. This was done by Scweigert (2001) and Ivlev *et al.* (2004b, 2005). The result is

$$\chi_i(\omega,k) = \frac{(k\lambda_{Di})^{-2}}{1+i(k\mu\nu/k^2 v_{T_i})M_T}\left[\frac{1+\langle\mathscr{F}(\xi_2)\rangle}{1+\dfrac{i\nu}{\omega+i\nu}\mathscr{F}(\xi_1)}\right], \tag{2.76}$$

where the variables $\xi_{1,2}$ are

$$\xi_1 = \frac{(\omega+i\nu)/\sqrt{2}kv_{T_i}}{\sqrt{1+i(k\mu\nu/k^2v_{T_i})M_T}}, \qquad \xi_2 = \frac{(\omega-k\mu v_{T_i}M_T x+i\nu)/\sqrt{2}kv_{T_i}}{\sqrt{1+i(k\mu\nu/k^2v_{T_i})M_T}},$$

$\mathscr{F}(\xi)$ is the dispersion function of the *Maxwellian* plasma (Fried and Conte 1961), and the average is $\langle\ldots\rangle = \int_0^\infty \ldots e^{-x}dx$. Using plasma permittivity with the ion response Equation (2.76), one can numerically calculate the ion drag force for arbitrary Mach number and ion collision frequency. For small Mach numbers, χ_i tends to that of Maxwellian plasmas and we recover Equation (2.73). For large Mach numbers Ivlev *et al.* (2004b, 2005) obtained the following approximate expression for the ion drag force

$$F_{\mathrm{id}} \simeq \sqrt{\frac{2}{\pi}}\left(\frac{T_i}{e}\right)^2 \ln\left(4\frac{\ell_i}{\lambda_D}\frac{M_T}{\beta_T}\right)\frac{\beta_T^2}{M_T} + O(M_T^{-2}). \tag{2.77}$$

Figure 2.20 shows the ion drag force normalized to $\beta_T^2(T_i/e)^2$ versus the Mach number for different values of β_T and λ_D/ℓ_i. One can see that analytic asymptotes agree fairly well with the numerical results – depending on the value of λ_D/ℓ_i, the discrepancy is ≤ 10 % at $M_T \leq$0.2–0.3 [Equation (2.73)] and $M_T \geq$10 –20 [Equation (2.77)].

At large M_T the kinetic approach yields the force which scales as $F_{\mathrm{id}} \propto M_T^{-1}$, in contrast to the scaling $\propto M_T^{-2}$ in the binary collision approach [see Equation (2.72)]. This is because the ion distribution (2.75) deviates significantly from the Maxwellian form in the superthermal regime. The scaling $F_{\mathrm{id}} \propto M_T^{-1}$ is not affected by a particular dependence of ν on the ion velocity, and hence, it is a generic feature of the self-consistent approach at large Mach numbers. Another feature which follows from the kinetic consideration is the dependence of the force on the ion mean free path. Figure 2.20 shows that frequent ion–neutral collisions ($\ell_i \ll \lambda_D$) enhance the force at small M_T. This is due to the ion focusing (Ivlev *et al.* 2004a): Each collision "eliminates" the angular momentum the ion had (with respect to the particle) before the collision. The motion of the flowing ions becomes more radial due to attraction towards the negatively charged particle. Thus, with increasing ion–neutral collision

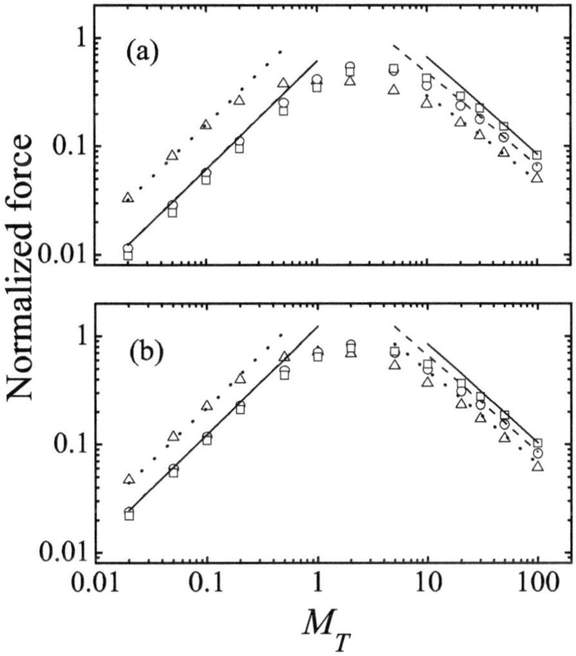

FIGURE 2.20
Normalized ion drag force versus the thermal Mach number of the ion flow M_T (Ivlev *et al.* 2005). The force depends on two parameters: Scattering parameter β_T, and ratio of the screening length to the thermal mean free path, λ_D/ℓ_i. The data points correspond to numerical calculations with the ion susceptibility from Equation (2.76), for $\beta_T^{-1} = 10$ (a) and $\beta_T^{-1} = 100$ (b). Symbols represent $\lambda_D/\ell_i = 0.1$ (square), $\lambda_D/\ell_i = 1$ (circle), and $\lambda_D/\ell_i = 10$ (triangle). Analytic asymptotes at small and large Mach numbers [Equations (2.73) and (2.77), respectively] correspond to the same values of λ_D/ℓ_i (solid, dashed, and dotted lines, respectively).

frequency (decreasing ion mean free path), the focusing center moves closer to the particle and the magnitude of the local enhancement of the ion density at the focusing center increases (see Figure 2.21a in section 2.5.1.3). This additional positive space charge attracts the negatively charged particle enhancing the ion drag force acting in the direction of the ion flow. This mechanism, however, can operate only if the field of the charged particle is stronger than the global field E. Otherwise, if E is relatively strong (Mach number is large), it should de-focus the ion trajectories: After each collision, the ions should accelerate mostly along E. Increase of collisionality (decrease of ℓ_i) at constant $M_T \propto E\ell_i$ implies increase of the global electric field and, hence, stronger de-focusing. In turn, the latter implies the decrease of the ion drag force with increasing collisionality which we see in Figure 2.20 for superthermal ion flows.

The linear kinetic approach is not valid in the immediate vicinity of the charged particle, where the electrostatic perturbations are too strong and the ion–particle interactions are nonlinear. The criterium of the applicability of this approach is the requirement that the *actual* contribution to the force from the "nonlinear" region is relatively small. Usually this requirement is satisfied when the ion–particle scattering parameter β is sufficiently small. For $M_T \ll 1$, the applicability of Equation (2.73) is $\beta_T \ll 1$. In this limit, the collisionless part of Equation (2.73) coincides with the results of the binary collision approach [see Equation (2.70)]. Larger Mach numbers imply better applicability – similar to the results of the binary collision approach – since the effective value of β decreases with increasing M_T.

Another important issue is the effect of ion absorption, neglected in the kinetic consideration. This effect is considered below in the case of a highly collisional plasma.

2.5.1.3 Hydrodynamic approach

In contrast to the analytic results of the kinetic approach predicting an *increase* of the ion drag force with ion collisionality at low flow velocities, some numerical simulations performed for collisional situations (Schweigert *et al.* 2004; Maiorov 2005) demonstrated a *decrease* of the ion drag force compared to the collisionless situation, and even *negative values* were reported (i.e., ion drag force was directed oppositely to the ion flow). Since one of the most important differences between these two considerations is the effect of plasma absorption on the particle – neglected in theory, but accounted for in simulations – it is natural to assume that this effect is responsible for the observed discrepancy. Although, Patacchini and Hutchinson (2008) later doubted the correctness of the numerical results by Schweigert *et al.* (2004) and Maiorov (2005), the above mentioned discrepancy between theory and simulations motivated Khrapak *et al.* (2007c) to perform a detailed analysis of the highly collisional limit in the hydrodynamic approximation, where the effects of plasma absorption on the particle surface can be easily accounted for. This consideration is merely of methodological interest, since highly collisional regime is seldom met in practical complex plasmas. Nevertheless, taking into account the interesting and unexpected character of the obtained results and possible applications to other fields it is worth to discuss this issue briefly.

The problem is formulated as follows: A small individual stationary point-like grain of charge Q is immersed in a highly collisional plasma. Quasi-neutral bulk plasma conditions are assumed, where the ions exhibit subthermal drift while the electrons form a stationary background (ambipolar plasma regime). The neutral component is stationary as well. There are no plasma sources and sinks in the vicinity of the particle (except for the particle surface). The collisional ion component is described by the continuity and momentum equation in the hydrodynamic approximation,

$$\nabla(n_i \mathbf{v}_i) = -J_i \delta(\mathbf{r}), \tag{2.78}$$

$$(\mathbf{v}_i\nabla)\mathbf{v}_i = -\frac{e}{m_i}\nabla\phi - \frac{\nabla n_i}{n_i}v_{T_i}^2 - \nu\mathbf{v}_i + \frac{\mathbf{f}}{m_i}, \quad (2.79)$$

where J_i is the ion flux to the particle and $\mathbf{f}$ is an external force responsible for the ion drift. In equilibrium the ion drift velocity is $\mathbf{u} = \mathbf{f}/m_i\nu$. In the considered case the ion drift is caused by a weak ambipolar electric field, but the final results are independent of the nature of the drift-generating term. The electron density satisfies the Boltzmann relation (2.41). The system is closed with the Poisson equation (2.43).

Standard linearization procedure yields

$$\phi(\mathbf{r}) = \frac{4\pi Q}{(2\pi)^3}\int\frac{\exp(i\mathbf{kr})d\mathbf{k}}{\chi_1(\mathbf{ku},k)} + \frac{4\pi e}{(2\pi)^3}\int\frac{\exp(i\mathbf{kr})d\mathbf{k}}{\chi_2(\mathbf{ku},k)}, \quad (2.80)$$

where

$$\chi_1(\mathbf{ku},k) = k^2 + k_{\mathrm{D}e}^2 + k_{\mathrm{D}i}^2\left[1 + \frac{\mathbf{ku}(i\nu - \mathbf{ku})}{k^2 v_{T_i}^2}\right]^{-1}, \quad (2.81)$$

and

$$\chi_2(\mathbf{ku},k) = -i\frac{k^2 v_{T_i}^2(k^2 + k_{\mathrm{D}}^2)}{J_i(i\nu - \mathbf{ku})} - i\frac{\mathbf{ku}(i\nu - \mathbf{ku})(k^2 + k_{\mathrm{D}e}^2)}{J_i(i\nu - \mathbf{ku})}. \quad (2.82)$$

The first term on the right-hand side of Equation (2.80) is the usual expression for the potential around a pointlike non-absorbing grain in the limit of high collisionality. The second term arises due to ion absorption [see Equation (2.78)]. The electron absorption is not accounted for since it yields only small corrections to the force (Khrapak *et al.* 2007c).

The ion drag force can be easily calculated in the limiting case of vanishing ion flow, $k\ell_i \gg u/v_{T_i} \equiv M_T$. The resulting expression is (Khrapak *et al.* 2007c)

$$F_{\mathrm{id}} \simeq (1/6)Q^2 k_{\mathrm{D}i}^2(\ell_i k_{\mathrm{D}})^{-1}M_T + (1/6)Qek_{\mathrm{D}i}^2(\ell_i k_{\mathrm{D}})^{-1}(J_i/k_{\mathrm{D}i}^2 v_{T_i}\ell_i)(1 + k_{\mathrm{D}e}^2/k_{\mathrm{D}}^2)M_T. \quad (2.83)$$

The first term on the right-hand side of Equation (2.83) yields the force acting on a non-absorbing particle. It coincides with the expression (2.74) obtained using kinetic approach [Note that (2.74) was written assuming $\lambda_{\mathrm{D}} = \lambda_{\mathrm{D}i}$)]. The second term (absent in the kinetic model) corresponds to the effect of absorption. It yields a *negative* contribution to the force, since $Q < 0$. Thus, in the highly collisional limit, plasma absorption on the particle *reduces* the absolute magnitude of the ion drag force.

The magnitude of this reduction is determined by the actual value of the ion flux J_i directed to the particle (Section 2.1). Using the continuum limit expression (2.27), we get

$$F_{\mathrm{id}} \simeq -(1/6)Q^2 k_{\mathrm{D}i}^2(\ell_i k_{\mathrm{D}})^{-1}(1+\tau)^{-1}M_T, \quad (2.84)$$

i.e., the force *reverses sign*! Clearly, this is a result of ion absorbtion on the particle surface.

Let us briefly discuss the physical reason for the ion drag force reduction and sign reversal. With no absorption taken into account, the ion–neutral collisions would *enhance* the ion drag force (compared with the collisionless case) due to collision induced ion focusing, as discussed in Section 2.5.1.2. In contrast, the absorption

An interesting consequence of the ion drag force sign reversal in the limit of a highly collisional plasma has been discussed by Vladimirov *et al.* (2008). When a particle moves in a plasma, it experiences the frictional (drag) force associated with all plasma components – positive ions, negative electrons, and neutral gas. The neutral drag force is likely to be dominant in most cases as long as the plasma is weakly ionized. However, there might exist a narrow parameter regime when plasma-related drag can overcome the neutral drag and, even more important, when the total drag force can be directed along the particle motion, causing the particle to accelerate until it reaches a free undamped frictionless motion.

Let us consider a stationary bulk quasi-neutral high pressure plasma, with motions of both electrons and ions dominated by collisions with neutrals. A small individual absorbing particle is slowly moving with respect to the plasma background. For simplicity let us also assume that the plasma is in equilibrium, $T_e = T_i = T_n = T$. Under the additional assumptions $u/v_{T_i} \ll \ell_i/\lambda_D < a/\lambda_D < \beta_T \ll 1$, we can use Equation (2.84) for the ion drag force acting on the particle. For a one-temperature plasmas it yields

$$F_{\text{id}} \simeq -(1/24)(Q^2/\lambda_D^2)(\lambda_D/\ell_i)(u/v_{T_i}), \tag{2.85}$$

where u is the particle velocity. The force acts in the direction of particle motion.

At high pressures considered here, it is reasonable to assume $\ell_n < a$, where ℓ_n is the mean free path of neutrals. In the limit of low Reynolds number corresponding to a slow particle motion, the neutral drag force is

$$F_n \simeq 6\pi\eta a u, \tag{2.86}$$

where $\eta \simeq n_n m_n \ell_n v_{T_n}$ is the gas viscosity. It is directed opposite to the particle motion.

The last contribution to the friction is associated with the electron component. However, since the ratio of the electron drag force to the ion (neutral) drag forces is $\propto [m_e/m_{i(n)}]^{1/2}$ in the considered case (Khrapak and Morfill 2004), the electron drag force can be neglected.

Thus, the direction of the total friction force is determined by the competition of the ion and neutral drag forces. Let us compare their absolute magnitudes. The force ratio is

$$|F_{\text{id}}/F_n| \simeq (1/18)(n_0/n_n)(v_{T_n}/v_{T_i})\beta_T(\lambda_D/\ell_i)(\lambda_D/\ell_n)z. \tag{2.87}$$

In a weakly ionized plasma we have $n_0 \ll n_n$; it is reasonable to assume that $v_{T_n} \sim v_{T_i}$; applicability of the linear approach requires $\beta_T \ll 1$; the normalized potential z is usually of the order of a few; the applicability of Equation (2.85) requires $\lambda_D \gg \ell_i$ and $\lambda_D \gg \ell_n$. The product of these three large factors can in principle compensate for the smallness in $(n_0/n_n)\beta_T$, reducing the total friction force compared to the pure neutral drag force or even making the total friction negative.

In this case the particle is accelerated until $u = u_{\text{cr}}$, when the balance between the ion drag and neutral drag is reached. The existence of such a balance follows from the fact that at high particle velocities only geometrical factors play a role and $F_{i(n)} \simeq \pi a^2 n_{0(n)} m_{i(n)} u^2$, and thus $F_{\text{id}} \ll F_n$ since $n_0 \ll n_n$. Although an exact value for $u_{\rm cr}$ cannot be obtained within the present approach, Vladimirov *et al.* (2008) suggested that $u_{\rm cr} \sim v_{T_i}(\ell_i/\lambda_{\rm D})$, which corresponds to the violation of the condition used to derive Equation (2.85). The final velocity $u_{\rm cr}$ is a stable equilibrium velocity and the particle exhibits free undamped motion.

The described effect can lead to the following behavior of the particle component in plasmas: A collection of absorbing particles can exhibit free frictionless motion in a highly collisional weakly ionized plasma. The main reason for such an unusual behavior is the openness of the plasma–particles system associated with continuous plasma loss on the particle surfaces. Apart from complex plasmas it would also be interesting to consider this effect in the context of colloidal suspensions, sand storms, volcanic plumes, etc. The mechanisms considered here might be effective in these systems, too.

Finally, let us point out that the effect of plasma production and loss on the ion drag force has been investigated by Chaudhuri *et al.* (2008) for the case of highly collisional plasmas. This has been done by simply adding the corresponding source and loss terms to the ion continuity equation (2.78). Electron impact ionization has been chosen to be the main plasma production mechanism. Two different scenarios of plasma loss have been investigated: Electron–ion volume recombination (which is relevant to high pressure plasmas) and ambipolar diffusion towards the discharge chamber walls and electrodes (which occurs in low- and moderate-pressure gas discharges).

The main results obtained by Chaudhuri *et al.* (2008) are as follows. For sufficiently low ionization rate the ion drag force is not sensitive to plasma production and loss in the vicinity of the particle. The force is directed opposite to the ion flow, as discussed above. The plasma production and loss processes start to affect the magnitude of the ion drag force when $\nu_I/\nu \sim (\ell_i/\lambda_{\rm D})^2$, where ν_I is the characteristic ionization frequency and ν is the frequency of ion–neutral collisions. With further increasing ionization rate, not only the magnitude of the force depends on ionization rate, but also its direction. In a plasma with sufficiently developed ionization the ion drag force always acts in the direction of the ion motion, independently of the plasma loss mechanism. Chaudhuri *et al.* (2008) identified the parameter regimes for the positive and negative ion drag forces for both plasma loss mechanisms considered.

2.5.1.4 Complementarity of the approaches

Comparing the results of these approaches, the most important conclusion to be drawn is that they are not really competitive but rather *complementary*: Binary collision approach is more suitable to describe highly nonlinear collisionless cases when both the characteristic length of ion–particle interaction and ion mean free path exceed considerably the plasma screening length. This situation is typical for (sub)thermal ion flows. Small Mach numbers also imply weak distortion of the potential around the charged particle and weak deviation of the ion distribution from the shifted Maxwellian function. In addition, binary collision formalism easily accounts for ion absorption on the particle surface. Therefore, in this regime binary collision approach is more reliable. Existing experiments in weakly collisional plasmas (Hirt *et al.* 2004; Nosenko *et al.* 2007) demonstrate reasonable agreement with the results of BC approach. Collisionless numerical simulations by Hutchinson (2005, 2006) also agree well with theory, although in some parameter regimes, discrepancies up to a factor of ~ 2 are reported.

On the other hand, for superthermal ions (when the characteristic length of ion–particle interaction decreases rapidly with the Mach number and, hence, the linear theory can be better applied!) both the particle potential and ion distribution function are highly anisotropic, and then the linear plasma response formalism is more reliable. For instance, the kinetic approach allowed to deduce how the *effective* screening length, λ, entering the expressions for the ion drag force in BC approach, depends on the ion flow velocity (Khrapak *et al.* 2005a). It was shown that at $M_T \lesssim 1$ the potential distribution around the particle is weakly affected by the flow, so that the screening is determined by the linearized screening length, $\lambda \simeq \lambda_{\rm D}$. For superthermal flows the ion contribution to the screening rapidly vanishes and the effective screening length tends to the asymptote $\lambda \simeq \lambda_{{\rm D}e}$. Figure 2.22 shows that this transition occurs in a fairly narrow range of velocities around $M_T \simeq$1–3. The exact analytic form of $\lambda(M_T)$ is rather complicated, but it can be approximated reasonably well (Fortov *et al.* 2005) with formula $\lambda^{-2} \simeq f(M_T)\lambda_{{\rm D}i}^{-2} + \lambda_{{\rm D}e}^{-2}$, where the fitting function is $f = \exp(-M_T^2/2)$ (shown in Figure 2.22) or $f = (1+M_T^2)^{-1}$.

An obvious advantage of the linear plasma response formalism is that it accounts consistently for the ion–neutral collisions. Land and Goedheer (2006) proposed to combine the results of BC and kinetic approaches to describe the ion drag force. Namely, they used Equation (2.73) derived using kinetic approach and containing the contribution from ion–neutral collisions, but with the Coulomb logarithm derived using BC approach and accounting for the effect of moderate ion–particle coupling [Equation (2.65)].

Note that it is rather complicated to describe self-consistently the effect of ion absorption on the particle surface within kinetic approach. On the other hand, this can be easily done in the limiting case of highly collisional plasmas using a fluid (hydrodynamic) formulation of Section 2.5.1.3. An interesting result of this approach is a possibility of the sign reversal of the ion drag force (Khrapak *et al.* 2007c). Essentially the same quantitative result was obtained from an analysis of individual ion trajectories in highly collisional plasmas under the assumption of an isotropic Coulomb interaction potential between the ions and the particle (Khrapak *et al.* 2007a).

One should emphasize, however, that often the experimental conditions are such that none of the approaches developed so far is directly applicable, e.g., when ion–particle coupling is not weak and the ion mean free path is comparable or shorter than the plasma screening length. There has been no approach proposed to treat this case analytically, and this issue remains the major challenge for the theory of ion drag. Presumably, in this regime the effect of nonlinear ion–particle coupling is more important than that of ion–neutral collisions. This is supported by experimental results obtained in weakly and moderately collisional plasmas (Khrapak *et al.* 2003a; Zafiu *et al.* 2003b; Klindworth *et al.* 2004; Yaroshenko *et al.* 2005) which demonstrated reasonable agreement with analytic results of the BC approach. An indirect evidence

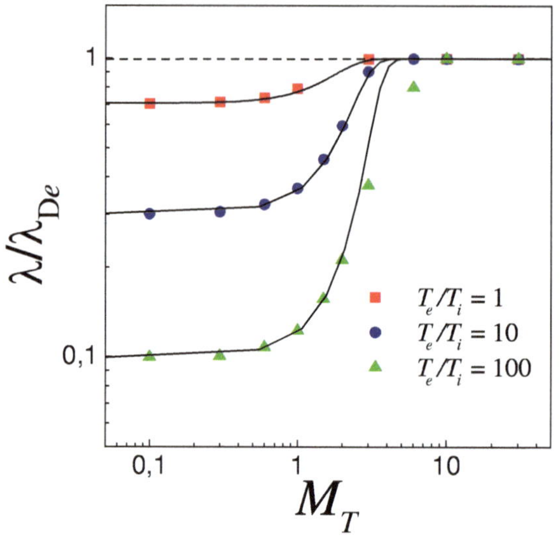

FIGURE 2.22
The effective screening length λ of the charged particle in the flowing plasma versus the thermal Mach number M_T (ion flow velocity normalized to the ion thermal velocity) for different electron-to-ion temperature ratios. Symbols are numerical calculations (Khrapak *et al.* 2005a), lines are simple analytic fits discussed in the text.

also comes from the good agreement between the theory and experimental observations related to the formation of voids – regions free of particles – in the central part of gas discharges under microgravity conditions, which are believed to be produced by the outwards pointing ion drag force (Khrapak *et al.* 2002; Kretschmer *et al.* 2005; Lipaev *et al.* 2007).

2.5.2 Other forces

Similar to the ion drag force, the *electron drag force* arises due to the momentum transfer from the electrons drifting relative to the charged particles. In the binary collision approximation, the electron drag force is $F_{\rm ed} = m_d \nu_{de} u_e$, where ν_{de} is given by Equation (2.63). Compared to the ion drag force, the effect of electron drag is usually ignored because the electron-to-ion mass ratio is small. This is correct when $u_e \sim u_i$, e.g., in rf discharges, where electrons and ions drift together due to the ambipolar diffusion. However, in the case of *independent* (mobility limited) drift (e.g., in the positive column of a dc discharge) the ratio of the ion-to-electron drag forces is *independent* of masses and can be approximately estimated as $F_{\rm id}/F_{\rm ed} \sim (T_e/T_i)^2(\sigma_{en}/\sigma_{in})$, where $\sigma_{e(i)n}$ is the transport cross section for electron (ion) collisions with neutrals. A detailed investigation performed by Khrapak and Morfill (2004) shows that the electron drag force can indeed dominate over the electric and ion drag force in most noble gases with relatively small electron temperatures ($T_e \lesssim 1$ eV).

In ground-based conditions the *gravitational force* $F_g = m_d g$ usually plays an important role. In order to levitate the particle, it should be counterbalanced by other forces. The *electric force* due to the electric field in the (pre)sheath or striation regions of discharges can provide the balance. The magnitude of the electric force is $F_{el} = QE$, where E is the electric field strength. A correction to F_{el} due to plasma polarization in the vicinity of the particle (of the order of a/λ_D, induced by the external electric field) was derived by Daugherty *et al.* (1993). This effect increases the absolute magnitude of the electric force. The external field also induces a dipole moment on a particle, $\sim a^3E$, which is pointed along the field. For a dielectric particle, an additional dipole moment can be induced due to anisotropy in charging (Ivlev *et al.* 1999). In the nonuniform electric field such a dipole will experience an additional force $\sim \frac{1}{2}a^3(E^2)'$. It is worth mentioning that the particle charge in the electric field is implicitly dependent on the field magnitude through, e.g., induced plasma and/or charging anisotropy, ion (electron) drift velocities. The problem of trapped ions is also an important issue related to the electric force acting on a particle in plasmas: Ions on trapped orbits can shield the particle from external electric field, leading to a decrease of the electric force.

If a temperature gradient is present in a neutral gas, then the particle experiences a *thermophoretic force*. The force is due to asymmetry in the momentum transfer from neutrals and is directed towards lower gas temperatures. In the case of full accommodation of neutrals colliding with the particle surface, Talbot *et al.* (1980) derived the following expression for the thermophoretic force:

$$F_{th} = -\frac{4\sqrt{2\pi}}{15}\frac{a^2}{v_{T_n}}\kappa_n \nabla T_n, \tag{2.88}$$

where κ_n is the thermal conductivity coefficient of gas. For atomic gases $\kappa_n \simeq 1.33(v_{T_n}/\sigma_{nn})$, where σ_{nn} is the cross section of neutral–neutral collisions (Raizer 1991). In this case $F_{th} \simeq -1.8(a^2/\sigma_{nn})\nabla T_n$. Rothermel *et al.* (2002) derived an expression with a different numerical factor, $F_{th} \simeq -3.33(a^2/\sigma_{nn})\nabla T_n$, and found good agreement between this expression and experimental results. Note that the thermophoretic force depends on the particle radius, gas type (through σ_{nn}), and temperature gradient, but does not depend on the gas pressure and temperature. For particles of about 1 μm radius and mass density ~ 1 g cm^{-3} in an argon plasma, the force is comparable to the force of gravity at temperature gradients $|\nabla T_n| \sim 10$ K cm^{-1}. The corrections to the force for the case when the particle is situated near the electrode or the walls of a discharge chamber, which basically change the numerical factor in Equation (2.88), were derived by Havnes *et al.* (1994). Experimental investigations of the effect of thermophoretic force on the behavior of particles in gas discharge plasmas were performed by Jellum *et al.* (1991), Balabanov *et al.* (2001), and Rothermel *et al.* (2002). In these works, it was shown that the thermophoretic force can be used for particle levitation in ground-based conditions as well as for controlled action on the ordered structures of particles.

And finally, the *neutral drag force* is the main mechanism responsible for friction when a particle is moving through a stationary plasma. This is because the ionization fraction is usually quite low, on the order of 10^{-7}–10^{-6}. Neutral drag can be also important when gas is flowing relative to the particles. When the Knudsen number $\mathrm{Kn} = \ell_n/a$ is large and the relative velocity between the particle and the gas u_d is small compared to the thermal velocity of neutrals v_{T_n}, then

$$F_{\mathrm{n}} = -m_d \nu_{dn} u_d, \tag{2.89}$$

where ν_{dn} is the momentum exchange rate given by Equation (2.69). The minus sign means that the force acts in the direction opposite to the relative velocity. For high relative velocities ($u_d \gg v_{T_n}$), the neutral drag force is proportional to the velocity squared (see, for example, Draine and Salpeter, 1979), $F_{\mathrm{n}} \simeq -\pi a^2 n_n m_n u_d^2$. In the opposite limit of small Knudsen numbers $\mathrm{Kn} \ll 1$, the Stokes expression applies, $F_{\mathrm{n}} = -6\pi\eta a u_d$, where η is the viscosity of neutral gas. In most practical cases Equation (2.89) is applicable to calculate the neutral drag force in complex plasmas.

2.6 Particle surface temperature

The particle surface temperature is determined by the balance of energy fluxes directed to/from the particle surface and is generally different from the temperature of surrounding gas. As discussed in Section 2.3.1 the temperature difference between the particle surface and the surrounding neutral gas gives rise to the so-called "neutral shadowing" interparticle interaction. Therefore, knowledge of the particle surface temperature is of considerable importance.

Two experiments were performed to measure the temperature of the particles suspended in plasmas. Daugherty and Graves (1993) have measured the temperature of phosphorescent manganese activated magnesium fluorogermanate (MFG) particles. They recorded the phosphorescent decay of MFG when the plasma was suddenly turned off and made use of the fact that the time constant for this decay is a well-defined function of temperature. Swinkels *et al.* (2000) employed laser–induced fluorescence of dyed melamine-formaldehyde spheres to measure their internal temperature. Both experiments were performed in low pressure rf discharges in argon and demonstrated that the particle temperature was somewhat hotter than the surrounding neutral gas temperature.

Theoretical model for the particle surface temperature in noble gases uses the following assumptions (Daugherty and Graves 1993; Swinkels *et al.* 2000; Ignatov 2002): The main heating mechanism is energy deposition from the collected ions, electrons, and their recombination on the particle surface, and the main cooling mechanisms are collisions with neutral atoms and radiation. Daugherty and Graves (1993), Swinkels *et al.* (2000) and Ignatov (2002) used collisionless OML theory to estimate plasma fluxes to the particle. Khrapak and Morfill (2006) refined this model

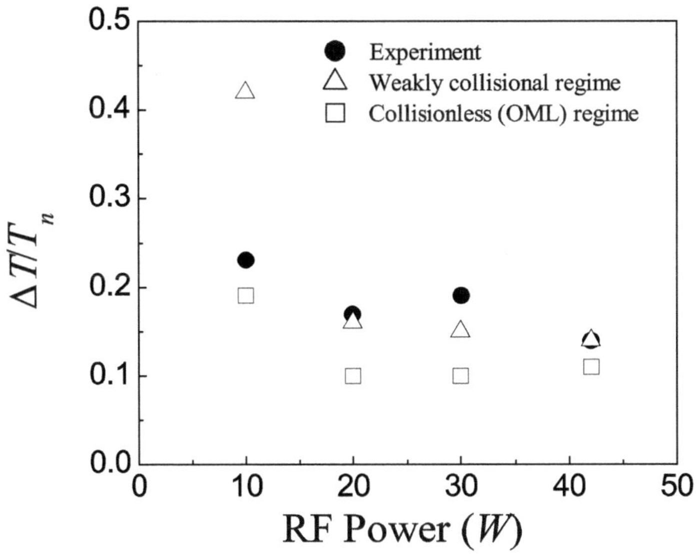

FIGURE 2.23
Relative difference between the particle and the neutral gas temperatures versus discharge power. In the figure experimental results by Swinkels *et al.* (2000) are compared with the results of collisional (CEC) and collisionless (OML) theoretical models of particle charging. Except for the lowest value of discharge power collisional model yields better agreement between theory and experiment. Relatively large discrepancy at low power correlates with dramatic increase in measured electron temperature in the experiment. This increase leads to the corresponding increase in the particle floating potential and ion–particle coupling, which can make CEC approximation for charging insufficiently accurate. This can explain the observed difference.

by taking into account collision induced enhancement of the ion flux to the particle using CEC approximation (see Section 2.1.2). This yields in general better agreement with the available experimental data as demonstrated in Figure 2.23. Although in the considered case the difference between the results of application of collisional and collisionless models of particle charging is not large, in other regimes one can expect much larger difference. Thus, the importance of the proposed modification should not be underestimated.

Typical values of the relative temperature difference between the particle surface and neutral gas are in the range $\Delta T/T_n \simeq 0.1$–0.2. Estimates by Khrapak and Morfill (2006) show that for usual plasma parameters the magnitude of neutral shadowing interaction is considerably smaller than that of ion shadowing and electric interactions.

References

Alexandrov, A. F., Bogdankevich, L. S., and Rukhadze, A. A. (1984). *Principles of plasma electrodynamics*. Springer-Verlag, New York.

Alexandrov, A. L., Schweigert, I. V., and Peeters, F. M. (2008). A non-Maxwellian kinetic approach for charging of dust particles in discharge plasmas. *New J. Phys.*, **10**, 093025/1–12.

Allen, J. E. (1992). Probe theory: The orbital motion approach. *Phys. Scripta*, **45**, 497–503.

Allen, J. E., Annaratone, B. M., and de Angelis, U. (2000). On the orbital motion limited theory for a small body at floating potential in a Maxwellian plasma. *J. Plasma Phys.*, **63**, 299–309.

Al'pert, Y. L., Gurevich, A. V., and Pitaevsky, L. P. (1965). *Space physics with artificial satellites*. Consultants Bureau, New York.

de Angelis, U. (2006). Dusty plasmas in fusion devices. *Phys. Plasmas*, **13**, 012514/1–10.

Balabanov, V. V., Vasilyak, L. M., Vetchinin, S. P., Nefedov, A. P., Polyakov, D. N., and Fortov, V. E. (2001). The effect of the gas temperature gradient on dust structures in a glow-discharge plasma. *JETP*, **92**, 86–92.

Barkan, A., D'Angelo, N., and Merlino, R. L. (1994). Charging of dust grains in a plasma. *Phys. Rev. Lett.*, **73**, 3093–3096.

Barnes, M. S., Keller, J. H., Forster, J. C., O'Neill, J. A., and Coultas, D. K. (1992). Transport of dust particles in glow-discharge plasmas. *Phys. Rev. Lett.*, **68**, 313–316.

Baroody, E. M. (1962). Classical scattering by some important repulsive potentials. *Phys. Fluids*, **5**, 925–932.

Baum, E. and Chapkis, R. L. (1970). Theory of spherical electroststic probe in a continuum gas: An exact solution. *AIAA J.*, **8**, 1073–1077.

Bhatnagar, P. L., Gross, E. P., and Krook, M. (1954). A model for collision processes in gases. I. Small amplitude proceses in charged and neutral one-component systems. *Phys. Rev.*, **94**, 511–525.

Bliokh, P., Sinitsin, V., and Yaroshenko, V. (1995). *Dusty and self-gravitational plasmas in space*. Kluwer, Dordrecht.

Boeuf, J. P. (1992). Characteristics of a dusty nonthermal plasma from a particle-in-cell Monte Carlo simulation. *Phys. Rev. A*, **46**, 7910–7922.

Bolotovskii, B. M. and Stolyarov, S. N. (1992). Radiation from and energy-loss by charged-particles in moving media. *Sov. Phys. Usp.*, **35**, 143–150.

Bouchoule, A. (1999). Technological impacts of dusty plasmas. In *Dusty plasmas: Physics, chemistry and technological impacts in plasma processing*, Bouchoule, A. (ed.), pp. 305–396. Wiley, Chichester.

Bryant, P. (2003). Floating potential of spherical probes and dust grains in collisional plasmas. *J. Phys. D: Appl. Phys.*, **36**, 2859–2868.

Bystrenko, O. and Zagorodny, A. (1999). Critical effects in screening of high-Z impurities. *Phys. Lett. A*, **255**, 325–330.

Bystrenko, O. and Zagorodny, A. (2003). Screening of dust grains in a weakly ionized gas: Effects of charging by plasma currents. *Phys. Rev. E*, **67**, 066403/1–5.

Chang, J.-S. and Laframboise, J. G. (1976). Probe theory for arbitrary shape in a large Debye length, stationary plasma. *Phys. Fluids*, **19**, 25–31.

Chaudhuri, M., Khrapak, S. A., and Morfill, G. E. (2007). Electrostatic potential behind a macroparticle in a drifting collisional plasma: Effect of plasma absorption. *Phys. Plasmas*, **14**, 022102/1–5.

Chaudhuri, M., Khrapak, S. A., and Morfill, G. E. (2008). Ion drag force on a small grain ih highly collisional weakly anisotropic plasmas: Effect of plasma production and loss mechanisms. *Phys. Plasmas*, **15**, 053703/1–7.

Chu, J. H. and I, L. (1994). Direct observation of Coulomb crystals and liquids in strongly coupled rf dusty plasmas. *Phys. Rev. Lett.*, **72**, 4009–4012.

Cui, C. and Goree, J. (1994). Fluctuations of the charge on a dust grain in a plasma. *IEEE Trans. Plasma Sci.*, **22**, 151–158.

D'Angelo, N. (1998). Dusty plasma ionization instability with ion drag. *Phys. Plasmas*, **5**, 3155–3160.

Daugherty, J. E. and Graves, D. B. (1993). Particulate temperature in radio frequency glow discharges. *J. Vac. Sci. Technol. A*, **11**, 1126–1134.

Daugherty, J. E., Porteous, R. K., Kilgore, M. D., and Graves, D. B. (1992). Sheath structure around particles in low-pressure discharges. *J. Appl. Phys*, **72**, 3934–3942.

Daugherty, J. E., Porteus, R. K., and Graves, D. B. (1993). Electrostatic forces on small particles in low-pressure discharges. *J. Appl. Phys.*, **73**, 1617–1620.

Delzanno, G. L., Lapenta, G., and Rosenberg, M. (2004). Attractive potential around a thermionically emitting microparticle. *Phys. Rev. Lett.*, **92**, 035002/1–4.

Delzanno, G. L., Bruno, A., Sorasio, G., and Lapenta, G. (2005). Exact orbital motion theory of the shielding potential around an emitting, spherical body. *Phys. Plasmas*, **12**, 062102/1–18.

Dimoff, K. and Smy, P. R. (1970). Dust induced quenching of an afterglow plasma. *Phys. Lett. A*, **32**, 13–14.

Draine, B. T. and Salpeter, E. E. (1979). On the physics of dust grains in hot gas. *Astrophys. J*, **231**, 77–94.

D'yachkov, L. G., Khrapak, A. G., Khrapak, S. A., and Morfill, G. E. (2007). Model of grain charging in collisional plasmas accounting for collisionless layer. *Phys. Plasmas*, **14**, 042102/1–6.

D'yachkov, L. G., Khrapak, A. G., and Khrapak, S. A. (2008). Effect of electron emission on charging and screening of a macroparticle in continuum plasma regime. *JETP*, **106**, 166–171.

Epstein, P. (1924). On the resistance experienced by spheres in their motion through gases. *Phys. Rev.*, **23**, 710–733.

Filippov, A. V., Dyatko, N. A., Pal', A. F., and Starostin, A. N. (2003). Development of a self-consistent model of dust grain charging at elevated pressures using the method of moments. *Plasma Phys. Rep.*, **29**, 190–202.

Filippov, A. V., Zagorodny, A. G., Momot A. I., Pal', A. F., and Starostin, A. N. (2007a). Charge screening in a plasma with an external ionization source. *JETP*, **104**, 147–161.

Filippov, A. V., Zagorodny, A. G., Pal', A. F., Starostin, A. N., and Momot A. I. (2007b). Kinetic description of the screening of the charge of macroparticles in a nonequilibrium plasma. *JETP Lett.*, **86**, 761–766.

Filippov, A. V., Zagorodniy, A. G., and Momot, A. I. (2008). Screening of a moving charge in a non-equilibrium plasma. *JETP Lett.*, **88**, 27–33.

Fortov, V. E., Nefedov, A. P., Petrov, O. F., Samarian, A. A., and Chernyschev, A. V. (1996a). Particle ordered structures in a strongly coupled classical thermal plasma. *Phys. Rev. E*, **54**, R2236–R2239.

Fortov, V. E., Nefedov, A. P., Petrov, O. F., Samarian, A. A., Chernyschev, A. V., and Lipaev, A. M. (1996b). Experimental observation of Coulomb ordered structure in sprays of thermal dusty plasmas. *JETP Lett*, **63**, 187–192.

Fortov, V. E., Nefedov, A. P., Torchinsky, V. M., Molotkov, V. I., Petrov, O. F., Samarian, A. A., Lipaev, A.M., and Khrapak, A. G. (1997). Crystalline structures of strongly coupled dusty plasmas in dc glow discharge strata. *Phys. Lett. A*, **229**, 317–322.

Fortov, V. E., Nefedov, A. P., Vaulina, O. S., Lipaev, A. M., Molotkov, V. I., Samaryan, A. A., Nikitskii, V. P., Ivanov, A. I., Savin, S. F., Kalmykov, A. V., Soloviev, A. Ya., and Vinogradov, P. V. (1998). Dusty plasma induced by solar radiation under microgravitational conditions: an experiment on board the Mir orbiting space station. *JETP*, **87**, 1087–1097.

Fortov, V. E., Nefedov, A. P., Molotkov, V. I., Poustylnik, M. Y., and Torchinsky, V. M. (2001). Dependence of the dust-particle charge on its size in a glow-discharge plasma. *Phys. Rev. Lett.*, **87**, 205002/1–4.

Fortov, V. E., Petrov, O. F., Usachev, A. D., and Zobnin, A. V. (2004). Micron-sized particle-charge measurements in an inductive rf gas-discharge plasma using gravity-driven probe grains. *Phys. Rev. E*, **70**, 046415/1–6.

Fortov, V. E., Ivlev, A. V., Khrapak, S. A., Khrapak A. G., and Morfill, G. E. (2005). Complex (dusty) plasmas: Current status, open issues, perspectives. *Phys. Reports*, **421**, 1–103.

Fried, B. D. and Conte, S. D. (1961). *The plasma dispersion function.* Academic Press, New York.

Gatti, M. and Kortshagen, U. (2008). Analytical model of particle charging in plasmas over a wide range of collisionality. *Phys. Rev. E*, **78**, 046402/1–6.

Ginzburg, V. L. (1996). Radiation by uniformly moving sources (Vavilov–Cherenkov effect, transition radiation, and other phenomena). *Phys. Usp.*, **39**, 973–982.

Goree, J. (1994). Charging of particles in a plasma. *Plasma Sources Sci. Technol.*, **3**, 400–406.

Goree, J., Morfill, G. E., Tsytovich, V. N. and Vladimirov, S. V. (1999). Theory of dust voids in plasmas. *Phys. Rev. E*, **59**, 7055–7067.

Goertz, C. K. (1989). Dusty plasmas in the solar system. *Rev. Geophys.*, **27**, 271–292.

Hahn, H.-S., Mason, E. A., and Smith, F. J. (1971). Quantum transport cross sections for ionized gases. *Phys. Fluids*, **14**, 278–287.

Havnes, O., Morfill, G. E., and Geortz, C. K. (1984). Plasma potential and grain charges in a dust cloud embedded in a plasma. *J. Geophys. Res.*, **89**, 999–1003.

Havnes, O., Nitter, T., Tsytovich, V., Morfill, G. E., and Hartquist, T. (1994). On the thermophoretic force close to walls in dusty plasma experiments. *Plasma Sources Sci. Technol.*, **3**, 448–457.

Hayashi, Y. and Tachibana, S. (1994). Observation of Coulomb-crystal formation from carbon particles grown in a methane plasma. *Jpn. J. Appl. Phys.*, **33**, L804–L806.

Hebner, G. A. and Riley, M. E. (2003). Measurement of attractive interactions produced by the ion wakefield in dusty plasmas using a constrained collision geometry. *Phys. Rev. E*, **68**, 046401/1–11.

Hebner, G. A., Riley, M. E., and Marder, B. M. (2003) Dynamic probe of dust wakefield interactions using constrained collisions. *Phys. Rev. E*, **68**, 016403/1–5.

Hirt, M., Block, D., and Piel, A. (2004). Measurement of the ion drag force on free falling microspheres in a plasma. *Phys. Plasmas*, **11**, 5690–5696.

Homann, A., Melzer, A., Peters, S., and Piel, A. (1997). Determination of the dust screening length by laser-excited lattice waves. *Phys. Rev. E*, **56**, 7138–7141.

Homann, A., Melzer, A., Peters, S., Madani, R., and Piel, A. (1998). Laser-excited dust lattice waves in plasma crystals. *Phys. Lett. A*, **242**, 173–180.

Homann, A., Melzer, A., and Piel, A. (1999). Measuring the charge on single particles by laser-excited resonances in plasma crystals. *Phys. Rev. E*, **59**, R3835–R3838.

Hou, L.-J., Wang, Y.-N., and Miskovic, Z. L. (2001). Interaction potential among dust grains in a plasma with ion flow. *Phys. Rev. E*, **64**, 046406/1–7.

Hou, L.-J., Wang, Y.-N., and Miskovic, Z. L. (2003). Induced potential of a dust particle in a collisional radio-frequency sheath. *Phys. Rev. E*, **68**, 016410/1–7.

Hutchinson, I. H. (2003). Ion collection by a sphere in a flowing plasma: 2. Non-zero Debye length. *Plasma Phys. Control. Fusion*, **45**, 1477–1500.

Hutchinson, I. H. (2005). Ion collection by a sphere in a flowing plasma: 3. Floating potential and drag force. *Plasma Phys. Control. Fusion*, **47**, 71–87.

Hutchinson, I. H. (2006). Collisionless ion drag force on a spherical grain. *Plasma Phys. Control. Fusion*, **48**, 185–202.

Hutchinson, I. H. and Patacchini L. (2007). Computation of the effect of neutral collisions on ion current to a floating sphere in a stationary plasma. *Phys. Plasmas*, **14**, 013505/1–9.

Ignatov, A. M. (1996). Lesage gravity in dusty plasmas. *Plasma Phys. Rep.*, **22**, 585–589.

Ignatov, A. M. (2002). Heat exchange in dusty plasma. *Plasma Phys. Rep.*, **28**, 847–857.

Ishihara, O. and Vladimirov, S. V. (1997). Wake potential of a dust grain in a plasma with ion flow. *Phys. Plasmas*, **4**, 69–74.

Ishihara, O., Vladimirov, S. V., and Cramer, N. F. (2000). Effect of dipole moment on the wake potential of a dust grain. *Phys. Rev. E*, **61**, 7246–7248.

Ishihara, O., Kamimura, T., Hirose, K. I., and Sato, N. (2002). Rotation of a two-dimensional Coulomb cluster in a magnetic field. *Phys. Rev. E*, **66**, 046406/1–6.

Ivlev, A. V., Morfill, G., and Fortov, V. E. (1999). Potential of a dielectric particle in a flow of a collisionless plasma. *Phys. Plasmas*, **6**, 1416–1420.

Ivlev, A. V., Konopka, U., and Morfill, G. (2000). Influence of charge variations on particle oscillations in the plasma sheath. *Phys. Rev. E*, **62**, 2739–2744.

Ivlev, A. V., Khrapak, S. A., Zhdanov, S. K., and Morfill, G. E. (2004a). Force on a charged test particle in a collisional flowing plasma. *Phys. Rev. Lett.*, **92**, 205007/1–4.

Ivlev, A. V., Zhdanov, S. K., Khrapak, S. A., and Morfill, G. E. (2004b). Ion drag force in dusty plasmas. *Plasma Phys. Control. Fusion*, **46**, B267–B279.

Ivlev, A. V., Zhdanov, S. K., Khrapak, S. A., and Morfill, G. E. (2005). Kinetic approach for the ion drag force in a collisional plasma. *Phys. Rev. E*, **71**, 016405/1–7.

Jellum, G. M., Daugherty, J. E., and Graves, D. B. (1991). Particle thermophoresis in low pressure glow discharges. *J. Appl. Phys.*, **69**, 6923–6934.

Kaw, P. K., Nishikawa, K., and Sato, N. (2002). Rotation in collisional strongly coupled dusty plasmas in a magnetic field. *Phys. Plasmas*, **9**, 387–390.

Kennedy, R. V. and Allen, J. E. (2003). The floating potential of spherical probes and dust grains. II: Orbital motion theory. *J. Plasma Phys.*, **69**, 485–506.

Kersten, H., Deutsch, H., Stoffels, E., Stoffels, W. W., Kroesen, G. M. W., and Hippler, R. (2001). Micro-disperse particles in plasmas: From disturbing side effects to new applications. *Contrib. Plasma Phys.*, **41**, 598–609.

Khrapak, S. A. and Morfill, G. E. (2002). Dust diffusion across a magnetic field due to random charge fluctuations. *Phys. Plasmas*, **9**, 619–623.

Khrapak, S. A. and Morfill, G. E. (2004). Dusty plasmas in a constant electric field: Role of the electron drag force. *Phys. Rev. E*, **69**, 066411/1–5.

Khrapak, S. A. and Morfill G. E. (2006). Grain surface temperature in noble gas discharges: Refined analytical model. *Phys. Plasmas*, **13**, 104506/1–4.

Khrapak, S. A. and Morfill, G. E. (2008a) A note on the binary interaction potential in complex (dusty) plasmas. *Phys. Plasmas*, **15**, 084502/1–4.

Khrapak, S. A. and Morfill, G. E. (2008b). An interpolation formula for the ion flux to a small particle in collisional plasmas. *Phys. Plasmas*, **15**, 114503/1–4.

Khrapak S. A. and Yaroshenko, V. V. (2003). Low-frequency waves in collisional plasmas with an ion drift. *Phys. Plasmas*, **10**, 4616–4621.

Khrapak, S. A., Nefedov, A.P., Petrov, O. F., and Vaulina, O. S. (1999). Dynamical properties of random charge fluctuations in a dusty plasma with different charging mechanisms., *Phys. Rev. E*, **59**, 6017–6022; and Erratum in *Phys. Rev. E*, **60**, 3450.

Khrapak, S. A., Ivlev, A. V., and Morfill, G. (2001). Interaction potential of microparticles in a plasma: Role of collisions with plasma particles. *Phys. Rev. E*, **64**, 046403/1–7.

Khrapak, S. A., Ivlev, A. V., Morfill, G. E., and Thomas, H. M. (2002). Ion drag force in complex plasmas. *Phys. Rev. E*, **66**, 046414/1–4.

Khrapak, S. A., Ivlev, A. V., Morfill, G. E., Thomas, H. M., Zhdanov, S. K., Konopka, U., Thoma, M. H., and Quinn, R. A. (2003a). Comment on "Measurement of the ion drag force on falling dust particles and its relation to the void formation in complex (dusty) plasmas" [*Phys. Plasmas* **10**, 1278 (2003)]. *Phys. Plasmas*, **10**, 4579–4581.

Khrapak, S. A., Ivlev, A. V., Morfill, G. E., and Zhdanov, S.K. (2003b). Scattering in the attractive Yukawa potential in the limit of strong interaction. *Phys. Rev. Lett.*, **90**, 225002/1–4.

Khrapak, S., Samsonov, D., Morfill, G., Thomas, H., Yaroshenko, V., Rothermel, H., Hagl, T., Fortov, V., Nefedov, A., Molotkov, V., Petrov, O., Lipaev, A., Ivanov, A., and Baturin, Y. (2003c). Compressional waves in complex (dusty) plasmas under microgravity conditions. *Phys. Plasmas*, **10**, 1–4.

Khrapak, S. A., Ivlev, A. V., and Morfill, G. E. (2004a). Momentum transfer in complex plasmas. *Phys. Rev. E*, **70**, 056405/1–9.

Khrapak, S. A., Ivlev, A. V., Morfill, G. E., Zhdanov, S.K., and Thomas, H. M. (2004b). Scattering in the attractive Yukawa potential: Application to the ion-drag force in complex plasmas. *IEEE Trans. Plasma Sci.*, **32**, 555–560.

Khrapak, S. A., Ivlev, A. V., Zhdanov, S. K., and Morfill, G. E. (2005a). Hybrid approach to the ion drag force. *Phys. Plasmas*, **12**, 042308/1–8.

Khrapak, S. A., Ratynskaia, S. V., Zobnin, A. V., Yaroshenko, V. V., Thoma, M. H., Kretschmer, M., Usachev, A. D., Höfner, H., Morfill, G. E., Petrov, O. F., and Fortov, V. E. (2005b). Particle charge in the bulk of gas discharges. *Phys. Rev. E*, **72**, 016406/1–10.

Khrapak, S. A., Morfill, G. E., Ivlev, A. V., Thomas, H. M., Beysens, D. A., Zappoli, B., Fortov, V. E., Lipaev, A. M., and Molotkov, V. I. (2006a). Critical point in complex plasmas. *Phys. Rev. Lett.*, **96**, 015001/1–4.

Khrapak, S. A., Morfill, G. E., Khrapak, A. G., and D'yachkov, L. G. (2006b). Charging properties of a dust grain in collisional plasmas. *Phys. Plasmas*, **13**, 052114/1–5.

Khrapak, S. A., Klumov, B. A., and Morfill, G. E. (2007a). Response to "Comment on 'Ion collection by a sphere in a flowing collisional plasma' ". *Phys. Plasmas*, **14**, 074702/1–2.

Khrapak, S. A., Morfill, G. E., Fortov, V. E., D'yachkov, L. G., Khrapak, A. G., and Petrov, O. F. (2007b). Attraction of positively charged particles in highly collisional plasmas. *Phys. Rev. Lett.*, **99**, 055003/1–4.

Khrapak, S. A., Zhdanov, S. K., Ivlev, A. V., and Morfill, G. E. (2007c). Drag force on an absorbing body in highly collisional plasmas. *J. Appl. Phys.*, **101**, 033307/1–4.

Khrapak, S. A., Klumov, B. A., and Morfill, G. E. (2008). Electric potential around an absorbing body in plasmas: Effect of ion–neutral collisions. *Phys. Rev. Lett.*, **100**, 225003/1–4.

Kilgore, M. D., Daugherty, J. E., Porteous, R. K., and Graves, D. B. (1993). Ion drag on an isolated particulate in low-pressure discharge. *J. Appl. Phys.*, **73**, 7195–7202.

Kilgore, M. D., Daugherty, J. E., Porteous, R. K., and Graves, D. B. (1994). Transport and heating of small particles in high density plasma sources. *J. Vac. Sci. Technol.*, **B12**, 486–493.

Klindworth M., Piel, A., Melzer, A., Konopka, U., Rothermel, H., Tarantik, K., and Morfill, G. E. (2004). Dust free regions around Langmuir probes in complex plasmas under microgravity. *Phys. Rev. Lett.*, **93**, 195002/1–4.

Kompaneets, R. (2007). Complex plasmas: Interaction potentials and non-Hamiltonian dynamics. Dissertation, Ludwig-Maximilians-Universität München, http://edoc.ub.uni-muenchen.de/7380.

Kompaneets, R., Konopka, U., Ivlev, A. V., Tsytovich, V., and Morfill, G. (2007). Potential around a charged dust particle in a collisional sheath. *Phys. Plasmas*, **14**, 052108/1–7.

Konopka, U., Ratke, L., and Thomas, H. M. (1997). Central collisions of charged dust particles in a plasma. *Phys. Rev. Lett.*, **79**, 1269–1272.

Konopka, U., Morfill, G. E., and Ratke, L. (2000a) Measurement of the interaction potential of microspheres in the sheath of a rf discharge. *Phys. Rev. Lett.*, **84**, 891–894.

Konopka, U., Samsonov, D., Ivlev, A. V., Goree, J., Steinberg, V., and Morfill, G. E. (2000b). Rigid and differential plasma crystal rotation induced by magnetic fields. *Phys. Rev. E*, **61**, 1890–1898.

Krasheninnikov, S. I., Tomita, Y., Smirnov, R. D., and Janev, R. K. (2004). On dust dynamics in tokamak edge plasmas. *Phys. Plasmas*, **11**, 3141–3150.

Kretschmer, M., Khrapak, S. A., Zhdanov, S. K., Thomas, H. M., Morfill, G. E., Fortov, V. E., Lipaev, A. M., Molotkov, V. I., Ivanov, A. I., and Turin, M. V. (2005). Force field inside the void in complex plasmas under microgravity conditions. *Phys. Rev. E*, **71**, 056401/1–6.

Lampe, M. (2001). Limits of validity for orbital-motion-limited theory for a small floating collector. *J. Plasma Phys.*, **65**, 171–180.

Lampe, M., Joyce, G., Ganguli, G., and Gavrishchaka, V. (2000). Interactions between dust grains in a dusty plasma. *Phys. Plasmas*, **7**, 3851–3861.

Lampe, M., Gavrishchaka, V., Ganguli, G., and Joyce, G. (2001a). Effect of trapped ions on shielding of a charged spherical object in a plasma. *Phys. Rev. Lett.*, **86**, 5278–5281.

Lampe, M., Joyce, G., and Ganguli, G. (2001b). Analytic and simulation studies of dust grain interaction and structuring. *Phys. Scripta*, **T89**, 106–111.

Lampe, M., Goswami, R., Sternovsky, Z., Robertson, S., Gavrishchaka, V., Ganguli, G., and Joyce, G. (2003). Trapped ion effect on shielding, current flow, and charging of a small object in a plasma. *Phys. Plasmas*, **10**, 1500–1513.

Land, V. and Goedheer, W. J. (2006). Effect of large-angle scattering, ion flow speed and ion–neutral collisions on dust transport under microgravity conditions. *New J. Phys.*, **8**, 8/1–23.

Landau, L. D. and Lifshitz, E. M. (1976). *Mechanics*. Pergamon, Oxford.

Lane, G. H. and Everhart, E. (1960). Calculations of total cross sections for scattering from Coulomb potentials with exponential screening. *Phys. Rev.*, **117**, 920–924.

Lapenta, G. (1999). Simulation of charging and shielding of dust particles in drifting plasmas. *Phys. Plasmas*, **6**, 1442–1447.

Lapenta, G. (2000). Linear theory of plasma wakes. *Phys. Rev. E*, **62**, 1175–1181.

Lapenta, G. (2002). Nature of the force field in plasma wakes. *Phys. Rev. E*, **66**, 026409/1–6.

Lemons, D. S., Murillo, M. S., Daughton, W., and Winske, D. (2000). Two-dimensional wake potentials in sub- and supersonic dusty plasmas. *Phys. Plasmas*, **7**, 2306–2313.

Lieberman, M. A. and Lichtenberg, A. J. (1994). *Principles of plasma discharges and materials processing*. Wiley, New York.

Lipaev, A. M., Molotkov, V. I., Nefedov, A. P., Petrov, O. F., Torchinskii, V. M., Fortov, V. E., Khrapak, A. G., and Khrapak, S. A. (1997). Ordered structures in a nonideal dusty glow-discharge plasma. *JETP*, **85**, 1110–1118.

Lipaev, A. M., Khrapak, S. A., Molotkov, V. I., Morfill, G. E., Fortov, V. E., Ivlev, A. V., Thomas, H. M., Khrapak, A. G., Naumkin, V. N., Ivanov, A. I., Tretschev, S. E., and Padalka, G. I. (2007). Void closure in complex plasmas under microgravity conditions. *Phys. Rev. Lett.*, **98**, 265006/1–4.

Liu, B., Goree, J., Nosenko, V. and Boufendi, L. (2003). Radiation pressure and gas drag forces on a melamine-formaidehyde microsphere in a dusty plasma. *Phys. Plasmas*, **10**, 9–20.

Maiorov, S. A. (2005). Influence of the trapped ions on the screening effect and frictional force in a dusty plasma. *Plasma Phys. Rep.*, **31**, 690–699.

Maiorov, S. A. (2006). Effect of ion collisions on the parameters of dust-plasmas structures. *Plasma Phys. Rep.*, **32**, 737–749.

Maiorov, S. A., Vladimirov, S.V., and Cramer, N. F. (2001). Plasma kinetics around a dust grain in an ion flow. *Phys. Rev. E*, **63**, 017401/1–4.

Massey, H. S. W., Burhop, E. H. S., and Gilbody, H. B. (1969). *Electronic and ionic impact phenomena*. Oxford University Press, Oxford.

Matsoukas, T. and Russell, M. (1995). Particle charging in low-pressure plasmas. *J. Appl. Phys.*, **77**, 4285–4292.

Matsoukas, T. and Russell, M. (1997). Fokker-Planck description of particle charging in ionized gases. *Phys. Rev. E*, **55**, 991–994.

Matsoukas, T., Russell, M., and Smith, M. (1996). Stochastic charge fluctuations in dusty plasmas. *J. Vac. Sci. Technol.*, **14**, 624–630.

Melandsø, F. and Goree, J. (1995). Polarized supersonic plasma flow simulation for charged bodies such as dust particles and spacecraft. *Phys. Rev. E*, **52**, 5312–5326.

Melzer, A., Trottenberg, T., and Piel, A. (1994). Experimental determination of the charge on dust particles forming Coulomb lattices. *Phys. Lett. A*, **191**, 301–307.

Melzer, A., Homann, A., and Piel, A. (1996a). Experimental investigation of the melting transition of the plasma crystal. *Phys. Rev. E*, **53**, 2757–2766.

Melzer, A., Schweigert, V. A., Schweigert, I. V., Homann A., Peters, S., and Piel, A. (1996b). Structure and stability of the plasma crystal. *Phys. Rev. E*, **54**, R46–R49.

Melzer, A., Schweigert, V. A., and Piel, A. (1999). Transition from attractive to repulsive forces between dust molecules in a plasma sheath. *Phys. Rev. Lett.*, **83**, 3194–3197.

Mendis, D. A. (2002). Progress in the study of dusty plasmas. *Plasma Sources Sci. Technol.*, **11**, A219–A228.

Montgomery, D., Joyce, G., and Sugihara, R. (1968). Inverse third power law for the shielding of test particles. *Plasma Phys.*, **10**, 681–685.

Morfill, G. E., Grün, E., and Johnson, T. V. (1980). Dust in Jupiter's magnetosphere – physical processes. *Planet. Space Sci.*, **28**, 1087–1100.

Morfill, G., Ivlev, A. V., and Jokipii, J. R. (1999a) Charge fluctuation instability of the dust lattice wave. *Phys. Rev. Lett.*, **83**, 971–974.

Morfill, G. E., Thomas, H. M., Konopka, U., Rothermel, H., Zuzic, M., Ivlev, A., and Goree, J. (1999b). Condensed plasmas under microgravity. *Phys. Rev. Lett.*, **83**, 1598–1602.

Morfill, G. E., Annaratone, B. M., Bryant, P., Ivlev, A. V., Thomas, H. M., Zuzic, M., and Fortov, V. E. (2002). A review of liquid and crystalline plasmas – New physical states of matter? *Plasma Phys. Control. Fusion*, **44**, B263–B277.

Morfill, G. E., Khrapak, S. A., Ivlev, A. V., Klumov, B. A., Rubin-Zuzic, M., and Thomas, H. M. (2004). From fluid flows to crystallization: New results from complex plasmas. *Phys. Scripta*, **T107**, 59–64.

Nambu, M., Vladimirov, S. V., and Shukla, P. K. (1995). Attractive forces between charged particulates in plasmas. *Phys. Lett. A*, **203**, 40–42.

Nefedov, A.P., Petrov, O. F., and Khrapak, S.A. (1998). Potential of electrostatic interaction in a thermal dusty plasma. *Plasma Phys. Reports*, **24**, 1037–1040.

Nefedov, A. P., Morfill, G. E., Fortov, V. E., Thomas, H. M., Rothermel, H., Hagl, T., Ivlev, A. V., Zuzic, M., Klumov, B. A., Lipaev, A. M., Molotkov, V. I., Petrov, O. F., Gidzenko, Y. P., Krikalev, S. K., Shepherd, W., Ivanov, A. I., Roth, M., Binnenbruck, H., Goree, J. A., and Semenov, Y. P. (2003). PKE–Nefedov: Plasma crystal experiments on the International Space Station. *New J. Phys.*, **5**, 33/1–10.

Nitter, T. (1996). Levitation of dust in rf and dc glow discharges. *Plasma Sources Sci. Technol.*, **5**, 93–99.

Nosenko V., Fisher, R., Merlino, R., Khrapak, S., Morfill, G., and Avinash, K. (2007). Measurement of the ion drag force in a collisionless plasma with strong ion-grain coupling. *Phys. Plasmas*, **14**, 103702/ 1–7.

Nunomura, S., Samsonov, D., and Goree, J. (2000) Transverse waves in a two-dimensional screened-Coulomb crystal (dusty plasma). *Phys. Rev. Lett.*, **84**, 5141–5144.

Nunomura, S., Goree, J., Hu, S., Wang, X., and Bhattacharjee, A. (2002). Dispersion relations of longitudinal and transverse waves in two-dimensional screened Coulomb crystals. *Phys. Rev. E*, **65**, 066402/1–11.

Pal', A. F., Starostin, A. N., and Filippov, A.V. (2001). Charging of dust grains in a nuclear-induced plasma at high pressures. *Plasma Phys. Rep.*, **27**, 143–152.

Pal', A.F., Sivokhin, D. V., Starostin, A. N., Filippov, A. V., and Fortov, V. E. (2002). Potential of a dust grain in a nitrogen plasma with a condensed disperse phase at room and cryogenic temperatures. *Plasma Phys. Rep.*, **28**, 28–39.

Patacchini, L. and Hutchinson, I. H. (2008). Fully self-consistent ion-drag-force calculations for dust is collisional plasmas with an external electric field. *Phys. Rev. Lett.*, **101**, 025001/1–4.

Peters, S., Homann, A., Melzer, A., and Piel, A. (1996). Measurement of dust particle shielding in a plasma from oscillations of a linear chain. *Phys. Lett. A*, **223**, 389–393.

Piel, A. and Melzer, A. (2002). Dynamical processes in complex plasmas. *Plasma Phys. Control. Fusion*, **44**, R1–R26.

Piel, A., Homann, A., Klindworth, M., Melzer, A., Zafiu, C., Nosenko, V., and Goree, J. (2003). Waves and oscillations in plasma crystals. *J. Phys. B: At. Mol. Opt. Phys.*, **36**, 533–543.

Quinn, R. A. and Goree, J. (2000). Single-particle Langevin model of particle temperature in dusty plasmas. *Phys. Rev. E*, **61**, 3033–3041.

Raizer, Yu. P. (1991). *Gas discharge physics*. Springer-Verlag, Berlin.

Rao, M. V. V. S., VanBrunt, R. J., and Olthoff, J. K. (1996). Resonant charge exchange and the transport of ions at high electric-field to gas-density ratios (E/N) in argon, neon, and helium. *Phys. Rev. E*, **54**, 5641–5656.

Ratynskaia, S., Khrapak, S., Zobnin, A., Thoma, M. H., Kretschmer, M., Usachev, A., Yaroshenko, V., Quinn, R. A., Morfill, G. E., Petrov, O., and Fortov, V. (2004a). Experimental determination of dust-particle charge in a discharge plasma at elevated pressures. *Phys. Rev. Lett.*, **93**, 085001/1–4.

Ratynskaia, S., Kretschmer, M., Khrapak, S., Quinn, R. A., Thoma, M. H., Morfill, G. E., Zobnin, A., Usachev, A., Petrov, O., and Fortov, V. (2004b). Dust mode in collisionally dominated complex plasmas with particle drift. *IEEE Trans. Plasma Sci.*, **32**, 613–616.

Ratynskaia, S., de Angelis, U., Khrapak, S., Klumov, B., and Morfill, G. E. (2006). Electrostatic interaction between dust particles in weakly ionized complex plasmas. *Phys. Plasmas*, **13**, 104508/ 1–4.

Rosenberg, M. and Mendis, D. A. (1995). UV-induced Coulomb crystallization in a dusty gas. *IEEE Trans. Plasma Sci.*, **23**, 177–179.

Rosenberg, M., Mendis, D. A., and Sheehan, D. P. (1996). UV-induced coulomb crystallization of dust grains in high-pressure gas. *IEEE Trans. Plasma Sci.*, **24**, 1422–1430.

Rosenberg, M., Mendis, D. A., and Sheehan, D. P. (1999). Positively charged dust crystals induced by radiative heating. *IEEE Trans. Plasma Sci.*, **27**, 239–242.

Rothermel, H., Hagl, T., Morfill, G. E., Thoma, M. H., and Thomas, H. M. (2002). Gravity compensation in complex plasmas by application of a temperature gradient. *Phys. Rev. Lett.*, **89**, 175001/1–4.

Rovagnati, B., Davoudabadi, M., Lapenta, G., and Mashayek, F. (2007). Effect of collisions on dust particle charging via particle-in-cell Monte-Carlo collision. *J. Appl. Phys.*, **102**, 073302/1–9.

Samarian, A. A. and Vladimirov, S. V. (2002). Comment on "Dependence of the dust-particle charge on its size in a glow-discharge plasma". *Phys. Rev. Lett.*, **89**, 229501/1.

Samarian, A. A. and Vladimirov, S. V. (2003). Charge of a macroscopic particle in a plasma sheath. *Phys. Rev. E*, **67**, 066404/1–5.

Schweigert, V. A. (2001). Dielectric permittivity of a plasma in an external electric field. *Plasma Phys. Reports*, **27**, 997–999.

Schweigert, I. V., Alexandrov, A. L., and Peeters, F. M. (2004). Negative ion-drag force in a plasma of gas discharge. *IEEE Trans. on Plasma Sci.*, **32**, 623–626.

Sickafoose, A. A., Colwell, J. E., Horanyi, M., and Robertson, S. (2000). Photoelectric charging of dust particles in vacuum. *Phys. Rev. Lett.*, **84**, 6034–6037.

Smirnov, B. M. (1982). A rigid sphere model in plasma and gas physics. *Usp. Fiz. Nauk*, **138**, 517–533.

Sodha, M. S. and Guha, S. (1971). Physics of colloidal plasmas. *Adv. Plasma Phys.*, **4**, 219–309.

Soo, S. L. (1990). *Multiphase fluid dynamics*. Gower Technical, Brookfield.

Spitzer, Jr., L. (1962). *Physics of fully ionized gases*. Wiley, New York.

Su, C. H. and Lam, S. H. (1963). Continuum theory of spherical electrostatic probes. *Phys. Fluids*, **6**, 1479–1491.

Swinkels, G. H. P. M., Kersten, H., Deutsch, H., and Kroesen, G. M. W. (2000). Microcalorimetry of dust particles in a radio-frequency plasma. *J. Appl. Phys.*, **88**, 1747–1755.

Takahashi, K., Oishi, T., Shimonai, K. I., Hayashi, Y., and Nishino, S. (1998). Analyses of attractive forces between particles in Coulomb crystal of dusty plasmas by optical manipulations. *Phys. Rev. E*, **58**, 7805–7811.

Talbot, L., Cheng, R. K., Schefer, R. W., and Willis, D. R. (1980). Thermophoresis of particles in a heated boundary layer. *J. Fluid Mech.*, **101**, 737–758.

Thomas, H., Morfill, G. E., Demmel, V., Goree, J., Feuerbacher, B., and Möhlmann, D. (1994). Plasma crystal: Coulomb crystallization in a dusty plasma. *Phys. Rev. Lett.*, **73**, 652–655.

Thomas, H. M. and Morfill, G. E. (1996). Melting dynamics of a plasma crystal. *Nature* (London), **379**, 806–809.

Thompson, W. B. and Hubbard, J. (1960). Long-range forces and the diffusion coefficients of a plasma. *Rev. Mod. Phys.*, **32**, 714–718.

Tomme, E. B., Law, D. A., Annaratone, B. M., and Allen, J. E. (2000). Parabolic plasma sheath potentials and their implications for the charge on levitated dust particles. *Phys. Rev. Lett.*, **85**, 2518–2521.

Trottenberg, T., Melzer, A., and Piel, A. (1995). Measurement of the electric charge on particulates forming Coulomb crystals in the sheath of a radiofrequency plasma. *Plasma Sources Sci. Technol.*, **4**, 450–458.

Tsytovich, V. N. (1997). Dust plasma crystals, drops, and clouds. *Phys. Usp.*, **40**, 53–94.

Tsytovich, V. N., Khodataev, Ya. K., and Bingham, R. (1996). Formation of a dust molecule in plasmas as a first step to super-chemistry. *Comments Plasma Phys. Control. Fusion*, **17**, 249–265.

Tsytovich, V. N., Khodataev, Ya. K., Morfill, G. E., Bingham, R. and Winter, D. J. (1998). Radiative dust cooling and dust agglomeration in plasmas. *Comments Plasma Phys. Control. Fusion*, **18**, 281–291.

Uglov, A. A. and Gnedovets, A. G. (1991). Effect of particle charging on momentum and heat-transfer from rarefied plasma-flow. *Plasma Chem. Plasma Process.*, **11**, 251–267.

Uhlenbeck, G. E. and Ornstein, L. S. (1930). On the theory of the Brownian motion. *Phys. Rev.*, **36**, 823–841.

Usachev, A., Zobnin, A., Petrov, O., Fortov, V., Thoma, M. H., Kretschmer, M., Ratynskaia, S., Quinn, R. A., Höfner, H., and Morfill, G. E. (2004). The project "Plasmakristall-4" – A dusty plasma experiment in a combined dc/rf(i) discharge plasma under microgravity conditions. *Czech. J. Phys.*, **54**, C639–C647.

Vaulina, O. S., Khrapak, S. A., Nefedov, A. P., and Petrov, O. F. (1999a). Charge fluctuations induced heating of dust particles in a plasma. *Phys. Rev. E*, **60**, 5959–5964.

Vaulina, O. S., Nefedov, A. P., Petrov, O. F. and Khrapak, S. A. (1999b). Role of stochastic fluctuations in the charge on macroscopic particles in dusty plasmas. *JETP*, **88**, 1130–1136.

Vaulina, O. S., Khrapak, S. A., Samarian, A. A., and Petrov, O. F. (2000). Effect of stochastic grain charge fluctuations on the kinetic energy of the particles in dusty plasma. *Phys. Scripta*, **T84**, 229–231.

Vaulina, O. S., Repin A. Yu., and Petrov, O. F. (2006). Empirical approximation for the ion current to the surface of a dust grain in a weakly ionized gas-discharge plasma. *Plasma Phys. Rep.*, **32**, 485–488.

Vladimirov, S. V. and Ishihara, O. (1996). On plasma crystal formation. *Phys. Plasmas* **3**, 444–446.

Vladimirov, S. V. and Nambu, M. (1995). Attraction of charged particulates in plasmas with finite flows. *Phys. Rev. E* **52**, R2172–R2174.

Vladimirov, S. V., Maiorov, S. A., and Ishihara, O. (2003). Molecular dynamics simulation of plasma flow around two stationary dust grains. *Phys. Plasmas*, **10**, 3867–3873.

Vladimirov, S. V., Khrapak, S. A., Chaudhuri, M., and Morfill, G. E. (2008). Superfluid-like motion of an absorbing body in a collisional plasma. *Phys. Rev. Lett.*, **100**, 055002/1–4.

Walch, B., Horanyi, M. and Robertson, S. (1995). Charging of dust grains in plasma with energetic electrons. *Phys. Rev. Lett.*, **75**, 838–840.

Wang, C. L., Joyce, G., and Nicholson, D. R. (1981). Debye shielding of a moving test charge in plasma. *J. Plasma Phys.*, **25**, 225–231.

Whipple, E. C. (1981). Potentials of surfaces in space. *Rep. Prog. Phys.*, **44**, 1197–1250.

Winske, D. (2001). Nonlinear wake potential in a dusty plasma. *IEEE Trans. Plasma Sci.*, **29**, 191–197.

Winter, J. (2000). Dust: A new challenge in nuclear fusion research? *Phys. Plasmas*, **7**, 3862–3866.

Xie, B., He, K., and Huang, Z. (1999). Attractive potential in weak ion flow coupling with dust-acoustic waves. *Phys. Lett. A*, **253**, 83–87.

Yakubov, I. T. and Khrapak, A. G. (1989). Thermophysical and electrophysical properties of low temperature plasma with condensed disperse phase. *Sov. Tech. Rev. B. Therm. Phys.*, **2**, 269–337.

Yaroshenko, V. V., Annaratone, B. M., Khrapak, S. A., Thomas, H. M., Morfill, G. E., Fortov, V. E., Lipaev, A. M., Molotkov, V. I., Petrov, O. F., Ivanov, A. I., and Turin, M. V. (2004). Electrostatic modes in collisional complex plasmas under microgravity conditions. *Phys. Rev. E*, **69**, 066401/1–7.

Yaroshenko V., Ratynskaia, S., Khrapak, S., Thoma, M. H., Kretschmer, M., Höfner, H., Morfill, G. E., Zobnin, A., Usachev, D., Petrov, O., and Fortov, V. (2005). Determination of the ion-drag force in a complex plasma. *Phys. Plasmas*, **12**, 093503/1–7.

Zafiu, C., Melzer, A., and Piel, A. (2003a). Measurement of the ion drag force on falling dust particles and its relation to the void formation in complex (dusty) plasmas. *Phys. Plasmas*, **10**, 1278–1282.

Zafiu, C., Melzer, A., and Piel, A. (2003b). Response to "Comment on 'Measurement of the ion drag force on falling dust particles and its relation to the void formation in complex (dusty) plasmas'" [*Phys. Plasmas*, **10**, 4579 (2003)]. *Phys. Plasmas*, **10**, 4582–4583.

Zakrzewski, Z. and Kopiczynski, T. (1974). Effect of collisions on positive ion collection by a cylindrical Langmuir probe. *Plasma Phys.* **16**, 1195–1198.

Zeuner, M. and Meichsner, J. (1995). Ion kinetics in collisional rf-glow discharge sheaths. *Vacuum* **46**, 151–157.

Zobnin, A. V., Nefedov, A. P., Sinelshchikov, V. A., and Fortov, V. E. (2000). On the charge of dust particles in a low-pressure gas discharge plasma. *JETP* **91**, 483–487.

Zobnin, A. V., Usachev, A. D., Petrov, O. F., and Fortov, V. E. (2008). Ion current on a small spherical attractive probe in a weakly ionized plasma with ion–neutral collisions (kinetic approach). *Phys. Plasmas*, **15**, 043705/1–6.

Zuzic, M., Thomas, H. M., and Morfill, G. E. (1996). Wave propagation and damping in plasma crystals. *J. Vac. Sci. Technol.*, **A14**, 496–500.

3

Particle dynamics

Alexey V. Ivlev

3.1 Vertical oscillations in an rf sheath

In most of the ground-based experiments, negatively charged dust particles can only levitate in the regions of sufficiently strong electric fields, where the electric force and other forces exerted in a plasma (e.g., ion drag) compensate for gravity (unless the particles are too heavy). This occurs, for example, in the pre-sheath and sheath regions of an rf discharge, where the electric field averaged over the oscillation period is directed along gravity force (due to the large mass, neither the dust particles nor the ions respond to the rf field at frequency 13.56 MHz). This is also true for striations in a dc discharge. The electric field E in these regions rapidly increases downwards. The particle charge Q varies with height, both due to the ion acceleration in the electric field (see Figure 2.5) and a decrease of the ratio $n_e/n_i < 1$ with E. Usually, the (negative) charge first somewhat decreases and attains a minimum, then it starts increasing and eventually can even reach positive values. Examples of numerical calculations of the dependence of the particle surface potential on the distance from the electrode in collisionless and collisional sheaths of rf and dc discharges were given by Nitter (1996) and Ikkurthi *et al.* (2008) for different sets of plasma parameters.

Let us first consider elementary dynamics of a single particle. If the vertical coordinate (height) $x = 0$ is assigned to the equilibrium particle position, then for small displacements around the equilibrium, the net force can be expanded into series,

$$F(x)/m_d = -\Omega_{\rm v}^2 x + \alpha_1 x^2 + \alpha_2 x^3 + \dots \tag{3.1}$$

where $\Omega_{\rm v}$ is the resonance frequency of vertical oscillations and coefficients α_i characterize nonlinearity. The major contribution to Equation (3.1) is often due to the electrostatic force $F_{\rm el} = QE$, and then the resonance frequency is determined by $m_d\Omega_{\rm v}^2 = -d(QE)/dx|_{x=0}$. It is well known (Ivlev *et al.* 2000b; Tomme et al. 2000) that at sufficiently high pressures (e.g., above $\simeq 20$ Pa for argon) the electric field in the sheath varies almost linearly. At lower pressures, however, the deviations from the linear profile can be significant. Therefore, depending on the discharge parameters and the particle mass, the nonlinearity in Equation (3.1) is determined either by the sheath field profile or by the charge variations with the height (Ivlev *et al.* 2000b; Zafiu *et al.* 2001).

Due to the relatively large mass of the dust particles, the magnitude of the resonance frequency Ω_v is rather low – it is typically in the range 1–100 Hz. Hence, it is convenient to use low-frequency excitations for determining the parameters of the force (3.1) which can then be expressed through the plasma and particle parameters. As the simplest example we refer to a harmonic excitation of particle oscillations. The oscillation amplitude $A(\omega)$ grows when ω approaches Ω_v. The amplitude reaches the maximum at $\omega = \sqrt{\Omega_v^2 - \frac{1}{2}\nu_{dn}^2}$; the width of the resonance peak is $\sim \nu_{dn}$. Hence, changing ω and measuring $A(\omega)$, one can determine Ω_v and ν_{dn}. As the excitation amplitude increases, the oscillations reveal all features peculiar to an unharmonic oscillator: hysteresis of the frequency response curve, shift of the resonance frequency, and secondary resonances (Ivlev *et al.* 2000b; Zafiu *et al.* 2001). Figure 3.1 shows evolution of the frequency response curve, $A(\omega)$, with the amplitude of the sinusoidal excitation voltage applied to the wire below the particle. Knowledge of the resonance frequency and of the nonlinear coefficients, recovered from the fitting of the measured curves with the analytical formulas, allows us to obtain the electric field and/or dust particle charge distributions in a relatively broad region across the sheath. The measurements can also be compared with the results of the numerical models, which take into account the dependencies of particle charge, electric field, and external force amplitude on the vertical coordinate, as well as the location of an excitation source with respect to the dust particle and force balance in the sheath (Wang *et al.* 2002).

3.2 Non-Hamiltonian dynamics

Complex plasmas are non-Hamiltonian systems, not only because of conventional friction of grains against the background neutral gas, but also due to specific plasma interactions that give rise to new classes of non-Hamiltonian dynamics. Under certain conditions these interactions result in spontaneous excitation of individual and collective particle motion (Zhakhovskii *et al.* 1997; Vaulina *et al.* 1999; Morfill *et al.* 1999, 2004; Ivlev *et al.* 2003). Below we consider a few interesting examples of such dynamics.

3.2.1 Role of variable charges

As we already mentioned in Section 2.1.6, the individual particle charges in complex plasmas fluctuate randomly with time around some equilibrium value which, in turn, is some function of the spatial coordinates (Fortov *et al.* 2005). This fact makes the particle dynamics nonconservative.

The simplest class of non-Hamiltonian dynamics is realized when the charge is a function of the coordinates (Zhakhovskii *et al.* 1997), $Q = Q(\mathbf{r})$: The force $Q\mathbf{E}$ acting on a particle in a potential electric field $\mathbf{E}(\mathbf{r}) = -\nabla\phi(\mathbf{r})$ *cannot* be expressed

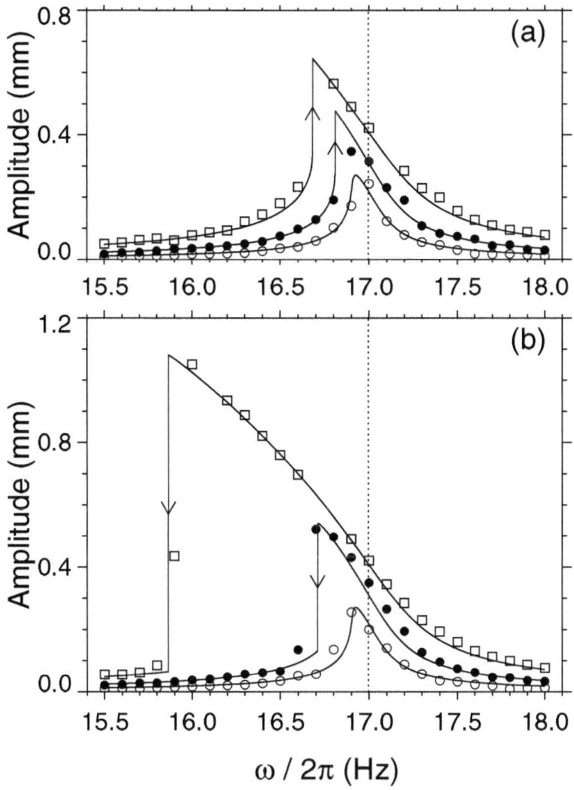

FIGURE 3.1
Amplitude of the vertical particle oscillations in an rf sheath (Ivlev *et al.* 2000b). The data for a 7.6 μm particle are shown near the primary resonance for increasing (a) and decreasing (b) frequency of excitation, ω, and for different amplitude of the sinusoidal excitation voltage: 50 mV (open circles), 100 mV (closed circles), and 200 mV (squares). Solid lines show the least-squares fit of the points to theory. The vertical dotted line indicates the position of the resonance frequency, Ω_v, obtained from the fit.

in terms of a gradient of a scalar function, because $\nabla \times (Q\nabla\phi) \equiv \nabla Q \times \nabla\phi$ is not equal to zero in the general case. The dynamics is Hamiltonian only when the charge gradient is collinear with the electric field (in this case, the force depends on a single longitudinal coordinate, and therefore, it can always be written as a derivative of some scalar function over the coordinate). Thus, particles with variable charges could gain energy from the ambient plasma.

Another example of nonconservative dynamics is due to the so-called "delayed charging" effect (Nunomura *et al.* 1999; Ivlev *et al.* 2000a; Pustylnik *et al.* 2006), which stems from the fact that the charging frequency Ω_{ch} [see Equation (2.16)] is finite. Therefore, the charge of a moving particle experiences some delay with respect to its equilibrium local value $Q(\mathbf{r})$, so that the dynamics is non-Hamiltonian also for 1D motion. Moreover, if $|Q(\mathbf{r})|$ increases along $\mathbf{E}$ (which is always the case for particles levitating in ground-based experiments), then the absolute value of the *momentary* charge is smaller than $|Q(\mathbf{r})|$ when the particle moves along $\mathbf{E}$ and larger when the particle moves back. Thus, the work done over the oscillation period is always positive – the particle acquires energy from the electric field. The oscillations grow exponentially provided the energy gain is higher than the friction dissipation, which requires (Ivlev *et al.* 2000a)

$$2\nu_{dn} \lesssim (\ell_E/\ell_Z)\Omega_{\rm v}^2/\Omega_{\rm ch},$$

where $\ell_E = |E|/|\nabla E|$ and $\ell_Z = |Z|/|\nabla Z|$ are the spatial scales of the field and charge inhomogeneity, respectively (usually, $\ell_E \ll \ell_Z$), and Z is the particle charge number, $Q = Ze$.

Now, let us consider random charge variations (fluctuations) assuming a constant (i.e., independent of $\mathbf{r}$) mean charge number Z (Vaulina *et al.* 1999; Ivlev *et al.* 2000a). In experiments, the vertical particle confinement is usually determined by the balance of electrostatic and gravity forces, $m_d g = ZeE$, and the equation of the vertical motion is determined by the charge fluctuation $Z_1(t)$,

$$\ddot{x} + \nu_{dn}\dot{x} + \Omega_{\rm v}^2[1 + Z_1(t)/Z]x = gZ_1(t)/Z. \tag{3.2}$$

Using the stochastic properties of the charge fluctuations [see Equations (2.37) and (2.38)], it can be easily shown (Vaulina *et al.* 1999) that for typical conditions $\nu_{dn} \ll \Omega_{\rm v} \ll \Omega_{\rm ch}$, the mean kinetic energy of vertical oscillations associated with the random force at the right-hand side of Equation (3.2) saturates at

$$\mathscr{E}_d \simeq \frac{\sigma_Z^4 |Z| m_d g^2}{2\nu_{dn}\Omega_{\rm ch}},$$

as it follows from the fluctuation–dissipation theorem. In accordance with Equation (2.38), the relative charge dispersion (due to charge discreteness) is $\sigma_Z^2 \sim |Z|^{-1}$. The neutral damping rate scales with gas pressure as $\nu_{dn} \propto p$, so that the mean energy decreases as $\mathscr{E}_d \propto p^{-1}$. Note also that, since $\Omega_{\rm ch} \propto a$, $\nu_{dn} \propto a^{-1}$, and $|Z| \propto a$, we have $\mathscr{E}_d \propto a^2 \propto m_d^{2/3}$; i.e., the mean energy of oscillations increases with the particle mass. For typical experimental conditions the energy can be of the order of a few eV or even higher.

In addition to this heating, the charge variations can trigger the parametric instability of the oscillations (Ivlev *et al.* 2000a), due to the random variations of the oscillation frequency in Equation (3.2). Then the mean energy grows exponentially with time – the instability condition is

$$2\nu_{dn} \lesssim \sigma_Z^2\Omega_{\rm v}^2/\Omega_{\rm ch}. \tag{3.3}$$

Note, however, that if the charge variations are due to the discreteness of plasma charges, then the magnitude of the dispersion is fairly small and the instability is only possible at pressures far below ~ 1 Pa.

For a set of particles, we have to define the mutual interactions. Let us again consider the role of spatial variations. The electrostatic potential created at $\mathbf{r}$ by a charge located at $\mathbf{r}_i$ is $\phi_i(\mathbf{r}) = Z(\mathbf{r}_i)\phi_{\text{unit}}(|\mathbf{r}-\mathbf{r}_i|)$, where $\phi_{\text{unit}}(r)$ is the (isotropic) potential of a *unit charge*. The resulting electric field is $\mathbf{E}_i(\mathbf{r}) = -Z(\mathbf{r}_i)(\partial/\partial\mathbf{r})\phi_{\text{unit}}(|\mathbf{r}-\mathbf{r}_i|)$. Hence, particles i and j interact via the force

$$\mathbf{F}_{ij} = -eZ(\mathbf{r}_i)Z(\mathbf{r}_j)(\partial/\partial\mathbf{r}_i)\phi_{\text{unit}}(|\mathbf{r}_i-\mathbf{r}_j|). \tag{3.4}$$

Note that the mutual interactions are reciprocal, $\mathbf{F}_{ij} = -\mathbf{F}_{ji}$, so that the total momentum of the system is conserved.

Principal features of non-Hamiltonian dynamics with interactions (3.4) can be understood by considering a 1D system of two charged grains which can move along the x-axis (Zhdanov *et al.* 2005). The repulsing particles have to be confined externally. Generally, the confinement is electrostatic and, hence, charge-dependent. However, since electrostatic forces are potential in the 1D case, we can always write the confinement force on a particle as $F_{\text{conf}} = -dU_{\text{conf}}/dx$. We now introduce a 2D space $\mathbf{x} = (x_1, x_2)$, with $x_{1,2}$ being the particle coordinates, and define the external confinement potential as $U_{\text{ext}}(\mathbf{x}) \equiv U_{\text{conf}}(x_1) + U_{\text{conf}}(x_2)$. Then the equations of two particle motion can be written in the following vector form:

$$m_d(\ddot{\mathbf{x}} + \nu_{dn}\dot{\mathbf{x}}) = -\partial U_{\text{ext}}/\partial\mathbf{x} - eZ_1Z_2(\partial\phi_{\text{unit}}/\partial\mathbf{x}), \tag{3.5}$$

where $\phi_{\text{unit}} = \phi_{\text{unit}}(|x_2 - x_1|)$ and $Z_{1,2} \equiv Z(x_{1,2})$. In addition to the confinement and interaction forces, we introduced a friction force with the damping rate ν_{dn}. One can see from Equation (3.5) that the 1D dynamics of two particles is mathematically identical to 2D dynamics of a single particle. The dynamics is *non-conservative*, because work W_{loop} done (due to mutual interactions) over a closed path (loop) ℓ in plane $\mathbf{x}$ is not equal to zero. Using Stokes' theorem, the work can be expressed via the integral over the surface S_ℓ bounded by the path:

$$W_{\text{loop}} = e\int_{S_\ell}(Z_1Z_2' + Z_1'Z_2)\phi_{\text{unit}}'\,dx_1dx_2,$$

where the prime denotes the derivative with respect to the argument.

The sign of W_{loop} is determined by the direction of motion along ℓ; i.e., the charge variations can serve either as a sink ($W_{\text{loop}} < 0$) or a source ($W_{\text{loop}} > 0$) of the energy. In the latter case the motion of interacting particles can be unstable. In dissipative systems one can expect that at the nonlinear stage motion converges asymptotically to a limit cycle, with the balance between the energy gain and frictional loss, $W_{\text{loop}} - 2\nu_{dn}\tau\mathcal{E}_d = 0$, where τ is the oscillation period and $\mathcal{E}_d$ is the mean kinetic energy averaged over τ. We see that the magnitude of the work done over path ℓ is determined by the area S_ℓ. This implies that when $\nu_{dn} \to 0$, the contour ℓ of periodic motion (if such motion is possible at all) should degenerate into some line, so that S_ℓ tends to zero as well.

The non-Hamiltonian dynamics of many particles with variable charges was investigated using Molecular Dynamics (MD) simulations by Zhdanov *et al.* (2005). The

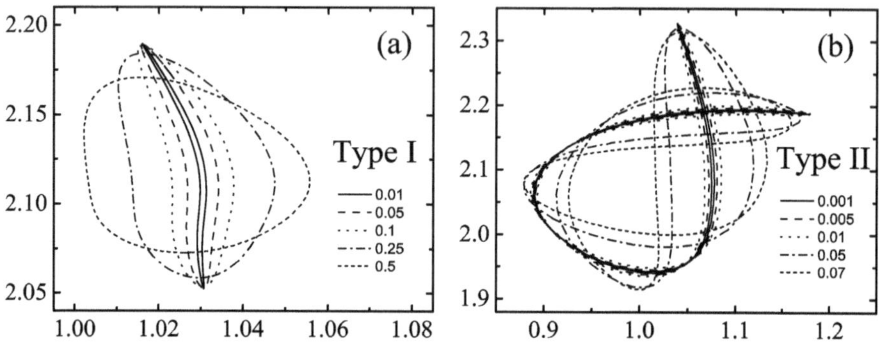

FIGURE 3.2
"Mutual" phase portraits of self-excited particle oscillations triggered by spatially varying charges (Zhdanov *et al.* 2005). Two particles perform a 1D motion, their coordinates (x_1, x_2) form periodic trajectories that are shown for several values of damping rate ν_{dn}; the highest values are chosen in the vicinity of the self-excitation cut-off. By varying initial momentums of the particles, one can obtain attractors of type I (a) or II (b). Coordinates are measured in units of the screening length λ and the damping rate is normalized by the dust-lattice frequency scale $\Omega_{\mathrm{DL}} = \sqrt{Q^2/m_d\lambda^3}$.

simplest case is the "2+2" particle system, where two outer particles are fixed and two inner particles movable. Particles interact via the Yukawa potential $\phi_{\mathrm{unit}}(x) = e\exp(-x/\lambda)/x$ with the screening length λ, and the spatial dependence of the charge is given by a step-wise function $\propto \tanh[(x - x_{\mathrm{jump}})/\sigma_{\mathrm{jump}}]$ varying along the x-axis by the magnitude ΔZ, where x_{jump} is the position of the charge "jump" and $\sigma_{\mathrm{jump}} \ll x_{\mathrm{jump}}$ is the width of the "jump". The initial interparticle distance Δ is determined from the equilibrium condition. It was found that spontaneous oscillations set in when the magnitude of the charge gradient exceeds a certain threshold: When the initial (equilibrium) coordinates of the particles are relatively far from x_{jump} (a few σ_{jump}), which implies relatively weak charge variations, the oscillations decay and the mean kinetic energy falls off as $\propto e^{-2\nu_{dn}t}$. When the initial position of one of the particles is sufficiently close to x_{jump} (about σ_{jump} or less), so that the charge variations are strong enough, the kinetic energy does not decay. On the contrary, $\mathcal{E}_d$ eventually saturates at a constant level and the oscillations become periodic, converging asymptotically to the attractors shown in Figure 3.2. Varying initial conditions it was found that two different types of attractors are possible, either type I (Figure 3.2a) or type II (Figure 3.2b), with no regular correspondence to the initial conditions (e.g., initial particle momenta). One can see that the oscillation contours become narrower and have a tendency to degenerate into a single line as the damping rate ν_{dn} decreases. On the other hand, there exists a critical friction beyond which self-sustaining oscillations are no longer possible. The "width" and "length" of the oscillation contours at the critical ν_{dn} are roughly the same, indicating that the area S_ℓ of the contour is about to achieve its maximum.

3.2.2 Role of plasma wakes

Another class of non-Hamiltonian dynamics occurs when charged grains (now we assume for simplicity that the charges are constant) are embedded in a flowing plasma, with ions moving relative to grains due to, e.g., ambipolar diffusion. Then the screening cloud around a charged grain is no longer spherically symmetric, which gives rise to higher (dipole, quadruple, etc.) moments in the mutual interaction. As we discussed in Section 2.2.2, the screening cloud in this case is usually referred to as a "plasma wake", instead of the "Debye sphere" in the isotropic case. We should point out the curious fact that in some cases the mathematical description of wakes is identical to the equations describing, e.g., hydrodynamic interactions of bubbles in conventional fluid flows (Beatus *et al.* 2006).

In order to understand the dynamics of such systems, one should note the following: Complete ensembles of elementary charges in complex plasmas can be conveniently subdivided into two distinct categories – a subsystem of "bound" charges at the grain surface and a subsystem of "free" plasma charges in the surrounding wakes. Plasma wakes play the role of a "third body" in the mutual grain–grain interaction and, hence, make the pair interaction *nonreciprocal* (Melzer *et al.* 1999): The force exerted by a wake of grain 1 on grain 2 is generally *not equal* to the force of wake 2 acting on grain 1. Thus, in contrast to the case of variable charges, the total particle momentum is no longer conserved. The center-of-mass motion is governed by equation (Kompaneets 2007)

$$m_d(\ddot{\mathbf{r}}_c + \nu_{dn}\dot{\mathbf{r}}_c) = \tfrac{1}{2}Q(\partial/\partial\mathbf{r}_r)[\phi(-\mathbf{r}_r) - \phi(\mathbf{r}_r)] + \mathbf{F}_{ext}. \tag{3.6}$$

Here $\mathbf{r}_c = \frac{1}{2}(\mathbf{r}_1 + \mathbf{r}_2)$ and $\mathbf{r}_r = \mathbf{r}_2 - \mathbf{r}_1$ are the center-of-mass and relative coordinates, respectively, and $\phi(\mathbf{r}_r)$ $[\neq \phi(-\mathbf{r}_r)]$ is an anisotropic wake potential [see, e.g., Equation (2.50)]. In addition to the mutual interactions, in Equation (3.6) we introduced a force $\mathbf{F}_{ext}$ which describes the interaction of grains with a (constant) external electric field (assuming constant charges, this field may be presented as a gradient of some scalar function). Equation (3.6) shows that one can easily construct such a loop ℓ for the motion of the center of mass that the work W_{loop} done by the nonreciprocal interaction over the loop is not equal to zero:

$$W_{loop} \propto \int_{S_\ell} (\partial/\partial\mathbf{r}_c) \times (\partial/\partial\mathbf{r}_r)[\phi(-\mathbf{r}_r) - \phi(\mathbf{r}_r)]\, dS_\ell.$$

This occurs when $\mathbf{r}_c$ and $\mathbf{r}_r$ are *correlated* (e.g., due to resonances), so that $(\partial/\partial\mathbf{r}_c) \times (\partial/\partial\mathbf{r}_r) \neq 0$ and the dynamics of interacting particles is non-conservative.

Such non-conservative dynamics results in spontaneous heating of complex plasmas. As an example, let us consider an experiment on melting of a crystalline monolayer (Ivlev *et al.* 2003). The microparticles were levitated in the strong vertical electric field of the sheath above a planar rf electrode and confined horizontally by a weak radial field. When the number of particles in the monolayer exceeded a certain

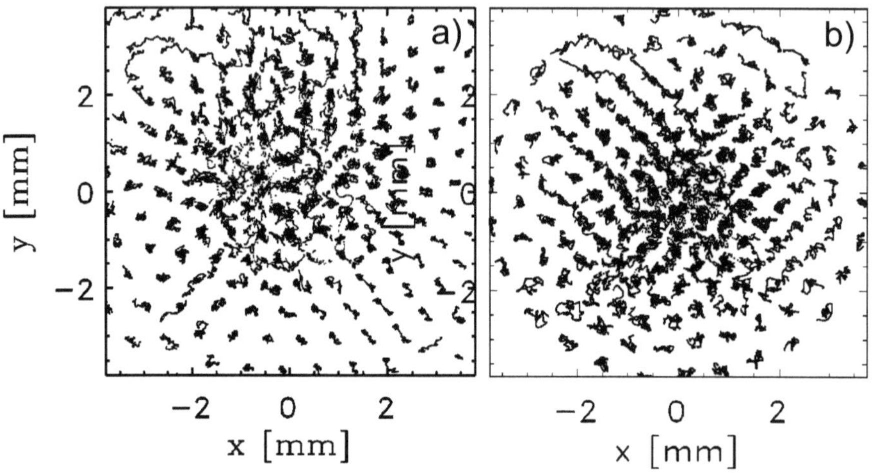

FIGURE 3.3
Melting of 2D crystal (Ivlev *et al.* 2003). (a) Top view of the particle monolayer in the experiment with 8.9 μm particles in argon at pressure of 2.8 Pa. The particles are in crystalline state, until the density exceeds a certain threshold and the monolayer melts. The figure shows trajectories over 3.1 s after the melting starts. (b) Top view of the particle monolayer in the MD simulation with parameters similar to the experiment. In both cases, the average kinetic energy of particles in the center saturated at $\sim$ 30 eV.

threshold (correspondingly, the interparticle distance decreased, with the smallest distance in the center due to the radial confinement) the monolayer started melting, from the center to the periphery. The horizontal particle trajectories during the melting process are shown in Figure 3.3a. Simultaneously, oscillations in a vertical direction were triggered. It was possible to stop the melting by increasing the pressure – the system always returned to a stable crystalline monolayer.

The experiment was simulated using an MD code which contains a first-principles representation of the short-range shielded Coulomb forces and the wake forces due to the streaming ions. In the numerical simulations the basic findings of the experiment could be reproduced: When the number of particles in the simulation exceeded a certain threshold the monolayer melted. In such unstable cases the interparticle distance in the center of the monolayer was close to the experimentally measured values. The melting developed in a manner similar to the experiment (see Figure 3.3b). Also, with increasing pressure (neutral friction) the system became stable. Finally, when the particle interaction was reduced to a spherically symmetric potential (without wakes), the system was always stable, indicating that the wakes indeed play the decisive role in the melting.

Quantitative conditions for the melting onset can be derived from the following simple model proposed by Ivlev *et al.* (2003): Negatively charged particles of charge Q and mass m_d are separated horizontally by a distance Δ and interact via a (spherically symmetric) screened potential. The excess positive charge of the wake, $\tilde{Q}$, is approximated by a pointlike charge located at distance $\tilde{\Delta}$ downstream from the particle. Vertically, the particles are confined in a potential well with their eigenfrequency $\Omega_{\rm v}$. It is known that in a crystalline monolayer, in addition to conventional in-plane acoustic modes (dust lattice waves, see Section 4.4.2), there exists also an out-of-plane optical mode associated with vertical particle oscillations (Samsonov *et al.* 2005). The in-plane and out-of-plane modes are coupled due to particle–wake interaction. The coupling is weak when the corresponding branches $\omega(k)$ are far away from each other, but becomes strong at the resonance, when they intersect. This happens when the number density of particles exceeds a threshold. Then the branches are modified and form a hybrid branch in the vicinity around the intersection point, where the resonance coupling can drive an instability with the growth rate

$$\mathrm{Im}\ \omega \sim |\tilde{Q}\tilde{\Delta}/Q\Delta|(\omega_0^2/\Omega_{\rm v}) - \nu_{dn}, \tag{3.7}$$

where $\omega_0 = \sqrt{Q^2/m_d\Delta^3}$ (see Section 4.4.2). One can see from Equation (3.7) that the coupling part is proportional to the dipole moment of the wake, $\tilde{Q}\tilde{\Delta}$, and the instability is suppressed when the damping is sufficiently high. It is remarkable that such a simple model not only recovers all qualitative features seen in the experiment, but also gives very good (within $\simeq 5$ %) quantitative agreement.

An additional mechanism contributing to the mode coupling can be due to spatial charge variations and/or variations of the screening, and this effect might significantly affect the instability (Kompaneets *et al.* 2005; Yaroshenko *et al.* 2005). Note that the non-Hamiltonian dynamics of dust grains due to nonreciprocity of the particle–wake interactions (and/or charge/screening variations) can also cause melting of 3D crystals (Melzer *et al.* 1996).

3.3 Kinetics of ensembles with variable charges

Since ensembles of particles with variable charges are generally non-Hamiltonian systems, the use of thermodynamic potentials to describe them is not really justified. An appropriate way to investigate the evolution of such systems is to use the kinetic approach (Ivlev *et al.* 2004, 2005). As long as properties of the charge variations are known, one can consider the dynamics and kinetics of the grains independently from the plasma kinetics.

Kinetic equation. In the absence of external fields, the kinetics of charged grains is governed by the mutual collisions and by the collisions with neutrals, so that the kinetic equation for the grain distribution function $f(\mathbf{p})$ is:

$$df/dt = \mathrm{St}_{dd}f + \mathrm{St}_{dn}f. \tag{3.8}$$

The grain-neutral collision integral does not depend on particle charges and can be written in usual Fokker–Planck form (equivalent to the Langevin equation).

As regards the grain–grain collisions (here we investigate dilute gaseous ensembles and hence focus on the binary interactions only), one should note a very important point (Ivlev *et al.* 2005): Generally, we cannot use the collision integral in the classical Boltzmann form, because its derivation employs the unitarity relation (Lifshitz and Pitaevskii 1981). This relation is not necessarily satisfied for ensembles with variable charges: Naturally, the grain–grain collision integral applies only for those transitions occurring (between different kinetic "states") in the *subsystem of charged grains*. Due to the exchange of energy with free plasma charges, the subsystem of grains is not conservative – the momentum exchange during a collision is affected by the charging processes. Therefore, the unitarity relation can be fulfilled only after the summation over the *complete* set of states, including those corresponding to the subsystem of the plasma charges. Thus, we have to write the collision integral in the most general form:

$$\mathrm{St}_{dd}f(\mathbf{p}) = \int \left[w(\mathbf{p}',\mathbf{p}_1';\mathbf{p},\mathbf{p}_1)f(\mathbf{p}')f(\mathbf{p}_1') - w(\mathbf{p},\mathbf{p}_1;\mathbf{p}',\mathbf{p}_1')f(\mathbf{p})f(\mathbf{p}_1)\right]d\mathbf{p}_1 d\mathbf{p}' d\mathbf{p}_1'. \tag{3.9}$$

Here, $w(\mathbf{p},\mathbf{p}_1;\mathbf{p}',\mathbf{p}_1')$ is a probability function for a pair of colliding particles with momenta $\mathbf{p}$ and $\mathbf{p}_1$ to acquire momenta $\mathbf{p}'$ and $\mathbf{p}_1'$, respectively, after the scattering. Equation (3.9) accounts for all possible transitions $(\mathbf{p}',\mathbf{p}_1') \to (\mathbf{p},\mathbf{p}_1)$ (sources) and $(\mathbf{p},\mathbf{p}_1) \to (\mathbf{p}',\mathbf{p}_1')$ (sinks), and then is averaged over $\mathbf{p}_1$. The function w can be determined by solving the mechanical problem of the binary scattering with given interaction between the particles.

The mechanics of binary grain collisions can be conveniently considered in terms of the center-of-mass and relative coordinates. (Below we consider grains of the same mass, although all results can be straightforwardly generalized for arbitrary mass ratio.) For a pair of particles with momenta $\mathbf{p}$ and $\mathbf{p}_1$, the center-of-mass and relative momenta are $\mathbf{p}_\mathrm{c} = \frac{1}{2}(\mathbf{p}+\mathbf{p}_1)$ and $\mathbf{p}_\mathrm{r} = \mathbf{p}_1 - \mathbf{p}$, respectively. In the absence of external forces, the center-of-mass momentum is conserved, and the relative momentum is changed during the collision,

$$\mathbf{p}_\mathrm{c}' = \mathbf{p}_\mathrm{c}, \quad \mathbf{p}_\mathrm{r}' = \mathbf{p}_\mathrm{r} + \mathbf{q}. \tag{3.10}$$

For constant charges, the absolute value of the relative momentum, $p_\mathrm{r} \equiv |\mathbf{p}_\mathrm{r}|$, is conserved, and only the direction changes (elastic scattering) (Landau and Lifshitz 1976). Charge variations also cause p_r variations (Ivlev *et al.* 2005). Hence, the exchange of the relative momentum can be divided into elastic and inelastic parts, $\mathbf{q} = \mathbf{q}_0 + \delta\mathbf{q}$: The elastic part keeps the magnitude of the relative momentum constant, $|\mathbf{p}_\mathrm{r} + \mathbf{q}_0| = p_\mathrm{r}$. The vector of inelastic momentum exchange, $\delta\mathbf{q}$, is parallel to $\mathbf{p}_\mathrm{r}'$, and its magnitude is $\delta q = p_\mathrm{r}' - p_\mathrm{r}$.

The kinetics of particles with variable charges has a very important *hierarchy of time scales* (Ivlev *et al.* 2005): Each interparticle collision is accompanied by (i) elastic momentum exchange $\mathbf{q}_0$, which provides the relaxation of the distribution function to the Maxwellian equilibrium (Lifshitz and Pitaevskii 1981) – while keeping the mean kinetic energy of the particles $\mathscr{E}_d$ constant, and (ii) inelastic momentum exchange $\delta\mathbf{q}$, which causes variation of $\mathscr{E}_d$. Due to the relative smallness of the charge variations, the resulting inelastic momentum exchange is small as well, $\delta q \ll q_0$. This implies that process (ii) is much slower than (i). Therefore, the velocity distribution remains close to the Maxwellian form, $f(\mathbf{p}) \simeq f_{\rm M}(\mathbf{p})$, with the temperature $T_d = \frac{2}{3}\mathscr{E}_d$.

Thus the temperature is the only parameter that determines the evolution of the ensemble. This implies that the system can be treated with fluid equations: The momentum equation [with the friction force $-\nu_{dn}\mathbf{v}$ added] remains unaffected since the charge variations conserve the net momentum. In the temperature equation, along with the friction (sink) term one has to add a source term due to charge variations. In accordance with Equation (3.8), the resulting combination of these terms is

$$\dot{T}_d = \int (p^2/3m_d)\,({\rm St}_{dd}f + {\rm St}_{dn}f)\,d\mathbf{p}. \tag{3.11}$$

For the grain–neutral collisions the integral is simply equal to $-2\nu_{dn}(T - T_n)$. For the grain–grain collisions, one can expand the integrand into a series over δq. Retaining the linear and quadratic terms and integrating in parts, we obtain (Ivlev *et al.* 2005),

$$\int p^2{\rm St}_{dd}f\,d\mathbf{p} \simeq \frac{1}{2}\int (p_{\rm r}\mathscr{A} + \mathscr{B})f_{\rm M}(\mathbf{p}_{\rm c})f_{\rm M}(\mathbf{p}_{\rm r})\,d\mathbf{p}_{\rm c}d\mathbf{p}_{\rm r}, \tag{3.12}$$

where $\mathscr{A}(\mathbf{p}_{\rm c}, \mathbf{p}_{\rm r}) = \int \delta q \tilde{w}\, d\delta q$ and $\mathscr{B}(\mathbf{p}_{\rm c}, \mathbf{p}_{\rm r}) = \frac{1}{2}\int (\delta q)^2 \tilde{w}\, d\delta q$ are the Fokker–Planck coefficients (Lifshitz and Pitaevskii 1981; van Kampen 1981). Here $\tilde{w}(\mathbf{p}_{\rm r}, \mathbf{p}_{\rm c}; \delta q) \equiv w(\mathbf{p}, \mathbf{p}_1; \mathbf{p}', \mathbf{p}'_1)$ and the momenta are related by Equation (3.10). The smallness of coefficients $\mathscr{A}$ and $\mathscr{B}$ is ensured by the smallness of the charge variations (for constant particle charges, the inelastic momentum exchange is equal to zero and, hence, $\mathscr{A} = \mathscr{B} \equiv 0$).

Heating. Irrespective of which type of charge variations plays the major role – charge inhomogeneity or fluctuations – the interparticle interaction can be distinguished in terms of the “interaction strength”: For particles interacting via a short-range screened electrostatic potential (with the screening length λ), the measure of the interaction strength is the “scattering parameter” $\beta_T^{dd} = Q^2/\lambda T_d$ (see Section 2.4). When β_T^{dd} is large enough the interaction is of the hard-spheres type. In the opposite case, when the ratio is small, the interaction is of the Coulomb type, similar to that between electrons and ions in usual plasmas. Below, these two limits are refereed to as the “low-temperature” and “high-temperature” regimes, respectively, with the transition temperature being $T_{\rm tr} = Q^2/\lambda$. Equations (3.11) and (3.12) result in the following equation for the particle temperature (Ivlev *et al.* 2005):

$$\dot{T}_d \sim \alpha T_d^{\gamma} - 2\nu_{dn}(T_d - T_n). \tag{3.13}$$

Coefficient α and exponent γ in the source term depend on the temperature regime. In the case of *inhomogeneous charges*, the exponent is $\gamma = 3/2$ for $T_d \ll T_{\rm tr}$ and $\gamma = 1/2$ for $T_d \gg T_{\rm tr}$. We see that in the low-temperature regime the temperature exhibits an explosion-like growth provided the friction rate ν_{dn} is low enough. At higher temperatures, however, the growth is always saturated (see Figure 3.4a). For

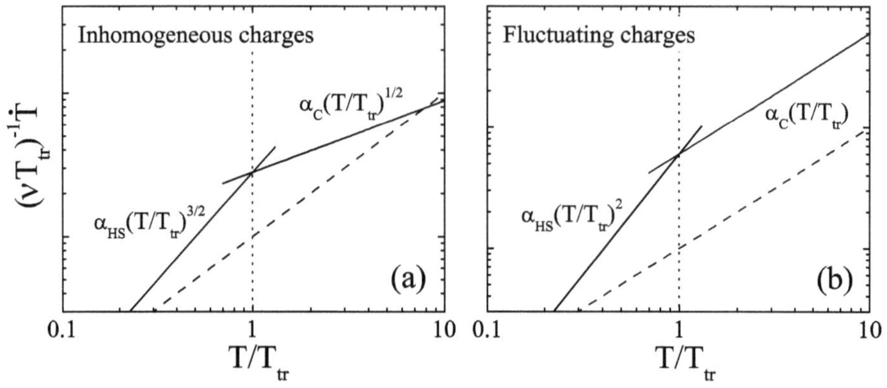

FIGURE 3.4
Self-heating in a system of particles with variable charges (Ivlev *et al.* 2005). Asymptotic behavior of the particle kinetic temperature is shown for inhomogeneous charges (a) and fluctuating charges (b). The transition temperature T_{tr} separates the "hard-spheres" (low-temperature, $T_d \ll T_{tr}$) and "Coulomb" (high-temperature, $T_d \gg T_{tr}$) collisional regimes. Solid lines correspond to the source (charge variation) term in Equation (3.13), symbols α_{HS} and α_C denote coefficients for the "hard-spheres" and "Coulomb" regimes, respectively. Dashed line correspond to the friction term.

the *fluctuating charges* we have, respectively, $\gamma = 2$ and $\gamma = 1$ for the low- and high-temperature regimes. This means that, unlike the case of the inhomogeneous charges, the temperature does not saturate but can grow exponentially at $T_d \gg T_{tr}$ (see Figure 3.4b). Numerical MD simulations fully support the theoretically predicted scalings.

References

Beatus, T., Tlusty, T., and Bar-Ziv, R. (2006). Phonons in a one-dimensional microfuildic crystal. *Nature Physics*, **2**, 743–748.

Fortov, V. E., Ivlev, A. V., Khrapak, S. A., Khrapak, A. G., and Morfill, G. E. (2005). Complex (dusty) plasmas: Current status, open issues, perspectives. *Phys. Rep.*, **421**, 1–103.

Ikkurthi, V. R., Matyash, K., Melzer, A., and Schneider, R. (2008). Computation of dust charge and potential on a static spherical dust grain immersed in rf discharges. *Phys. Plasmas*, **15**, 123704/1–9.

Ivlev, A. V., Konopka, U., and Morfill, G. (2000a). Influence of charge variations on particle oscillations in the plasma sheath. *Phys. Rev. E*, **62**, 2739–2744.

Ivlev, A. V., Suetterlin, R., Steinberg, V., Zuzic, M., and Morfill, G. (2000b). Nonlinear vertical oscillations of a particle in a sheath of rf discharge. *Phys. Rev. Lett.*, **85**, 4060–4063.

Ivlev, A. V., Konopka, U., Morfill, G., and Joyce, G. (2003). Melting of monolayer plasma crystals. *Phys. Rev. E*, **68**, 026405/1–4.

Ivlev, A. V., Zhdanov, S. K., Klumov, B. A., Tsytovich, V. N., de Angelis, U., and Morfill, G. E. (2004). Kinetics of ensembles with variable charges. *Phys. Rev E*, **70**, 066401/1–5.

Ivlev, A. V., Zhdanov, S. K., Klumov, B. A., and Morfill, G. E. (2005). Generalized kinetic theory of ensembles with variable charges. *Phys. Plasmas*, **12**, 092104/1–8.

Kompaneets, R. (2007). Complex plasmas: Interaction potentials and non-Hamiltonian dynamics. Dissertation, Ludwig-Maximilians-Universität München, http://edoc.ub.uni-muenchen.de/7380.

Kompaneets, R., Ivlev, A. V., Tsytovich, V., and Morfill, G. (2005). Dust-lattice waves: Role of charge variations and anisotropy of dust-dust interaction. *Phys. Plasmas*, **12**, 062107/1–6.

Landau, L. D. and Lifshitz, E. M. (1976). *Mechanics*. Pergamon, Oxford.

Lifshitz, E. M. and Pitaevskii, L. P. (1981). *Physical kinetics*. Pergamon, Oxford.

Melzer, A., Schweigert, V. A., Schweigert, I. V., Homann, A., Peters, S., and Piel, A. (1996). Structure and stability of the plasma crystal. *Phys. Rev. E*, **54**, R46–R49.

Melzer, A., Schweigert, V. A., and Piel, A. (1999). Transition from attractive to repulsive forces between dust molecules in a plasma sheath. *Phys. Rev. Lett.*, **83**, 3194–3197.

Morfill, G. E., Thomas, H. M., Konopka, U., Rothermel, H., Zuzic, M., Ivlev, A., and Goree, J. (1999). Condensed plasmas under microgravity. *Phys. Rev. Lett.*, **83**, 1598–1602.

Morfill, G. E., Rubin-Zuzic, M., Rothermel, H., Ivlev, A. V., Klumov, B. A., Thomas, H. M., and Konopka, U. (2004). Highly resolved fluid flows: "Liquid plasmas" at the kinetic level. *Phys. Rev. Lett.*, **92**, 175004/1–4.

Nitter, T. (1996). Levitation of dust in RF and DC glow discharges. *Plasma Sources Sci. Technol.*, **5**, 93-111.

Nunomura, S., Misawa, T., Ohno, N., and Takamura, S. (1999). Instability of dust particles in a Coulomb crystal due to delayed charging. *Phys. Rev. Lett.*, **83**, 1970–1973.

Pustylnik, M. Y., Ohno, N., Takamura, S., and Smirnov, R. (2006). Modification of the damping rate of the oscillations of a dust particle levitating in a plasma due to the delayed charging effect. *Phys. Rev. E*, **74**, 046402/1–9.

Samsonov, D., Zhdanov, S., and Morfill, G. (2005). Vertical wave packets observed in a crystallized hexagonal monolayer complex plasmas. *Phys. Rev. E*, **71**, 026410/1-7.

Tomme, E. B., Law, D.A., Annaratone, B. M., and Allen, J. E. (2000). Parabolic plasma sheath potentials and their implications for the charge on levitated dust particles. *Phys. Rev. Lett.*, **85**, 2518–2521.

van Kampen, N. G. (1981). *Stochastic processes in physics and chemistry*. Elsevier, Amsterdam.

Vaulina, O. S., Khrapak, S. A., Nefedov, A. P., and Petrov, O. F. (1999). Charge fluctuations induced heating of dust particles in a plasma. *Phys. Rev. E*, **60**, 5959–5964.

Wang, Y.-N., Hou, L.-J., and Wang, X. (2002). Self-consistent nonlinear resonance and hysteresis of a charged microparticle in a rf sheath. *Phys. Rev. Lett.*, **89**, 155001/1–4.

Yaroshenko, V. V., Ivlev, A. V., and Morfill, G. E., (2005). Coupled dust–lattice modes in complex plasmas. *Phys. Rev. E*, **71**, 046405/1–4.

Zafiu, C., Melzer, A., and Piel, A. (2001). Nonlinear resonances of particles in a dusty plasma sheath. *Phys. Rev. E*, **63**, 066403/1–8.

Zhakhovskii, V. V., Molotkov, V. I., Nefedov, A. P., Torchinskii, V. M., Khrapak, A. G., and Fortov, V. E. (1997). Anomalous heating of a system of dust particles in a gas-discharge plasma. *JETP Lett.*, **66**, 419–425.

Zhdanov, S. K., Ivlev, A. V., and Morfill, G. E. (2005). Non-Hamiltonian dynamics of grains with spatially varying charges. *Phys. Plasmas*, **12**, 072312/1–7.

4

Waves and instabilities

Alexey V. Ivlev and Sergey A. Khrapak

The charged dust grains embedded into plasmas not only change the electron–ion composition and thus affect conventional wave modes (e.g., ion–acoustic waves), but also introduce new low-frequency modes associated with the microparticle motion, alter dissipation rates, give rise to instabilities, etc. Moreover, the particle charges vary in time and space (see Section 3.1), which results in important qualitative differences between complex plasmas and usual multicomponent plasmas. Depending on the strength of interparticle interaction, complex plasmas can be in weakly coupled (gaseous-like) or strongly coupled (liquid-like) states, and form crystalline structures (see Section 5.1). This gives us a unique opportunity to investigate wave phenomena occurring in different phase states – in particular, nonlinear waves – at the kinetic level.

Complex plasmas observed in laboratory or space experiments in most cases form strongly coupled liquid or crystalline states. The uncorrelated gaseous-like phase can be seen when there is a strong energy influx into the sub-system of grains, which causes substantial increase of the grain temperature and, hence, decrease of the coupling strength. This heating can be due to the spatial and/or temporal charge variations (as discussed in Section 3.1), or induced by dust wave instabilities triggered in complex plasmas (as discussed below in Sections 4.2.2 and 4.4.4). At the same time, for ideal plasmas the theoretical analysis of the wave modes and major instabilities can be performed in the most simple form. Therefore, we first consider major wave properties of gaseous complex plasmas and then discuss features peculiar to the waves in strongly coupled plasmas.

The comprehensive kinetic approach to study waves in complex plasmas is accompanied by serious difficulties: One has to deal with the dust–dust and dust–ion collision integrals which, in contrast to usual plasmas, cannot be considered in the linear approximations for realistic experimental conditions. Also, the grain charge should be treated as a new independent variable in the kinetic equation, which makes the calculations much more complicated. There has been a series of publications where the substantial progress in the self-consistent kinetic theory of complex plasmas has been achieved by Tsytovich and de Angelis (1999, 2000, 2001, 2002, 2004). One should admit, however, that this problem is still far from being solved. On the other hand, in many cases the (relatively) simple hydrodynamic approach based on the analysis of the fluid equations allows us to catch essential physics of the processes and, hence, to understand major dynamical properties of complex plasmas. Therefore, the analysis of major wave modes and instabilities can be done with the fluid model. Of course, in some cases the applicability of the results of the hydrodynamic approach have certain limitations, especially where the damping and/or the growth rates of the modes are concerned, and then the kinetic approach has to be employed.

In this chapter we first briefly describe different experimental techniques used to excite the waves, and then discuss in detail the wave properties of complex plasmas in gaseous, liquid, and crystalline states.

4.1 Wave excitation technique

The methods used for the wave excitation in complex plasmas conditionally divide the experiments into two categories: passive and active. The former employ "natural" perturbations which are triggered spontaneously, e.g., wave instabilities (Chu *et al.* 1994; Barkan *et al.* 1995; Molotkov *et al.* 1999; Fortov *et al.* 2000, 2003; Misawa *et al.* 2001; Zobnin *et al.* 2002; see Sections 4.2.2, 4.4.4, and 4.5.2) or Mach cones (Samsonov *et al.* 1999, 2000; see Section 4.5.3), whereas the latter use methods of controlled action produced with the specially designed devices. Generally, the active experiments provide much better flexibility, but in some particular cases the passive experiments yield remarkably good results, e.g., natural spectrum of waves in plasma crystals (Nunomura *et al.* 2002b, 2005; Zhdanov *et al.* 2003a; see also Section 4.4.2).

The methods of the active (controlled) wave excitation in complex plasmas are very diverse. First of all, this can be the electrical action produced in plasmas with biased Langmuir probes (Peters *et al.* 1996; Nakamura *et al.* 1999; Nakamura and Sarma 2001; Nakamura 2002), wires (Zuzic *et al.* 1996; Pieper and Goree 1996; Samsonov *et al.* 2002, 2004a,b, 2005), or electrodes (Thompson *et al.* 1997; Merlino *et al.* 1998; Luo *et al.* 1999, 2000; Khrapak *et al.* 2003; Yaroshenko *et al.* 2004), which allows the excitation of both ion and dust waves. The electrical methods are very effective in creating waves of large amplitude and arbitrary geometry (see Section 4.5). On the other hand, the major drawback of these methods is that the electrical perturbations cannot be localized in small regions, and that this action can strongly affect global plasma parameters (Samsonov *et al.* 2001). To produce the local action on dust particles, which does not affect the discharge plasma, the laser radiation is the most widely used method (Homann *et al.* 1997, 1998; Melzer *et al.* 2000; Nunomura *et al.* 2000, 2002a, 2003; Piel *et al.* 2002; Nosenko *et al.* 2002a, 2004; Liu *et al.* 2003). This method is employed for manipulation by single particles and small particle sets, and allows us to excite the Coulomb cluster rotation (see Section 1.4.5), vertical oscillations of individual particles (see Section 3.1) and low-frequency waves in plasma crystals (see Section 4.4), and to generate Mach cones (see Section 4.5.3). This technique, however, requires very high laser power when waves of large amplitude are needed, e.g., Mach cones, solitons (Melzer *et al.* 2000; Nunomura *et al.* 2003; Nosenko *et al.* 2004), or when the perturbation should be simultaneously produced over an extended area, e.g., to excite planar waves in three-dimensional plasmas. The alternative method which provides local action on microparticles and, at the same time, the sufficient strength of the perturbations is the use of the electron beams (Vasilyak *et al.* 2002, 2003). Recently, this method was successfully employed to cause local excitation, melting, and disruption of plasma crystals, by changing the magnitude of the beam current. This technique, however, requires further development to be widely used in complex plasma experiments. Large amplitude waves can also be produced using perturbations of a neutral gas pressure (density) (Samsonov *et al.* 2003; Fortov *et al.* 2004; see Section 4.5.2).

4.2 Waves in ideal (gaseous) complex plasmas

Considering the dust species as an ideal gas, one can write the continuity and momentum equations for the dust density n_d and velocity $\mathbf{v}_d$ in the following forms:

$$\frac{\partial n_d}{\partial t} + \nabla(n_d \mathbf{v}_d) = 0, \tag{4.1}$$

$$\frac{\partial \mathbf{v}_d}{\partial t} + (\mathbf{v}_d \cdot \nabla)\mathbf{v}_d = -\frac{Q}{m_d}\nabla\phi - \frac{\nabla(n_d T_d)}{m_d n_d} - \sum_\beta \nu_{d\beta}(\mathbf{v}_d - \mathbf{v}_\beta). \tag{4.2}$$

The last term in Equation (4.2) describes the momentum transfer force ("drag") on the dust particles caused by the collisions with the "light" species – electrons, ions, and neutrals ($\beta = e, i, n$). The corresponding momentum exchange rates derived in the binary collision approximation, $\nu_{d\beta}$, are given in Section 2.4.2. As long as the flow velocities of the light species are much smaller than the thermal velocities, $\nu_{d\beta}$ does not depend on v_β. An important difference between the drag force due to collisions with neutrals ("neutral drag") and the force caused by the collisions with the charged species ("ion drag" and "electron drag") is that the latter includes both the direct collisions with the grain surface ("collection" part) and elastic scattering by the grain electrostatic potential ("orbital" part), i.e., $\nu_{d\beta} = \nu_{d\beta}^{\rm coll} + \nu_{d\beta}^{\rm orb}$. The dust viscosity usually does not play a noticeable role in the gaseous phase and, therefore, is not included in Equation (4.2).

The fluid equations for electrons and ions are ($\alpha = e, i$)

$$\frac{\partial n_\alpha}{\partial t} + \nabla(n_\alpha \mathbf{v}_\alpha) = Q_{\rm I\alpha} - Q_{\rm L\alpha} - J_\alpha n_d, \tag{4.3}$$

$$\frac{\partial \mathbf{v}_\alpha}{\partial t} + (\mathbf{v}_\alpha \cdot \nabla)\mathbf{v}_\alpha = -\frac{e_\alpha}{m_\alpha}\nabla\phi - \frac{\nabla(n_\alpha T_\alpha)}{m_\alpha n_\alpha} - \sum_\beta \nu_{\alpha\beta}^{\rm orb}(\mathbf{v}_\alpha - \mathbf{v}_\beta) - \left(\frac{Q_{\rm L\alpha}}{n_\alpha} + \nu_{\alpha d}^{\rm coll}\right)\mathbf{v}_\alpha. \tag{4.4}$$

The continuity equations include source terms, $Q_{\mathrm{I}\alpha}$, and two types of sink – "discharge" loss $Q_{\mathrm{L}\alpha}$ and the "dust" loss $J_\alpha n_d$. The source of electrons and ions which sustains the discharge is usually the volume ionization in electron–neutral collisions (Raizer 1991; Lieberman and Lichtenberg 1994), and then $Q_{\mathrm{I}e} = Q_{\mathrm{I}i} = \nu_{\mathrm{I}} n_e$, where ν_{I} is the ionization frequency. (In some cases the secondary electron emission, thermionic emission, and photoelectric emission from the surface of the grains can also provide contributions to $Q_{\mathrm{I}e}$, see Section 2.1.4). The "discharge" loss term is usually due to the diffusion towards the discharge chamber walls and can be estimated as $Q_{\mathrm{L}e} = Q_{\mathrm{L}i} \sim (D_{\mathrm{a}i}/L^2)n_i$, where $D_{\mathrm{a}i}$ is the ambipolar (ion) diffusion coefficient and L is the spatial scale of the "global" plasma inhomogeneity (i.e., distance between the rf electrodes or the radius of the dc discharge tube). The "dust" loss terms are due to electron and ion absorption on the grain surface (see Section 2.1.5) and are determined by the corresponding fluxes on a grain, J_α, described in Section 2.1.

The representation of the momentum transfer force in the form $\nu_{\alpha\beta}(\mathbf{v}_\alpha - \mathbf{v}_\beta)$ is valid as long as the mean free path of the species is shorter than the spatial scale of the perturbations (e.g., the inverse wave vector k^{-1}). Note that the change of the electron or ion momentum due to absorption by the dust grains does not depend on the grain velocity, and this is also taken into account in Equation (4.4). The reciprocal momentum transfer rates are related to each other as follows:

$$m_\alpha n_\alpha \nu_{\alpha\beta} = m_\beta n_\beta \nu_{\beta\alpha}. \tag{4.5}$$

Variability of the grain charges implies that the fluid equations for the density and momentum should be coupled to the charge transport equation which has the following form:

$$\frac{\partial Z}{\partial t} + \mathbf{v}_d \cdot \nabla Z = J_i - J_e, \tag{4.6}$$

where $Z = Q/e$ is the particle charge number. The system of equations is closed by the Poisson equation,

$$\nabla^2 \phi = -4\pi e(n_i - n_e + Z n_d). \tag{4.7}$$

One can also take into account the temperature variation of the species caused by the wave perturbations. There are two limiting cases: Isothermal variations – when the time scale of the perturbations exceeds the time scale of temperature relaxation due to the thermal conductivity – and adiabatic variations in the opposite case. Then the partial pressure of each species, $n_\alpha T_\alpha$, scales as $\propto n_\alpha^{\gamma_\alpha}$, where γ_α is the effective polytropic index.

4.2.1 Major wave modes

In *ideal* unmagnetized plasmas only longitudinal wave modes can be sustained. The dispersion relations of these modes can be written as a sum of the partial susceptibilities (plasma responses),

$$\varepsilon(\omega, \mathbf{k}) = 1 + \chi_e + \chi_i + \chi_d = 0, \tag{4.8}$$

where the electron and ion responses are expressed via density and potential perturbations as $\chi_{e,i} = \pm 4\pi e k^{-2}\delta n_{e,i}/\delta\phi$. The dust response depends also on the charge variations, so that $\chi_d = -4\pi k^{-2}(Z\delta n_d + n_d\delta Z)/\delta\phi$. By linearizing Equations (4.1)–(4.7) one can obtain the partial responses in the hydrodynamic approximation (for the results of the kinetic theory, see, e.g., Tsytovich *et al.* 2001, 2002).

In order to retrieve the wave modes existing in complex plasmas, let us first consider the case of the *multi-component plasmas* – when the variations of the grain charges are neglected. At this point we also neglect collisions and assume the equilibrium plasma ionization and loss: This approach allows us to obtain satisfactory results for the *real part* of the dispersion relations $\omega(k)$, unless the actual damping (growth) rate of the waves is comparable with ω. The partial plasma responses in this case are

$$\chi_\alpha = -\frac{\omega_{\mathrm{p}\alpha}^2}{\omega^2 - \gamma_\alpha k^2 v_{T_\alpha}^2}, \tag{4.9}$$

where $\omega_{\mathrm{p}\alpha}$ is the plasma frequency of the corresponding species. (When the flow is present with the drift velocities $\mathbf{u}_\alpha$, one should simply substitute $\omega \to \omega - \mathbf{k}\cdot\mathbf{u}_\alpha$.) For the plasma waves (plasmons with $\omega \gg k v_{T_e}$) the microparticles remain at rest and, therefore, the functional form of the dispersion relation, $\omega^2 = \omega_{\mathrm{p}e}^2 + 3k^2 v_{T_e}^2$, is not affected by the presence of the dust grains (in the hydrodynamic approach one can treat plasma waves as one-dimensional oscillations with $\gamma_e = 3$). However, the electron plasma frequency $\omega_{\mathrm{p}e}$ is changed because the charged grains affect the quasineutrality condition for unperturbed densities, $n_i = n_e + |Z| n_d$, and hence the electron density. A similar effect is also observed for the ion–acoustic (IA) waves, where the electrons provide equilibrium neutralizing background and dust remains at rest, $k v_{T_i} \ll \omega \ll k v_{T_e}$. For electrons we have $\chi_e = (k\lambda_{\mathrm{D}e})^{-2}$ and then Equations (4.8) and (4.9) yield

$$\frac{\omega^2}{k^2} = \gamma_i v_{T_i}^2 + \frac{\omega_{\mathrm{p}i}^2 \lambda_{\mathrm{D}e}^2}{1 + \lambda_{\mathrm{D}e}^2 k^2}. \tag{4.10}$$

The first term represents ordinary ion thermal sound mode which can exist when the ion mean free path is much smaller than the wavelength k^{-1}. Usually this term is relatively small and can be neglected compared to the second term, which actually represents the IA mode. This IA term depends on the ion-to-electron density ratio, which can be conveniently characterized by the "Havnes parameter" P (Havnes *et al.* 1987),

$$n_i/n_e - 1 = |Z| n_d/n_e \equiv P. \tag{4.11}$$

Generally, when $P \ll 1$ the effect of dust on the conventional (plasma and ion–acoustic) modes can be neglected. Otherwise, for $P \gtrsim 1$ the role of dust can be significant and then, in order to highlight this effect, the IA waves are referred to as the *dust ion–acoustic* (DIA) mode. In the long-wavelength limit $k\lambda_{\mathrm{D}e} \ll 1$ the phase velocity of the DIA mode can be conveniently expressed via the ion thermal velocity,

$$C_{\mathrm{DIA}} = \omega_{\mathrm{p}i}\lambda_{\mathrm{D}e} \equiv \sqrt{(1+P)\tau}\, v_{T_i}, \tag{4.12}$$

where $\tau = T_e/T_i$ is the electron-to-ion temperature ratio, which is much larger than unity for typical rf and dc discharges, so that $C_{\rm DIA}$ exceeds significantly the ion thermal velocity (note that $C_{\rm DIA}$ does not depend on T_i). The role of the dust species on the IA waves was first considered by Shukla and Silin (1992). The DIA waves were studied in a series of experiments (see, e.g., Barkan *et al.* 1996a,b; Merlino *et al.* 1998) where the increase of the phase velocity with the grain density was clearly demonstrated.

Charged dust particles give rise to another acoustic mode associated with the motion of charged grains, whereas both the electrons and ions provide equilibrium neutralizing background. For $kv_{T_d} \ll \omega \ll kv_{T_i}$, we have $\chi_{e,i} = (k\lambda_{{\rm D}e,i})^{-2}$, and then Equations (4.8) and (4.9) yield

$$\frac{\omega^2}{k^2} = \gamma_d v_{T_d}^2 + \frac{\omega_{\rm pd}^2 \lambda_{\rm D}^2}{1+\lambda_{\rm D}^2 k^2}, \qquad (4.13)$$

where $\lambda_{\rm D}^{-2} = \lambda_{{\rm D}e}^{-2} + \lambda_{{\rm D}i}^{-2}$ is the linearized Debye length. In analogy with the DIA waves [Equation (4.10)], the first term represents the dust thermal mode and the second one corresponds to the *dust–acoustic* (DA) mode. The phase velocity of the DA mode does not depend on the dust temperature and in the long-wavelength limit $k\lambda_{\rm D} \ll 1$ can be written as

$$C_{\rm DA} = \omega_{\rm pd}\lambda_{\rm D} \equiv \sqrt{\frac{P\tau}{1+(1+P)\tau}}\sqrt{|Z|\frac{T_i}{T_d}}\, v_{T_d}. \qquad (4.14)$$

There is a clear similarity between the ion and dust acoustic modes: Equations (4.12) and (4.14) show that for both modes the ratio of the phase velocity to the thermal velocity is determined by the temperature ratio of the light-to-heavy species – T_e/T_i for DIA waves and T_i/T_d for DA waves. A peculiarity of the DA waves is that the charge-to-mass ratio of the dust grains is typically 10^8–10^{10} times smaller than that of the ions, and therefore, the dust waves have relatively low frequencies, $\sim$10–100 Hz. The dispersion relation for the DA wave was first derived by Rao *et al.* (1990). Since the typical values of $|Z|$ are on the order of thousands, the phase velocity of DA waves can be much larger than v_{T_d}, even if T_d exceeds T_i (of course, the Havnes parameter should not be too small).

The first reported observation of spontaneously excited dust waves was in an rf magnetron discharge at a frequency $\simeq$ 12 Hz (Chu *et al.* 1994). Later the dust waves, either self-sustained or excited externally, were seen in numerous experiments under quite different experimental conditions: For example, in Q-machine at $\simeq$ 15 Hz with the phase velocity $\simeq$ 9 cm s^{-1} (Barkan *et al.* 1995), in dc discharges in the range $\simeq$6-30 Hz with the phase velocity $\simeq$ 12 cm s^{-1} (Thompson *et al.* 1997; Merlino *et al.* 1998), and at $\simeq$ 60 Hz with the phase velocity $\simeq$ 1 cm s^{-1} (Molotkov *et al.* 1999; Fortov *et al.* 2000). There were numerous experiments performed in rf discharges where self-excited waves (Zobnin *et al.* 2002; Fortov *et al.* 2003; Piel *et al.* 2006; Schwabe *et al.* 2007) or externally driven waves (Khrapak *et al.* 2003; Yaroshenko *et al.* 2004; Thomas *et al.* 2007; Annibaldi *et al.* 2007) were investigated. Figure 4.1

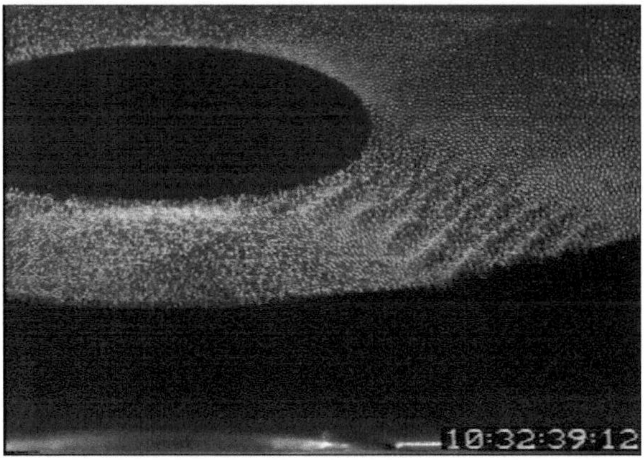

FIGURE 4.1
Wave structure observed in PKE-Nefedov experiments by Annibaldi *et al.* (2007). The experiments were performed under microgravity conditions in argon rf-discharge plasma at a pressure 12 Pa with particles of 3.4 μm diameter. The particles were visualized by the vertical laser sheet of about 150 μm thickness. The waves were excited by applying a low-frequency modulation voltage to the lower rf electrode. The snapshot (side view) shows the response of the particles at a frequency of 23 Hz.

shows the waves excited by a periodic modulation of the dc bias of the (lower) rf electrode obtained under microgravity conditions by Annibaldi *et al.* (2007). Most of such experiments, however, were done with strongly coupled plasmas. The weak coupling was probably achieved only when the excitation amplitude was fairly large or the waves were unstable (e.g., Barkan *et al.* 1995; Fortov *et al.* 2003; Schwabe *et al.* 2007), which eventually provided sufficiently high "temperature" of grains. A quantitative comparison with the linear dispersion relations is not really justified in these cases. Therefore, an accurate experiment to verify the DA dispersion relation in gaseous complex plasmas is still necessary, though the first experimental observations of the dust thermal mode was recently reported by Nunomura *et al.* 2005.

Note that when the variations of the grain charges induced by waves are taken into account, the DIA phase velocity remains the same, but the DA velocity is changed (Melandsø *et al.* 1993; Fortov *et al.* 2000, Ivlev and Morfill 2000). However, this change does not exceed the factor $\simeq \sqrt{(2+z)/(1+z)}$ compared to Equation (4.14), which is typically less than $\simeq 15\%$. Therefore, for practical use Equation (4.14) is quite sufficient.

4.2.2 Damping and instabilities

The wave modes can exist only when the damping is weak, so that the actual imaginary part of the dispersion relation, $|\omega_{\rm i}|$, is much smaller than the real part $\omega_{\rm r}$ – only then one can speak about the wave propagation. The waves can also be unstable, because of various mechanisms operating in complex plasmas – we discuss these mechanisms below. As long as $|\omega_{\rm i}|$ is much smaller than $\omega_{\rm r}$, the latter is approximately determined by the real part of the permittivity (4.8), i.e., Re $\varepsilon(\omega_{\rm r},\mathbf{k}) \simeq 0$, and the former is given by (Alexandrov *et al.* 1984)

$$\omega_{\rm i} \simeq -\left.\frac{\mathrm{Im}\ \varepsilon(\omega,\mathbf{k})}{\partial \mathrm{Re}\ \varepsilon(\omega,\mathbf{k})/\partial\omega}\right|_{\omega=\omega_{\rm r}}.$$

This is very convenient formula for the practical use.

First we discuss the kinetic effects – namely, the role of the Landau damping. For each wave mode, the Landau damping can be due to wave resonance with "heavy" species (i.e., ions for DIA waves and dust for DA waves) and with "light" species (electrons for DIA waves and ions for DA waves). The damping caused by heavy species scales as $|\omega_{\rm i}/\omega_{\rm r}| \propto \exp(-\frac{1}{2}C^2/v_T^2)$, where C and v_T are the corresponding phase velocity and the thermal velocity of heavy species, respectively (Lifshitz and Pitaevskii 1981; Rosenberg 1993). From Equations (4.12) and (4.14) we see that even in isothermal complex plasmas, the C/v_T ratios can be quite large: $C_{\rm DIA}/v_{T_i}$ is large when $P \gg 1$, and $C_{\rm DA}/v_{T_d}$ is large because $|Z| \gg 1$. This makes substantial difference compared to usual plasmas, where $C_{\rm IA}/v_{T_i}$ can be large and, thus, the IA waves can propagate only when $\tau \gg 1$ [DIA waves were studied in the double plasma device, e.g., by Nakamura and Sarma (2001), and the Q-machine, e.g., by Barkan *et al.* (1996a), where $\tau \simeq 1$]. Thus, the damping on heavy species is usually small in (dense) complex plasmas.

In the absence of the plasma flows the Landau damping on light species is relatively weak as well, because of the small charge-to-mass ratios: For DIA waves the (relative) damping rate is $|\omega_{\rm i}/\omega_{\rm r}| \lesssim \sqrt{(1+P)m_e/m_i}$, whereas for DA waves $|\omega_{\rm i}/\omega_{\rm r}| \lesssim \sqrt{P(1+P)^{-1}|Z|m_i/m_d}$ (Lifshitz and Pitaevskii 1981; Rosenberg 1993). Nevertheless, in experiments with the DIA waves performed at very low pressures, $p \sim 10^{-2}$–10^{-3} Pa, the electron Landau damping can be an important mechanism of dissipation (Popel *et al.* 2004). For DA waves, however, it does not play a noticeable role. The Landau damping is, of course, modified when a stream of light species exists in a plasma (see below).

Now let us dwell upon the other mechanisms responsible for the damping and instabilities of the DIA and DA waves. Below we assume that the electron-to-ion temperature ratio τ is large, as it usually is in experiments. Such assumption allows us to simplify formulas substantially (note that even for $\tau \simeq 1$, the resulting expressions yield fairly good quantitative agreement with the exact formulas).

DIA mode. Along with the Landau damping the major dissipation mechanisms are the collisions with neutrals and variations of the grain charges (Jana *et al.* 1993; Varma *et al.* 1993; Ma and Yu 1994; D'Angelo 1994; Ivlev and Morfill 2000). In addition, there is a counterplay between ionization and loss – this can cause either damping or instability, depending on the value of P (D'Angelo 1997; Ivlev *et al.* 1999; Wang *et al.* 2001b). All three contributions to the imaginary part $\omega_{\rm i}$ (assuming that it is much smaller than ω_r) can be derived from the fluid approach which yields

$$2\omega_i \simeq -\nu_{in} - \frac{1-P}{1+P}\nu_I - \frac{P^2\Omega_{ch}}{(1+P)(1+z)},$$

where ν_{in} is the frequency of the ion–neutral collisions and Ω_{ch} is the charging frequency [see Equation (2.16)]. In sufficiently dense complex plasmas (when $P \gg 1$), the major damping mechanism can be due to the "coherent" charge variations induced by waves (the mechanism is effective because the DIA frequency can be comparable to Ω_{ch}). The ion–neutral collisions as well as ionization do not usually play any significant role in the DIA wave experiments, since the gas pressure is low enough (see, e.g., Barkan *et al.* 1996a; Luo *et al.* 1999; Nakamura and Sarma 2001).

As regards the DIA instabilities, the major mechanism operating in experiments is associated with the electron drift relative to ions (Barkan *et al.* 1996b; Merlino 1997) – the so-called "current-driven instability" which is well-known for the IA waves in usual plasmas. Essentially, this instability is the reversed electron Landau damping – the energy exchange due to the resonance electron-wave interaction changes the sign when the drift velocity u_e exceeds the phase velocity of the DIA waves C_{DIA}. The growth rate associated with this instability can be estimated as (Rosenberg 1993)

$$\frac{\omega_i}{\omega_r} \simeq \sqrt{\frac{\pi}{8}\frac{m_e}{m_i}(1+P)}\frac{(u_e/C_{DIA}-1)}{(1+k^2\lambda_{De}^2)^{3/2}}.$$

When the damping rates discussed above (including the ion Landau damping) are low enough, the current-driven instability sets on (Merlino 1997; Merlino *et al.* 1998). The charge variations can somewhat modify the growth rate (Annou 1998).

DA wave mode. The major damping mechanism operating in experiments with complex plasmas is certainly neutral gas friction. The resulting damping, $2\omega_i \simeq -\nu_{dn}$, is determined by the corresponding momentum exchange rate [see Equation (2.69)]. However, along with the damping there are a number of instability mechanisms which turn out to be quite important in experiments. Below we mention the most important types of the DA instability:

(i) Ion streaming instability: This can be triggered when ion currents are present in a plasma (e.g., due to electric fields in rf sheaths and dc striations). The mechanism of the (DA) ion streaming instability is completely identical to that of the (DIA) current-driven instability. The ion streaming instability is often observed in complex plasma experiments performed in different discharges (see, e.g., Barkan *et al.* 1995; Molotkov *et al.* 1999; Ratynskaia *et al.* 2004), and it has been studied theoretically in numerous publications, e.g., Rosenberg 1993, 1996, 2002; D'Angelo and Merlino 1996; Kaw and Singh 1997; Mamun and Shukla 2000. Figure 4.2 shows an example of such instability observed in a rf capacitively coupled discharge by Schwabe *et al.* (2007).

The presence of the ion flux modifies properties of the DA mode. This can be appropriately taken into account by using the kinetic expression for the ion susceptibility. Also, the kinetic approach allows us to include properly the effect of the ion–neutral collisions: The collisions in discharges are mostly of the charge-exchange

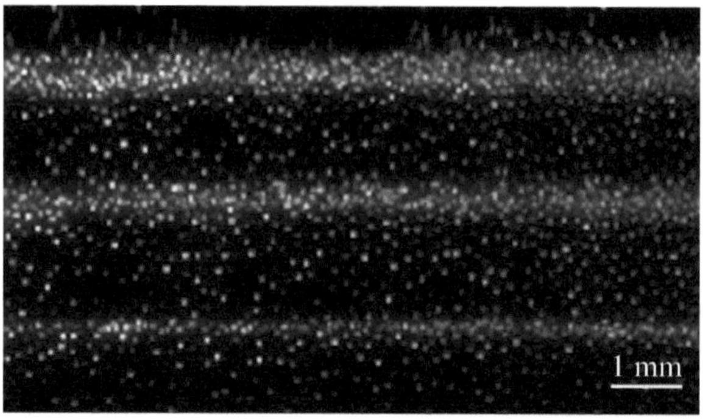

FIGURE 4.2
Spontaneous excitation of dust waves observed in laboratory experiments by Schwabe *et al.* (2007). The experiments were carried out in an rf discharge in argon at a pressure of 23 Pa with particles of 1.28 μm diameter. The instability sets on when the particle density in the cloud exceeds a threshold, provided the gas pressure is low enough.

type, which makes it possible to employ the model BGK (Bhatnagar, Gross and Krook) form of the ion collision integral. Assuming *shifted Maxwellian* distribution, the ion response is (Alexandrov *et al.* 1984)

$$\chi_i(\omega,\mathbf{k}) = \frac{1}{(k\lambda_{Di})^2}\left[\frac{1+\mathscr{F}(\xi)}{1+\dfrac{i\nu_{in}}{\omega-\mathbf{k}\cdot\mathbf{u}_i+i\nu_{in}}\mathscr{F}(\xi)}\right], \quad \xi = \frac{\omega-\mathbf{k}\cdot\mathbf{u}_i+i\nu_{in}}{\sqrt{2}kv_{T_i}}, \tag{4.15}$$

where $\mathscr{F}(\xi)$ is the Maxwellian dispersion function (Fried and Conte 1961). In limiting cases Equation (4.15) can be substantially simplified: For $|\xi| \ll 1$, the power series for the dispersion function is $\mathscr{F}(\xi) \simeq -2\xi^2 + i\sqrt{\pi}\xi$, and for $|\xi| \gg 1$ the asymptotic expansion is $\mathscr{F}(\xi) \simeq -1 - \frac{1}{2}\xi^{-2} + i\sqrt{\pi}\xi e^{-\xi^2}$. Therefore, when $|\omega - \mathbf{k}\cdot\mathbf{u}_i + i\nu_{in}| \ll kv_{T_i}$ we obtain

$$\chi_i(\omega,\mathbf{k}) \simeq \frac{1}{(k\lambda_{Di})^2}\left[1 + i\sqrt{\frac{\pi}{2}}\frac{\omega-\mathbf{k}\cdot\mathbf{u}_i}{kv_{T_i}}\right], \tag{4.16}$$

where $\mathbf{u}_i$ is the drift velocity of ions. The real part in Equation (4.16) coincides with the results of the fluid approach in this limit [see Equation (4.9)], the imaginary part is due to the Landau damping. In the opposite limit $|\omega - \mathbf{k}\cdot\mathbf{u}_i + i\nu_{in}| \gg kv_{T_i}$ the resulting susceptibility can be written in the following form:

$$\chi_i(\omega,\mathbf{k}) \simeq -\frac{\omega_{pi}^2}{(\omega-\mathbf{k}\cdot\mathbf{u}_i)(\omega-\mathbf{k}\cdot\mathbf{u}_i+i\nu_{in}) - k^2v_{T_i}^2}. \tag{4.17}$$

This limit denotes either a strongly collisional case (when the ion mean free path is shorter than k^{-1}) or the case of "cold hydrodynamics" (when $u_i \gg v_{T_i}$, so that the thermal motion can be neglected). In both cases the fluid approach is applicable and, hence, Equation (4.17) can be directly obtained from Equations (4.3) and (4.4), assuming equilibrium ionization/recombination and neglecting other collisions.

Note that Equation (4.15) is derived assuming the shifted Maxwellian function for the ion velocity distribution. This assumption, however, is only justified when the ion flow is (sub)thermal – otherwise deviations from the Maxwellian form become too strong [see Equation (2.75)] which, in turn, strongly affects the expression for the ion susceptibility χ_i. Therefore, for the superthermal flow one should use Equation (2.76) (Schweigert 2001; Ivlev *et al.* 2004, 2005).

The threshold for the ion streaming instability is determined from the (numerical) solution of Equation (4.8), by using Equations (4.16) or (4.17) for the ion response and substituting $\chi_e \simeq (k\lambda_{\mathrm{D}e})^{-2}$ and $\chi_d \simeq -\omega_{\mathrm{pd}}^2/\omega(\omega+i\nu_{dn})$. Experiments show that by increasing the neutral gas pressure up to sufficiently high values (typically, dozens of Pa) the instability can be suppressed, apparently because the neutral gas friction increases as well. The theoretical analysis [which can be somewhat simplified for the subthermal (Fortov *et al.* 2000) and superthermal (Joyce *et al.* 2002) limits of the ion drift] yields the pressure threshold which is in a good agreement with the experiments.

(ii) Ionization instability: Unlike the DIA waves, ionization cannot directly cause the instability of the DA waves – because the ionization creates new ions, but not dust grains. Nevertheless, ionization can in fact trigger the DA instability, because the ions can effectively transfer their momentum to the grains via the ion drag force (D'Angelo 1998; Ivlev *et al.* 1999; Ostrikov *et al.* 2000). The whole instability mechanism operates as follows: When the dust density fluctuates – say, decreases – in some region, ionization increases (because the electron density grows keeping quasi-neutrality), which creates additional ion outflow from the region. This flux exerts an additional ion drag force pushing the grains away, and thus, the dust density decreases further. Obviously, this instability is of the aperiodic type (i.e., $\omega_{\mathrm{r}} = 0$) and, thus, is independent of ν_{dn}. The instability condition $\omega_{\mathrm{i}} > 0$ is satisfied when (D'Angelo 1998; Ivlev *et al.* 1999; Wang *et al.* 2001b)

$$[P^{-1}\nu_{id} - (1+P)^{-1}\nu_{in}]\nu_{\mathrm{I}} \gtrsim k^2 v_{T_i}^2.$$

Here, ν_{id} is the effective frequency of the ion–dust collisions, which is related to the "ion drag" rate ν_{di} introduces in Section 2.4.2.2 via $m_i n_i \nu_{id} = m_d n_d \nu_{di}$. The larger the dust grains are, the higher the value of ν_{id} is and, hence, the condition for the instability is more relaxed. There are grounds to believe that this instability is responsible for the onset of the void formation in complex plasmas (D'Angelo 1998; Ivlev *et al.* 1999; Samsonov and Goree 1999; Ostrikov *et al.* 2000; Wang *et al.* 2001b).

(iii) Charge variation instability: This is due to the grain charge variations induced by the DA wave. In contrast to the DIA waves, now the charges are very close to the momentary equilibrium (because $|\omega| \ll \Omega_{\mathrm{ch}}$), and therefore, their variations alone are unlikely to be a reason for an instability or damping. However, in the presence of an external electric field **E** (e.g., ambipolar fields or the fields in rf sheaths and dc striations), the wave-correlated charge variations result in non-zero (average) work done by the electric force (Fortov *et al.* 2000, 2003; Zobnin *et al.* 2002). The sign of this work is determined by the orientation of the wave vector **k** with the respect to the electric field. The dust susceptibility that takes into account these effects is (Fortov *et al.* 2000)

$$\chi_d(\omega,k) \simeq -\frac{\omega_{\rm pd}^2}{\omega(\omega+i\nu_{dn})}\left[1+\frac{ie\mathbf{E}\cdot\mathbf{k}}{(1+z)T_ik^2}\right].$$

This expression should be used together with Equation (4.16) or (4.17) for the flowing ions, because the electric field causes an ion drift with a velocity which is usually determined by the ion mobility μ_i, via $u_i = \mu_i E$. Numerical solution of Equation (4.8) by Fortov *et al.* (2000) yields the instability threshold which takes into account both the ion stream and the charge variations. It was shown that the charge variations can relax the conditions for the instability onset significantly, resulting into lower values of the threshold pressure. Then the instability can be triggered when the density of dust particles exceeds a critical value (Fortov *et al.* 2000; Zobnin *et al.* 2002).

4.3 Waves in strongly coupled (liquid) complex plasmas

In the beginning of this chapter we mentioned that complex plasmas are normally observed in experiments forming strongly coupled states. The pair correlation function of microparticles usually exhibits short-range order indicating that plasmas are in liquid-like states or that the particles form ordered crystalline structures. Dispersion properties of strongly coupled plasmas significantly deviate from those of ideal gaseous plasmas discussed above. There are many different theoretical approaches to study waves in strongly coupled systems: These are, e.g., the "quasi-localized charge approximation" (Kalman and Golden 1990) employed for complex plasmas by Rosenberg and Kalman (1997) and Kalman *et al.* (2000), the "multicomponent kinetic approach" by Murillo (1998, 2000), and the "generalized hydrodynamic approach" applied by Kaw and Sen (1998), Kaw (2001), and Xie and Yu (2000). The last is probably the most physically "transparent" approach which allows us to track evolution of the dispersion properties of complex plasmas in a broad range of coupling parameter Γ, from the ideal gaseous state ($\Gamma \lesssim 1$) up to the strongly coupled state ($\Gamma \gg 1$) – when the system crystallizes. There have also been numerical MD simulations of the wave modes in strongly coupled complex plasmas (Winske *et al.* 1999; Ohta and Hamaguchi 2000), which are in reasonably good agreement with the results of the above mentioned theoretical approaches.

Following the model of "very viscous liquids" originally proposed by Maxwell and generalized by Frenkel (1946), in the framework of generalized hydrodynamics (GH), the ensemble of strongly coupled dust grains is treated as a continuous medium which reveals properties of viscous liquids in response to slow perturbations, but behaves as an elastic body when the perturbation time scales are short enough (Kaw and Sen 1998; Kaw 2001). The transition between these two regimes occurs at the so-called "Maxwellian relaxation time" $\tau_{\rm M}$. The fluid equation of motion for the velocity perturbation $\delta\mathbf{v}_d$ has the following form:

$$\frac{\partial\delta\mathbf{v}_d}{\partial t} = -\frac{Q}{m_d}\nabla\phi - \frac{\nabla p_d}{m_d n_d} - \nu_{dn}\delta\mathbf{v}_d - \int_{-\infty}^{t} dt' \int d\mathbf{r}'\, \eta_d(\mathbf{r}-\mathbf{r}', t-t')\delta\mathbf{v}_d(\mathbf{r}', t'). \qquad (4.18)$$

The integral term is the linear viscoelastic operator in a homogeneous stationary medium written in a general form. It takes into account both spatial and temporal correlations of stresses exerted in strongly-coupled systems, in addition to the local homogeneous stress – the pressure term $\propto \nabla p_d$. By using the simplest form of the viscoelastic operator with the exponentially decaying memory effects, we have for the viscosity kernel:

$$\eta_d(\omega,\mathbf{k}) \simeq \frac{\eta k^2 + (\frac{1}{3}\eta + \zeta)\mathbf{k}(\mathbf{k}\cdot\ )}{1 - i\omega\tau_{\rm M}}.$$

[Here $(\mathbf{k}\cdot\ )$ denotes a scalar product with $\delta\mathbf{v}_d$.] Parameters of the stress operator – viscosities η and ζ and relaxation time $\tau_{\rm M}$, as well as pressure p_d – are determined by the correlation part of the energy of the electrostatic interaction $u(\Gamma,\kappa)$ (normalized by the dust temperature). In a weakly coupled regime, $\Gamma \lesssim 1$, the Debye–Hückel approximation yields $u \simeq -\frac{\sqrt{3}}{2}\Gamma^{3/2}$ (here and below in this section Γ corresponds to the Wigner-Seitz radius). For the liquid phase in the range $1 \lesssim \Gamma \lesssim 200$, the normalized correlation energy can be well approximated by the scaling (Farouki and Hamaguchi 1994) $u = a\Gamma + b\Gamma^{1/4} + c + d\Gamma^{-1/4}$, where coefficients a, b, c, and d are some functions of the lattice parameter κ. In the one-component plasma (OCP) limit, $\kappa = 0$, the Monte Carlo (MC) simulations by Slattery *et al.* (1980) yield $a \simeq -0.89$, $b \simeq 0.94$, $c \simeq -0.80$, and $d \simeq 0.18$. For $\kappa \lesssim 1$ the dependence of the coefficients on the lattice parameter is rather weak (Farouki and Hamaguchi 1994), $a(\kappa) \simeq -0.89 - 0.10\kappa^2 + 0.0025\kappa^4 + \ldots$, which means that the OCP results are quite applicable for this range of κ. The relaxation time is expressed as follows (Ichimaru *et al.* 1987): $\tau_{\rm M} v_{T_d}^2 = \eta_*(1 - \gamma_d\mu_d + \frac{4}{15}u)^{-1}$, where $\eta_* = \frac{4}{3}\eta + \zeta$ and $\mu_d = T_d^{-1}(\partial p_d/\partial n_d)_{T_d} = 1 + \frac{1}{3}u + \frac{1}{9}\Gamma\partial u/\partial\Gamma$ is the isothermal compressibility. The first and second viscosity coefficients, η and ζ, can also be deduced from the results of numerical simulations and experiments by Saigo and Hamaguchi (2002), Salin and Caillol (2002, 2003), Nosenko and Goree (2004).

Using the Fourier transformation of Equation (4.18) together with the continuity equation (4.1), one can derive the dispersion relations of different wave modes.

4.3.1 Longitudinal waves

The dust susceptibility is ($\delta\mathbf{v}_d$ parallel to $\mathbf{k}$)

$$\chi_d(\omega,k) = -\frac{\omega_{pd}^2}{\omega(\omega+i\nu_{dn}) - \gamma_d\mu_d k^2 v_{T_d}^2 + \dfrac{i\eta_*\omega k^2}{1-i\omega\tau_M}}. \tag{4.19}$$

The dispersion relation is determined by Equation (4.8) by substituting $\chi_{e,i} \simeq (k\lambda_{De,i})^{-2}$ together with Equation (4.19). Naturally, two limits can be distinguished: The "hydrodynamic regime" $\omega\tau_M \ll 1$ and the "strongly coupled regime" $\omega\tau_M \gg 1$ (following the terminology adopted from the OCP literature). For $\Gamma \gg 1$ the correlation energy is mostly determined by the Madelung part, $u \simeq -0.89\Gamma$, and then the results are reduced to the following simple form:

$$\begin{aligned}
\omega\tau_M \ll 1: \quad & \omega\left[\omega + i(\nu_{dn} + \eta_* k^2)\right] \simeq \left[\frac{1}{1+k^2\lambda_D^2} - 0.4\left(\frac{v_{T_d}}{C_{DA}}\right)^2\Gamma\right] C_{DA}^2 k^2, \\
\omega\tau_M \gg 1: \quad & \omega(\omega + i\nu_{dn}) \simeq \left[\frac{1}{1+k^2\lambda_D^2} - 0.24\left(\frac{v_{T_d}}{C_{DA}}\right)^2\Gamma\right] C_{DA}^2 k^2.
\end{aligned} \tag{4.20}$$

The right-hand side of Equation (4.20) is independent of T_d, and the second terms represent the "coupling correction" to the gaseous DA dispersion relation (4.13). At sufficiently large Γ these terms play a very important role – the dispersion can even change the sign, so that the group velocity becomes negative ($\partial\omega/\partial k < 0$) at large k. This feature is peculiar to the longitudinal modes in plasma crystals (see Section 4.4), which indicates that there is no *qualitative* difference between the dispersion properties of (strongly coupled) liquid and crystalline complex plasmas. It is noteworthy that the other approaches yield essentially the same results for the real part of the dispersion relation (see Figure 4.3). The important difference revealed in the GH approach is only for the imaginary part – in addition to the neutral friction, the viscosity contributes to the dissipation in the hydrodynamic regime. However, for typical experimental conditions the neutral gas friction prevails and the viscosity can provide the major contribution to the dissipation only if the gas pressure is low enough. Also, the neutral gas friction can hamper the role of the "coupling correction", and when ν_{dn} becomes comparable to ω_{pd}, the difference between the dispersion relations of ideal and strongly coupled complex plasmas can be washed away completely (Rosenberg and Kalman 1997; Kaw and Sen 1998; Kaw 2001). Presumably, that was the reason why in experiments where the coupling parameter was quite high, the dispersion properties were nevertheless found to be very similar to those derived for ideal plasmas (see, e.g., Pieper and Goree 1996). Note that the longitudinal waves in strongly coupled plasmas are subject to the ion streaming instability, similar to that discussed in Section 4.2.2, now with the thresholds and the growth rate functions of Γ (e.g., Kaw and Sen 1998).

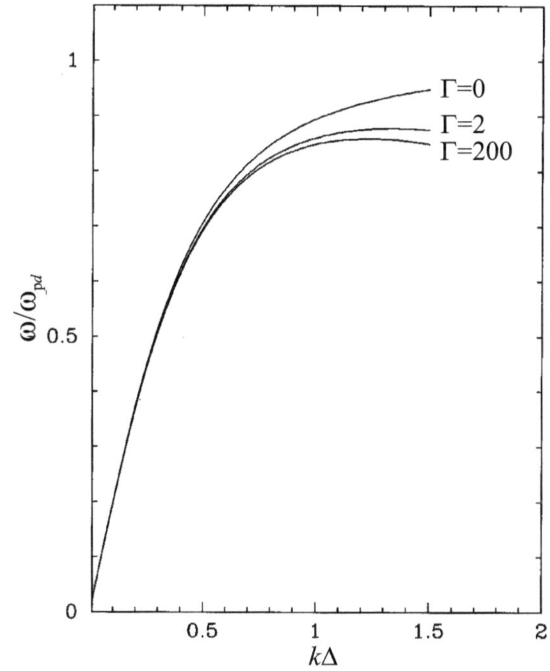

FIGURE 4.3
Dispersion relation of dust waves in strongly coupled complex plasmas, as derived from the quasi-localized charge approximation by Rosenberg and Kalman (1997). The frequency is normalized to the dust plasma frequency ω_{pd} and the wave vector is normalized to the particle separation Δ. The results are for the screening parameter $\kappa = 0.5$ and different values of the coupling parameter Γ. The curve $\Gamma = 0$ recovers the dispersion relation for the DA waves in gaseous complex plasmas [Equation (4.13) with $T_d = 0$].

4.3.2 Transverse waves

For the transverse waves ($\delta \mathbf{v}_d$ perpendicular to $\mathbf{k}$), the dispersion relation can be directly obtained from the Fourier transformed Equation (4.18),

$$\omega \left[\omega + i(\tau_{\rm M}^{-1} + \nu_{dn})\right] = \frac{\eta_* k^2 + \nu_{dn}}{\tau_{\rm M}}. \tag{4.21}$$

This mode is also strongly affected by the neutral gas friction – the effects of strong coupling disappear when $\nu_{dn} \gtrsim \eta_* k^2$. If the neutral friction can be neglected then in the hydrodynamic regime $\omega\tau_{\rm M} \ll 1$, Equation (4.21) reduces to the ordinary damped mode for a shear flow in viscous liquids, $\omega \simeq -i\eta_* k^2$. In the opposite regime we obtain a non-dispersive acoustic mode, $\omega \simeq \sqrt{\eta_*/\tau_{\rm M}}\, k$, which is analogous to elastic shear waves in solids. The shear waves triggered in liquid complex plasmas due to instability have been observed experimentally by Pramanik *et al.* (2002) [however, the mechanism responsible for the instability is not fully understood (Mishra *et al.* 2000; Kaw 2001; Pramanik *et al.* 2002)].

Thus, the GH approach allows us to track evolution of the wave dispersion properties as the coupling parameter Γ increases and, thus, to cover the transition from the ideal gaseous to the strongly coupled state. However, the phenomenological *hydrodynamic* approach can provide us with only the qualitative convergence to the wave modes in crystals and, of course, cannot retrieve features peculiar to a particular lattice type (especially when the wavelength becomes comparable to the interparticle distance). Therefore, for a quantitative study of waves in plasma crystals, one should employ a different approach which is discussed in the next section.

4.4 Waves in plasma crystals

The theoretical model of waves in crystals – the so-called "dust–lattice" (DL) waves – is based on the analysis of the equation of motion for individual particles. For a particle having the coordinate $\mathbf{r}$, the equation of motion is

$$m_d\ddot{\mathbf{r}} + m_d \nu_{dn}\dot{\mathbf{r}} = -\nabla U_{dd} + \mathbf{F}_{\text{ext}}. \tag{4.22}$$

Here $U_{dd} = Q^2 \sum_i |\mathbf{r}-\mathbf{r}_i|^{-1} \exp(-|\mathbf{r}-\mathbf{r}_i|/\lambda_{\rm D})$ is the total energy of the electrostatic dust–dust coupling (interaction potential is assumed to be of the Debye–Hückel form), the summation is over all particles with $\mathbf{r}_i \neq \mathbf{r}$. The force $\mathbf{F}_{\text{ext}}$ includes all "external" forces (except for the neutral drag force which is explicitly included to the left-hand side), e.g., confinement, excitation (lasers, electric pulses, beams, etc.), thermal noise (Langevin force). Such diversification of the forces is convenient because the eigenmodes of the system do not depend on $\mathbf{F}_{\text{ext}}$. When the particles in a crystalline state are sufficiently far from the melting line (see Figure 5.1, Section 5.1), as a "first iteration" one can neglect the influence of the thermal motion in the dispersion properties. Then the waves can be considered as perturbations of cold particles which form an ideal lattice in the equilibrium.

4.4.1 One-dimensional strings

The simplest model for studying waves in crystals is the one-dimensional "particle string" (Kittel 1976). The motion is allowed only along the string, which formally corresponds to the infinite transverse confinement. This model was adopted to study waves in plasma crystals by Melandsø (1996). The string model shows very good agreement with the first experiments performed with one-dimensional plasma crystals (Peters *et al.* 1996; Homann *et al.* 1997). Moreover, when particle separation Δ exceeds the screening length $\lambda_{\rm D}$, so that only the interaction with the *nearest neighbors* is important, the string model turns out to be appropriate to describe DL waves also in two-dimensional plasma crystals (Zuzic *et al.* 1996; Homann *et al.* 1998). The string model yields the following dispersion relation in the nearest neighbor approximation:

$$\omega(\omega + i\nu_{dn}) = 4\Omega_{\rm DL}^2 e^{-\kappa}(\kappa^{-1} + 2\kappa^{-2} + 2\kappa^{-3}) \sin^2 \tfrac{1}{2}k\Delta, \tag{4.23}$$

where $\Omega_{\rm DL}^2 = Q^2/m_d\lambda_{\rm D}^3$ is the DL frequency scale and $\kappa = \Delta/\lambda_{\rm D}$ is the lattice parameter. The experimentally observed wave frequencies usually vary from a few Hz for strings up to a few dozens of Hz for monolayers (e.g., Peters *et al.* 1996; Zuzic *et al.* 1996; Nunomura *et al.* 2002b), which is in agreement with the estimated magnitude of $\Omega_{\rm DL}$. Simple formula (4.23) is very convenient to evaluate spectra of the longitudinal DL waves.

4.4.2 Two-dimensional triangular lattice

Most of the experiments on the DL waves have been performed so far in two-dimensional complex plasmas – crystalline monolayers suspended in rf electrode sheaths (Zuzic *et al.* 1996; Homann *et al.* 1998; Nunomura *et al.* 2000, 2002a,b, 2005; Zhdanov *et al.* 2003a). Particles in the monolayers form a hexagonal (triangular) lattice. The dispersion relation for the *in-plane* DL modes in such lattices was derived by Peeters and Wu (1987), Dubin (2000), Wang *et al.* (2001a), and Zhdanov *et al.* (2003a). The perturbations are determined by the following equation (Zhdanov *et al.* 2003a): $\omega(\omega + i\nu_{dn})\delta\mathbf{r}_{\omega,\mathbf{k}} = \mathbf{D}_{\omega,\mathbf{k}}\delta\mathbf{r}_{\omega,\mathbf{k}}$, where the components of the dynamics matrix are $D_{\omega,\mathbf{k}}^{xx} = \alpha - \beta$, $D_{\omega,\mathbf{k}}^{yy} = \alpha + \beta$, and $D_{\omega,\mathbf{k}}^{xy} = D_{\omega,\mathbf{k}}^{yx} = \gamma$, and the coefficients α, β, and γ are represented by the following sums over all neighbors (m and n are integers):

$$\begin{aligned}
\alpha &= \Omega_{\rm DL}^2 \sum_{m,n} e^{-K}(K^{-1} + K^{-2} + K^{-3}) \sin^2 \tfrac{1}{2}\mathbf{k}\cdot\mathbf{R}, \\
\beta &= \Omega_{\rm DL}^2 \sum_{m,n} e^{-K}(K^{-1} + 3K^{-2} + 3K^{-3})[(R_y^2 - R_x^2)/R^2] \sin^2 \tfrac{1}{2}\mathbf{k}\cdot\mathbf{R}, \\
\gamma &= \Omega_{\rm DL}^2 \sum_{m,n} e^{-K}(K^{-1} + 3K^{-2} + 3K^{-3})[2R_xR_y/R^2] \sin^2 \tfrac{1}{2}\mathbf{k}\cdot\mathbf{R}.
\end{aligned} \tag{4.24}$$

Here $K = R/\lambda_{\rm D}$ is the lattice parameter for the neighbor separated by vector $\mathbf{R} = (R_x, R_y)$, with the components

$$R_x = m\tfrac{\sqrt{3}}{2}\Delta, \quad R_y = (\tfrac{1}{2}m + n)\Delta. \tag{4.25}$$

The dispersion relation for the in-plane DL modes is determined by the eigenvalues of the dynamics matrix,

$$\omega_\pm(\omega_\pm + i\nu_{dn}) = \alpha \pm \sqrt{\beta^2 + \gamma^2}. \tag{4.26}$$

The two branches, $\omega_+(\mathbf{k})$ and $\omega_-(\mathbf{k})$, represent the "high-frequency" and "low-frequency" modes, respectively. These modes are shown in Figure 4.4 for different

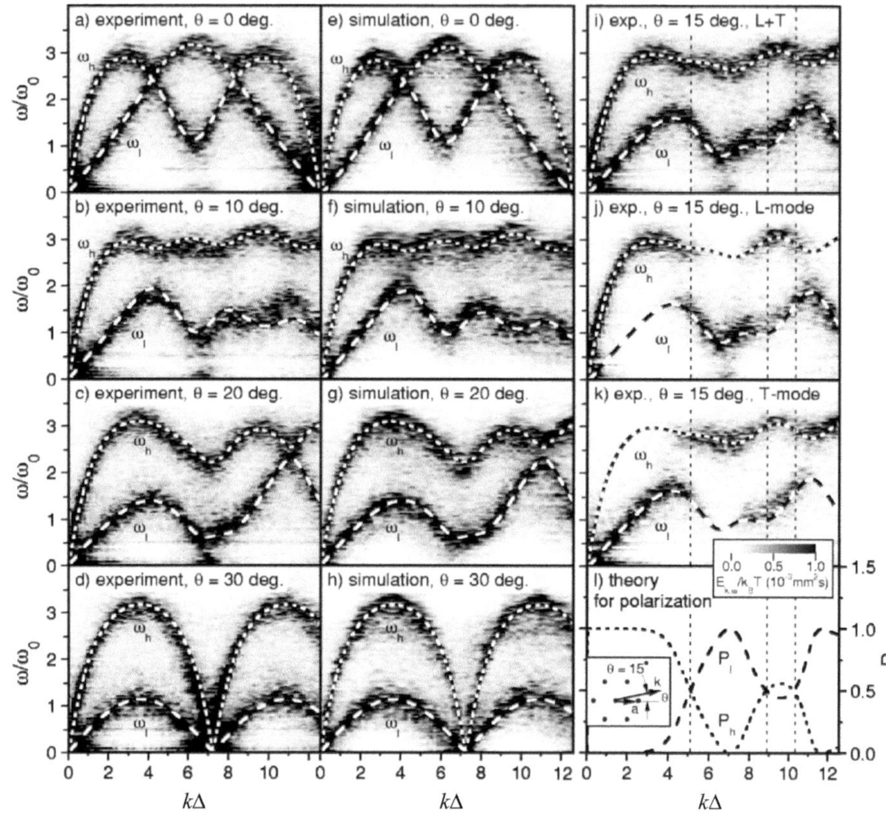

FIGURE 4.4
Dispersion relations of the DL waves in a monolayer hexagonal lattice at different propagation angles θ (Zhdanov *et al.* 2003a). The phonon spectra of thermally excited waves is shown, as measured (a)–(d) in experiments (argon gas pressure about 1 Pa, plastic particles of 8.9 μm diameter) and (e)–(h) in MD simulations. The theoretical dispersion relations are superposed: dotted line for the high-frequency mode ω_h and dashed line for the low-frequency mode ω_l (in the text $\omega_\pm$ are adopted, respectively). The frequency is normalized to $\omega_0 = \sqrt{Q^2/m_d\Delta^3} \equiv \kappa^{-3/2}\Omega_{DL}$ and the wave vector is normalized to the interparticle distance Δ. (i) – (k) Phonon spectra of the waves propagating at $\theta = 15^\circ$ measured experimentally. The high- and low-frequency branches have mixed longitudinal (L) and transverse (T) polarization. (l) Polarization of the high-frequency (P_h) and low-frequency (P_l) modes predicted by the theory. In the long-wavelength limit the high-frequency mode is purely longitudinal ($P_h = 1$), and the low-frequency mode is purely transverse ($P_l = 0$).

orientations of the wave vector with respect to the lattice (i.e., different propagation angle θ) (Zhdanov *et al.* 2003a). In the long-wavelength limit branches $\omega_\pm(\mathbf{k})$ are isotropic. The dependence on the propagation angle is revealed only at larger k, where the dispersion can be negative ($\partial\omega/\partial k < 0$). Along with the theoretical curves and the results of MD simulations (Figure 4.4e–h), the experimental data are shown in Figure 4.4a–d. The experimental results were obtained by employing a very effective technique proposed by Nunomura *et al.* (2002b): Instead of using external excitation and measuring the particle response, the naturally excited (thermal) particle motion is recorded and a Fourier transform of the velocities, $V_{\omega,\mathbf{k}}$, is computed. The highest values of the energy density, $\propto |V_{\omega,\mathbf{k}}|^2$, are concentrated in close proximity to distinct curves in $(\omega,\mathbf{k})$-space, which are identified as dispersion curves.

One should be reminded here that $\omega(\mathbf{k})$ is *completely* defined by the wave vectors from the first Brillouin zone (Kittel 1976). For a hexagonal lattice the first zone is a hexagon determined by the basis reciprocal vectors $\mathbf{A},\mathbf{B} = 2\pi\Delta^{-1}(\frac{1}{\sqrt{3}},\pm 1)$, which corresponds to $k \leq \frac{2}{\sqrt{3}}\pi\Delta^{-1}$ for $\theta = 0°$, and to $k \leq \frac{4}{3}\pi\Delta^{-1}$ for $\theta = 30°$. If $\mathbf{k}$ is beyond the first zone, then vector $\mathbf{G} = i\mathbf{A} + j\mathbf{B}$ (with some integer i and j) should be subtracted, so that $\mathbf{k}' = \mathbf{k} - \mathbf{G}$ lies in the first zone and, thus, $\omega(\mathbf{k})$ is equal to $\omega(\mathbf{k}')$.

Branches $\omega_\pm(\mathbf{k})$ are often referred to as the "longitudinal" and "transverse" modes. Such a distinction, however, is not always appropriate (Zhdanov *et al.* 2003a): The branches are purely longitudinal and transverse only when the propagation angle is $\theta = 0°$ and $30°$. Otherwise, for an arbitrary θ, the longitudinal polarization can be prescribed to ω_+, and the transverse one to ω_- (i.e., perturbation $\delta\mathbf{r}_{\omega,\mathbf{k}}$ is parallel or perpendicular to $\mathbf{k}$, respectively) only when the wave vectors are sufficiently small, i.e., in the long-wavelength limit. As $\mathbf{k}$ approaches the first Brillouin zone the polarization of branches $\omega_\pm(\mathbf{k})$ starts alternating between longitudinal and transverse, as shown in Figure 4.4i–l. This is because for arbitrary θ the short-wavelength longitudinal perturbations cause the transverse ones and vice versa: The modes become coupled, so that one cannot distinguish between them. The coupling disappears only in "symmetrical" cases – when $\theta = 0°$ and $30°$. Thus, the more general division into the "high-frequency" and "low-frequency" branches seems to be more suitable.

Note that the one-dimensional dispersion relation for the particle string can be recovered from Equations (4.24)–(4.26) by setting $m = 0$. This corresponds to the summation along the primitive translation vector $\Delta(0,1)$ (mode ω_+ represents perturbations parallel to the string, and mode ω_- is prohibited). It is remarkable that in the nearest neighbor approximation ($\kappa \gtrsim 2$) *and* in the long-wavelength limit, branch $\omega_+(\mathbf{k})$ almost coincides with Equation (4.23) (Dubin 2000; Samsonov *et al.* 2000; Zhdanov *et al.* 2003a). Moreover, for the propagation angle about $\theta = 30°$, the high-frequency branch is well approximated by the string model at *any* k.

In the long-wavelength limit the DL branches (which are purely longitudinal or transverse in this case) have an acoustic dispersion, with the phase velocities $C_{\mathrm{DL}}^{\mathrm{l,t}} = \lim_{k\to 0} \omega_\pm/k$ (superscripts "l" and "t" denote the longitudinal and transverse polarization, respectively). The velocities can be written as $C_{\mathrm{DL}}^{\mathrm{l,t}} = C_{\mathrm{DL}}F_{\mathrm{l,t}}(\kappa)$ (Peeters and

Wu 1987; Wang *et al.* 2001a; Zhdanov *et al.* 2003b), where

$$C_{\rm DL} = \Omega_{\rm DL}\lambda_{\rm D} \equiv \sqrt{\frac{Q^2}{m_d\lambda_{\rm D}}} \qquad (4.27)$$

is the DL velocity scale. The magnitude of $C_{\rm DL}$ is on the order of a few cm $\rm s^{-1}$ for typical experimental conditions (Nunomura *et al.* 2000, 2002a). In order to compare $C_{\rm DL}$ with the phase velocity of DA waves in gaseous complex plasmas, one can use the following convenient expression: $C_{\rm DL} \equiv \sqrt{\kappa^3/3}\tilde{C}_{\rm DA}$, where $\tilde{C}_{\rm DA}$ is given by Equation (4.14) calculated for dust density $\tilde{n}_d = (\frac{4}{3}\pi\Delta^3)^{-1}$. Exact formulas for functions $F_{\rm l,t}(\kappa)$, which can be derived from Equations (4.24)–(4.26), are rather complicated. However, for a practical range of κ, the functions can be very well approximated by simple polynomial expansions. For $\kappa \leq 5$, we have with accuracy better than 1% (Peeters and Wu 1987; Zhdanov *et al.* 2003b):

$$F_{\rm l} \simeq 2.69\kappa^{-1}(1-0.096\kappa-0.004\kappa^2), \quad F_{\rm t} \simeq 0.51\kappa^{-1/2}(1-0.039\kappa^2). \qquad (4.28)$$

Note that in the OCP regime ($\kappa \ll 1$), the scaling of the longitudinal velocity with κ as well as the magnitude of the velocity ($C_{\rm DL}^{\rm l} \simeq 2.7\kappa^{-1}C_{\rm DL}$) is different from the results for the one-dimensional string ($C_{\rm DL}^{\rm l} \simeq 1.4\sqrt{-\kappa^{-1}\ln\kappa}\, C_{\rm DL}$) (Wang *et al.* 2001a). Recently, the long-wavelength DL modes in crystalline monolayers were investigated in active experiments (Nunomura *et al.* 2000, 2002a), where the waves were excited with chopped laser radiation. The measured dispersion relations were found to be in remarkably good agreement with the theoretical results.

In addition to the in-plane waves, the particles in crystalline monolayers can also sustain the *vertical* (out-of-plane) DL wave mode, which is shown in Figure 4.5. The vertical mode is due to the balance between gravity and strongly inhomogeneous vertical electric force on a particle (e.g., in rf sheaths). This implies the (lowest-order) vertical parabolic confinement characterized by the frequency of a single particle oscillation, $\Omega_{\rm v}$. Employing the one-dimensional string model with parabolic transverse confinement, one can derive the dispersion relation for the vertical DL mode (nearest neighbor approximation) (Vladimirov *et al.* 1997)

$$\omega(\omega + i\nu_{dn}) = \Omega_{\rm v}^2 - 4\Omega_0^2 e^{-\kappa}(\kappa^{-2}+\kappa^{-3})\sin^2\tfrac{1}{2}k\Delta. \qquad (4.29)$$

This is the optical branch, $\lim_{k\to 0}\omega = \Omega_{\rm v}$, which has a negative dispersion, so that the group and phase velocities have opposite signs. An analytical dispersion relation for the vertical mode in a two-dimensional hexagonal lattice has been derived by Qiao and Hyde (2003) and Samsonov *et al.* (2005). In the nearest neighbor approximation *and* in the long-wavelength limit, it agrees well with Equation (4.29). However, for **k** close to the first Brillouin zone, the dispersion of the two-dimensional vertical mode becomes positive for any propagation angle, and then the vertical mode cannot be approximated by Equation (4.29). The theoretical dispersion relations are in reasonable agreement with the experimental results obtained for long waves by Liu *et al.* (2003) and Samsonov *et al.* (2005). Nevertheless, deeper experimental investigations of the vertical DL mode (similar to what has been done for the in-plane modes) are still required in order to perform comprehensive quantitative comparison with the theory.

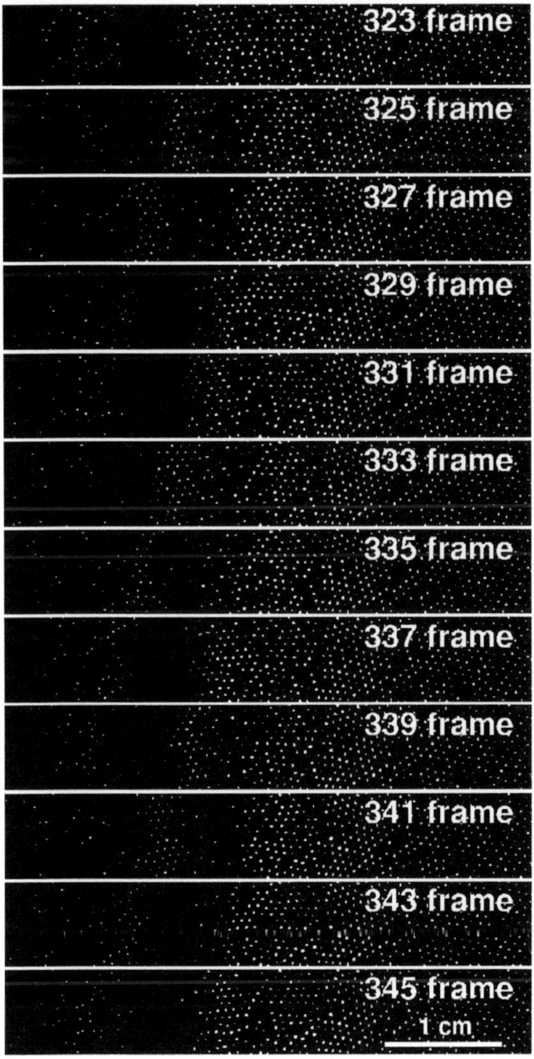

FIGURE 4.5
Vertical DL waves observed in a monolayer hexagonal lattice by Samsonov *et al.* (2005). Experiments were performed in a GEC rf chamber in argon gas at pressure about 1–2 Pa, with plastic particles of 8.9 μm diameter. Particles were illuminated by a horizontal laser sheet of about 200–300 μm thickness. The waves were excited by applying a (negative) voltage pulse to the horizontal wire positioned at the left edge. Top view of the lattice is shown at time 1.4–1.5 s after the excitation, when the wave packet was well formed. The particles are visible only if they are in the plane of the illuminating laser sheet. The stripes of particles apparently move from right to left due to the vertical wave motion, revealing the lines of constant phase. Individual particles do not move horizontally. The numbers on the images indicate the frame number (at 230.75 fps).

4.4.3 Three-dimensional plasma crystals

So far, no reliable experimental results on the DL waves in three-dimensional plasma crystals have been reported. Basically, there are two reasons for this:

(i) The major problem of the wave investigation in three-dimensional complex plasmas is the lack of effective 3D diagnostics. The laser sheets which are employed to render the particle motion cannot precisely reveal the information (viz., particle velocity) in the direction perpendicular to the sheet. The technique that is currently available for 3D diagnostics [based on the particle color coding of the third dimension, see brief description by Annaratone *et al.* (2004)] restricts the analysis to a relatively small volume and also is very complicated technically. Therefore, the major experimental efforts (to study waves and other dynamical phenomena) have been focused so far on the crystalline monolayers.

(ii) In the ground-based experiments it is rather difficult to obtain 3D plasma crystals of "good quality": The particle clouds are very stressed in the vertical direction (unless the particles are about 1 μm or smaller, but then their recognition becomes difficult) – the inhomogeneity scale can be comparable to the particle separation (e.g., Zuzic *et al.* 2000; Nefedov *et al.* 2003). In contrast, the plasma crystals produced under microgravity conditions occur in the most normal, isotropic, stress-free state which can be obtained in complex plasmas. Recent experiments performed with the PK-3 Plus rf setup onboard the ISS by Thomas *et al.* (2008) allowed us to form fairly homogeneous crystals with a relatively small number of dislocations, which is very important for the comparison with theory.

Below we briefly mention the main theoretical results for the DL waves in three-dimensional plasma crystals: The number of acoustic modes which can be sustained in crystals is 3. Since the number of particles per elementary lattice cell, r, can be more than one (e.g., $r = 2$ for the body centered cubic (bcc) and $r = 3$ for the face centered cubic (fcc) lattices), the remaining $3(r-1)$ modes have an optical dispersion (although these modes can be degenerate) (Kittel 1976). In the long-wavelength limit, the phase velocities of the (acoustic) modes are isotropic. When $\kappa \ll 1$ (OCP regime), the κ-scaling of the longitudinal phase velocity, $C_{\rm DL}^{\rm l} \simeq 5.0(7.0)\kappa^{-3/2}C_{\rm DL}$ for a bcc (fcc) lattice, is different from that for one- and two-dimensional model, whereas the transverse acoustic velocity, $C_{\rm DL}^{\rm t} \simeq 0.19\kappa^{-1/2}C_{\rm DL}$ (both for bcc and fcc lattices), has the same scaling (Wang *et al.* 2001a). Note that in comparison to monolayers, Equations (4.27) and (4.28), the magnitude of the phase velocity in three-dimensional crystal is larger for the longitudinal mode and is smaller for the transverse mode. For arbitrary κ one can obtain the phase velocities of all modes in the long-wavelength limit by using the results for the elastic constants of Yukawa crystals (e.g., Robbins *et al.* 1988).

It is noteworthy that the wave modes in three-dimensional plasma crystals are similar to those in solids. Therefore, the comprehensive investigation of particle dynamics in plasma crystals can give us an excellent opportunity to study generic wave phenomena – mode interaction, umklapp processes, phonon scattering on defects, etc. – at the kinetic level.

4.4.4 Instabilities in plasma crystals

A number of different mechanisms can trigger wave instabilities and cause eventual melting of plasma crystals. Some of these instabilities can operate irrespective of the phase state (although, the parameters of the instabilities depend on the coupling parameter Γ), some are peculiar to plasma crystals, and some can set in only when the crystal has a particular dimensionality (e.g., the instability can be triggered in monolayers only). The common type is the ion streaming instability, which is similar to that discussed in Section 4.2.2. For strongly coupled and crystalline states, the instability threshold was calculated by Kalman and Rosenberg (2003). It was found that the strong coupling generally leads to an enhancement of the growth rates. The major wave instabilities peculiar to plasma crystals are as follows:

(i) Wake-induced instability in three-dimensional crystals: The charged grains suspended in rf sheaths or dc striations often assemble themselves into the so-called "vertically aligned" hexagonal lattices. Such structures can be stable only at sufficiently high pressures. When the pressure (and, thus, the damping rate ν_{dn}) decreases below a certain threshold, the particles start oscillating horizontally, which indicates the instability onset (Melzer *et al.* 1996). The further (relatively slight) pressure decrease leads to an increase of the oscillation amplitude and melting of the crystal (Melandsø 1997; Schweigert *et al.* 1998). This instability occurs because the presence of wakes makes the interparticle interaction non-reciprocal, and hence, the total energy of the particle system is not conserved. The source of the energy is the ion flux. The instability was first analyzed theoretically by Melzer *et al.* (1996) using the model of a point-like dipole downstream from the grain. This model yields very good qualitative agreement with experiments.

(ii) Coupling instability in monolayers: This mechanism was considered in detail in Section 3.2.2. It might be one of the main reasons for the monolayer melting at pressures below ~ 10 Pa.

(iii) Instability due to defects: Another instability mechanism which can be especially important in bilayer crystals is associated with the so-called "strong defects" – the particles which are located above and below the "complete" layers (Schweigert *et al.* 2000). These particles were shown to be very effective sources of the local heating. The instability due to strong defects starts somewhat before the wake-induced instability sets on, and makes the melting transition more smoothed as the pressure decreases.

(iv) Instability due to charge fluctuations: Stochastic variations of the grain charges can trigger another instability in plasma crystals (Morfill *et al.* 1999). The mechanism of energy gain in this case is similar to stochastic heating considered in Section 3.2.1: The charge variations not only provide an additional Langevin-like term in the equations of the particle motion, but also result in a multiplicative effect, inducing a parametric instability. The instability can be triggered when the neutral damping is below a threshold,

$$\nu_{dn} \lesssim \sigma_Z^2 \Omega_{\rm DL}^2 / \Omega_{\rm ch},$$

where σ_Z^2 is the dimensionless charge dispersion [see also the condition for the instability of a single particle, Equation (3.3)]. If the variations of the grain charges are due to the discreteness of plasma (electron and ion) charges, then the magnitude of the dispersion is fairly small, $\sigma_Z^2 \sim |Z|^{-1}$, and the instability is only possible at pressures far below ~ 1 Pa (Morfill *et al.* 1999). However, in sufficiently dense complex plasmas, the charge fluctuations might be due to the dust grain discreteness (Tsytovich and de Angelis 2002), which yields substantially larger values of σ_Z^2, and hence, the instability can be possible at much larger pressures. Nevertheless, one should note that so far there have been no reliable experiments where this type of instability has been clearly identified.

4.5 Nonlinear waves

Complex plasmas, as any other plasmas are nonlinear media where the waves of finite amplitude cannot be generally considered independently. Nonlinear phenomena in complex plasma are very diverse, due to a large number of different wave modes which can be sustained. The wave amplitude can reach a nonlinear level because of different processes: This is not necessarily an external forcing or the wave instabilities – it can also be a regular collective process of nonlinear wave steepening. In the absence of dissipation (or when the dissipation is small enough), nonlinear steepening can be balanced by wave dispersion which, in turn, can result in the formation of *solitons*. When the dissipation is large, it can overcome the role of dispersion and then the balance of nonlinearity and dissipation can generate *shock waves*. In many cases the lowest-order nonlinear terms are quadratic, and then the weakly nonlinear soliton dynamics is governed by the Korteweg-de Vries (KdV) equation (Karpman 1975). For solitons of arbitrary amplitude, the method of the Sagdeev pseudopotential is very convenient (Sagdeev 1966): In particular, this method allows us to determine the upper value of the Mach number beyond which the dispersion is no longer sufficient to balance the nonlinearity, and thus, the collisionless shock is formed due to "collective" dissipation. The "conventional" dissipation is often determined by viscosity, and then the shock waves can be described by the KdV-Burgers equation (Karpman 1975; Shukla 2003). However, in complex plasmas there is a rich variety of mechanisms which determine nonlinear and dispersive properties of the medium. This generally makes the description of nonlinear waves in complex plasmas more complicated.

4.5.1 Ion solitons and shocks

The theory predicts that in complex plasmas (as well as in electronegative plasmas) both compressive and rarefactive dust ion–acoustic solitons are possible (e.g., Bharuthram and Shukla 1992; Pillay and Bharuthram 1992). It was shown that the properties of the DIA solitons (profile and the range of Mach numbers where the solitons can exist) are strongly affected by the form of the electron and ion distribution function, in particular, by the presence of "cold" and "hot" populations (e.g., in space environment) and trapped electrons (e.g., in laboratory plasmas). The compressive DIA solitons were observed in experiments by Nakamura and Sarma (2001) performed in a dusty plasma device at very low pressures ($p \sim 10^{-2}$ Pa), whereas in plasmas with a negative ion component the rarefactive solitons were also reported (Ludwig *et al.* 1984; Nakamura 1987). At such pressures the collisions with neutrals play almost no role, and a weak dissipation does not destroy the profile of DIA solitons as long as the dissipation time scales are longer than the duration of the soliton existence (Popel *et al.* 2003, 2004). As regards the DIA shocks, they were observed in different experiments (Luo *et al.* 1999; Nakamura *et al.* 1999; Nakamura 2002) performed at pressures $p \sim 10^{-2}$–10^{-3} Pa. Depending on the parameter regime (in particular, number density of grains), different dissipation mechanisms can play the major role (Popel *et al.* 2001, 2004): Along with the ion viscosity (due to collisions with grains), these are grain charge variations (ion absorbtion) and Landau damping. The general trend is that in the absence of dust the shock front exhibits pronounced oscillatory structure (Nakamura 2002) typical for collisionless ion–acoustic shocks (Taylor *et al.* 1970). As the dust density increases, the peaks become smoothed and eventually disappear, leaving the monotonic front profile, as shown in Figure 4.6.

4.5.2 Dust solitons and shocks

Longitudinal dust solitons of moderate amplitude were observed in experiments by Samsonov *et al.* (2002) and Nosenko *et al.* (2002b). Both experiments were performed in rf discharges at low pressures ($p \simeq 1.8$–2 Pa). The solitons were excited in crystalline monolayers by electrical pulses or by the laser beams. Figure 4.7 shows the evolution of the soliton propagating along the crystal (Samsonov *et al.* 2002). Theoretical study of the soliton dynamics is based on the analysis of Equation (4.22). Defining $\mathbf{x}$ as the propagation vector and retaining the lowest-order nonlinearity and dispersion terms, the resulting equation for the nonlinear wave dynamics is (Samsonov *et al.* 2002; Zhdanov *et al.* 2002)

$$\frac{\partial^2 u}{\partial t^2} + \nu_{dn}\frac{\partial u}{\partial t} = C^2 \frac{\partial^2}{\partial x^2}\left(u + \ell^2 \frac{\partial^2 u}{\partial x^2} + \frac{1}{2}\Lambda u^2\right). \tag{4.30}$$

Here $u = \partial \delta \mathbf{r}/\partial \mathbf{x} \simeq -\delta n_d/n_d$ is the particle density modulation expressed via the longitudinal derivative of the (in-plane) displacement, C is the long-wavelength DL phase velocity (which is independent of the direction of propagation), ℓ^2 is the dispersion coefficient which generally can have *either* sign (it has the dimension of squared length), and Λ is the nonlinear coefficient. Without the frictional dissipation, Equation (4.30) is readily reduced to the KdV equation by employing the stretched coordinates $(x - Ct, t)$. The soliton can only exist when ℓ^2 and Λ have opposite signs, so that the following relations can be fulfilled: $-\frac{1}{3}\Lambda A = 4\ell^2/L^2 = M^2 - 1$, where A and L are the soliton amplitude and width, respectively, and $M = V/C$ is the Mach number for the soliton velocity. The Mach number is a convenient control parameter which defines the soliton profile, $-u = A\cosh^{-2}(\xi/L)$, with $\xi = x - Vt$.

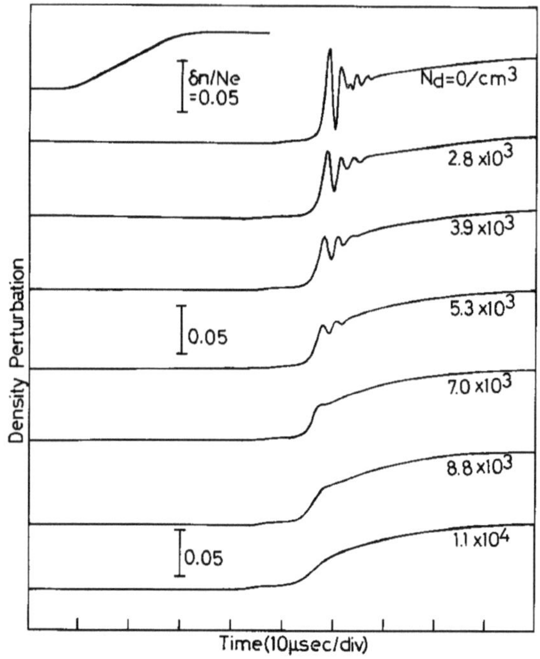

FIGURE 4.6
DIA shock observed in a double plasma device (Nakamura 2002). Experiments were performed with argon gas at pressure about $(2-4)\times 10^{-2}$ Pa, for different densities n_d of the dust particles of about 8.9 μm diameter. The DIA waves were excited with a positive ramp voltage applied to the source anode, and the signals were detected by the movable Langmuir probe. The electron density perturbations were recovered from the perturbations of the electron saturation current on the probe.

In two-dimensional hexagonal lattices (Zhdanov *et al.* 2002), ℓ^2 is always positive and has a very weak dependence on the direction of propagation; Λ is always negative and can depend on the direction substantially, especially at $\kappa \gtrsim 1$. Such a combination of signs implies that only compressive ($A > 0$) supersonic ($M > 1$) solitons can propagate in crystalline monolayers, as is observed in experiments. For $\kappa \gtrsim 1$, one can calculate parameters of Equation (4.30) by using the results for a one-dimensional string (Samsonov *et al.* 2002),

$$\begin{aligned} C^2 &= C_{\rm DL}^2\kappa^2\left[G(\kappa)/\kappa\right]'', \\ \ell^2 &= \tfrac{1}{12}\lambda_{\rm D}^2\kappa^2\left[G''(\kappa)/\kappa\right]''/\left[G(\kappa)/\kappa\right]'', \\ \Lambda &= \kappa\left[G(\kappa)/\kappa\right]'''/\left[G(\kappa)/\kappa\right]'', \end{aligned} \tag{4.31}$$

where $G(\kappa) = -\ln(e^{\kappa}-1)$. This relatively simple theoretical model provides re-

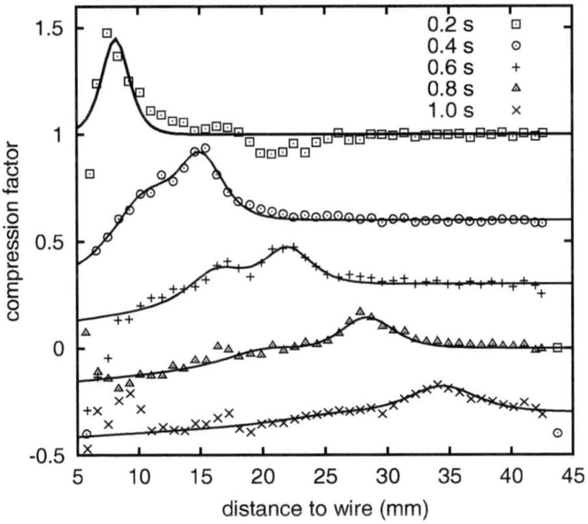

FIGURE 4.7
Dust soliton observed in experiments with a monolayer hexagonal lattice by Samsonov *et al.* (2002). Experimental conditions and the wave excitation technique are described in Figure 4.5. Compression factor $1 + \delta n_d / n_d$ versus distance to the wire is plotted at different times. The solid lines show the theoretical fits to the experimental data. The fits and experimental points at later times are offset down (by 0.4, 0.7, 1.0, and 1.3).

markably good agreement with the experiments. If the neutral gas pressure is low enough, the friction does not destroy the soliton (Samsonov *et al.* 2002). The perturbation simply slows down, approaching the asymptote $V = C$, and the form of the soliton changes in accordance with the analytical solution (i.e., the amplitude decreases and the width increases, keeping the "soliton relation" $AL^2 = const$, see Figure 4.7). Thus, one can speak about a "weakly dissipative soliton" when the dissipation time scale, $\sim \nu_{dn}^{-1}$, exceeds the time scale of the wave itself, $\sim \Omega_{\rm DL}^{-1}$.

The theory predicts in-plane transverse (shear) solitons in two-dimensional lattices (Zhdanov *et al.* 2002), as well as the solitons due to the coupling between longitudinal in-plane and vertical out-of-plane modes (Ivlev *et al.* 2003). Such solitons, however, have not yet been observed in experiments. There have also been a number of theoretical papers on properties of the DA solitons in gaseous complex plasma (Rao *et al.* 1990; Verheest 1992; Mamun *et al.* 1996; Ma and Liu 1997; Xie *et al.* 1999; Ivlev and Morfill 2001), but no experiments have been done so far.

As regards the dust shock waves, this topic still needs to be explored, both theoretically and experimentally. The theory of shocks in weakly coupled complex plasmas has been studied, e.g., by Melandsø and Shukla (1995) and Popel *et al.* (1996). It was suggested that the major dissipation mechanism providing the shock formation

(a)
(b)
(c)
(d)

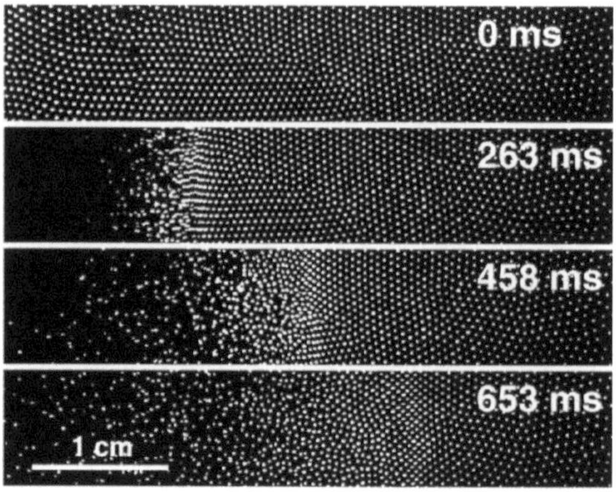

FIGURE 4.8
Dust shock wave propagating in a monolayer hexagonal lattice (Samsonov *et al.* 2004b). Experimental conditions and the the wave excitation technique are described in Figure 4.5. Initially undisturbed particles (a) were swept from left to right (b) and (c) forming a shock with a sharp front. The lattice melted behind the front. At later times (d) the shock weakened due to the neutral drag and a soliton was formed.

can be the dust charge variations, and the weak shocks can be described by the KdV-Burgers equation. These results, however, have better applicability to the space environment, where the complex plasmas can be found in the gaseous state and where the the charge variation effects are not inhibited by the gas friction (see Section 4.2.2). For the strongly coupled plasmas, the generalized hydrodynamic approach (see Section 4.3) was proposed by Kaw and Sen (1998) and Shukla (2003). This approach suggests that weak shocks cannot be described by the KdV-Burgers equation in general case (Kaw and Sen 1998). In experiments, "pure" shocks have been observed so far only in two-dimensional crystals by Samsonov *et al.* (2004a,b). (The term "pure" implies here that the momentum exchange in dust–dust collisions prevails over the momentum loss due to neutral gas friction, $\nu_{dd} \gg \nu_{dn}$, so that charged grains have properties of one-phase fluids, see Section 2.4.3.) These shocks [generated by electrical pulses, like the solitons in experiments of Samsonov *et al.* (2002)] caused melting of the crystal behind the front, as shown in Figure 4.8. As the shock propagated and weakened it was seen that the melting ceased. Further propagation of the pulse was in the form of a soliton, as described above. Other examples of strong dust discontinuities were observed in microgravity experiments with an rf discharge by Samsonov *et al.* (2003) and in ground-based dc experiments by Fortov *et al.* (2004). In both cases the shock-like structures were triggered in three-dimensional complex plasmas using gas pulses. The experiments were performed at high gas pressures about 50–120 Pa, when the friction dissipation is very strong, ν_{dn} ~100–300 s^{-1}.

Nevertheless, the "shocks" were observed during a few seconds almost undamped, which suggests that there must be a mechanism of strong energy "influx" into the structures (e.g., modulational instability). Therefore, the dust discontinuities observed by Samsonov *et al.* (2003) and Fortov *et al.* (2004) should rather be referred to as "dissipative structures" where pure hydrodynamic effects presumably play a minor role.

4.5.3 Mach cones

Dispersion relations of dust modes in complex plasmas suggest that, irrespective of the plasma state [see Equations (4.13), (4.20), (4.21), and (4.26)], the phase velocity attains the maximal value in the long-wavelength limit. For acoustic modes this velocity – the "sound speed" C – is finite and, therefore, similar to conventional media, the supersonic perturbations (i.e., with Mach number $M = V/C > 1$) are always localized behind the object which produces these perturbations (this can be a rapidly moving charged particle or a bunch of particles, biased probe, etc.). The perturbation front has a conical form in a three-dimensional case and therefore it is called a "Mach cone". In a two-dimensional case the same name is adopted, although the front is a planar V-shaped perturbation. The opening angle μ of the front at large distances from the object (where the nonlinearity should not play important role) is determined by the well-known relation $\sin\mu = C/V \equiv M^{-1}$.

Originally, it was suggested that the Mach cones (wakes) can be excited in space dusty plasmas – e.g., in planetary rings by big boulders (Havnes *et al.* 1995, 1996) moving through the dust at a velocity that is somewhat higher than C. It was expected that the discovery of Mach cones and the measurements of the opening angles during the Cassini mission to Saturn will yield new information on the dusty plasma conditions in planetary rings. Unfortunately, no Mach cones detected during this mission were reported. The Mach cones in laboratory complex plasmas were discovered by Samsonov *et al.* (1999, 2000) in two-dimensional plasma crystals. They were generated by single particles spontaneously moving beneath the monolayer along straight trajectories. [The physical mechanism which drives such motion is still an open issue; for one of the possible explanations, see Schweigert *et al.* (2002).] In experiments of Melzer *et al.* (2000), the Mach cones were excited by the radiation pressure of a focused laser beam. The wake reveals a multiple cone structure behind the front, as shown in Figure 4.9. Generally, the wake structure is determined by the dispersion and nonlinear properties of particular wave modes excited behind the front (Zhdanov *et al.* 2002; Nosenko *et al.* 2003). The formation of the second cone behind the first one, with the opening angle smaller for the second cone can be ascribed to the shear (transverse) wave front (Nosenko *et al.* 2002a, 2003), because the (longitudinal) sound speed is larger than the shear phase velocity [see Equations (4.27) and (4.28)]. Also, the shape of the cone wings can be affected by the inhomogeneity of the particle density, as suggested by Zhdanov *et al.* (2004). It was proposed by Havnes *et al.* (1996) and Melzer *et al.* (2000) to use the Mach cones as a tool to determine local parameters of complex plasmas, e.g., particle charge and the screening length, making use of the measured sound speed.

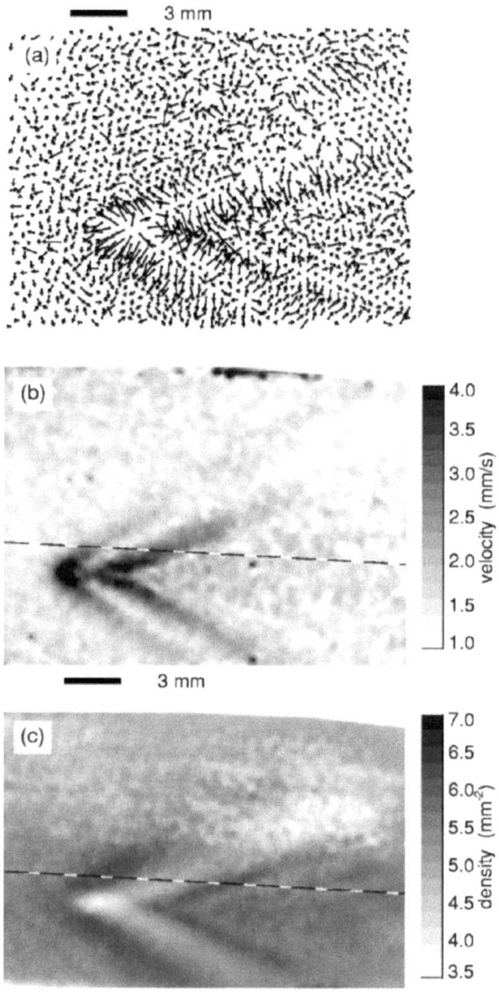

FIGURE 4.9
Mach cone observed in a monolayer hexagonal lattice by Samsonov *et al.* (2000). Experiments were performed in a GEC rf chamber in krypton gas at pressure about 1.2 Pa, with plastic particles of 8.9 μm diameter. The cone was excited by a supersonic particle which moved spontaneously beneath the monolayer. (a) Particle velocity vector map derived from particle positions in two consecutive video fields, (b) gray-scale speed map, and (c) gray-scale number density map. The first cone consists of particles moving forward, and it coincides with the high density region. The second cone has particles moving backward, and it coincides with the low density region.

References

Alexandrov, A. F., Bogdankevich, L. S., and Rukhadze, A. A. (1984). *Principles of plasma electrodynamics*. Springer-Verlag, New York.

Annaratone, B. M., Antonova, T., Goldbeck, D. D., Thomas, H. M., and Morfill, G. E. (2004). Complex-plasma manipulation by radiofrequency biasing. *Plasma Phys. Control. Fusion*, **46**, B495–B510.

Annibaldi, S. V., Ivlev, A. V., Konopka, U., Ratynskaia, S., Thomas, H. M., Morfill, G. E., Lipaev, A. M., Molotkov, V. I., Petrov, O. F., and Fortov, V. E. (2007). Dust–acoustic dispersion relation in three-dimensional complex plasmas under microgravity. *New J. Phys.*, **9**, 327/1–9.

Annou, B. (1998). Current-driven dust ion–acoustic instability in a collisional dusty plasma with a variable charge. *Phys. Plasmas*, **5**, 2813–2814.

Barkan, A., Merlino, R. L., and D'Angelo, N. (1995). Laboratory observation of the dust–acoustic wave mode. *Phys. Plasmas*, **2**, 3563–3565.

Barkan, A., D'Angelo, N., and Merlino, R. L. (1996a). Experiments on ion–acoustic waves in dusty plasmas. *Planet. Space Sci.*, **44**, 239–242.

Barkan, A., D'Angelo, N., and Merlino, R. L. (1996b). Potential relaxation instability and ion acoustic waves in a single-ended Q-machine dusty plasma. *Phys. Lett. A*, **222**, 329–332.

Bharuthram, R. and Shukla, P. K. (1992). Large amplitude ion–acoustic solitons in a dusty plasma. *Planet Space Sci.*, **40**, 973–977.

Chu, J. H., Du, J. B., and I, L. (1994). Coulomb solids and low-frequency fluctuations in RF dusty plasmas. *J. Phys. D: Appl. Phys.*, **27**, 296–300.

D'Angelo, N. (1994). Ion–acoustic waves in dusty plasmas. *Planet. Space Sci.*, **42** , 507–511.

D'Angelo, N. (1997). Ionization instability in dusty plasmas. *Phys. Plasmas*, **4**, 3422–3426.

D'Angelo, N. (1998). Dusty plasma ionization instability with ion drag. *Phys. Plasmas*, **5**, 3155–3160.

D'Angelo, N. and Merlino, R. L. (1996). Current-driven dust–acoustic instability in a collisional plasma. *Planet. Space Sci.*, **44**, 1593–1598.

Dubin, D. H. E. (2000). The phonon wake behind a charge moving relative to a two-dimensional plasma crystal. *Phys. Plasmas*, **7**, 3895–3903.

Farouki, R. T. and Hamaguchi, S. (1994). Thermodynamics of strongly-coupled Yukawa systems near the one-component-plasma limit. II. Molecular dymanics simulations. *J. Chem. Phys.*, **101**, 9885–9893.

Fortov, V. E., Khrapak, A. G., Khrapak, S. A., Molotkov, V. I., Nefedov, A. P., Petrov, O. F., and Torchinsky, V. M. (2000). Mechanism of dust–acoustic instability in a current glow discharge plasma. *Phys. Plasmas*, **7**, 1374–1380.

Fortov, V. E., Usachev, A. D., Zobnin, A. V., Molotkov, V. I., and Petrov, O. F. (2003). Dust–acoustic wave instability at the diffuse edge of radio frequency inductive low-pressure gas discharge plasma. *Phys. Plasmas*, **10**, 1199–1207.

Fortov, V. E., Petrov, O. F., Molotkov, V. I., Poustylnik, M. Y., Torchinsky, V. M., Khrapak, A. G. and Chernyshev, A. V. (2004). Large-amplitude dust waves excited by the gas-dynamic impact in a dc glow discharge plasma. *Phys. Rev. E*, **69**, 016402/1–5.

Frenkel, Ya. I. (1946). *Kinetic theory of liquids*. Clarendon, Oxford.

Fried, B. D. and Conte, S. D. (1961). *The plasma dispersion function*. Academic Press, New York.

Havnes, O., Goertz, C. K., Morfill, G. E., Grun, E., and Ip, W. (1987). Dust charges, cloud potential, and instabilities in a dust cloud embedded in a plasma. *J. Geophys. Res. A*, **92**, 2281–2293.

Havnes, O., Aslaksen, T., Hartquist, T. W., Li, F., Melandsø, F., Morfill, G. E., and Nitter, T. (1995). Probing the properties of planetary ring dust by the observation of Mach cones. *J. Geophys. Res.*, **100**, 1731–1734.

Havnes, O., Li, F., Melandsø, F., Aslaksen, T., Hartquist, T. W., Morfill, G. E., Nitter, T., and Tsytovich, V. (1996). Diagnostic of dusty plasma conditions by the observation of Mach cones caused by dust acoustic waves. *J. Vac. Sci. Technol. A*, **14**, 525–528.

Homann, A., Melzer, A., Peters, S., and Piel, A. (1997). Determination of the dust screening length by laser-excited lattice waves. *Phys. Rev. E*, **56**, 7138–7141.

Homann, A., Melzer, A., Peters, S., Madani, R., and Piel, A. (1998). Laser-excited dust lattice waves in plasma crystals. *Phys. Lett. A*, **242**, 173–180.

Ichimaru, S., Iyetomi, H., and Tanaka, S. (1987). Statistical physics of dense plasmas: Thermodynamics, transport coefficients and dynamic correlations. *Phys. Rep.*, **149**, 91–205.

Ivlev, A. V. and Morfill, G. (2000). Acoustic modes in a collisional dusty plasma: Effect of the charge variation. *Phys. Plasmas*, **7**, 1094–1102.

Ivlev, A. V. and Morfill, G. (2001). Dust acoustic solitons with variable particle charge: Role of ion distribution. *Phys. Rev. E*, **63**, 026412/1-5.

Ivlev, A. V., Samsonov, D., Goree, J., Morfill, G., and Fortov, V. E. (1999). Acoustic modes in a collisional dusty plasma. *Phys. Plasmas*, **6**, 741–750.

Ivlev, A. V., Zhdanov, S. K., and Morfill, G. E. (2003). Coupled dust-lattice solitons in monolayer plasma crystals. *Phys. Rev. E*, **68**, 066402/1–4.

Ivlev, A. V., Zhdanov, S. K., Khrapak, S. A., and Morfill, G. E. (2004). Ion drag force in dusty plasmas. *Plasma Phys. Control. Fusion*, **46**, B267–B279.

Ivlev, A. V., Zhdanov, S. K., Khrapak, S. A., and Morfill, G. E. (2005). Kinetic approach for the ion drag force in a collisional plasma. *Phys. Rev. E*, **71**, 016405/1–7.

Jana, M. R., Sen, A., and Kaw, P. K. (1993). Collective effects due to charge-fluctuations dynamics in a dusty plasma. *Phys. Rev. E*, **48**, 3930–3933.

Joyce, G., Lampe, M., and Ganguli, G. (2002). Instability-triggered phase transition to a dusty-plasma condensate. *Phys. Rev. Lett.*, **88**, 095006/1–4.

Kalman, G. and Golden, K. I. (1990). Response function and plasmon dispersion for strongly coupled Coulomb liquids. *Phys. Rev. A*, **41**, 5516–5527.

Kalman, G. J. and Rosenberg, M. (2003). Instabilities in strongly coupled plasmas. *J. Phys. A: Math. Gen.*, **36**, 5963–5969.

Kalman, G., Rosenberg, M., and DeWitt, H. E. (2000). Collective modes in strongly correlated Yukawa liquids: Waves in dusty plasmas. *Phys. Rev. Lett.*, **84**, 6030–6033.

Karpman, V. I. (1975). *Nonlinear waves in dispersive media*. Pergamon, Oxford.

Kaw, P. K. (2001). Collective modes in a strongly coupled dusty plasma. *Phys. Plasmas*, **8**, 1870–1878.

Kaw, P. K. and Sen, A. (1998). Low frequency modes in strongly coupled dusty plasmas. *Phys. Plasmas*, **5**, 3552–3559.

Kaw, P. and Singh, R. (1997). Collisional instabilities in a dusty plasma with recombination and ion-drift effects. *Phys. Rev. Lett.*, **79**, 423–426.

Khrapak, S., Samsonov, D., Morfill, G., Thomas, H., Yaroshenko, V., Rothermel, H., Hagl, T., Fortov, V., Nefedov, A., Molotkov, V., Petrov, O., Lipaev, A., Ivanov, A., and Baturin Y. (2003). Compressional waves in complex (dusty) plasmas under microgravity conditions. *Phys. Plasmas*, **10**, 1–4.

Kittel, C. (1976). *Introduction to Solid State Physics*. Wiley, New York.

Lieberman M. A. and Lichtenberg, A. J. (1994). *Principles of plasma discharges and materials processing*. Wiley, New York.

Lifshitz E. M. and Pitaevskii, L. P. (1981). *Physical kinetics*. Pergamon, Oxford.

Liu, B., Avinash, K., and Goree, J. (2003). Transverse optical mode in a one-dimensional Yukawa chain. *Phys. Rev. Lett.*, **91**, 255003/1–4.

Ludwig, G. O., Ferreira, J. L., and Nakamura, Y. (1984). Observation of ion–acoustic rarefaction solitons in a multicomponent plasma with negative ions. *Phys. Rev. Lett.*, **52**, 275–278.

Luo, Q.-Z., D'Angelo, N., and Merlino, R. L. (1999). Experimental study of shock formation in a dusty plasma. *Phys. Plasmas*, **6**, 3455–3458.

Luo, Q.-Z., D'Angelo, N., and Merlino, R. L. (2000). Ion acoustic shock formation in a converging magnetic field geometry. *Phys. Plasmas*, **7**, 2370–2373.

Ma, J. X. and Liu, J. (1997). Dust–acoustic soliton in a dusty plasma. *Phys. Plasmas*, **4**, 253–255.

Ma, J.-X. and Yu, M. Y. (1994). Self-consistent theory of ion acoustic waves in a dusty plasma. *Phys. Plasmas*, **1**, 3520–3522.

Mamun, A. A. and Shukla, P. K. (2000). Streaming instabilities in a collisional dusty plasma. *Phys. Plasmas*, **7**, 4412–4417.

Mamun, A. A., Cairns, R. A., and Shukla, P. K. (1996). Solitary potentials in dusty plasmas. *Phys. Plasmas*, **3**, 702–704.

Melandsø, F. (1996). Lattice waves in dust plasma crystals. *Phys. Plasmas*, **3**, 3890–3901.

Melandsø, F. (1997). Heating and phase transitions of dust-plasma crystals in a flowing plasma. *Phys. Rev. E*, **55**, 7495–7506.

Melandsø, F. and Shukla, P. K. (1995). Theory of dust–acoustic shocks. *Planet. Space Sci.*, **43**, 635–648.

Melandsø, F., Aslaksen, T., and Havnes, O. (1993). A new damping effect for the dust–acoustic wave. *Planet. Space Sci.*, **41**, 321–325.

Melzer, A., Schweigert, V. A., Schweigert, I. V., Homann, A., Peters, S., and Piel, A. (1996). Structure and stability of the plasma crystal. *Phys. Rev. E*, **54**, R46–R49.

Melzer, A., Nunomura, S., Samsonov, D., Ma, Z. W., and Goree, J. (2000). Laser-excited Mach cones in a dusty plasma crystal. *Phys. Rev. E*, **62**, 4162–4168.

Merlino, R. L. (1997). Current-driven dust ion–acoustic instability in a collisional dusty plasma. *IEEE Trans. Plasma Sci.*, **25**, 60–65.

Merlino, R. L., Barkan, A., Thompson, C., and D'Angelo, N. (1998). Laboratory studies of waves and instabilities in dusty plasmas. *Phys. Plasmas*, **5**, 1607–1614.

Misawa, T., Ohno, N., Asano, K., Sawai, M., Takamura, S., and Kaw, P. K. (2001). Experimental observation of vertically polarized transverse dust-lattice wave propagating in a one-dimensional strongly coupled dust chain. *Phys. Rev. Lett.*, **86**, 1219–1222.

Mishra, A., Kaw, P. K., and Sen, A. (2000). Instability of shear waves in an inhomogeneous strongly coupled dusty plasma. *Phys. Plasmas*, **8**, 3188–3193.

Molotkov, V. I., Nefedov, A. P., Torchinskii, V. M., Fortov, V. E., and Khrapak, A. G. (1999). Dust acoustic waves in a dc glow-discharge plasma. *JETP*, **89**, 477–480.

Morfill, G., Ivlev, A. V., and Jokipii, J. R. (1999). Charge fluctuation instability of the dust lattice wave. *Phys. Rev. Lett.*, **83**, 971–974.

Murillo, M. S. (1998). Static local field correction description of acoustic waves in strongly coupling dusty plasmas. *Phys. Plasmas*, **5**, 3116–3121.

Murillo, M. S. (2000). Longitudinal collective modes of strongly coupled dusty plasmas at finite frequencies and wave vectors. *Phys. Plasmas*, **7**, 33–38.

Nakamura, Y. (1987). Observation of large-amplitude ion acoustic solitary waves in a plasma. *J. Plasma Phys.*, **38**, 461–471.

Nakamura, Y. (2002). Experiments on ion–acoustic shock waves in a dusty plasma. *Phys. Plasmas*, **9**, 440–445.

Nakamura, Y. and Sarma, A. (2001). Observation of ion–acoustic solitary waves in a dusty plasma. *Phys. Plasmas*, **8**, 3921–3926.

Nakamura, Y., Bailung, H., and Shukla, P. K. (1999). Observation of ion–acoustic shocks in a dusty plasma. *Phys. Rev. Lett.*, **83**, 1602–1605.

Nefedov, A. P., Morfill, G. E., Fortov, V. E., Thomas, H. M., Rothermel, H., Hagl, T., Ivlev, A. V., Zuzic, M., Klumov, B. A., Lipaev, A. M., Molotkov, V. I., Petrov, O. F., Gidzenko, Y. P., Krikalev, S. K., Shepherd, W., Ivanov, A. I., Roth, M., Binnenbruck, H., Goree, J. A., and Semenov, Yu. P. (2003). PKE-Nefedov: Plasma crystal experiments on the International Space Station. *New J. Phys.*, **5**, 33/1–10.

Nosenko, V., and Goree, J. (2004). Shear flows and shear viscosity in a two-dimensional Yukawa system (dusty plasma). *Phys. Rev. Lett.*, **93**, 155004/1–4.

Nosenko, V., Goree, J., Ma, Z. W., and Piel, A. (2002a). Observation of shear-wave Mach cones in a 2D dusty-plasma crystal. *Phys. Rev. Lett.*, **88**, 135001/1–4.

Nosenko, V., Nunomura, S., and Goree, J. (2002b). Nonlinear compression pulses in a 2D crystallized dusty plasma. *Phys. Rev. Lett.*, **88**, 215002/1–4.

Nosenko, V., Goree, J., Ma, Z. W., Dubin, D. H. E., and Piel, A. (2003). Compressional and shear wakes in a two-dimensional dusty plasma crystal. *Phys. Rev. E*, **68**, 056409/1–15.

Nosenko, V., Avinash, K., Goree, J., and Liu, B. (2004). Nonlinear interaction of compressional waves in a 2D dusty plasma crystal. *Phys. Rev. Lett.*, **92**, 085001/1–4.

Nunomura, S., Samsonov, D., and Goree, J. (2000). Transverse waves in a two-dimensional screened-Coulomb crystal (dusty plasma). *Phys. Rev. Lett.*, **84**, 5141–5144.

Nunomura, S., Goree, J., Hu, S., Wang, X., and Bhattacharjee, A. (2002a). Dispersion relations of longitudinal and transverse waves in two-dimensional screened Coulomb crystals. *Phys. Rev. E*, **65**, 066402/1–11.

Nunomura, S., Goree, J., Hu, S., Wang, X., Bhattacharjee, A., and Avinash, K. (2002b). Phonon spectrum in a plasma crystal. *Phys. Rev. Lett.*, **89**, 035001/1–4.

Nunomura, S., Zhdanov, S., Morfill, G. E., Goree, J. (2003). Nonlinear longitudinal waves in a two-dimensional screened Coulomb crystal. *Phys. Rev. E*, **68**, 026407/1–7.

Nunomura, S., Zhdanov, S., Samsonov, D., and Morfill, G. (2005). Wave spectra in solid and liquid complex (dusty) plasmas. *Phys. Rev. Lett.*, **94**, 045001/1–4.

Ohta, H. and Hamaguchi, S. (2000). Wave dispersion relations in Yukawa fluids. *Phys. Rev. Lett.*, **84**, 6026–6029.

Ostrikov, K. N., Vladimirov, S. V., Yu, M. Y., and Morfill, G. E. (2000). Low-frequency dispersion properties of plasmas with variable-charge impurities. *Phys. Plasmas*, **7**, 461–465.

Peeters, F. M. and Wu, X. (1987). Wigner crystal of a screened-Coulomb-interaction colloidal system in two dimensions. *Phys. Rev A*, **35**, 3109–3114.

Peters, S., Homann, A., Melzer, A., and Piel, A. (1996). Measurement of dust particle shielding in a plasma from oscillations of a linear chain. *Phys. Lett. A*, **223**, 389–393.

Piel, A., Nosenko, V., and Goree, J. (2002). Experiments and molecular-dynamics simulation of elastic waves in a plasma crystal radiated from a small dipole source. *Phys. Rev. Lett.*, **89**, 085004/1–4.

Piel, A., Klindworth, M., Arp, O., Melzer, A., and Wolter, M. (2006). Obliquely propagating dust-density plasma waves in the presence of an ion beam. *Phys. Rev. Lett.*, **97**, 205009/1–4.

Pieper, J. B. and Goree, J. (1996). Dispersion of plasma dust acoustic waves in the strong-coupling regime. *Phys. Rev. Lett.*, **77**, 3137–3140.

Pillay, R. and Bharuthram, R. (1992). Large amplitude solitons in a multi-species electron-positron plasma. *Astrophys. Space Sci.*, **198**, 85–93.

Popel, S. I., Yu, M. Y., and Tsytovich, V. N. (1996). Shock waves in plasmas containing variable-charge impurities. *Phys. Plasmas*, **3**, 4313–4315.

Popel, S. I., Golub', A. P., Losseva, T. V. (2001). Dust ion–acoustic shock-wave structures: Theory and laboratory experiments. *JETP Lett.*, **74**, 362–366.

Popel, S. I., Golub', A. P., Losseva, T. V., Ivlev, A. V., Khrapak, S. A., and Morfill, G. (2003). Weakly dissipative dust ion–acoustic solitons. *Phys. Rev. E*, **67**, 056402/1–5.

Popel, S. I., Andreev, S. N., Gisko, A. A., Golub', A. P., and Losseva, T. V. (2004). Dissipative processes during the propagation of nonlinear dust ion–acoustic perturbations. *Plasma Phys. Rep.*, **30**, 314–329.

Pramanik, J., Prasad, G., Sen, A., and Kaw, P. K. (2002). Experimental observations of transverse shear waves in strongly coupled dusty plasmas. *Phys. Rev. Lett.*, **88**, 175001/1–4.

Qiao, K. and Hyde, T. W. (2003). Dispersion properties of the out-of-plane transverse wave in a two-dimensional Coulomb crystal. *Phys. Rev. E*, **68**, 046403/1–5.

Raizer, Yu. P. (1991). *Gas discharge physics*. Springer-Verlag, Berlin.

Rao, N. N., Shukla, P. K., and Yu, M. Y. (1990). Dust–acoustic waves in dusty plasmas. *Planet. Space Sci.*, **38**, 543–546.

Ratynskaia, S., Khrapak, S., Zobnin, A., Thoma, M. H., Kretschmer, M., Usachev, A., Yaroshenko, V., Quinn, R. A., Morfill, G. E., Petrov, O., and Fortov, V. (2004). Experimental determination of dust particle charge at elevated pressures. *Phys. Rev. Lett.*, **93**, 085001/1–4.

Robbins, M. O., Kremer, K., and Grest, G. S (1988). Phase diagram and dynamics of Yukawa systems. *J. Chem. Phys.*, **88**, 3286–3312.

Rosenberg, M. (1993). Ion- and dust–acoustic instabilities in dusty plasmas. *Planet. Space Sci.*, **41**, 229–233.

Rosenberg, M. (1996). Ion-dust streaming instability in processing plasmas. *J. Vac. Sci. Technol. A*, **14**, 631–633.

Rosenberg, M. (2002). A note on ion-dust streaming instability in a collisional dusty plasma. *J. Plasma Phys.*, **67**, 235–242.

Rosenberg, M. and Kalman, G. (1997). Dust acoustic waves in strongly coupled dusty plasmas. *Phys. Rev. E*, **56**, 7166–7173.

Sagdeev, R. Z. (1966). Cooperative phenomena and shock waves in collisionless plasmas. In *Reviews of plasma physics, Vol. 4*, Leontovich M. A. (ed.), pp. 23–93. Consultants Bureau, New York.

Saigo, T. and Hamaguchi, S. (2002). Shear viscosity of strongly coupled Yukawa systems. *Phys. Plasmas*, **9**, 1210–1216.

Salin, G. and Caillol, J.-M. (2002). Transport coefficients of the Yukawa one-component plasma. *Phys. Rev. Lett.*, **88**, 065002/1–4.

Salin, G. and Caillol, J.-M. (2003). Equilibrium molecular dynamics simulations of the transport coefficients of the Yukawa one component plasma. *Phys. Plasmas*, **10**, 1220–1230.

Samsonov, D. and Goree, J. (1999). Instabilities in a dusty plasma with ion drag and ionization. *Phys. Rev. E*, **59**, 1047–1058.

Samsonov, D., Goree, J., Ma, Z. W., Bhattacharjee, A., Thomas, H. M., and Morfill, G. E. (1999). Mach cones in a coulomb lattice and a dusty plasma. *Phys. Rev. Lett.*, **83**, 3649–3652.

Samsonov, D., Goree, J., Thomas, H. M., and Morfill, G. E. (2000). Mach cone shocks in a two-dimensional Yukawa solid using a complex plasma. *Phys. Rev. E*, **61**, 5557–5572.

Samsonov, D., Ivlev, A. V., Morfill, G. E., and Goree, J. (2001). Long-range attractive and repulsive forces in a two-dimensional complex (dusty) plasma. *Phys. Rev. E*, **63**, 025401(R)/1–4

Samsonov, D., Ivlev, A. V., Quinn, R. A., Morfill, G., and Zhdanov, S. (2002). Dissipative longitudinal solitons in a two-dimensional strongly coupled complex (dusty) plasma. *Phys. Rev. Lett.*, **88**, 095004/1–4.

Samsonov, D., Morfill, G., Thomas, H., Hagl, T., Rothermel, H., Fortov, V., Lipaev, A., Molotkov, V., Nefedov, A., Petrov, O., Ivanov, A., and Krikalev, S. (2003). Kinetic measurements of shock wave propagation in a three-dimensional complex (dusty) plasma. *Phys. Rev. E*, **67**, 036404/1–5.

Samsonov, D., Zhdanov, S., and Morfill, G. (2004a). Shock waves and solitons in complex (dusty) plasmas. In *Shock compression of condensed matter-2003*, Furnish, M. D., Gupta, Y. M., and Forbes, J. W. (eds.), pp. 111–114. AIP, New York.

Samsonov, D., Zhdanov, S., Quinn, R. A., Popel, S. I., and Morfill, G. E. (2004b). Sock melting of a two-dimensional complex (dusty) plasma. *Phys. Rev. Lett.*, **92**, 255004/1–4.

Samsonov, D., Zhdanov, S., and Morfill, G. (2005). Vertical wave packets observed in a crystallized hexagonal monolayer complex plasmas. *Phys. Rev. E*, **71**, 026410/1–7.

Schwabe, M., Rubin-Zuzic, M., Zhdanov, S., Thomas, H. M., and Morfill, G. E. (2007). Highly resolved self-excited density waves in a complex plasma. *Phys. Rev. Lett.*, **99**, 095002/1–4.

Schweigert, I. V., Schweigert, V. A., Melzer, A., and Piel, A. (2000). Melting of dust plasma crystals with defects. *Phys. Rev. Lett.*, **62**, 1238–1244.

Schweigert, V. A. (2001). Dielectric permittivity of a plasma in an external electric field. *Plasma Phys. Rep.*, **27**, 997–999.

Schweigert, V. A., Schweigert, I. V., Melzer, A., Homann, A., and Piel, A. (1998). Plasma crystal melting: A nonequilibrium phase transition. *Phys. Rev. Lett.*, **80**, 5345–5348.

Schweigert, V. A., Schweigert, I. V., Nosenko, V., and Goree, J. (2002). Acceleration and orbits of charged particles beneath a monolayer plasma crystal. *Phys. Plasmas*, **9**, 4465–4472.

Shukla, P. K. (2003). Nonlinear waves and structures in dusty plasmas. *Phys. Plasmas*, **10**, 1619–1627.

Shukla P. K. and Silin, V. P. (1992). Dust ion–acoustic wave. *Phys. Scripta*, **45**, 508.

Slattery, W. L., Doolen, G. D., and DeWitt, H. E. (1980). Improved equation of state for the classical one-component plasma. *Phys. Rev. A*, **26**, 2087–2095.

Taylor, R. J., Baker, D. R., Ikezi, H. (1970). Observation of collisionless electrostatic shocks. *Phys. Rev. Lett.*, **24**, 206–209.

Thomas, H. M., Morfill, G. E., Fortov, V. E., Ivlev, A. V., Molotkov, V. I., Lipaev, A. M., Hagl, T., Rothermel, H., Khrapak, S. A., Suetterlin, R. K., Rubin-Zuzic, M., Petrov, O. F., Tokarev, V. I., and Krikalev, S. K. (2008). Complex plasma laboratory PK-3 Plus on the International Space Station. *New J. Phys.*, **10**, 033036/1–14.

Thomas, Jr., E., Fisher, R., and Merlino, R. L. (2007). Observations of dust acoustic waves driven at high frequencies: Finite dust temperature effects and wave interference. *Phys. Plasmas*, **14**, 123701/1–6.

Thompson, C., Barkan, A., D'Angelo, N., and Merlino, R. L. (1997). Dust acoustic waves in a direct current glow discharge. *Phys. Plasmas*, **4**, 2331–2335.

Tsytovich, V. N. and de Angelis, U. (1999). Kinetic theory of dusty plasmas. I. General approach. *Phys. Plasmas*, **6**, 1093–1106.

Tsytovich, V. N. and de Angelis, U. (2000). Kinetic theory of dusty plasmas. II. Dust-plasma particle collision integrals. *Phys. Plasmas*, **7**, 554–563.

Tsytovich, V. N. and de Angelis, U. (2001). Kinetic theory of dusty plasmas. III. Dust-dust collision integrals. *Phys. Plasmas*, **8**, 1141–1153.

Tsytovich, V. N. and de Angelis, U. (2002). Kinetic theory of dusty plasmas. IV. Distribution and fluctuations of dust charges. *Phys. Plasmas*, **9**, 2497–2506.

Tsytovich, V. N. and de Angelis, U. (2004). Kinetic theory of dusty plasmas. V. The hydrodynamic equations. *Phys. Plasmas*, **11**, 496–506.

Tsytovich, V. N., de Angelis, U., and Bingham, R. (2001). Low-frequency responses and wave dispersion in dusty plasmas. *Phys. Rev. Lett.*, **87**, 185003/1–4.

Tsytovich, V. N., de Angelis, U., and Bingham, R. (2002). Low frequency responses, waves and instabilities in dusty plasmas. *Phys. Plasmas*, **9**, 1079–1090.

Varma, R. K., Shukla, P. K., and Krishan, V. (1993). Electrostatic oscillations in the presence of grain-charge perturbations in dusty plasmas. *Phys. Rev. E*, **47**, 3612–3616.

Vasilyak, L. M.,Vetchinin, S. P., Polyakov, D. N., and Fortov, V. E. (2002). Cooperatove formation of dust structures in a plasma. *JETP*, **94**, 521–524.

Vasilyak, L. M., Vasil'ev, M. N., Vetchinin, S. P., Polyakov, D. N., and Fortov, V. E. (2003). The action of an electron beam on dust structures in a plasma. *JETP*, **96**, 440–443.

Verheest, F. (1992). Nonlinear dust–acoustic waves in multispecies dusty plasmas. *Planet Space Sci.*, **40**, 1–6.

Vladimirov, S. V., Shevchenko, P. V., and Cramer, N. F. (1997). Vibrational modes in the dust-plasma crystal. *Phys. Rev. E*, **56**, R74–R76.

Wang, X., Bhattacharjee, A., and Hu, S. (2001a). Longitudinal and transverse waves in Yukawa crystals. *Phys. Rev. Lett.*, **86**, 2569–2572.

Wang, X., Bhattacharjee, A., Gou, S. K., and Goree, J. (2001b). Ionization instabilities and resonant acoustic modes. *Phys. Plasmas*, **8**, 5018–5024.

Winske, D., Murillo, M. S., and Rosenberg, M. (1999). Numerical simulation of dust–acoustic waves. *Phys. Rev. E*, **59**, 2263–2272.

Xie, B. S. and Yu, M. Y. (2000). Dust acoustic waves in strongly coupled dissipative plasmas. *Phys. Rev. E*, **62**, 8501–8507.

Xie, B. S., He, K., and Huang, Z. (1999). Dust–acoustic solitary waves and double layers in dusty plasma with variable dust charge and two-temperature ions. *Phys. Plasmas*, **6**, 3808–3816.

Yaroshenko, V. V., Annaratone, B. M., Khrapak, S. A., Thomas, H. M., Morfill, G. E., Fortov, V. E., Lipaev, A. M., Molotkov, V. I., Petrov, O. F., Ivanov, A. I., and Turin, M. V. (2004). Electrostatic modes in collisional complex plasmas under microgravity conditions. *Phys. Rev. E*, **69**, 066401/1–7.

Zhdanov, S. K., Samsonov, D., and Morfill, G. E. (2002). Anisotropic plasma crystal solitons. *Phys. Rev. E*, **66**, 026411/1–11.

Zhdanov, S., Nunomura, S., Samsonov, D., and Morfill, G. (2003a). Polarization of wave modes in a two-dimensional hexagonal lattice using a complex (dusty) plasma. *Phys. Rev. E*, **68**, 035401(R)/1–4.

Zhdanov, S., Quinn, R. A., Samsonov, D., and Morfill, G. E. (2003b). Large-scale steady-state structure of a 2D plasma crystal. *New J. Phys.*, **5**, 74/1–11.

Zhdanov, S. K., Morfill, G. E., Samsonov, D., Zuzic, M., and Havnes, O. (2004). Origin of the curved nature of Mach cone wings in complex plasmas. *Phys. Rev. E*, **69**, 026407/1–7.

Zobnin, A. V., Usachev, A. D., Petrov, O. F., and Fortov, V. E. (2002). Dust–acoustic instability in an inductive gas-discharge plasma. *JETP*, **95**, 429–439.

Zuzic, M., Thomas, H. M., and Morfill, G. E. (1996). Wave propagation and damping in plasma crystals. *J. Vac. Sci. Technol. A*, **14**, 496–500.

Zuzic, M., Ivlev, A. V., Goree, J., Morfill, G. E., Thomas, H. M., Rothermel, H., Konopka, U., Sütterlin, R., and Goldbeck, D. D. (2000). Three-dimensional strongly coupled plasma crystal under gravity conditions. *Phys. Rev. Lett.*, **85**, 4064–4067.

5

Kinetic studies of fluids and solids with complex plasmas

Alexey V. Ivlev, Gregor E. Morfill, and Sergey A. Khrapak

How relevant are liquid-like or solid-like complex plasmas to the study of classic phenomena in conventional condensed media? The implication is clear – if they are relevant, this opens up a completely new kinetic approach, which will then have a major impact in a field of great future potential. As was pointed out in Section 2.4.3, one of the interesting aspects of strongly coupled complex plasmas is that although they are intrinsically multispecies systems, the rate of momentum exchange through mutual (electrostatic) interactions between the microparticles can exceed that of interactions with the background neutral gas significantly – thus providing an essentially single-species system for kinetic studies. Moreover, comparison in terms of similarity parameters – e.g., Reynolds, Rayleigh, or Weber numbers for fluids – suggests that liquid complex plasmas can be like conventional liquids (e.g., water) – but observed at the atomistic level!

Because of these unique properties, complex plasmas can indeed serve as a powerful new tool for investigating fluid flows on (effectively) nanoscales, including the all-important mesoscopic transition from collective hydrodynamic behavior to the dynamics of individual particles, as well as nonlinear processes on scales that have not been accessible for studies so far. Of particular interest could be kinetic investigations of the onset and nonlinear development of hydrodynamic instabilities. Individual particle observations can provide crucial new insights – e.g., whether the coarse-grained concept of basic hydrodynamical instabilities (Kelvin–Helmholtz, Rayleigh–Taylor, Richtmyer–Meshkov, etc.) is still adequate on interparticle distance scales, whether there are any microscopic origins of instabilities (in particular, what trajectories can trigger instabilities), etc. Another important issue is, of course, the atomistic structure and dynamics of fluids – in particular, what critical changes occur in the (atomic) structure of solids that give them the ability to flow, are there any characteristic patterns in microscopic dynamics associated with that transition, what conditions form supercooled liquids and glassy states, etc.

Regarding the solid phases, the current interest where highly resolved dynamical measurements in complex plasmas may bring significant advances lies in domain boundaries and defects – associated with excited crystal lattice states and even grain boundary melting and the pre-melting phenomenon (Gleiter 1989; Phillpot and Wolf 1990; Alsayed *et al.* 2005; Pusey 2005). This is of relevance in understanding possible kinetic scenarios of both crystal–crystal phase transitions (in particular, in the context of externally constrained systems) and crystal–liquid transitions (especially in 2D). Other areas of interest are annealing, phonons, shock melting and various nonlinear phenomena.

5.1 Phase diagram of complex plasma

One of the fundamental characteristics of an interacting many-particle system is the coupling strength between the particles. It is measured in units of the potential energy of interaction between neighboring particles normalized by their mean kinetic energy. As discussed in Sections 2.2.1 and 2.3.1 there is a wide parameter range where interparticle interaction (at least its short- and middle-range parts) can be well approximated with the Debye–Hückel (Yukawa) form. For the Debye–Hückel interaction potential, the coupling strength is characterized by two parameters because the interaction has a length scale. These are the coupling parameter Γ determined by the magnitude of the bare Coulomb interaction and the screening parameter κ:

$$\Gamma = Q^2/T_d\Delta, \qquad \kappa = \Delta/\lambda, \tag{5.1}$$

where T_d characterizes mean kinetic energy (temperature) of the particles, λ is the appropriate screening length (e.g., for small grains in isotropic plasmas $\lambda \simeq \lambda_D$), and $\Delta = n_d^{-1/3}$ is the interparticle distance. The coupling strength is characterized by the "screened" coupling parameter $\Gamma_S = \Gamma\exp(-\kappa)$, and the system is usually called "strongly coupled" when $\Gamma_S \gtrsim 1$. Note that the coupling parameter is related to the grain–grain "scattering parameter" β_{dd} introduced in Section 2.4 via $\beta_{dd} = \Gamma\kappa$.

Most theories developed so far to describe the properties of complex plasmas employ the following model: Negatively charged particles are trapped within the plasma volume due to some confining force (usually of electrostatic character) and interact with each other via the isotropic Yukawa repulsive potential, with the screening determined by the plasma electrons and ions. This model gives a rather simplified picture of complex plasma behavior and is not always applicable, especially when plasma anisotropy plays an important role. Moreover, this model does not take into account variations of particle charges, long-range interactions, non-reciprocity, etc. (see Section 3.1). However, the model was shown to be useful in providing qualitative results which are in many cases confirmed by experiments, and hence, it should be considered as a reasonable basis from which more sophisticated models might be constructed.

Besides complex plasmas, particles interacting with the Yukawa potential have been extensively studied in different physical systems ranging from elementary particles to colloidal suspensions. Not surprisingly, their phase diagrams have received considerable attention. Various numerical methods (usually MC or MD simulations) have been employed by Kremer *et al.* (1986), Robbins *et al.* (1988), Meijer and Frenkel (1991), Stevens and Robbins (1993), Hamaguchi *et al.* (1997), Vaulina

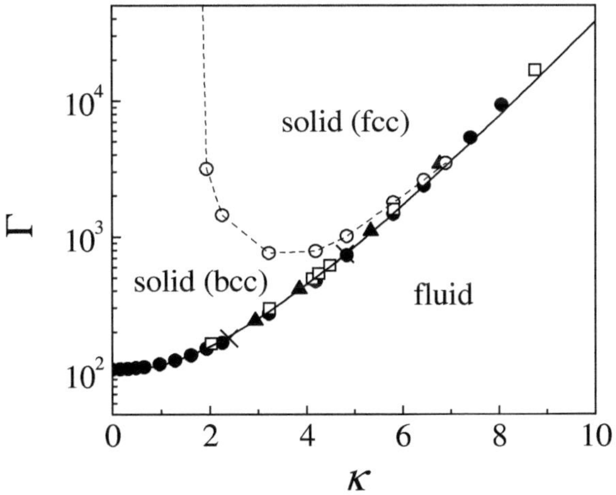

FIGURE 5.1
Phase diagram of Yukawa systems, obtained from numerical modeling. Open circles correspond to the bcc–fcc phase boundary (Hamaguchi *et al.* 1997). The fluid–solid phase boundary (melting line) is marked by squares (Stevens and Robbins 1993), solid circles (Hamaguchi *et al.* 1997), and triangles (Meijer and Frenkel 1991). The crosses correspond to jumps in the diffusion constant, observed in the simulations of dissipative Yukawa systems (Vaulina and Khrapak 2001; Vaulina *et al.* 2002). The solid line represents the analytic approximation of the melting line [Equation (5.2)], the dashed line is the fit to the numerical data judged by eye.

and Khrapak (2001),Vaulina *et al.* (2002). Figure 5.1 shows the phase diagram of Yukawa systems in the (Γ, κ)-plane, summarizing available numerical results. For sufficiently strong coupling, $\Gamma > \Gamma_{\rm M}$, where $\Gamma_{\rm M}(\kappa)$ denotes the melting curve, there are solid face-centered cubic (fcc) and body-centered cubic (bcc) phases, whereas for $\Gamma < \Gamma_{\rm M}$, the system is in a fluid state. The bcc phase is stable at small κ, while fcc is stable at larger κ. The triple point is at $\Gamma \simeq 3.47 \times 10^3$ and $\kappa \simeq 6.90$ (Hamaguchi *et al.* 1997).

Of particular interest for plasma crystallization experiments is the form of the melting (crystallization) curve $\Gamma_{\rm M} = \Gamma_{\rm M}(\kappa)$ (Ikezi 1986; Robbins *et al.* 1988; Meijer and Frenkel 1991; Stevens and Robbins 1993; Hamaguchi *et al.* 1997; Ohta and Hamaguchi 2000; Vaulina and Khrapak 2000). Results obtained for one-component plasma (OCP) systems ($\kappa = 0$) indicate that the crystallization occurs at $\Gamma \simeq 106$ [if the distance is measured in units of the Wigner-Seitz radius, $(4\pi n_d/3)^{-1/3}$, then $\Gamma \simeq 172$, see Ichimaru (1982), Dubin (1990), Farouki and Hamaguchi (1993)]. Different analytical approximations for $\Gamma_{\rm M}(\kappa)$ were proposed (see, e.g., Fortov *et al.* 2004). Vaulina and Khrapak (2000) suggested employing the Lindemann criterion where the dust–lattice frequency for a 1D string (see Section 4.4) is used for the characteristic timescale. The melting line obtained is

$$\Gamma_{\mathrm{M}} e^{-\kappa}\left(1+\kappa+\tfrac{1}{2}\kappa^2\right) \simeq 106, \tag{5.2}$$

which yields remarkably good agreement with the results of numerical simulations at $\kappa \lesssim 10$ (see Figure 5.1). One should note, however, that the arguments used in deriving Equation (5.2) are not really rigorous (for instance, there are no clear physical arguments to justify the choice of the dust–lattice frequency instead of, e.g., the Einstein frequency).

As discussed in Section 2.2.1, plasma absorption on the particles, nonlinearities in ion–particle interaction, and ion–neutral collisions can lead to considerable deviations of the actual electric interaction potential from the Debye–Hückel form. The deviations are especially pronounced at long distances, where the interaction is not exponentially screened, but has a power law asymptote ($\propto 1/r^2$ in collisionless plasmas and $\propto 1/r$ in collisional plasmas). This can have important consequences for the phase diagram of complex plasmas. For example, in the extreme case of strongly collisional plasmas, combining Equations (2.47) and (2.47) with Equation (2.27) for J_i in the SC limit, we get $U(r) \simeq (Q^2/r)(k_{\mathrm{D}i}/k_{\mathrm{D}})^2[1+(T_i/T_e)\exp(-k_{\mathrm{D}}r)]$. Since usually $T_e > T_i$, the interaction potential is very close to the unscreened Coulomb form for all r. Thus, in this case the phase diagram of Yukawa systems is completely irrelevant. Complex plasmas behave as a Coulomb system of particles with effective charge $Q(k_{\mathrm{D}i}/k_{\mathrm{D}})$ somewhat smaller than the actual charges due to partial plasma screening. The crystallization/melting condition is $\Gamma_{\mathrm{M}} \simeq 106(k_{\mathrm{D}}/k_{\mathrm{D}i})^2$. For $T_e \gg T_i$ it reduces to $\Gamma_{\mathrm{M}} \gtrsim 106$, while for a one-temperature plasma ($T_e = T_i$) we get $\Gamma_{\mathrm{M}} \gtrsim 212$.

There are different phenomenological criteria for the crystallization (melting), which are often practically independent of the exact form of the interparticle interaction, and therefore, many of them are applicable to complex plasmas. Best known is the Lindemann criterion (Lindemann 1910), according to which melting of the crystalline structure occurs when the ratio of the root-mean-square particle displacement to the mean interparticle distance reaches a value of $\simeq$0.1–0.2. Another criterion was suggested by Hansen and Verlet (1969) who observed that in 3D hard-sphere systems the first maximum of the static structure factor is equal to $\simeq 2.85$ at the melting curve [for inverse-power-law interaction potentials $\propto r^{-n}$, this value varies in the range from $\simeq 2.6$ for $n = 1$ to $\simeq 3.0$ for $n = 12$, Hansen and Schiff (1973)]. For 2D systems, to our knowledge, this criterion has not been systematically tested so far. There also exists a crystallization criterion for the pair correlation function proposed by Raveche *et al.* (1974): For inverse-power-law interactions, the critical ratio of the first (nonzero) minimum to the first maximum lies in the range from $\simeq 0.1$ (for $n = 1$) to $\simeq 0.26$ (for hard spheres). A simple dynamic crystallization criterion, similar to some extent to the Lindemann criterion, was proposed by Löwen *et al.* (1993). According to this criterion, crystallization occurs when the properly normalized diffusion constant reduces below a certain value. This critical value depends on the dissipation ratio $\nu_{dn}/\omega_{\mathrm{E}}$. In highly dissipative limit, the ratio of the diffusion constant to that of non-interacting particles is ~ 0.1 on the crystallization line. In the

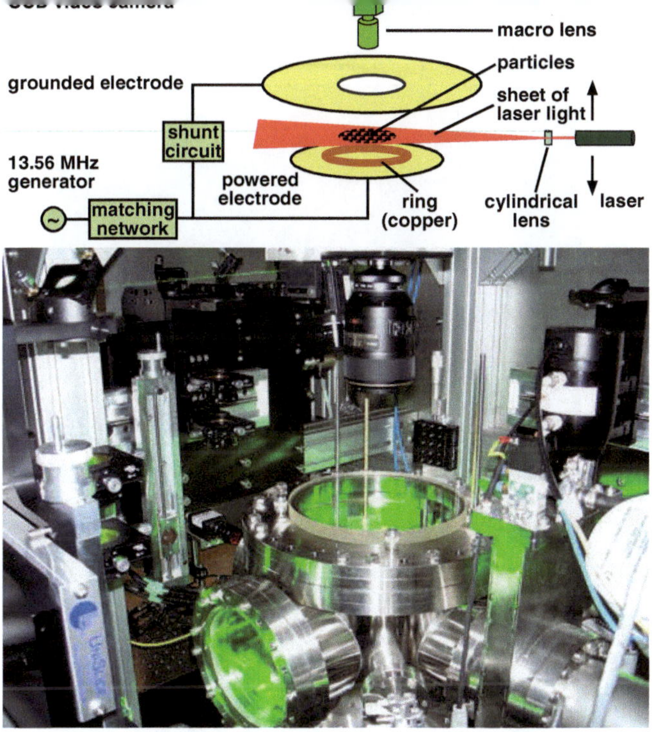

FIGURE 1.3
The sketch shows a schematic drawing of a typical rf-electrode system with lower driven electrode and the grounded upper ring electrode. The microparticles are injected by a dispenser (not shown) and levitate in the sheath electric field, additionally trapped horizontally by a parabolic potential formed by a ring positioned on the electrode. The microparticles are illuminated by a laser beam expanded into a sheet parallel to the electrode, and the reflected light is observed at 90° by a CCD camera. The image shows a typical assembly of the GEC-RF-Reference Cell.

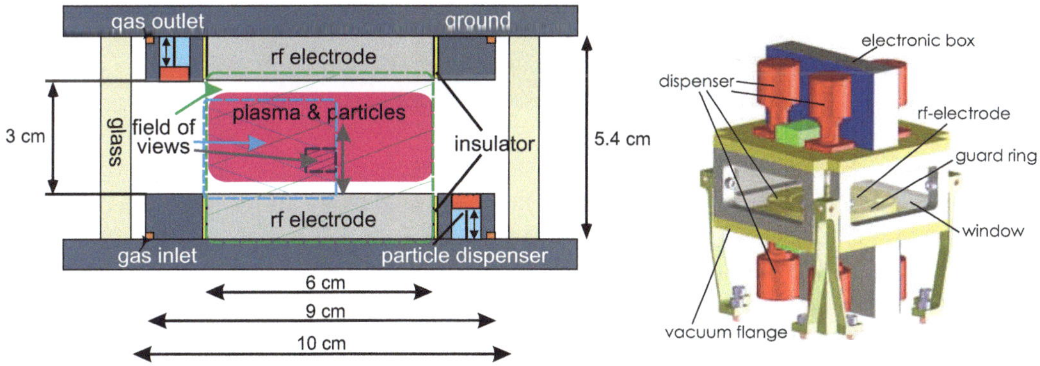

FIGURE 1.11
The sketches show the 2D (left) and 3D view of the plasma chamber (right) (Thomas *et al.* 2008).

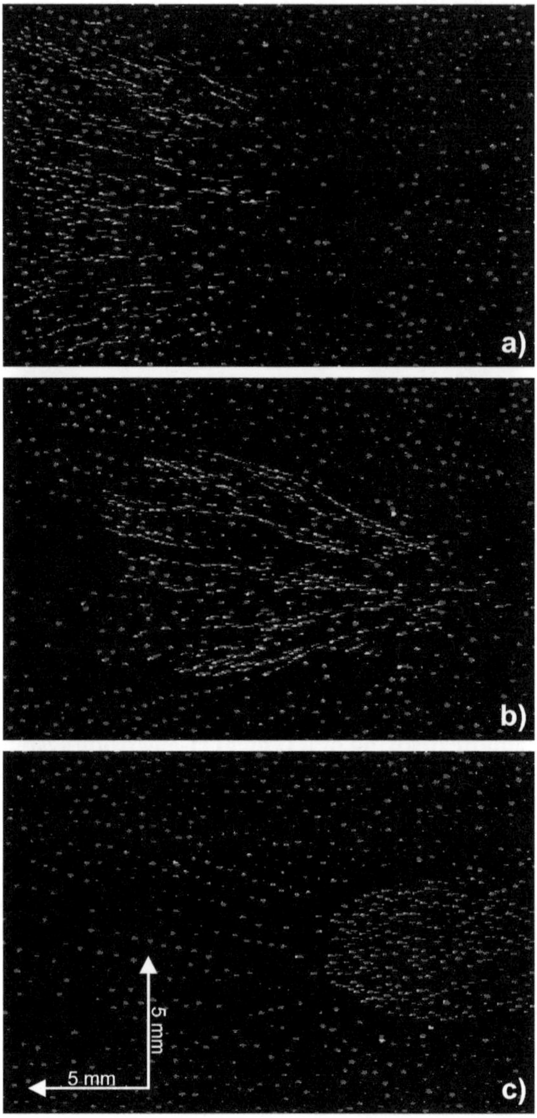

FIGURE 5.6
Lane formation in complex plasmas. A short burst of small (3.4 μm) particles injected into a cloud of large (9.2 μm) background particles are driven from left to right. Stages of (a) initial lane formation, (b) merging of lanes into larger streams, and (c) eventual droplet formation are shown. Each figure is a superposition of two consecutive color-coded images (1/50th s apart, green to red), entire sequence is about 2.5 s long. Images are kindly provided by M. Rubin-Zuzic.

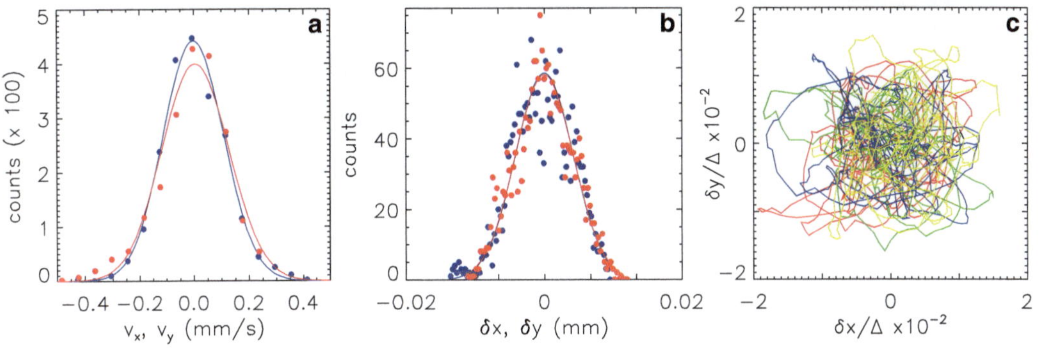

FIGURE 5.12
Dynamics of particles in a lattice of a 2D crystal *et al.* 2007). (a) Distribution of velocities v_x (red dots) and v_y (blue dots) with Maxwellian fits (solid lines). (b) Distribution of displacements x (red dots) and y (blue dots) of particles in their nearest-neighbor cage, solid lines are Gaussian fits. (c) Particle trajectories in their respective nearest-neighbor cells during the measurement time of $\simeq 12.3$ s (colors correspond to the progression of time). Particles are of $9.19\ \mu$m diameter.

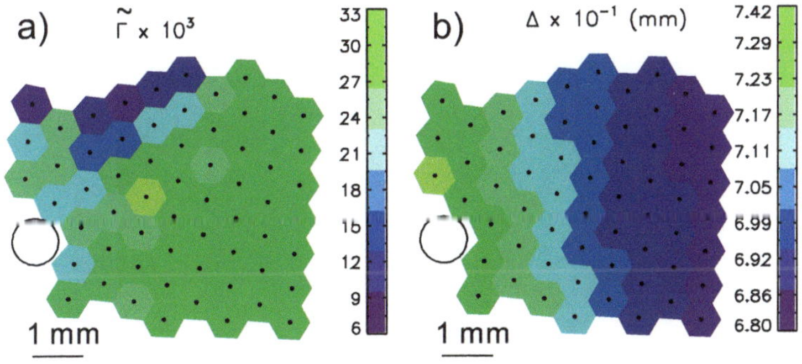

FIGURE 5.13
2D maps of local crystal parameters (Knapek *et al.* 2007). Distribution of (a) effective coupling parameter $\tilde{\Gamma}$ and (b) interparticle distance Δ is shown, the Voronoi cell around each particle is color coded according to the value of the measured quantity. The circles indicate the position of a sevenfold-fivefold pair defect, blue cells seen at the upper edge of (a) are due to the particle cage-escape event (see Section 5.2.1).

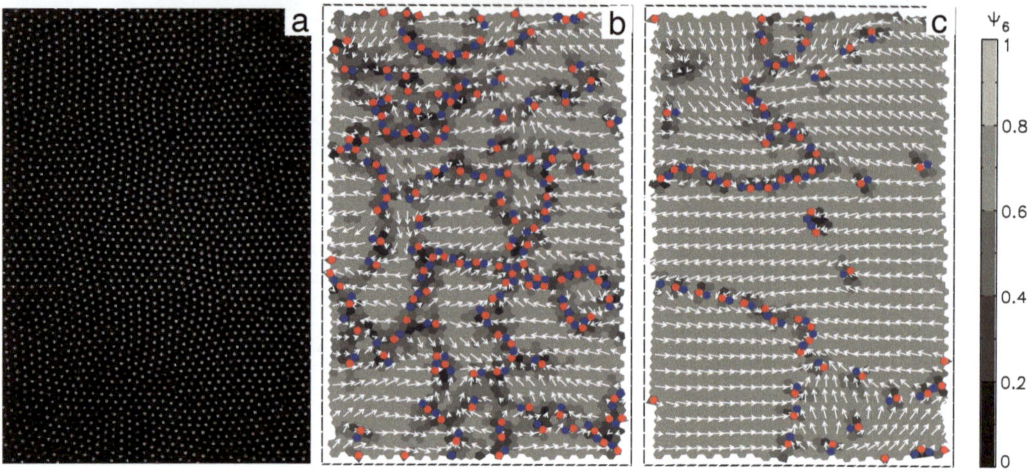

FIGURE 5.14
Recrystallization in 2D complex plasmas (Knapek *et al.* 2007b). (a) Snapshot showing intermediate structure of $9.19\ \mu$m particles during the recrystallization. (b,c) Color-coded 2D maps for two consecutive stages of recrystallization (about 10 s apart, map b corresponds to snapshot a). The background gray scale corresponds to the local value of the bond-orientational function $|\psi_6|$, the arrows represent the vector field of ψ_6 on the complex plane, defects are marked by red (fivefold) and blue (sevenfold) dots.

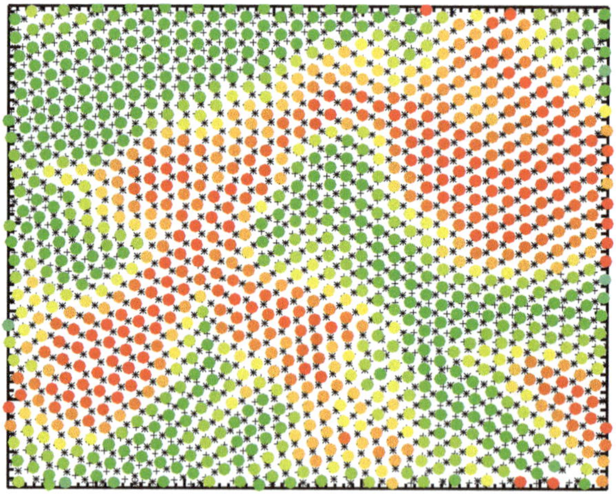

FIGURE 5.16
Domain structure of a 3D plasma crystal (Zuzic *et al.* 2000). Particles are of $3.38\ \mu$m diameter three consecutive lattice planes are shown, each particle in the middle plane is color-coded in accordance with the local order (red corresponds to the fcc lattice cell and green to hcp) particle in two adjacent planes are indicated by crosses and stars.

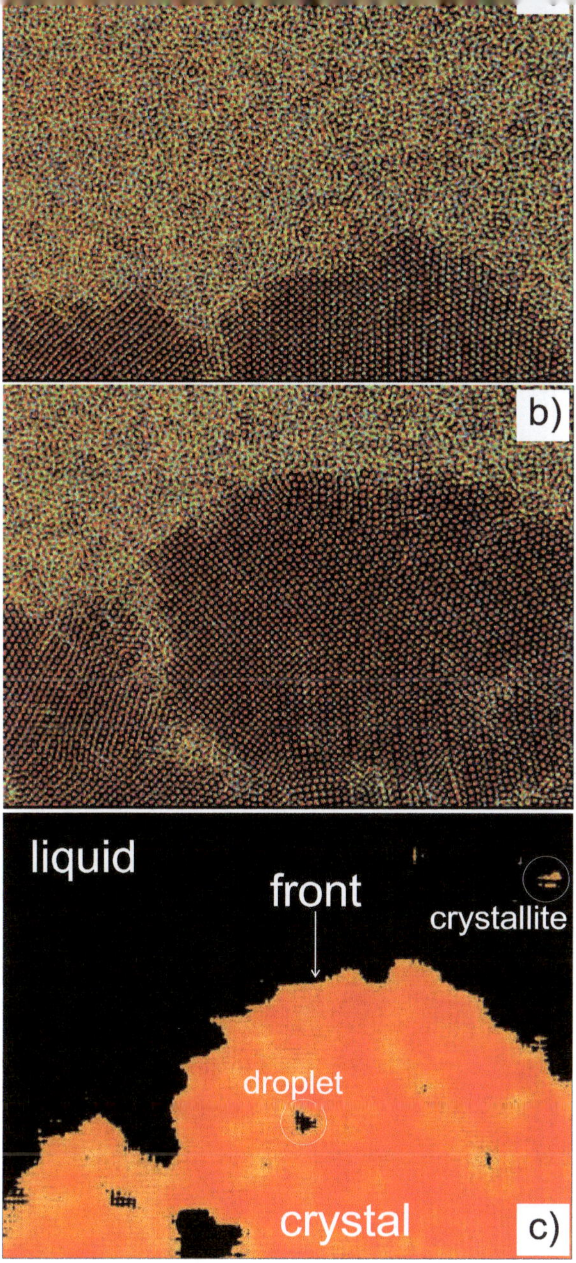

FIGURE 5.17
Crystallization front in a 3D complex plasma (Rubin-Zuzic *et al.* 2006). Figures (a) and (b) illustrate the front propagating upwards (images are about 16 s apart from each other). Each figure is a superposition of 10 consecutive video frames (about 0.7 s), particle positions are color-coded from green to red, i.e., "caged" particles appear redder, "fluid" are multicolored. (c) The local order for figure (b), where red implies high crystalline order, black denotes the fluid phase, and yellow indicates transitional regions. Along with the crystallization front, droplets and crystallites are seen that may grow and then dissolve again. Particles are of 1.28 μm diameter.

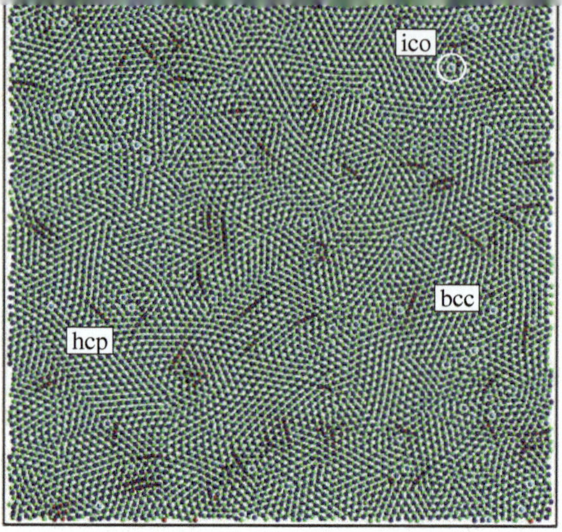

FIGURE 7.18
Crystallized Yukawa system in the narrow channel for the parabolic confinement. The particles form three layers A, B, and C. The particles of layers A, B, and C are marked by blue, green, and red, respectively. It is seen that layer C is almost completely screened by layer A. It means that the hcp lattice is dominant in that case. The bcc phase is also seen. A small number of clusters ($\sim 1\%$) have icosahedral-like (fivefold) symmetry.

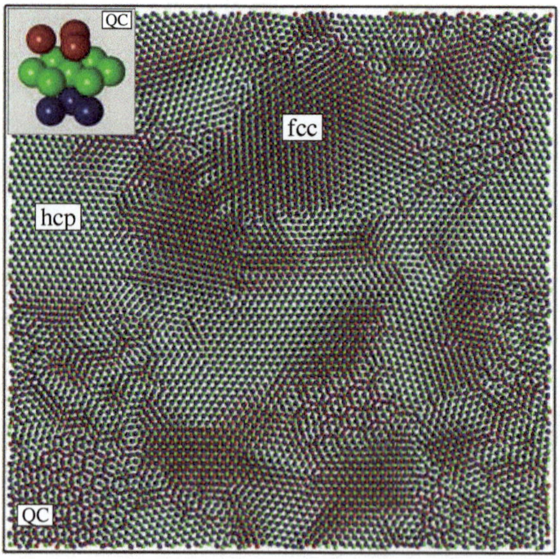

FIGURE 7.19
Crystallization of the particles in the narrow channel for the hard wall confinement. The domains having hcp and fcc lattice types are clearly seen. A significant number of clusters have a quasicrystalline (QC) phase. The inset shows the unit cell of the QC phase, which is a distorted hcp/fcc unit cell.

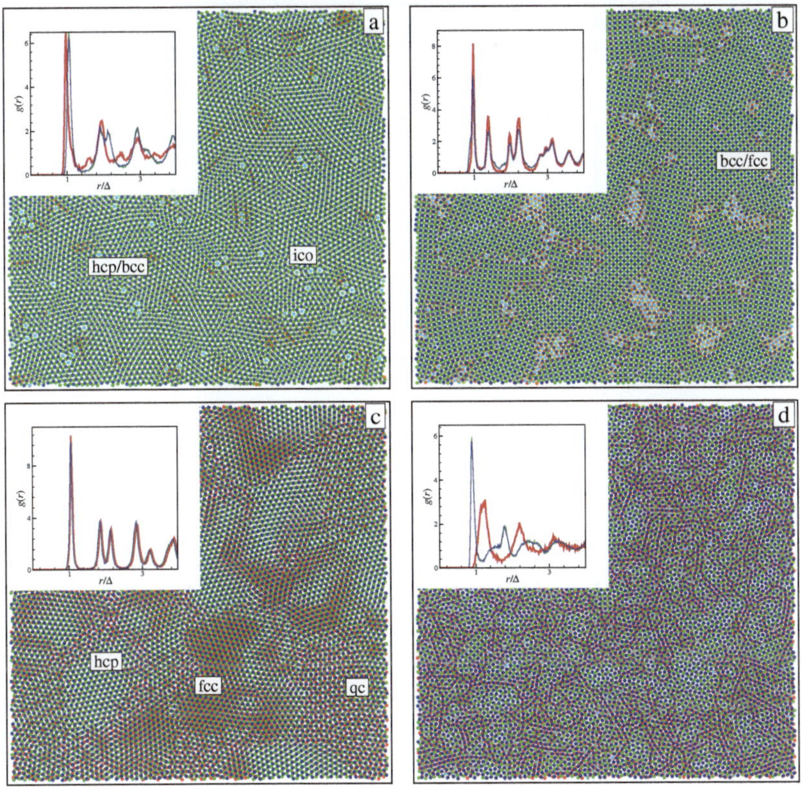

FIGURE 7.20
Crystallization of the particles in the narrow channel for the parabolic (a, b) and hard wall (c, d) confinements. Particles are color-coded by z-coordinate. Stable three-layers configurations of the Yukawa system (a, c) are presented. These systems close to the two-layer configuration (b, d) are also shown. The insets show pair correlation function $g(r/\Delta)$ for each layer, including the central one (solid line).

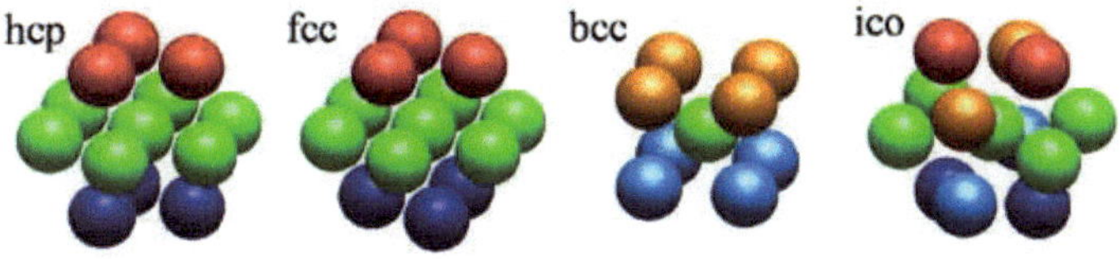

FIGURE 7.27
The lattice types we try to identify: hexagonal close packing (hcp), face centered cubic (fcc), body centered cubic (bcc) and icosahedron (ico) (from left to right).

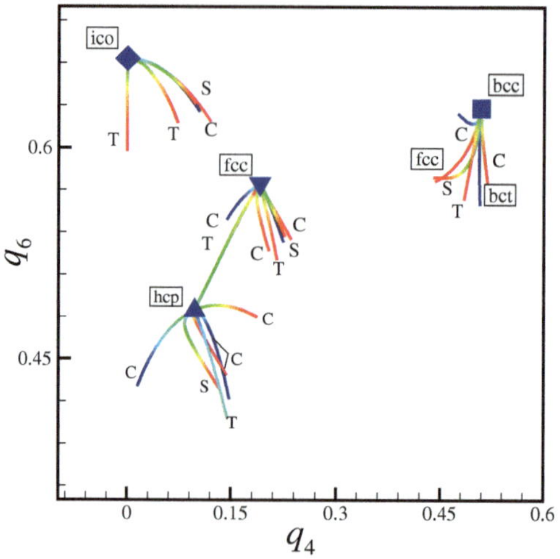

FIGURE 7.28
Variations of the rotational invariants q_4 and q_6 for different lattice types at different distortions: compression/extraction (C), shear (S) and torsion (T). The curves are color-coded by relative deformation values (S, T). We also plotted here the invariants for the fcc lattice (calculated by using 8 nearest neighbors) and body centered tetragonal (bct) - the compressional modification of bcc.

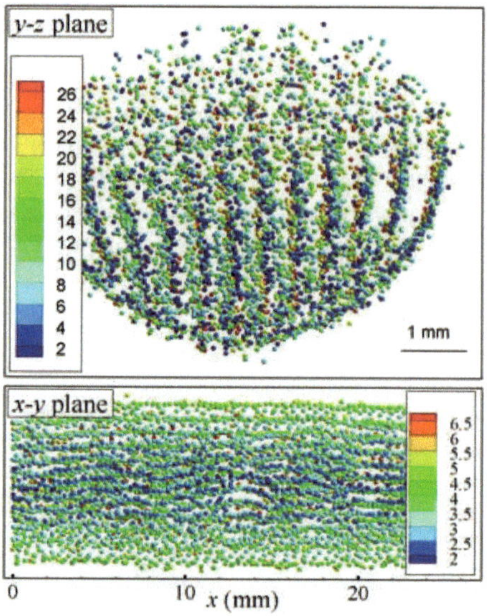

FIGURE 7.29
Experimentally recorded particle positions in the $y-z$ (top) and $x-y$ (bottom) planes. Particles are color-coded by the corresponding third coordinate presented in mm. About 6000 particles were detected.

"single-species" (non-dissipative) regime the diffusion constant normalized to $\omega_E \Delta^2$ has a distinct asymptote of $\simeq 0.0032$ at crystallization (Ohta and Hamaguchi 2000; Vaulina *et al.* 2002). This dynamical criterion holds for both 2D and 3D systems (Löwen 1996).

So far we have considered only the repulsive electric interaction between like-charged particles. However, as discussed in Section 2.3.1, other interaction mechanisms can operate in complex plasmas as a result of their thermodynamic openness. For example, the so-called ion shadowing effect can provide the long-range attractive branch of the interaction potential. This can further complicate the phase diagram of complex plasmas. For instance, Khrapak *et al.* (2006) considered the possibility for the liquid–vapor phase transition and, in particular, a liquid–vapor critical point occurrence in complex plasmas. Their analysis was based on the qualitative similarities in the binary interaction potentials compared to conventional gases and liquids (electrical repulsion at short distances and attraction due to ion shadowing at larger distances). The main requirements for the observation of the critical point formulated by Khrapak *et al.* (2006) are as follows: Isotropic plasma conditions (in order for ion shadowing to operate ion drifts should be at least subthermal), weak external confinement (interparticle separation should be large enough so that the particles can "feel" attraction), low neutral gas pressure (ion shadowing operates and attraction exists), and the critical temperature should be higher than the neutral gas temperature (which is the lower limit of the particle kinetic energy due to Brownian motion). Theoretical estimation of the critical point parameters yielded the result that it could be observable in microgravity experiments for realistic plasma conditions (Khrapak *et al.* 2006). Unfortunately, calculations by Khrapak *et al.* neglected any direct effect of ion–neutral collisions on the interaction potential. More recent calculations by Khrapak *et al.* (2008) and Khrapak and Morfill (2008) taking into account collisional effects demonstrated that much lower pressures (on the order of 1 Pa) than previously expected are required to approach the critical point. This can complicate the search for the liquid–vapor critical point for the following reasons: First, gas discharge operation at such low pressures might be unstable. Second, the problem of a void formation may become crucial at low pressures. Obviously, the whole issue of the effect of "natural" attraction on the phase states and phase transitions in complex plasmas needs further detailed investigation. The possibility of external manipulation of the interparticle interactions and the emerging new properties of the phase diagram of complex (electrorheological) plasmas are discussed in Section 5.2.6.

Some experiments are preformed with *multispecies* complex plasmas, when microparticles of different sizes are injected in the discharge chamber – such investigations have been regularly carried out under microgravity conditions (Nefedov *et al.* 2003; Thomas *et al.* 2008). Fluid phase transitions in this case are governed by the mechanisms that are very different from those operating in one-component complex plasmas: It is well known that a tendency for particles of different sizes to mix or demix is basically determined by the relative strengths of their interactions (Hansen and McDonald 1986). The fluid–fluid phase separation in such multicomponent systems *does not* require an attraction in the interparticle interactions – the necessary condition for the fluid phase transition in single-species systems. The phase equilibrium is governed by relations that are often referred to as the Lorentz-Berthelot mixing rules (Maitland *et al.* 1981). It is remarkable that in the vicinity of the critical point the behavior of binary mixtures (with short-range, e.g., Yukawa, interactions) belongs to the same universality class as that of a conventional liquid–vapor phase transition in simple fluids (Fisher 1974).

It has been shown by Ivlev *et al.* (2009) that the asymmetry of interparticle interactions in binary complex plasmas always stimulates the phase separation in the isotropic (bulk) regime. This tendency does not depend on a particular shape of the interaction potential. For typical conditions of experiments with binary complex plasmas the regime of the spinodal decomposition is easily achievable. Apparently, this process is illustrated in Section 5.2.4 (see Figure 5.6c), where the appearance of a smooth-surface droplet is the clear manifestation of a positive surface tension. This conclusion provides us with strong grounds to believe that binary complex plasmas could be ideal model systems to study atomistic dynamics of fluid phase transitions and the associated phenomena, such as critical behavior or effects of the surface tension.

In concluding this section it is worth noting that depending on the phase state – liquid or solid – one can choose different timescales to characterize collective dynamics of microparticles. For liquid and amorphous solid complex plasmas the dust plasma frequency, $\omega_{\rm pd} = \sqrt{4\pi Q^2 n_d/m_d}$, can be used as the measure, whereas for crystals the modes depend on a particular lattice structure, so that the Einstein frequency $\omega_{\rm E}$ is the more appropriate scale. The ratio $\omega_{\rm E}/\omega_{\rm pd}$ is a (monotonously decreasing) function of the screening parameter κ, and for different lattices it typically varies between a few units and a few tenths (Robbins *et al.* 1988; Knapek *et al.* 2007a). To avoid confusion, in this chapter we decided to use $\omega_{\rm E}^{-1}$ as the characteristic dynamical timescale for both liquids and solids.

5.2 Strongly coupled fluids

In this section we discuss various generic processes that can be studied at the kinetic level with fluid complex plasmas. We start with the consideration of physical mechanisms underlying the atomistic dynamics of supercooled fluids, proceed with the detailed discussion of elementary processes governing momentum and energy transport, and address the "discreteness issue" of hydrodynamics – what happens to basic hydrodynamic phenomena when the relevant spatial scales become comparable to the interparticle distance. Finally, we focus on processes occurring in finite-sized systems, by considering confined fluids and discuss novel types of "electrorheological plasmas" where the binary interactions can be tuned by an external ac electric field, similar to that in "regular" electrorheological fluids.

5.2.1 Atomistic dynamics in fluids

Liquid complex plasmas can be considered as one of the best model systems to investigate fundamental long-standing problems of the classical physics of fluids. In particular, understanding the properties of supercooled fluids (especially in the vicinity of the glass transition) is one of the most controversial issues (Jäckle 1986; March and Tosi 2002), where a number of mutually exclusive interpretations of various aspects of the complex behavior are still under discussion. Weak neutral damping of complex plasmas plays a constructive role for such investigations, allowing us to control the cooling rate and, therefore, to bring the fluid to a desirable degree of overcooling [and, hence, vary the glass transition temperature, see, e.g., March and Tosi (2002)]. Kinetic investigations of supercooled fluids with complex plasmas may help us to get a deeper insight into other major issues, e.g., which elementary mechanisms determine the stability of supercooled fluids against crystallization (Jäckle 1986), what is the kinetics of the glass transition and how do the relevant processes like arrest of the structural relaxation and loss of ergodicity evolve (Fischer 1993; Sillescu 1999), what microscopically determines the variation of the transport properties (especially, self-diffusion) in the supercooled state (Saltzman and Schweizer 2006). Liquid complex plasmas are apparently the only available model system where the dynamics of rapid relaxation can be studied at the kinetic level.

Atomistic behavior in liquid states has been observed in numerous experiments with 2D and quasi-2D strongly coupled complex plasmas (Juan *et al.* 2001; Lai and I 2002; Woon and I 2004; Nunomura *et al.* 2006; Ratynskaia *et al.* 2006; Huang and I 2007; Liu and Goree 2007, 2008). As an example, let us consider one particular experiment performed by Juan *et al.* 2001. Figure 5.2a shows a snapshot where most particles are mutually confined by (quasi-ordered) neighbors, and exhibit caged motion with small amplitude oscillations. However, there is a certain fraction of particles that are in a rearrangement state. Spatially, one can observe coherent cage-escape events – strings or vortices surrounding crystallites (ordered domains) with the size of a few Δ. Usually, a local rearrangement ceases after the involved particles jump a distance of $\simeq \Delta$ and then reenter the new caged state. Particles may start coherent rearrangement only after accumulation of sufficient "constructive" perturbations, and then transfer the excess energy to the neighbors through mutual interactions. The coherent motion is rapidly smeared out unless further constructive perturbation occurs at a timescale smaller than the momentum relaxation time.

Introducing external stress greatly enhances the formation of micro-vortices. Figures 5.2b and c show an experiment by Juan *et al.* 2001, where external stress in liquid complex plasmas was introduced by a laser. One can see that the intensity of micro-vortices gradually decays with distance from the shear source to the remoter regions. The observations can be reasonably explained by the following phenomenology: Even in a stress-free cool liquid, thermal agitation can distort the caging potential of neighboring particles through changing the particles' relative positions, transfer energy to particles, and induce vortex-like escape over caging barriers. But introducing an external stress breaks the symmetry and further promotes

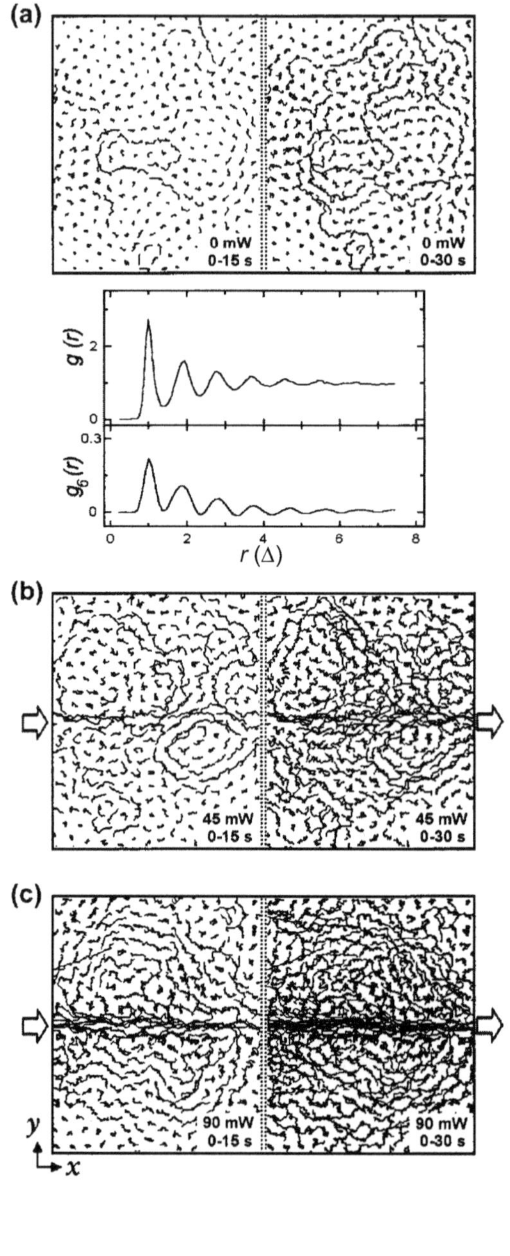

FIGURE 5.2
Dynamical heterogeneity seen in quasi-2D liquid complex plasmas (Juan *et al.* 2001). (a) Trajectories of 7 μm diameter particles with 15 and 30 s exposure times for the laser-free liquid state. Shown below are the pair correlation functions of particle positions, $g(r)$, and bond orientations, $g_6(r)$. Figures (b) and (c) show particle trajectories in the laser-enhanced vortex motion under 45 and 90 mW laser power, respectively. The arrows indicate the position and direction of the laser beam.

forward jumping. In the low stress regime, when any rearrangement occurs at a low rate, the motion is still strongly constrained by caging. Increasing the stress level further usually promotes the rearrangement, although the advection can sometimes be jammed, forming local solid-like regions. Assisted by thermal fluctuations, the stressed particles will find the easiest "percolation" paths for rearrangement and branch off from the stressed zone. The vacancy left behind can be filled by the trailing particles or by the particles in the neighborhood of the laser beam, thus forming the vortices originated from the laser zone. Under the strong mutual particle interaction, these vortices quickly relax through cascaded excitations of new vortices with decaying strength in remote regions.

As we already mentioned above, observation at the individual particle level may help us to shed light on what "elementary processes" determine the rich variety of unusual properties peculiar to supercooled fluids. In addition, knowledge of the fully resolved particle kinetics would allow us to calculate basic transport properties of the system from first statistical principles and then compare the results with existing models (Hansen and McDonald 1986). The common approach is to employ the Green–Kubo formalism that yields transport coefficients expressed in terms of time integrals over the relevant microscopic autocorrelation functions (of, e.g., velocity, shear stress, and energy flux for self-diffusion, shear viscosity, and heat conduction, respectively). This standard theory, however, is based on the assumption that the time integrals converge and, therefore, excludes an important class of processes called "fractional Gaussian noises", which lead to particle trajectories described in terms of the fractional Fokker-Planck dynamics. For these processes the mean square displacement (MSD) scales as $\propto t^{2H}$, where H is the Hausdorff exponent: For $H = 1/2$ we have standard diffusion, for $H < 1/2$ the resulting motion is subdiffusive, and for $H > 1/2$ the motion is superdiffusive. Standard diffusion theory also fails if the velocity probability distribution function is non-Gaussian but has algebraic tails, so that the velocity variance diverges.

Statistical analysis of individual particle trajectories in complex plasmas at sufficiently low temperatures (high densities) usually reveals subdiffusion at short and intermediate timescales (which are at the same time much longer than the "in-cage" oscillation time $\sim \omega_E^{-1}$), with the crossover to normal diffusion at much longer times (Lai and I 2002; Nunomura *et al.* 2006), as shown in Figure 5.3. (The crossover can be mediated by a relatively short superdiffusive stage.) On the other hand, several experiments (Ratynskaia *et al.* 2006; Liu and Goree 2007, 2008) demonstrate persisting superdiffusive long-time behavior. One should bear in mind that the long-time superdiffusion in complex plasmas has been observed either in relatively small and inhomogeneous systems (e.g., in experiment by Ratynskaia *et al.* (2006), where superdiffusion might be triggered by boundary/confinement effects), or in systems with noticeable large-scale flow [which enhances transport and and therefore increases asymptotical long-time value of H, e.g., an experiment by Liu and Goree (2008)]. One should also mention that in other systems (e.g., colloids) superdiffusive behavior apparently has never received reliable confirmation (Reichman *et al.* 2005).

It is generally accepted that above the glass transition fluids have properties of a

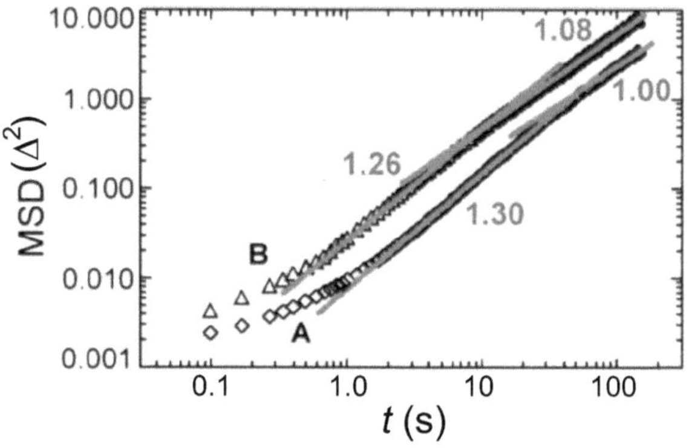

FIGURE 5.3
Mean squared displacement of particles in quasi-2D strongly coupled complex plasmas (Lai and I 2002). A and B indicate experiments performed at different temperatures with particles of 7 μm diameter. One can see a transition from subdiffusive behavior (dominated by in-cage motion) to the long-time normal diffusion, mediated by a short superdiffusive stage (associated with cage-escape events).

viscoelastic medium (March and Tosi 2002; Fischer 1993; Jäckle 1986). The simplest model that can be employed to describe diffusion in such media is based on a linear viscoelastic Langevin equation with a single relaxation time,

$$\dot{\mathbf{v}} + (\nu_{dn}/\tau)\int_{-\infty}^{t} e^{-(t-t')/\tau}\mathbf{v}(t')dt' = \mathbf{f}(t),$$

where $\mathbf{f}$ is a random force (per particle) that satisfies the fluctuation–dissipation theorem and τ is the memory relaxation time (Hansen and McDonald 1986; van Zanten and Rufener 2000). Assuming "long-range" memory, $\nu_{dn}\tau \gg 1$, the qualitative behavior of the MSD derived from this simple approach has all the distinct features observed in experiments: Short ballistic motion, MSD $\simeq 3v_T^2 t^2$; for $t \lesssim \sqrt{\tau/\nu_{dn}}$ with crossover to a plateau, MSD $\simeq$ *const*; and eventual transition to normal diffusion, MSD $\simeq 6v_T^2 \nu_{dn}^{-1} t$, at $t \gtrsim \tau$. Of course, quantitative agreement can only be received with more sophisticated models, e.g., by employing the nonlinear Langevin equation based on the formalism of the dynamic density functional theory (Saltzman and Schweizer 2006).

A similar approach can also be employed to describe the flow of supercooled fluids under external stress. The essential feature of a viscoelastic flow is that it displays elastic deformation on short temporal and spatial scales but looks more like a viscous flow on larger scales. In the framework of the linear Maxwell model (see, e.g., Landau and Lifshitz 1986), the strain γ is a superposition of two components: The elastic contribution responds to the stress through Hooke's law, $\sigma = G\gamma$, and the viscous contribution through Newton's relation, $\sigma = \eta\dot{\gamma}$, where G and η are the high-frequency Young's modulus and static shear viscosity, respectively. From these limiting relations follows the differential equation

$$G^{-1}\dot{\sigma} + \eta^{-1}\sigma = \dot{\gamma}. \tag{5.3}$$

The general solution expresses the stress as a linear response on the time history of the strain rate with an exponentially decaying response function with the Maxwell timescale $\tau_{\rm M} = \eta/G$. This classical model implies a separation between the elastic and hydrodynamic responses controlled by the Deborah number $\dot{\gamma}\tau_{\rm M}$: The response is viscous at timescales $t \gg \tau_{\rm M}$ and elastic for $t \ll \tau_{\rm M}$, at intermediate timescales we have a complex Young's modulus (or viscosity). By measuring the response to external stresses at different frequencies one can obtain the complex Young's modulus, derive $\tau_{\rm M}$, and compare it with the results retrieved from the diffusion measurements.

5.2.2 Kinetics of stable shear flows

Macroscopically, the hydrodynamic behavior of fluid complex plasmas can be described in the framework of continuous media. The momentum (Navier-Stokes) and energy equations should be modified appropriately, in order to take into account the frictional interaction with the background neutral gas (Fortov *et al.* 2005). This interaction is characterized by the dust–neutral momentum exchange rate ν_{dn} [see Equation (2.69)], and the resulting equations for the mean velocity $\mathbf{v}$ and kinetic temperature T of microparticles (complemented with the continuity equation for the mass density $\rho = m_d n_d$) are

$$\begin{aligned} &\partial_t \mathbf{v} + (\mathbf{v}\cdot\nabla)\mathbf{v} + \nu_{dn}\mathbf{v} = -\nabla p/\rho + (\eta/\rho)\nabla^2\mathbf{v}, \\ &\partial_t T + \mathbf{v}\cdot\nabla T + 2\nu_{dn}(T - T_n) = \chi\nabla^2 T + Q_H. \end{aligned} \tag{5.4}$$

Here, η and χ are, respectively, the dynamical shear viscosity and thermal diffusivity of the microparticle component (see this and next sections; note that η and $\rho\chi$ are functions of ρ and T and, therefore, in the general case the operator ∇ should act on them as well). Also, p is the pressure of dust species determined by an appropriate equation of state and Q_H is the heat source per particle (e.g., due to external heating or viscous heating).

The obvious necessary condition for implementing Equations (5.4) is that all relevant length scales should be much larger than the discreteness scale Δ, so that the model of continuous media can be well applied. Yet another essential assumption is that the background gas remains at rest – this allows us to consider complex plasmas as a single-species fluid with a weak background friction proportional to the local velocity (Ivlev *et al.* 2007b): Collisions with microparticles do not affect diffusive motion of neutrals as long as the diffusion length at timescales $\sim \ell_{nd}/v_{T_n}$ exceeds the complex plasma size L, which yields $\ell_{nn}\ell_{nd} \gtrsim L^2$. The mean free path of neutrals due to collisions with micron-size particles, $\ell_{nd} = (\pi a^2 n_d)^{-1}$, is usually about a few meters, whereas the mean free path of neutral–neutral collisions (say, at pressures ~ 3 Pa) is $\ell_{nn} \sim$0.3–1 cm. Thus, the assumption that neutrals remain unaffected is

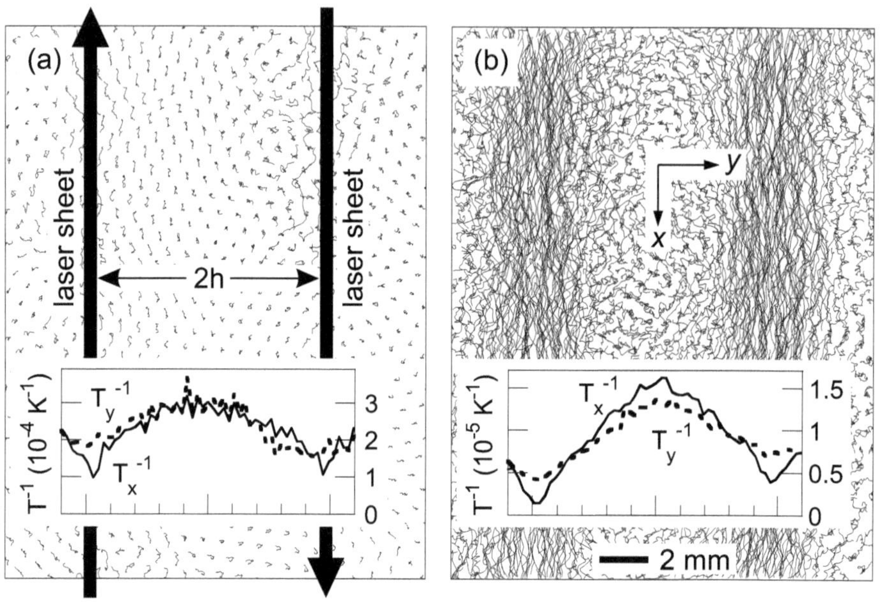

FIGURE 5.4
Planar Couette flow in a 2D complex plasma (Nosenko and Goree 2004). Initially crystalline microparticles of 8.09 μm diameter are sheared by two counter-propagating laser sheets. At the onset of plastic flow (a), the particles hop between equilibrium lattice sites. In a fully developed shear flow (b), the particle motion is highly irregular on smaller scales comparable with the interparticle spacing, but on larger scales it is like a laminar flow in a fluid. Trajectories over $\simeq$1.7 s are shown.

very well satisfied for typical system sizes $L \lesssim 10$ cm. Note, however, that sometimes spontaneous global flows in neutral gas might be triggered, e.g., due to strong temperature gradients (Mitic *et al.* 2008) or magnetic fields (Carstensen *et al.* 2009).

Let us now take a closer look at the individual particle trajectories in a fluid complex plasma that exhibits a macroscopic flow. In complex plasmas, one can easily induce various types of flows with controllable characteristics by applying laser beams or creating controllable flows in the neutral gas (see Section 4.1). A clear advantage of such methods of particle manipulations is that the background plasma, and hence parameters of the interparticle interaction, remain unchanged, yet the characteristics of the particle flow – especially the flow shear rate – can be varied over an exceptionally broad range.

In fact, shear flows appear as an almost inevitable ingredient of more complicated flows. Even in the simplest case of laminar shear flows, many fundamental questions immediately arise: What is the kinetic structure of the flow (e.g., how does the transverse momentum relaxation occur)? What is the kinetics of non-Newtonian fluids (e.g., what determines the relevant timescales in the viscoelastic fluids)? What happens at shear fluid boundaries (e.g., how good is the Navier ansatz for the slip velocity and what is the corresponding slip length)? In the case of multiphase flows, many more fundamental problems turn up, especially those related to the shear boundaries. Probably, the most "obvious" one is the contact-line singularity problem: a movable intersection of the fluid–fluid interface with the solid wall is incompatible with the no-slip boundary condition (Qian *et al.* 2006).

The simplest experimental configuration that allows us to encompass most of the issues mentioned above is a 2D monolayer of particles, and the easiest way to create shear flows in this case is to use laser manipulation. Figure 5.4 shows an example of such an experiment performed by Nosenko and Goree (2004), when particles formed an (almost perfect) hexagonal crystalline monolayer, and the shear flow was created by applying two counter-propagating laser sheets. By increasing the laser power and, hence, the level of shear stress, the authors observed that the particle suspension passed through four distinct stages: elastic deformation, defect generation while in a solid state, onset of plastic flow, and fully developed shear flow. Figure 5.4 presents data for the latter two stages. At the onset of plastic flow, Figure 5.4a, the particles hopped between equilibrium lattice sites. Domain walls developed, and they moved continuously. The crystalline order of the lattice in the shearing region deteriorated, broadening the peaks in the static structure factor (not shown here). At still higher levels of shear stress, the lattice fully melted everywhere, and a shear flow developed, Figure 5.4b.

In terms of the applied laser power (and hence the resulting stress), the onset of the plastic flow is a rather distinct phenomenon with well-defined yield stress, suggesting that simplest rheological models [e.g., modifications of the Bingham plastic model, see Meyers and Chawla (1998)] are quite appropriate to describe the shear-induced melting. On the other hand, the individual trajectories of "percolating" particles that identify the onset of the plastic flow are quite peculiar: They have a zigzag-like shape, jumping along the local principal vector of the hexagonal lattice, i.e., in the direction where the macroscopic lattice has the least yield stress.

At the stage of fully developed shear flow, the particle motion is highly irregular on a small scale compared to the interparticle spacing, but on a larger scale, it is like a laminar flow in a fluid. In this case, the liquid-like order of the particle suspension can be clearly identified from the diffusiveness of the structure factor. Particles are confined so that after flowing out of the field of view on one side, they circulate around the suspension's perimeter and reenter on the opposite side. Within the field of view, more than 95% of the time-averaged flow velocity is directed in the x-direction, with less than 5% of the flow velocity diverted in the y-direction. It is worth noting that for all values of the laser power used in the experiment the local velocity distribution of particles is (with very good accuracy) a Maxwellian one, although at highest shear rates the mismatch between the longitudinal and transverse temperatures is as high as $\sim 30\%$. This means that the internal momentum and energy equilibration in the particle ensemble is fast enough to balance the heat released due to the shear flow, and hence, the concept of equilibrium viscosity (as a function of self-consistent temperature corresponding to a given flow regime) is well justified.

Numerical simulations (Saigo and Hamaguchi 2002; Salin and Caillol 2002) predict that the shear viscosity of complex plasmas depends on the concentration of microparticles, which is one of the essential features of complex fluids. Moreover, experiments and simulations by Nosenko and Goree (2004), Gavrikov *et al.* (2005), Donko *et al.* (2006), and Ivlev *et al.* (2007a) verified that the viscosity can exhibit significant shear thinning and/or thickening. This non-Newtonian behavior of complex plasmas occurs because the viscosity η is a strong function of the particles' kinetic temperature which, in turn, is determined by the local viscous heat released due to shear flow and is proportional to $\eta\dot{\gamma}^2$. Based on this simple rheological model (Ivlev *et al.* 2007a), one can identify three distinct regimes for a qualitative dependence of the viscosity and the shear stress $\sigma = \eta\dot{\gamma}$ on the shear rate $\dot{\gamma}$: (i) At sufficiently low $\dot{\gamma}$, the viscosity remains constant and stress grows linearly with $\dot{\gamma}$, which corresponds to Newtonian fluids; (ii) above a certain critical value of $\dot{\gamma}$, shear-thinning is observed, which can be quite significant – the viscosity can decrease by an order of magnitude; (iii) at even higher $\dot{\gamma}$, the crossover to the shear-thickening occurs. A remarkable rheological feature is that the viscosity decrease in the second regime can be so rapid that the $\sigma(\dot{\gamma})$ dependence may have an anomalous N-shaped profile. In this case the part of the curve with $d\sigma/d\dot{\gamma} < 0$ becomes unstable and the flow is accompanied by a discontinuity in $\dot{\gamma}$. This causes the formation of shear bands – a phenomenon often observed in complex fluids (Salmon *et al.* 2003). Thus, liquid complex plasmas can exhibit essential rheological features peculiar to "classic" non-Newtonian fluids.

Moreover, by combining different methods to induce shear flows – e.g., inhomogeneous gas flows and laser beams – one can directly measure the shear viscosity in the entire range of shear rates – all the way to the limit where the discreteness enters and a fluid cannot be formally considered as a continuous medium. Probably, the most surprising result of such an investigation was that at "extreme" shear rates (up to $\dot{\gamma} \sim U/\Delta$, where U is the magnitude of the flow velocity and Δ is the interparticle distance), the formal hydrodynamic description with the Navier-Stokes equation still provides fairly good agreement with the experiment (Ivlev *et al.* 2007a).

It is worth mentioning that the transport coefficients of fluid complex plasmas, including the viscosity, could be calculated numerically for an arbitrary rate of the frictional dissipation (Vaulina *et al.* 2002; Vaulina and Dranzhevskii 2007). However, in contrast to steady-state structural properties (see Section 5.1), the kinetics of individual particles in strongly dissipative systems (say, when $\nu_{dn}/\omega_{\rm E} \gtrsim 1$) would inevitably be different from that in conventional single-species fluids. Therefore, such systems would probably not be relevant for investigating, e.g., the kinetics of the momentum transfer in shear flows. In the next subsection, where we focus on the kinetics of the energy transport, the importance of weak frictional dissipation becomes particularly clear.

5.2.3 Kinetics of heat transport

Thermal conductivity is an important property of matter that is essential in many engineering applications. At the same time, the behavior of thermal conductivity in various situations is governed by diverse fundamental processes that occur at the atomistic (kinetic) level. Measurements of the thermal conductivity in regular matter are only possible at a macroscopic scale and, therefore, cannot resolve the details of the heat transfer processes at their atomistic level due to the lack of experimental techniques to study the motion of individual atoms. Therefore, in this case also, liquid or solid complex plasmas occupy an invaluable position of an experimental model system where the motion of individual "atoms" can be observed in real time.

Analysis of the heat transport, especially in 2D crystalline systems, is a controversial problem that has a long history: Some authors claim that the thermal conductivity of such systems diverges in the thermodynamic limit. Liquid systems are far less studied – one can mention a simulation of frictionless hard disks by Shimada *et al.* (2000), where the thermal conductivity slowly diverged as well, and a theoretical study by Ernst *et al.* (1970), where the lack of a valid thermal conductivity was conjectured. Systems undergoing a phase transition, to our knowledge, were not studied at all.

Kinetics of the heat transport in liquid and solid complex plasmas was experimentally investigated by Fortov *et al.* (2007) and Nosenko *et al.* (2008). Below we focus on the experiment performed by Nosenko *et al.* (2008) in a 2D complex plasma that is undergoing a phase transition and therefore constitutes a mixture of crystalline and liquid phases. To melt the lattice locally and to control the temperature of the resulting liquid complex plasma, the laser-heating method has been employed, so that particles were pushed randomly by the radiation pressure force. To produce a quasi-1D temperature gradient, with temperature varying mostly in the y direction, a narrow area which extended fully across the particle suspension in the x direction was heated, as shown in Figure 5.5a. Under these conditions, heat was transferred mainly by thermal conductivity in the region where the temperature gradient was high.

Figure 5.5b shows the resulting profiles of the kinetic particle temperature, $T(y)$, measured for different values of the laser power. The particle suspension was melted in this temperature range, as can be seen from the analysis of the pair correlation function $g(r)$ (see Figure 5.5c) : Far from the laser-heated area, $g(r)$ has the characteristic appearance of the solid phase with notably many peaks, whereas inside the laser-heated area, $g(r)$ is typical for a liquid phase with a few peaks. (Note that the background temperature of the crystal, T_b, was naturally increasing with the applied laser power P_{laser}.) Also, according to the KTHNY theory (see Section 5.3.2), a 2D solid melts via two second-order phase transitions: Estimates show that the two temperatures corresponding to the transitions lie well within the temperature range achieved in the experiment. Irrespective of the applied heating power P_{laser}, all the measured temperature profiles are very well fitted by the exponential function $T(y) \propto \exp(y/L_{heat})$, where the heat transport length L_{heat} turned out to be practically constant. In the framework of the continuous approach to the heat transport, such scaling implies that L_{heat} is identical to the friction length $\sqrt{\chi/2\nu_{dn}}$ [see Equation (5.4)], and hence the thermal diffusivity χ is practically independent of T.

Thus, the heat transport in a 2D system that undergoes a phase transition turns out to be quite interesting: On the one hand, the experiment yielded the expected

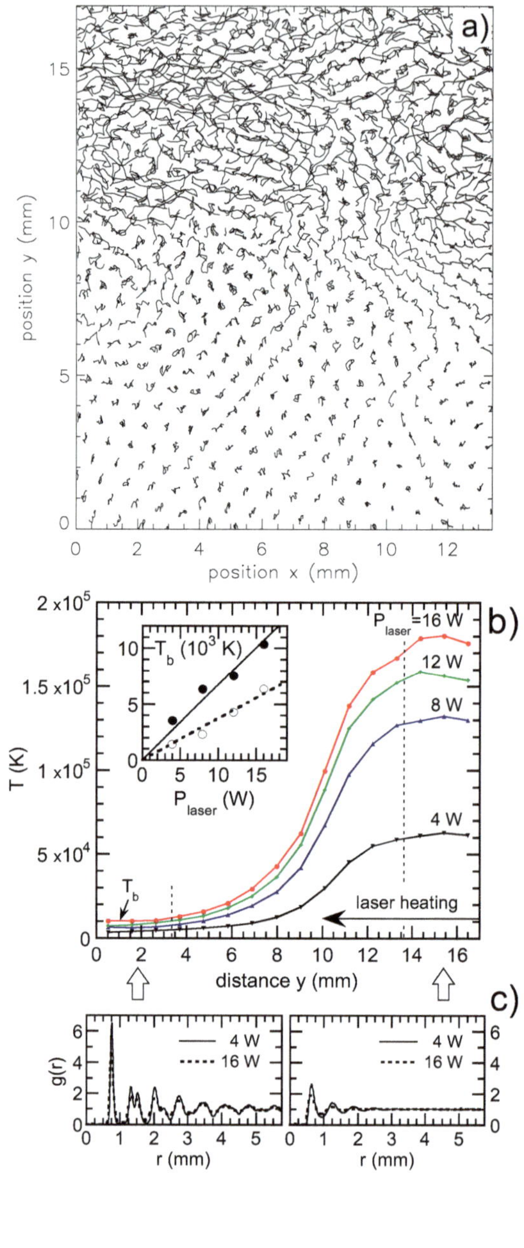

FIGURE 5.5

Heat transport in 2D complex plasmas (Nosenko *et al.* 2008). (a) Example of the particle trajectories (8.09 μm diameter, duration $\simeq$ 1.7 s, image is kindly provided by V. Nosenko) in a 2D plasma crystal heated by a laser at a power of P_{laser} = 16 W (heated region $y > 13.6$ mm). (b) Profiles of the kinetic particle temperature as a function of the transverse coordinate y, for different values of P_{laser}. The inset shows the background particle temperature T_b. (c) Pair correlation function $g(r)$ far from the heated region (left) and inside the heated region (right), suggesting crystalline and liquid states, respectively.

result that the thermal conductivity χ does not exhibit any major discontinuity at the liquid–solid phase boundary – such behavior is well known in regular matter (March and Tosi 2002). On the other hand, the values of χ obtained for different particle temperatures are almost the same. This result is not trivial, since individual phases, solid and liquid, are expected to have a temperature-dependent thermal conductivity. Nosenko *et al.* (2008) suggested that the dominant mechanism of thermal conduction in such systems is phonon scattering on heterogeneous fluctuations that occur in the melting region.

It is important to emphasize that although the measured temperature profile (viz., the value of L_{heat}) is determined by friction, the thermal conductivity itself is solely determined by internal generic properties of the medium (Yukawa system in our case) and does not depend on the damping: It was shown that the effective length of the phonon scattering ℓ_{ph} that actually determines the heat conduction ($\chi \simeq \frac{1}{2}C_{\text{l}}\ell_{\text{ph}}$) is at least an order of magnitude smaller than the length of the frictional phonon decay ($\simeq C_{\text{l}}/\nu_{dn}$), where C_{l} is the measured longitudinal acoustic velocity (see Section 3.3). This allows us to extrapolate the knowledge about the kinetics of heat transport (gained with weakly damped complex plasmas) to regular condensed matter and, thus, to understand more about generic atomistic processes governing the thermal conductivity.

5.2.4 Hydrodynamics at the discreteness limit

The "discreteness issue" of continuous media can be formulated as follows: "What is the smallest scale at which the conventional hydrodynamic description breaks down?" Apparently, the answer depends on the particular problem under consideration: It is determined by the similarity variables (and hence the related physical parameters) that play the major role in the description of the macroscopic problem. For instance, for a planar shear flow this is, primarily, the Reynolds and Mach numbers, whereas for a flow past an obstacle or a droplet breakup this can be the Weber number. (Of course, one should remember that the basic parameters entering hydrodynamics such as viscosity or surface tension are quantities which are well defined only for sufficiently large systems.)

To get the quantitative characteristics of hydrodynamic behavior at the discreteness limit, let us discuss the progress achieved so far in exploring interfacial instabilities occurring in large systems of discrete particles.

Wysocki and Löwen (2004) performed MD simulations of the Rayleigh-Taylor (RT) instability in the fully damped (Brownian) regime peculiar to colloidal suspensions. In these simulations, two different scenarios were observed that occur for either "high" or "low" surface tension. The high-surface-tension scenario is characterized by interfacial instability which is similar in spirit to the classical Rayleigh-Taylor instability (Chandrasekhar 1961). AS in the regular undamped case, the classical threshold value for the wavelength of unstable interface perturbations is confirmed. In contrast, when the interfacial surface tension is low enough, a completely different development is observed: The particles penetrate the interface easily as a result of the driving field and form microscopic lanes. The structure of these lanes is very similar to that seen in numerical simulations (Chakrabarti *et al.* 2004) and experiments (Leunissen *et al.* 2005) with driven colloidal suspensions. These results are obtained in the regime when the classical RT threshold for the unstable wavelength (calculated for given values of the surface tension and driving force) is smaller than the interparticle distance and hence a breakdown of hydrodynamics is expected.

Thus, the microscopic appearance of the RT instability might be completely different as the discreteness enters, and this conclusion is rather intuitive: The surface tension is the only stabilizing mechanism of the instability, and once this mechanism becomes negligible, and hence, allows growth at hydrodynamic scales smaller than the discreteness limit, the hydrodynamics itself becomes meaningless. On the other hand, the instability should develop in some form anyway, and the only imaginable picture for that is the interpenetrating strings, as observed in the simulations and experiments.

In addition to colloidal suspensions (Leunissen *et al.* 2005) and pedestrian zones (Helbing *et al.* 2000), the lane formation can be easily triggered in complex plasmas (Morfill *et al.* 2006; Sütterlin 2009). As we have already pointed out, complex plasmas provide a very important intermediate dynamical regime that is between classic undamped fluids and fully damped colloidal suspensions: In complex plasmas, the "internal" dynamics associated with the interparticle interaction can be undamped whereas the large-scale hydrodynamics can be strongly affected by friction. Nevertheless, the mesoscopic appearance of the lane formation in colloids and in complex plasmas is quite similar, which gives us grounds to believe that this phenomenon constitutes an ultimate generic form of the RT instability in any driven (strongly coupled) fluid.

Figure 5.6 shows an example of lane formation observed in complex plasmas with particles of different sizes (Sütterlin 2009). The net force acting on particles in a discharge plasma (a combination of the electric and ion drag forces; see Section 2.4) depends on the diameter and plays the role of an effective gravity pointed to the right (the force is relatively strong at the left edge and almost vanishing at the right edge of the figure). Initially, the large particles formed a (practically homogeneous) "background" fluid. When a small fraction of individual small particles entered the system from the left, their sedimentation towards the right edge of the figure was accompanied by a remarkable self-organization sequence: First, the particles form strings flowing along the force field (Figure 5.6a); then, as the field decreases, strings organize themselves into larger mesoscopic streams (Figure 5.6b); and at the later stage, when the field almost vanishes, streams merge to form a spheroidal droplet with well-defined surface (Figure 5.6c), indicating the transition to the regime when the effective surface tension should play the primary role.

In order to investigate the RT instability in further detail, let us consider examples of highly resolved shear flows observed in complex plasmas (Morfill *et al.* 2004a) and shown in Figure 5.7. Different flow topologies were observed, with the (average) flow lines being either straight (Figure 5.7a) or curved with a radius of curvature of about 80–100 Δ (Figure 5.7b). The lower part of the microparticle cloud is at rest. The observations suggest that the width and the structure of the transition (mixing)

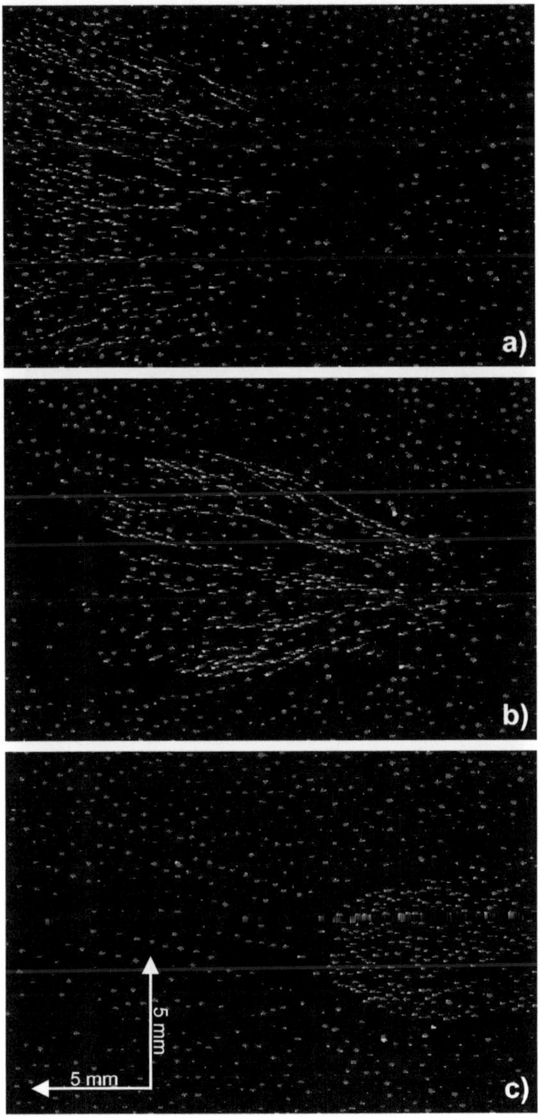

FIGURE 5.6

(See color insert following page 242). Lane formation in complex plasmas. A short burst of small (3.4 μm) particles injected into a cloud of large (9.2 μm) background particles is driven from left to right. Stages of (a) initial lane formation, (b) merging of lanes into larger streams, and (c) eventual droplet formation are shown. Each figure is a superposition of two consecutive color-coded images (1/50th s apart, green to red), entire sequence is about 2.5 s long. Images are kindly provided by M. Rubin-Zuzic.

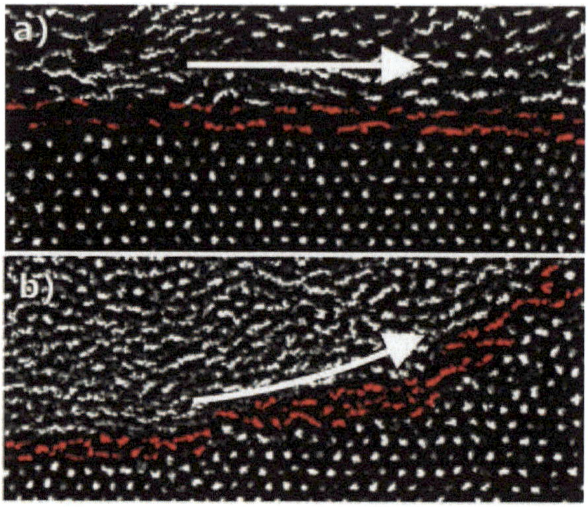

FIGURE 5.7
Two examples of highly resolved complex plasma flows (Morfill *et al.* 2004a). The figures show (a) a shear flow over a flat-surface plasma crystal and (b) a flow over a curved-surface plasma crystal. Note the small angle perturbations in the particle trajectories in (a), and the considerably larger scattering in the curved flow in (b). Particles are of 6.8 μm diameter, the flow velocity is $\sim$1 mm s^{-1}.

layer strongly depends on the geometry. For the planar flow the interface is quite smooth, with the flow along a particular monolayer. The trajectories of individual flowing particles experience only weak deflections and the overall flow appears to be stable and laminar. In contrast, the curved flow interface has a curious rough structure; the flow is not laminar; a "mixing layer" is formed. It is also apparent that the mixing layer becomes unstable at the individual particle level. The microscopic behavior may be interpreted as the centrifugally driven RT instability. Analyzing a whole sequence of such images, one can quantify "elementary" (discrete) perturbations in two ways – the fraction of interpenetrating (say, $\gtrsim \Delta$) particles, and the fraction of particles undergoing "large angle" (say, $\gtrsim 30^\circ$) collisions in the surface layer. For instance, for the straight flow the quantities are (almost) 0 %, and $\sim$ 3 %, for the curved flow $\sim$ 3 %, and $\sim$ 30 %. The latter can be understood kinetically in terms of the higher collision frequency with smaller impact parameter due to particle inertia at a curved surface. This has also been confirmed by numerical simulations conducted for similar geometry and flow conditions as in the experiments. The topology of the mixing layer found in the simulations corresponds closely to the measurements, which supports the kinetic interpretation.

Following these considerations, one can argue that the Kelvin–Helmholtz (KH) instability at the discreteness limit also has a different appearance. In order to illustrate this point, we consider another example of the hydrodynamic behavior of liquid complex plasmas (Morfill *et al.* 2004b) shown in Figure 5.8. Particles were flowing around an "obstacle" – the void of size $\sim 100\,\Delta$. One can see stable laminar shear flow around the obstacle, the development of a downstream "wake" exhibiting stable vortex flows, and a mixing layer between the flow and the wake. The enlargement of the mixing layer (Figure 5.8b) shows that the flow is quite unstable at the kinetic level, with instabilities becoming rapidly nonlinear. The width of the mixing layer grows monotonically with distance from the border where the laminar flow becomes detached from the obstacle. The growth length scale is on the order of a few Δ, i.e., much smaller than the hydrodynamic scales $n(dn/dx)^{-1}$ or $u(du/dx)^{-1}$, which would be expected macroscopically in fluids and which refer to the RT or KH instability, respectively. This rapid onset of surface instabilities followed by mixing and momentum exchange at scales $\sim \Delta$, i.e., the smallest interaction length scale available, is not consistent, therefore, with conventional macroscopic fluid instability theories. While this could not rightfully be expected at the kinetic level, it clearly points to new physics and, possibly, a hierarchy of processes that is necessary to describe interacting fluid flows: First, binary collision processes provide particle and momentum exchange on discreteness scales (a few Δ); then collective effects (due to the correlations defining fluid flows) take over and propel this "discrete" instability to macroscopic scales, creating cascades of growing clumps characterized by increased vorticity.

Although the onset of the instability shown in Figure 5.8 occurs at scales $\sim \Delta$, its further development is in amazing agreement with the simplest conceptual picture of the continuous jet turbulence: It is well known that the mixing between a jet and its surroundings occurs in two stages (see, e.g., Tennekes and Lumley 1972). During the first stage (which is a distinct peculiarity of jets), a shear layer is formed immediately downstream of the jet source, between jet stream and surroundings. As one moves downstream, there is an early linear-instability regime, involving exponential growth of small perturbations introduced at the jet source. Then, there is a gradual transition to the second stage associated with the nonlinear regime of the KH instability, where the dynamics of large-scale vortex formation and merging becomes the defining feature of the transitional shear flow. Apparently, the observed clump cascading fully mimics this scenario, which suggests – again – that the similarity of the coarse-grained hydrodynamics is preserved down to the physical discreteness limit.

Unfortunately, so far in experiments with complex plasmas, it has been impossible to observe the second stage typical to any developed turbulence – when vortices (clumps) break down leading to a more disorganized flow regime characterized by smaller-scale vortices. The spectral energy content at this stage should be consistent with the Kolmogorov's inverse cascade theory of turbulence. These processes develop at much longer timescales, when the neutral friction starts playing an important role and simply "freezes out" free hydrodynamic motion. In order to observe this turbulent stage in experiments, one needs to decrease the neutral gas pressure substantially and to increase the size of the complex plasmas.

These examples suggest a naive microscopic picture of the hydrodynamic instabilities: It is not unreasonable to conclude that many instabilities have a kinetic analog or trigger and that the most effective trigger mechanism is provided by binary large

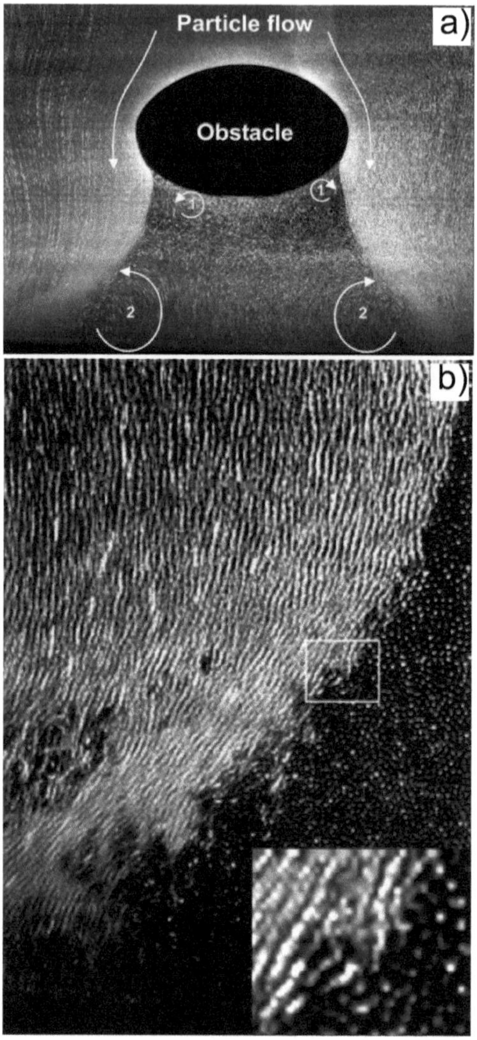

FIGURE 5.8
Flow past an obstacle in fluid complex plasmas (Morfill *et al.* 2004b). (a) Overall topology of the 3.7 μm particle flow, the system is approximately symmetric around the vertical axis (exposure time 1 s). The flow leads to a compressed laminar layer, which becomes detached at the outer perimeter of the wake. The steady vortex flow patterns in the wake are illustrated. The boundary between the laminar flow and wake becomes unstable; a mixing layer is formed, which grows in width with distance downstream. (b) An example of the mixing layer (an enlargement of the left side, exposure time 0.05 s). The points (lines) represent traces of slow (fast) moving microparticles. The inset shows trajectories of individual particles in the mixing layer.

angle scattering in localized structures and/or inhomogeneities of scales comparable to the particle correlation length. However, the mathematical techniques required to quantify the kinetic behavior and to transfer this to macroscopic scales still need to be developed.

5.2.5 Confined fluids

As fluid systems are engineered to smaller and smaller scales, down to the atomic size, the special effects associated with the confinement of fluids become increasingly important. The behavior of such systems is a fundamental problem in technology (including areas such as lubrication, adhesion, nanofluidics, microchannel spectrometry, and surface functionalization). The general consensus is that the smaller the system, the more important the confinement even for intrinsic properties (Hummer *et al.* 2001; de Mello 2006; Heller *et al.* 2006; Whitby and Quirke 2007). It is inevitable that there will be new physics associated with finite size effects, due to surface interactions and reduced dimensionality. From the application point of view, understanding the functionalization of nanoflow surfaces to achieve the desired form of hydrophobic or hygroscopic behavior (for a given fluid) in the absence or presence of, e.g., external fields (which would give rise to nano-electrorheology or electro-osmotic flows) is clearly one of the aims – and no doubt there are many others (Miller *et al.* 2001; Vaitheeswaran *et al.* 2004).

There have been many studies of confined flow systems, e.g., nanoporous materials (ordered or disordered), thin fluid films, and microchannels. Amongst the areas of interest are topics such as demixing (segregation) of biological fluid components, flows in nano-capillaries, and the effects of confinement on the fluid structure and on freezing and melting [for a review, see Whitby and Quirke (2007) and Alba-Simionesco *et al.* (2006) and references therein]. The optimum way to study the basic (generic) physics is to employ a system where kinetic measurements are possible at all relevant length and time scales. Currently the only systems capable of satisfying all these requirements are complex plasmas. We will see in Section 5.2.6 that complex plasmas have electrorheological properties under certain conditions and that it is possible to "design" the binary interaction potential between the particles using external fields (Ivlev *et al.* 2008). This provides great opportunities for future basic and applied research in a number of fields in condensed matter physics and beyond, and in particular for confined (nano) systems.

In this section we concentrate specifically on the first studies involving liquid complex plasmas, their "kinetic structure" in confined channel flows, and their dependence on the confinement potential. All confined flow experiments with complex plasmas so far have been conducted on the ground, i.e., the microparticles are suspended against gravity in the sheath region above the lower electrode. Horizontal confinement is effected either by non-conducting glass walls (which then attain floating potential), by conducting segmented electrodes (that can be actively powered and used to transport the particles), or by conducting metal channels placed on the lower electrode.

Teng *et al.* (2003) reported on the microscopic observation of the confinement-induced layering in quasi-2D complex plasma liquids. Two parallel vertical plates were put on a horizontal rf electrode surface to laterally confine particles and, hence, to form mesoscopic channels down to a few interparticle spacings in width. Microscopically, the particle mutual interaction tends to generate ordered triangular lattice-type domains with small amplitude position oscillations, which can be reorganized through string- or vortex-type hopping activated by thermal noise. However, the boundaries suppress the nearby transverse hopping. Figure 5.9 shows some snapshots of particle configurations and the corresponding transverse density distribution for different number of "layers", N. Basically, at larger N, the density profiles with their decaying oscillation from both boundaries manifest the confinement-induced (two to three) almost frozen outer layers near each boundary, which sandwich the more disordered isotropic liquid with a flat density profile in the center region. The transition to the layered structure up to the center at $N \lesssim 7$ is evidenced by the appearance of sharp peaks of the density profile. Similar structure was observed in a series of experiments with the so-called "dusty balls" – 3D spheroidal clusters consisting of a few thousand particles, which have a shell structure (of 3–4 layers) near the surface and a liquid (amorphous) state in the central bulk (Arp *et al.* 2004).

Using the same experimental setup as in Teng *et al.* (2003), the atomistic dynamics of the shear flow in a quasi-2D mesoscopic complex plasma liquid has been studied by Chan *et al.* (2004). Due to the formation of the nearby layered structure shown in Figure 5.9, the persistent and directional slow drive from the external stress along the boundary enhances cage-escape structural rearrangements which cascade into the liquid through many-body interaction. It was found that the flow consists of two outer shear bands, about three interparticle distances in width, adjacent to the boundaries and a central small-shear zone. The former has higher levels of both longitudinal and transverse velocity fluctuations. The shear banding phenomenon originates from the local stress release through the local rearrangement events adjacent to the boundary.

In a different experiment, converging and diverging ("nano") flows were investigated by Fink *et al.* (2005). One of the interests here was the determination of possible "selection rules" for the flow – e.g., how in detail the system evolves kinetically from N flow lines to $N-1$ flow lines when N becomes small. A second interest was to find out if there was a preferred instantaneous "structure" of the fluid particles during the flow line transitions. An example is shown in Figure 5.10a. The flow converges by one interparticle spacing Δ over a distance of typically six Δ (i.e., reduction of one flow line), so that the convergence angle is about 10 degrees. The figure shows the following features: (i) The typical structure of the fluid is hexagonal – i.e., the same as the 2D crystalline ground state. (ii) The transition from 4 to 3 flow lines goes via a localized 5/7 dislocation. (iii) The transition from 3 to 2 flow lines goes via alternating jumps ("zipping") of particles from the "central" flow line (which disappears) to the two outer ones. The characteristic "structures" observed are shown schematically in Figure 5.10b.

The results by Fink *et al.* (2005) confirm the observations shown in Figure 5.9 for a plane non-converging channel. As the system becomes smaller (in terms of flow lines), it begins to look instantaneously like a solid. This is, of course a consequence of the channel surface, which in these experiments is a "slip surface" (i.e., the

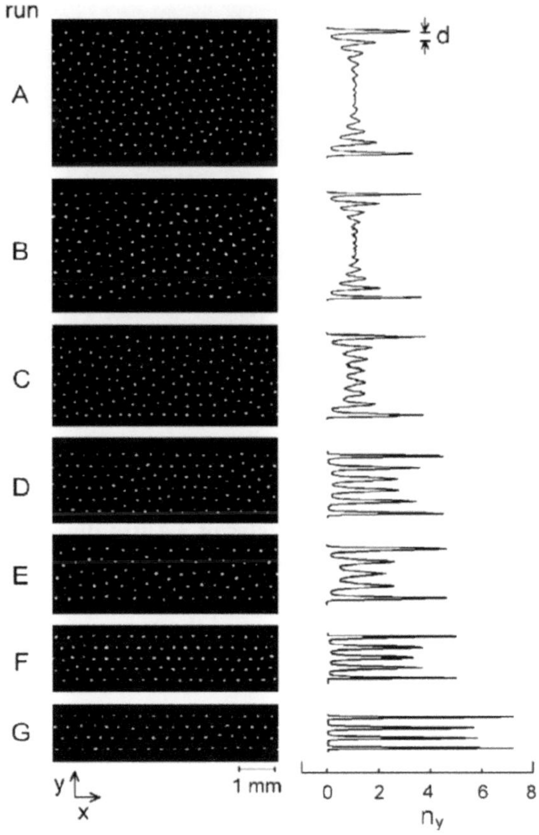

FIGURE 5.9
Liquid complex plasmas in narrow channels (Teng *et al.* 2003). The typical snapshots of the 7 μm particle configurations and the transverse particle density distributions, n_y, for different experiments with decreasing "number of layers" N (width measured in units of the interparticle spacing), from 11 to 3.

complex plasma does not have any "wetting" properties). Experiments with rough surfaces (on the scale of the interparticle separation) have not been performed yet. In such a case we would expect surface friction to play a role with associated modification of the flow structure and dynamics.

5.2.6 Electrorheological fluids

In this section we focus on an interesting class of so-called "electrorheological" (ER) fluids which have presently acquired significant attention. "Conventional" ER fluids consist of suspensions of microparticles in (usually) nonconducting fluids with a

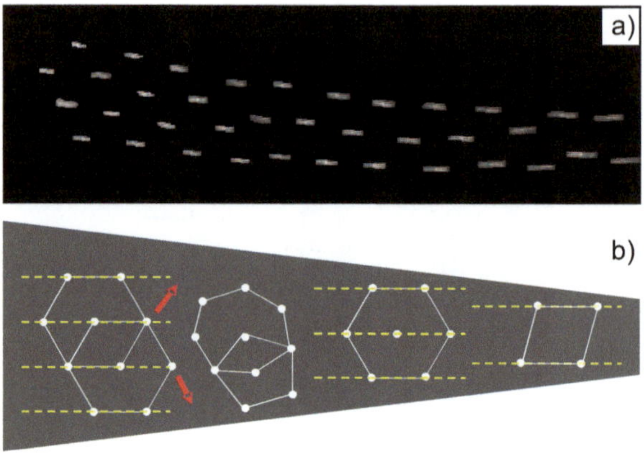

FIGURE 5.10

Converging 2D complex plasma flow in the limit of very few flow lines (Fink *et al.* 2005). (a) The convergence of particles of 3.4 μm diameter goes from 4 to 2 lines. The experiment was designed to investigate fluid structure and dynamical selection rules during the (discrete) flow line reduction. (b) Characteristic "fluid structures" observed in different regimes of the converging flow.

different dielectric constant (Chen *et al.* 1992; Dassanayake *et al.* 2000). The interparticle interaction, and hence the rheology of ER fluids, is determined by an external electric field, which polarizes microspheres and thus induces additional dipole–dipole coupling. The electric field plays the role of a new degree of freedom that allows us to "tune" the interaction between particles. This makes the phase diagram of ER fluids remarkably diversified (Yethiraj and van Blaaderen 2003; Hynninen and Dijkstra 2005).

The term "electrorheological fluid" is self-explaining (Stangroom 1983; Carlson *et al.* 1990): At low electric fields microparticles may be fully disordered and then (provided their concentration is low as well) ER fluids may be just normal Newtonian fluids. At larger fields, however, the situation can change dramatically – due to the increased dipole–dipole attraction particles arrange themselves into strongly coupled chains ("strings", or even "sheets") along the field. This naturally changes the rheology – e.g., at low shear stresses ER fluids can behave like elastic solids, while at stresses greater than a certain yield stress they are viscous liquids again. ER fluids have a significant industrial application potential – they can be used in hydraulics, photonics, display production, etc. (Stangroom 1983; Carlson *et al.* 1990; Yethiraj *et al.* 2004).

In contrast to conventional ER fluids (e.g., colloids) where the induced dipoles are due to polarization of the microparticles themselves, in complex plasmas the primary role is played by clouds of compensating plasma charges (mostly, excess ions) surrounding negatively charged microparticles (see Section 2.2). Without an external field the cloud is spherical ("Debye sphere"); when a field is applied, the cloud (which then acquires a fairly complicated shape and is called "plasma wake") is shifted downstream from the particle, along the field-induced ion drift. In this case the pair interaction between charged microparticles is generally non-reciprocal (i.e., non-Hamiltonian, see Section 3.2.2). The non-reciprocity of the interaction could be eliminated only if the wake potential were an even function of coordinates, i.e., $\varphi(\mathbf{r}) = \varphi(-\mathbf{r})$. A simple "recipe" to create such a reciprocal wake potential is as follows (Ivlev *et al.* 2008): One has to apply an ac field of a frequency that is (i) much lower than the inverse timescale of the ion response (ion plasma frequency, typically $\sim 10^7\ \mathrm{s}^{-1}$) and, at the same time, (ii) much higher than the inverse dust response time (dust plasma frequency, typically $\sim 10^2\ \mathrm{s}^{-1}$ or less). Then the ions react instantaneously to the field whereas the microparticles do not react at all. The effective interparticle interaction in this case is determined by the *time-averaged* wake potential. The resulting interaction is rigorously reciprocal (Hamiltonian), so that one can directly apply the formalisms of statistical physics to describe ER plasmas.

Quantitatively, the (field-induced) interparticle interaction in ER plasmas can be determined from the linear dielectric response formalism (see Section 2.2.2). For subthermal ion drift the interaction potential is given by Equation (2.50), which basically represents the far-field asymptotics for the potential expanded into a series over small u_i (with the angular dependence of the first three coefficients being proportional to that of the corresponding multipoles, i.e., charge, dipole, quadrupole). Furthermore, all "odd" terms ($\propto u_i^j$ with odd j) are proportional to linear combinations of the odd-order Legendre polynomials whereas "even" terms are combinations of the even-order polynomials. Thus, for an ac field $E(t)$ with $\langle E\rangle_t = 0$, all odd-order terms disappear in the time-averaged potential $\langle\varphi\rangle_t$, which becomes an even function of coordinates. The effective energy $Q\langle\varphi\rangle_t$ of the time-averaged pair interaction is (Ivlev *et al.* 2008)

$$W(r,\theta) \simeq Q^2\left[\frac{e^{-r/\lambda}}{r} - 0.43\frac{M_T^2\lambda^2}{r^3}(3\cos^2\theta - 1)\right]. \tag{5.5}$$

Thus, the effective interaction consists of two principal contributions: The first "core" term represents the spherically symmetric Debye–Hückel part, whereas the second term is due to the interaction between the charge of one particle and the quadrupole part of the wake produced by another particle. The charge–quadrupole interaction is identical to the interaction between two equal and parallel dipoles of magnitude $\simeq 0.65 M_T Q\lambda$. This implies that for small M_T the interactions in ER plasmas are *equivalent* to dipolar interactions in conventional ER fluids.

One can compare ER colloids and ER plasmas in terms of the dipole–dipole coupling (Tao 1993; Gulley and Tao 1997; Hynninen and Dijkstra 2005). Since the magnitude of the induced dipole is proportional to the volume of the "polarizable sphere", the field necessary to achieve a given coupling in ER colloids will be much larger than that in ER plasmas. In colloids, microparticles of radius a acquire dipoles $\sim a^3 E_{\mathrm{coll}}$ and are separated by distance $\sim a$, whereas the interaction in plasmas is determined by Equation (5.5) with typical separation $\sim \lambda$. The equivalent field for colloids is then $E_{\mathrm{coll}} \sim M_T (a/\lambda)^{1/2}(|Q|/a^2)$. For typical experimental conditions, the electric field $E \sim 3$ V/cm in plasmas (which corresponds to $M_T \sim 1$) is equivalent to $E_{\rm coll} \sim 3$ kV/cm in colloids.

The investigated phase diagram of ER colloids reveals a remarkable variety of crystalline states (Chen *et al.* 1992; Yethiraj and van Blaaderen 2003; Hynninen and Dijkstra 2005). In addition to "isotropic" bcc and fcc lattices, the hexagonal close-packed (hcp) structure can be a ground state in a fairly broad range of phase variables. Moreover, unusual "anisotropic" crystalline states become possible, like body-centered orthorhombic (bco) and body-centered tetragonal (bct); the phase transition between them is of the second order. On the other hand, relatively little research has been done on the fluid phase. In particular, the dynamics and details of the phase transition between "isotropic" and "string" fluids is practically unexplored (Tao 1993; Gulley and Tao 1997).

The "isotropic-to-string" phase transition in ER plasmas was investigated in experiments under microgravity conditions by Ivlev *et al.* (2008). Particles remained in a disordered fluid state as long as the amplitude of the applied AC (alternating current) field was below a certain threshold. Increasing the field further triggered rearrangement of particles: They became more and more ordered, until eventually well-defined particle strings were formed along the direction of the field. The transition between isotropic and string fluid states was fully reversible – decreasing the field brought the particles back into their initial isotropic state. The trend to form strings increased with particle size. The molecular dynamic (MD) simulations performed with similar parameters gave remarkable agreement with the experiment.

In order to quantify the isotropic-to-string phase transition, a suitable order parameter that is sensitive to the changing particle structures has to be employed (Ivlev *et al.* 2008). Conventional approaches, e.g., binary correlation or bond orientation functions, and Legendre polynomials turned out to be too insensitive. Much more satisfactory results were obtained by implementing the α (see, e.g., Räth *et al.* 2002) – a local nonlinear measure for structure characterization, with which any symmetry changes can be quantified by using the longitudinal and transverse distributions $P_{\parallel}(\alpha)$ and $P_{\perp}(\alpha)$. For the onset of the isotropic-to-string transition, the difference between the transverse and longitudinal scaling indices averaged over the ensemble, $\Delta\alpha = \int \alpha P_{\perp} d\alpha - \int \alpha P_{\parallel} d\alpha$, was used as a scalar order parameter, whereas M_T played the role of the control parameter. The obtained data were quite well approximated with a two-parametric fit $\Delta\alpha \propto (M_T - M_T^{\rm cr})^{\gamma}$ for $M_T > M_T^{\rm cr}$ and $\Delta\alpha = 0$ for $M_T \leq M_T^{\rm cr}$, which might suggest a second-order or a weak first-order phase transition between isotropic and string fluids (Tao 1993). Note that for a weakly coupled ER plasma, which implies gaseous-like ensembles of particles where triple interactions play a minor role, a simple analytical criterion for the isotropic-to-string phase transition can be derived from the analysis of the second virial coefficient, $B = \pi \int_0^{\infty} \int_{-1}^{1} (1 - e^{-W/T_d}) r^2 dr dx$ (see, e.g., Landau and Lifshitz 1978), where $x = \cos\theta$ and $W(r,x)$ is given by Equation (5.5).

Figure 5.11 summarizes the experimental results and the comparison with the MD simulations (Ivlev *et al.* 2008). The structural order of the well-developed strings is quite evident in both experimental and simulation data shown in the top two rows. The lower two rows show the corresponding distributions $P_{\parallel}(\alpha)$ and $P_{\perp}(\alpha)$.

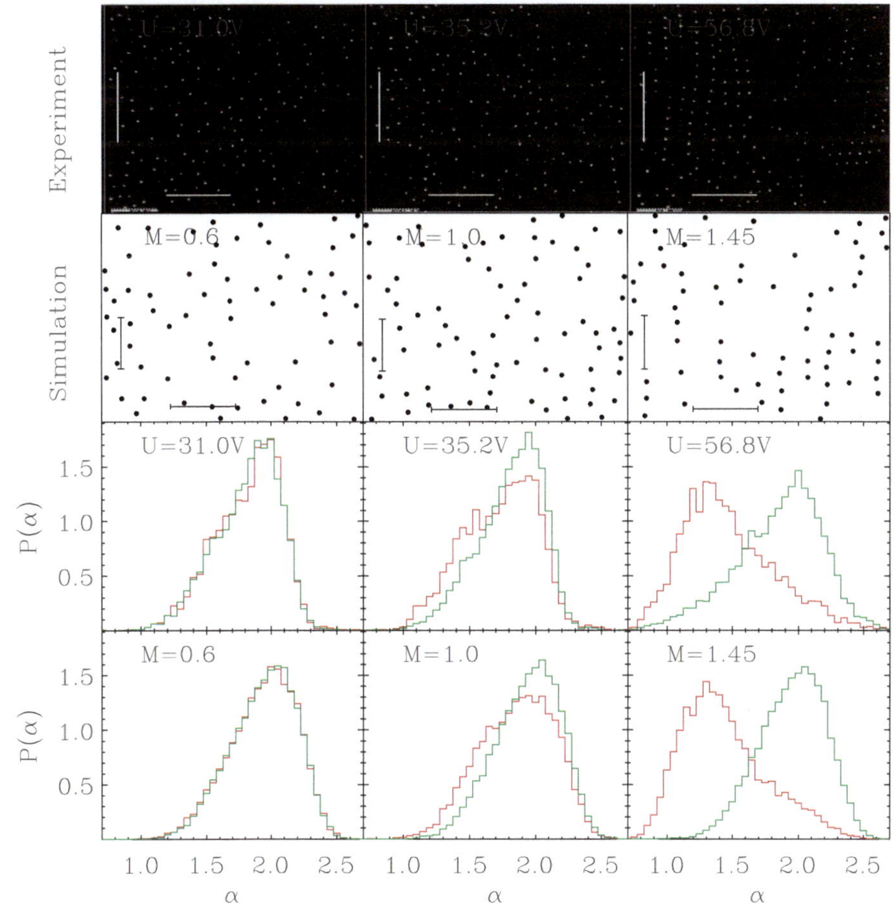

FIGURE 5.11

Formation of strings in ER plasmas (Ivlev *et al.* 2008). First row: Microgravity experiments (6.8 μm particles, raw data), microparticles are illuminated by a thin (less than mean interparticle distance) laser sheet parallel to the applied ac electric field. Examples of "low" (first column), "intermediate" (second column), and "high" (third column) fields are shown, the peak-to-peak voltage of the ac signal (applied to two parallel horizontal electrodes) is indicated. Second row: MD simulations, the same configuration as in the experimental setup, the field is measured in units of the thermal Mach number M_T (scale bars correspond to 2 mm). Third and forth rows: Histograms for longitudinal (left curve) and transverse (right curve) distributions of the scaling indices, $P_{\parallel}(\alpha)$ and $P_{\perp}(\alpha)$, calculated for the experiment and simulation, respectively. (Note that at higher densities the particle positions in neighboring strings became highly correlated.)

So far, colloidal suspensions have been the major focus for ER studies, providing a wealth of information (Chen *et al.* 1992; Dassanayake *et al.* 2000; Yethiraj and van Blaaderen 2003; Hynninen and Dijkstra 2005). The discovery that complex plasmas also have electrorheological properties adds a new dimension to such research – in terms of time/space scales and for studying new phenomena: An essentially single-species system of microparticles in complex plasmas enables us to investigate previously inaccessible rapid elementary processes that govern the dynamical behavior of ER fluids – at the level of individual particles. In particular, such investigations may allow us to study critical phenomena accompanying second-order phase transitions (Khrapak *et al.* 2006; Kompaneets 2009).

5.3 Solids

In this section we concentrate specifically on the kinetic description of the crystalline state in complex plasmas, with the focus on various dynamical aspects that may have generic nature and therefore play an important role in regular solids. This section starts with the kinetic characterization of crystals – the approach which (in principle) is equally appropriate for 2D and 3D cases. Then we proceed with different crystallization scenarios peculiar to 2D and 3D systems. We also discuss creation and dynamics of dislocations – the process that is absolutely relevant for 3D crystals as well, but has been properly investigated so far only in 2D plasma crystals.

5.3.1 Atomistic dynamics in crystals

Transitions between solid and fluid phases as well as between different crystalline states, rheological and transport properties of the fluid phase, energy relaxation and hierarchy of metastable states are determined by the magnitude of the coupling parameter Γ [see Equation (5.1)], which can be also considered as the measure of (inverse) temperature. In turn, Γ depends sensitively on local variations in crystal structure and provides information about the occurrence of localized excited states and nonstationary processes.

The value of Γ can be determined experimentally (Knapek *et al.* 2007a), by linking the individual particle dynamics with the local density and crystal structure using the Einstein frequency $\omega_{\rm E}$, which refers to linear oscillations of individual particles (atoms) in a lattice. In local equilibrium, the dynamics of individual particles in each lattice cell is statistically equivalent and can be described by a Langevin equation (see, e.g., VanKampen 1981). Therefore, cells represent a canonical ensemble with the Maxwell-Boltzmann distribution, $\propto \exp[-m_d(v^2 + \omega_{\rm E}^2 r^2)/2T_d]$, as illustrated in Figure 5.12. Then one can deduce thermodynamic characteristics locally, from the independent Gaussian fit of the velocity and displacement distribution: The velocity dispersion is T_d/m_d and the displacement dispersion is $T_d/m_d\omega_{\rm E}^2 \equiv \Delta^2/\tilde{\Gamma}$, where

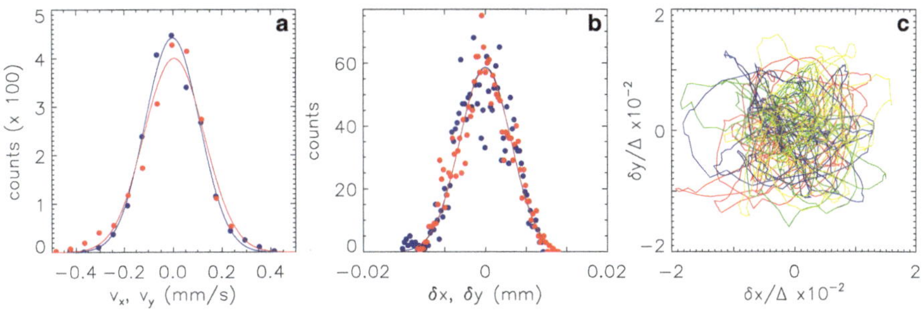

FIGURE 5.12
(See color insert following page 242). Dynamics of particles in a lattice of a 2D crystal (Knapek *et al.* 2007a). (a) Distribution of velocities v_x (red dots) and v_y (blue dots) with Maxwellian fits (solid lines). (b) Distribution of displacements x (red dots) and y (blue dots) of particles in their nearest-neighbor cage, solid lines are Gaussian fits. (c) Particle trajectories in their respective nearest-neighbor cells during the measurement time of $\simeq$12.3 s (colors correspond to the progression of time). Particles are of 9.19 μm diameter.

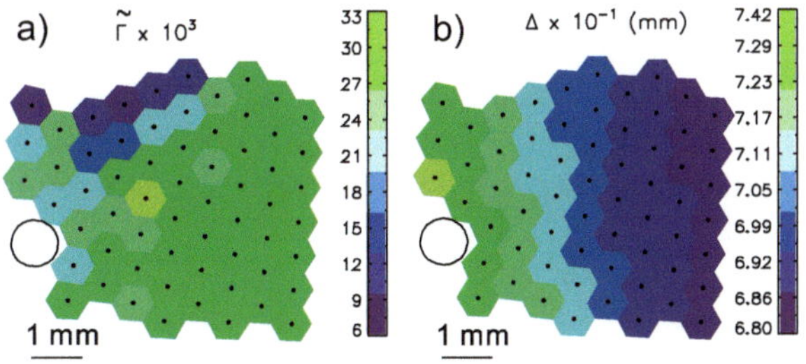

FIGURE 5.13
(See color insert following page 242). 2D maps of local crystal parameters (Knapek *et al.* 2007a). Distribution of (a) effective coupling parameter $\tilde{\Gamma}$ and (b) interparticle distance Δ is shown; the Voronoi cell around each particle is color coded according to the value of the measured quantity. The circles indicate the position of a sevenfold-fivefold pair defect; blue cells seen at the upper edge of (a) are due to the particle cage-escape event (see Section 5.2.1).

$\tilde{\Gamma} = (\omega_{\rm E}/\omega_{pd})^2\Gamma$ is the effective coupling parameter modified by the screening [the ratio of the Einstein to the dust-plasma frequency is a function of the screening parameter κ only, see Robbins *et al.* (1988) and Knapek *et al.* (2007a)]. For such a linear description to hold, it is essential that particles in the lattice perform sufficiently small oscillations. The role of anharmonic effects can then be neglected, so that the oscillations of the neighboring particles are uncoupled and can be treated independently.

An exemplary map of $\tilde{\Gamma}$ measured in a 2D plasma crystal is shown in Figure 5.13a. This map can be used to probe correlations with various local processes occurring in a crystal. For instance, there is a 5/7 dislocation just outside the regime analyzed (the position is marked by a circle), and there is some indication that the coupling strength in the vicinity of a non-stationary cage-escape event is substantially decreased. At the same time, comparison with the density map (see Figure 5.13b) shows no correlation.

The interparticle spacing shown in Figure 5.13b varies by about 0.5% per cell, so that the 2D density inhomogeneities are about 1% per cell. By performing independent measurements of the longitudinal and transverse acoustic modes (viz., acoustic velocities $C_{l,t}$, see Section 3.3) that are very sensitive to the screening parameter κ (Fortov *et al.* 2005), one can obtain a map of the coupling parameter Γ (rather than $\tilde{\Gamma}$) and thus define the state of the crystal within the phase diagram (see Figure 5.1).

A straightforward application of the method described above could be to determine the *local* Lindemann criterion of crystal melting, viz., what is the critical magnitude of the mean squared displacement, what are the characteristic patterns of the caged particle motion in the vicinity of the melting transition, what is the role of dynamical heterogeneity, etc.

5.3.2 Scalings in 2D crystallization

The characterization of solid, supercooled (glassy) and liquid states is, in general, not straightforward. Different models for the solid–liquid phase transition have been put forward. For 2D systems, models of particular relevance are the dislocation theory of melting – the Kosterlitz–Thouless–Halperin–Nelson–Young (KTHNY) theory [which involves two phase transitions – with an intermediate, so called "hexatic phase" in between – one associated with the unbinding of dislocation pairs and the other with the unbinding of disclination pairs, see Kosterlitz and Thoules (1973), Halperin and Nelson (1978, 1979), Young (1979), and Nelson (2002)], and the theory of grain-boundary induced melting (see Chui 1982, 1983).

Apparently, one of the central questions in understanding phase transitions in 2D strongly coupled systems is what the critical parameters are that determine which melting scenario will be realized in a particular experiment (i.e., whether the melting occurs in accordance with the KTHNY scenario, or the transition is preempted by grain-boundary-induced melting). The accompanying questions are whether the correlation functions associated with the crystal and hexatic phases have the appropriate scaling behavior, and what the order is of the observed phase transitions in the "thermodynamic limit". These issues have been discussed extensively (see, e.g., reviews by Alba-Simionesco *et al.* (2006) and Strandburg 1988).

It is generally believed that the value of the defect core energy plays a critical role in the realization of the melting scenario (Strandburg 1988). The KTHNY mechanism should operate when the core energy exceeds $\simeq 2.8T_{M1}$ (where T_{M1} is the temperature of unbinding of dislocation pairs), otherwise grain-boundary-induced melting should occur. Below we focus on two experiments that illustrate the kinetics accompanying these melting mechanisms.

The experiments by Zahn and Maret (2000) performed with colloidal particles are

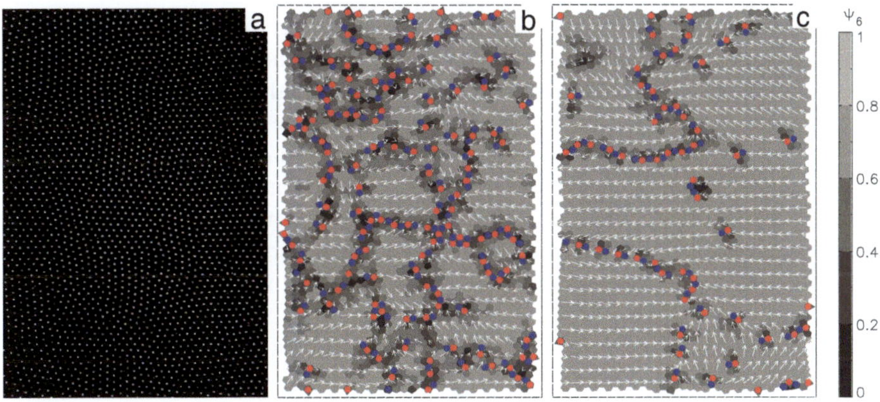

FIGURE 5.14
(See color insert following page 242). Recrystallization in 2D complex plasmas (adapted from Knapek *et al.* 2007b). (a) Snapshot showing intermediate structure of 9.19 μm particles during the recrystallization. (b,c) Color-coded 2D maps for two consecutive stages of recrystallization (about 10 s apart, map b corresponds to snapshot a). The background gray scale corresponds to the local value of the bond-orientational function $|\psi_6|$; the arrows represent the vector field of ψ_6 on the complex plane; defects are marked by red (fivefold) and blue (sevenfold) dots.

an excellent example of the KTHNY scenario [for other examples, see, e.g., Murray and Winkle (1987), and Marcus and Rice (1997)]. Supermagnetic spherical colloids were confined by gravity to a horizontal flat water/air interface. A vertical magnetic field was applied, which induced a magnetic moment, so that a (repulsive) dipole–dipole potential dominated the interaction. Thus changing the magnetic field strength allows external tuning of the coupling parameter and the study of phase transitions in a controlled way. The data obtained with video microscopy (about 2000 particles) were analyzed in terms of the bond order correlation function, $g_6(t) = \langle e^{i6\theta(t)} \rangle$, as a function of time, where $\theta(t)$ denotes the angle fluctuation of a fixed bond. Three regimes can be identified: The crystalline regime at large Γ, where $g_6 = const$, the isotropic liquid regime at low Γ, where $g_6(t)$ decays exponentially; and an intermediate regime, where $g_6(t)$ decays as a power law, indicating the hexatic phase. These findings are in quite good agreement with the KTHNY theory, supporting the two-stage melting for systems with an r^{-3} interaction potential (the core energy occurs above the critical value of $2.8T_{\rm M1}$).

The melting via grain boundaries was apparently seen in several experiments with complex plasmas (see, e.g., Quinn *et al.* 1996; Melzer *et al.* 1996a; Knapek *et al.* 2007b; Nosenko *et al.* 2009). Let us consider the experiment by Knapek *et al.* (2007b), where a 2D monolayer of about 3400 particles was first allowed to crystallize, and then it was perturbed and melted by an electric impulse. The subsequent recrystallization was recorded with high spatial and temporal resolution (see Figure 5.14). To make a comparison with the KTHNY theory, the local variation of orientational ordering was investigated by calculating the bond-orientational function, $\psi_6 = \frac{1}{n}\sum_j e^{i6\theta_j}$, over the n nearest neighbors for each particle, with θ_j being the angle between the nearest-neighbor bond and a reference axis. The modulus $|\psi_6|$ of this complex quantity yields the bond order parameter, which is unity for an ideal hexagonal structure, and the argument $\arg(\psi_6)$ is a measure for cell orientations with respect to the reference axis. The kinetic temperature of the system was defined from the velocity distribution of the particles (by fitting with a Maxwell-Boltzmann distribution, see Figure 5.12).

Figure 5.14b,c shows color-coded maps of $|\psi_6|$ for two consecutive stages of recrystallization. The location of jumps in bond orientation is clearly correlated with the lines of (fivefold/sevenfold) defect locations. After melting, as the system cools down, the crystalline domains grow and merge with neighboring regions, as illustrated in Figure 5.14b, causing the bonds to tilt to the (single) orientation of the growing region. Eventually, a metastable state shown in Figure 5.14c is reached which is characterized by highly ordered adjoined crystalline domains.

The dynamic evolution of the lattice defects in 2D complex plasmas can be summarized as follows: (i) The instantaneous 2D structure revealed mainly hexagons, pentagons and septagons at all temperatures sampled. (ii) The fraction of pentagons and septagons was identical within the statistical uncertainties – they practically always appear in pairs. (iii) The hexagonal "ground state" (also the lowest energy state) dominated at all temperatures sampled. (iv) The local disorder, identified as the fraction of pentagons or septagons, could be approximated by a power-law dependence on temperature, as shown in Figure 5.15.

These facts indicate that unlike 3D liquids, which may have their own distinct local order [pentagon-like, see Frank (1952); Reichert *et al.* (2000)] – quite different from the crystalline state – 2D liquids do not exhibit a special local order. They can appear as a crystal with different amounts of lattice dislocations, which depend on the temperature. As the temperature decreases, these dislocations may partially annihilate (anneal) and they can also have a tendency to form strings, which act as domain boundaries separating homogeneous ordered regions.

The implications of these experimental findings are obvious:

- they show that the fundamental stability principles of condensed matter depend on the external constraints – in such a way that for 2D systems the self-organization favors mixtures of the ground state and the next most excited states;

- furthermore, the power-law behavior with respect to the order parameter "temperature" shows that there is no characteristic scale;

- if these findings are generic (perhaps for a certain class of materials) then they are significant for characterizing physical properties and – ultimately – for monolayer, membrane and nano-engineering;

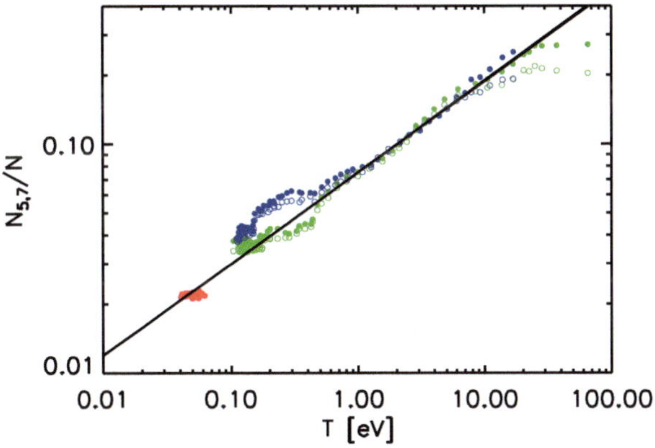

FIGURE 5.15
Fraction of the five- and sevenfold defects, $N_{5,7}/N$, during the recrystallization (adapted from Knapek *et al.* 2007b). As temperature decreases, the number of defects in a hexagonal lattice obeys a power-law dependence, $N_{5,7} \propto T_d^{0.4}$ (solid line), revealing a classic scale-free behavior. Three different experiments are shown.

- to understand whether the findings have generic implications, we need to develop a kinetic theory that has a sufficiently general character to allow extrapolation to other systems.

Temperature scaling for domain boundaries. Given the above findings one can develop a simple theory that describes this process of self-organization. This theory is based on the early work of Frenkel (1955).

At a given temperature, a 2D system of N particles is divided into $\zeta = N/\bar{N}_\mathrm{d}$ homogeneous domains, each containing $\bar{N}_\mathrm{d}$ particles on average, with boundaries made up of pairs of pentagons and septagons. We assume that the structural order in the individual domains is uncorrelated. While it is clear that there will be a spectrum of domain sizes, let us for the moment consider only averages (the justification for this approach will become apparent later).

If the mean separation between the particles is Δ, then the mean domain radius is determined by $\pi\bar{r}^2 = \pi(\Delta/2)^2(N/\zeta)$, i.e., $\bar{r} = \frac{1}{2}(N/\zeta)^{1/2}\Delta$. The interfaces have an additional amount of "line energy" $\bar{E} = 2\pi\bar{r}\zeta\sigma$, where σ is the "line tension" (interaction between domains is neglected). Substituting for $\bar{r}$ gives $\bar{E} = \pi\Delta(N\zeta)^{1/2}\sigma$. As a result of the domain structure, the system entropy increases. The measure of disorder is characterized by the number of different ways in which the particles may be organized (assuming homogeneity inside each domain), $P = N!/[(N/\zeta)!]^\zeta$. If N and N/ζ are sufficiently large, then using Stirling's approximation yields $P \simeq \zeta^N$. The entropy is $S = \ln P$ and the mean free Helmholtz energy is accordingly $\bar{F} = \pi\Delta(N\zeta)^{1/2}\sigma - NT_d \ln\zeta$. Assuming that during recrystallization the system remains always in thermodynamic equilibrium, from $\partial \bar{F}/\partial \zeta = 0$ we have

$$\zeta = \left(\frac{2T_d}{\pi \Delta \sigma}\right)^2 N.$$

Remarkably, ζ does not depend on $\bar{N}_\mathrm{d}$, the mean particle population of a domain. We have now established a relationship between ζ and T_d. At this stage we can introduce the fractal nature of the domains as a hypothesis. This hypothesis is also intended to account for the size distribution of the domains, which we have not explicitly discussed. We write $\bar{N}_\mathrm{d}\Delta^2 = const(\bar{N}_\mathrm{s}\Delta)^{1+\alpha}$, where $\bar{N}_\mathrm{s}$ is the average number of particles in a domain wall. (Note that with the above definition we would have $\alpha = 1$ if the domain was circular, whereas for long narrow strip domains $\alpha \to 0$, which suggests that $0 < \alpha < 1$.) Substituting this scaling yields finally the total number of particles in all the domain boundaries, $N_\mathrm{tot}(= N_5 + N_7) \equiv \zeta \bar{N}_\mathrm{s}$, which obeys the scaling $N_\mathrm{tot}/N \propto T_d^{\frac{2\alpha}{1+\alpha}}$. From the measurements we have $2\alpha/(1+\alpha) \simeq 0.4$, which gives $\alpha \simeq 0.25$ if σ is temperature independent (obviously, if σ is temperature dependent then α becomes larger than 0.25). The obtained fractal exponent lies in the expected range $0 < \alpha < 1$, but no physical argument has been obtained so far regarding its specific value. This could imply that α may be material-dependent.

Thus, it is possible to explain the observations of the recrystallization of a 2D plasma crystal with simple thermodynamic arguments, provided the following major assumptions are satisfied: (i) The system is instantaneously in thermodynamic equilibrium. (ii) The evolution takes the form of uncorrelated domains with size and number depending on temperature. (iii) Domain boundaries are always of the same type (here 5/7 dislocations). (iv) The domain lines satisfy on average a constant fractal relationship, independent of temperature. (v) The line tension of the domain boundaries is temperature independent (or has a power law dependence on T_d). (vi) The free energy of the domain walls dominates the system evolution.

5.3.3 Dynamics of dislocations

Even far above the melting line, dislocations are ubiquitous in both 2D and 3D crystals. Dislocations are essential for understanding such properties as plasticity, yield stress, susceptibility to fatigue, and fracture. Their generation and motion is of interest in materials science (Kittel 1961), the study of earthquakes and snow avalanches (Kirchner *et al.* 2002), colloidal crystals (Schall *et al.* 2004), 2D foams (Abd el Kader and Earnshaw 1999), and various types of shear cracks (Abraham and Gao 2000; Rosakis *et al.* 1999).

In elastic theory, a dislocation's core is treated as a singularity in an otherwise continuous elastic material. Obviously, such a simplified approach is often too crude to capture essential quantitative characteristics of dislocations, whose scales are usually on the order of the lattice constant. In regular solids dislocation dynamics is almost impossible to study experimentally at an atomistic level (Murayama *et al.* 2002) because of the small distances between the atoms (or molecules), high characteristic

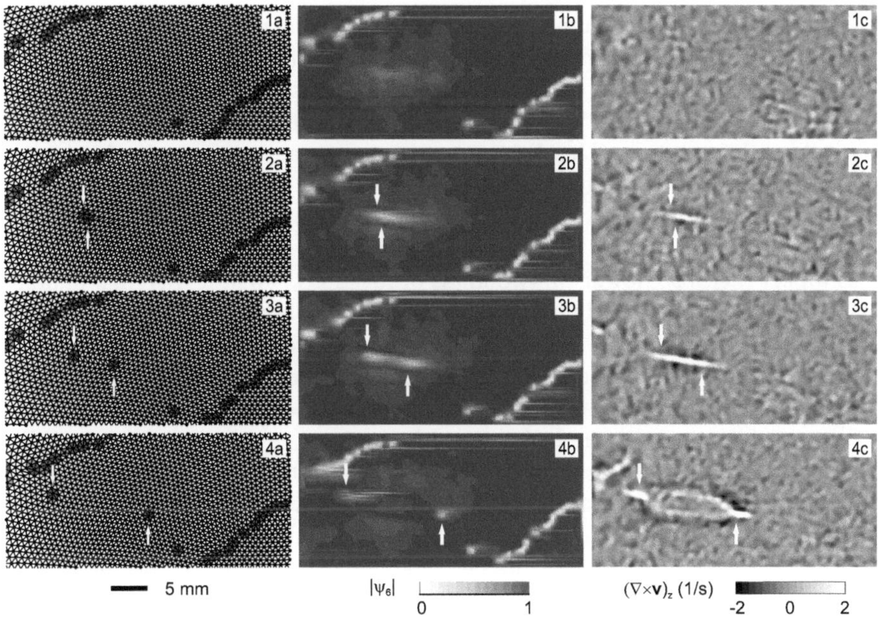

FIGURE 5.16
Generation and dynamics of dislocation pairs in a 2D plasma crystal (Nosenko *et al.* 2008). Maps of (a) triangulation of the particle positions, (b) bond-orientational function $|\psi_6|$, and (c) vorticity $(\nabla \times \mathrm{v})$ are shown for four different moments of time: (1) 0.33, (2) 0.57, (3) 0.70, and (4) 1.00 s. A pair of dislocations is indicated in (a) by arrows. Particles are 8.09 μm in diameter.

frequencies, and the lack of experimental techniques of visualizing the motion of individual atoms.

In contrast to regular solids, complex plasmas turned out to be an exceptionally suitable model system for experimental study of the discrete structure and dynamics of dislocations. In the experiment by Nosenko *et al.* (2007), a 2D plasma crystal was heavily stressed due to inhomogeneous (parabolic) radial confinement. That was the reason for the strong variation of the number density across the crystal and, as a consequence, for the appearance of topological defects (indicated in Figure 5.16a). Most of the defects formed linear chains that constitute domain boundaries in the crystal. During the course of the experiment, dislocations (i.e., isolated pairs of five-fold and sevenfold defects) were continuously generated due to the shear introduced by a slow rotation of the crystal. They then moved around and finally annihilated with each other or at domain boundaries.

In order to characterize dislocations at the discreteness limit, one has to relate discrete and continuous measures of shear deformation. The most appropriate discrete measure (which is, at the same time, insensitive to uniform compressions, rotations, and translations, etc.) is the modulus of the bond-orientational function $|\psi_6|$ shown in Figure 5.16b. In the limit of weak simple shear, the following relation can be used: $|\psi_6| \simeq 1 - 9\gamma^2$ where γ is the shear strain (Nosenko *et al.* 2007). For weak pure shear, $|\psi_6| \simeq 1 - 2.25\varepsilon^2$, where ε is the elongation, which is the measure of pure shear deformation. The dislocation dynamics can be conveniently characterized in terms of 2D vorticity, $(\nabla \times \mathbf{v})$, shown in Figure 5.16c (where $\mathbf{v}$ is the particle velocity).

Figure 5.16b shows that the shear strain had quite a nonuniform distribution. It was higher (i.e., $|\psi_6|$ lower) in two kinds of locations. First, it was high in domain boundaries – the two nearly parallel bright stripes in Figure 5.16b (or equivalently the chains of fivefold and sevenfold defects in Figure 5.16a). Second, a "diffuse background" of shear strain appeared between the domain boundaries. The diffuse shear strain increased with time. When it locally exceeded a certain threshold, a pair of edge dislocations was created in that location, as one can see in the second row of Figure 5.16; these dislocations appear as bright spots in (b) or as pairs of fivefold and sevenfold defects in (a), all indicated by arrows. Once a pair of dislocations was created, they moved rapidly apart (third and fourth rows). The Burgers vectors in such a pair were oppositely directed and equal in magnitude, so that the total Burgers vector was naturally conserved.

Creation of dislocation pairs is characterized by several distinct stages in the evolution of the shear strain: First, the shear strain builds up gradually in a certain location. Second, when the shear strain in this location exceeds a threshold, a pair of dislocations is born. Third, the shear stress is rapidly relaxed when the dislocations separate, and gradually drops to the background level. This cycle then starts over again, perhaps in a different location.

Dislocations that move supersonically create clear signatures – Mach cones that can be seen in Figure 5.16c, fourth row (see also Section 4.5.3). The Mach cones were composed of shear waves and not of compressional waves, because they were excited by dislocations moving faster than the transverse acoustic velocity, C_{t}, but slower than the longitudinal one, C_{l}. The average speed of supersonic dislocations in the experiment was about $2C_{\mathrm{t}}$. In fact, linear elastic theory predicts that a gliding edge dislocation cannot overcome the sound speed of shear waves C_{t} because the energy radiated by a moving dislocation becomes infinite at this speed. However, gliding edge dislocations moving at the speed of $1.3C_{\mathrm{t}}$ to $1.6C_{\mathrm{t}}$ were observed in atomistic computer simulations (Gumbsch and Gao 1999). To the best of our knowledge, the results reported by Nosenko *et al.* (2007) provide the first experimental evidence that dislocations can indeed move faster than C_{t}.

5.3.4 3D crystallization

Growth of 3D crystals is a very important branch of industry, with numerous applications ranging from semiconductors, substrates for high temperature superconductors, piezo sensors, ferroelectric memories, optical elements to nano-structures, quantum dots and organic systems. There are different facets to crystal growth, homogeneous nucleation, heterogeneous nucleation, epitaxial growth, molecular beam epitaxy, chemical vapor deposition, etc. While techniques for visualization (and quality control) of crystal growth have improved greatly, the detailed kinetic understanding of dynamical growth processes is still far from complete. The same holds for nano- and microparticle contamination in production processes.

For a deeper understanding of the kinetics of crystal growth, the use of model systems that allow visualization in real space and time at the individual particle level is desirable. It is no surprise, therefore, that colloidal suspensions have been widely studied in the past in order to learn more about the generic properties of self-organization [see, e.g., Vlasov *et al.* (2001), Alsayed *et al.* (2005) and references therein]. The only essential limitation of colloids for this purpose is the damping by the suspension fluid, which makes it practically impossible to investigate atomistic dynamics.

With the discovery of plasma crystals, a new system became available for studying the fully resolved dynamics of self-organization processes. Research into 3D crystallization may benefit from this, and consequently a number of studies have been conducted, beginning with the investigation of basic crystal properties [3D crystal structure, acoustic modes, etc., see Zuzic *et al.* (2000) and Zhdanov (2003)] and the liquid-solid phase transitions (see Thomas and Morfill 1996; Rubin-Zuzic *et al.* 2006).

Structural properties of steady-state 3D plasma crystals were investigated in numerous experiments (Chu and I 1994; Quinn *et al.* 1996; Hayashi 1999; Zuzic *et al.* 2000; Arp *et al.* 2004). Crystalline structures such as bcc, fcc, and hcp, as well as their coexistence, were found for certain plasma and particle parameters, as illustrated in Figure 5.17. We see the co-existence of the (presumable) ground state (fcc) and a metastable state (hcp), which seems to mark the domain borders (Zuzic *et al.* 2000). Such borders are also seen between domains of the same structure but different lattice orientation. Note that in addition to these "isotropic" structures, also the vertically aligned hexagonal lattices – when particles form consecutive hexagonal layers in the horizontal direction, but vertically they are aligned in strings – were observed (see, e.g., Melzer *et al.* 1996b). Such lattices usually form in rf sheaths or dc striations, where significant ion flow is present. They can exist because the lower particles are attracted by the wake potentials of the upper ones, so that this attraction overcomes the mutual particle repulsion (see Section 2.3.2). The vertically aligned lattices are quite common for ground based experiments and are usually formed by particles of a few microns in diameter. For smaller particles the wake effect presumably becomes too weak, so that the particles form conventional close-packed crystals.

Based on available experimental data one can claim that there are (at least) two distinct macroscopic scenarios of crystallization in 3D complex plasmas. These can be referred to as "uniform nucleation" and "crystallization front", and which pathway is realized in the experiment depends heavily on the boundary conditions. If the (initially) liquid complex plasma is brought into the regime corresponding to a solid phase in Figure 5.1 then in the bulk region, where boundaries play no role, the system usually develops towards the uniform nucleation (although sometimes particles form a "visibly" amorphous solid; whether this is a 3D glassy state still needs

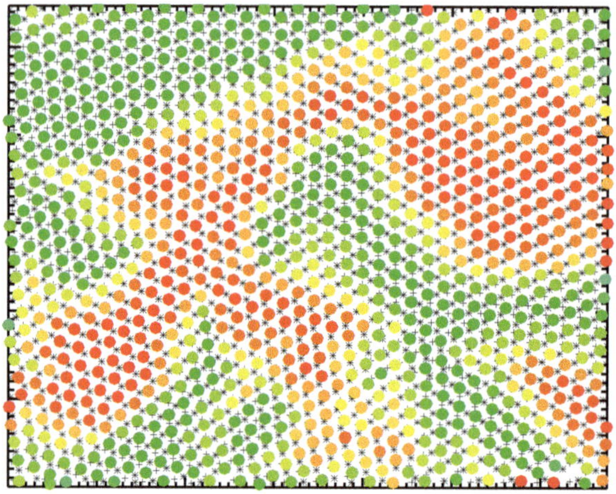

FIGURE 5.17
(See color insert following page 242). Domain structure of a 3D plasma crystal (Zuzic *et al.* 2000). Particles are of 3.38 μm diameter; three consecutive lattice planes are shown; each particle in the middle plane is color-coded in accordance with the local order (red corresponds to the fcc lattice cell and green to hcp); particle in two adjacent planes are indicated by crosses and stars.

to be clarified). In this case one normally observes coexistence of mesoscopic crystalline domains of different structure and orientation (see Figure 5.17), similar to nanostructured regular solids (Gleiter 1989, 2000). However, closer to the complex plasma boundaries, when a steep potential well exists (e.g., plasma sheaths close to rf electrodes), the crystallization often develops in the form of a front propagating from boundaries inwards into the particle cloud. Apparently, a steep boundary in this case facilitates formation of a hexagonal "substrate" which then triggers the propagating layer-by-layer crystallization process (Rubin-Zuzic *et al.* 2006).

Below we discuss the measurements of the dynamical evolution and kinetic structure of a 3D crystallization front (Rubin-Zuzic 2006) and relate this to theoretical models. The first two images of Figure 5.18 show a slice through a 3D complex plasma crystallization front. One can see a number of features, such as the detailed (kinetic) structure of the front and different crystal domains with different structure and/or orientation.

Let us focus on two particular features, which could be quite generic for a certain class of substances – both in the liquid and solid phases. These are the discovery of a distribution of small "droplets" in the crystal phase and small "crystallites" in the fluid phase (henceforth called *phaselets*), that are seen in Figure 5.18c, and a narrow (few lattice distance extent) premelted region in the crystalline regime (perpendicular to the front) where particles exhibit enhanced mobility signifying *interfacial melting*.

Phaselets. Figure 5.19 summarizes the measured characteristics of phaselets. Due

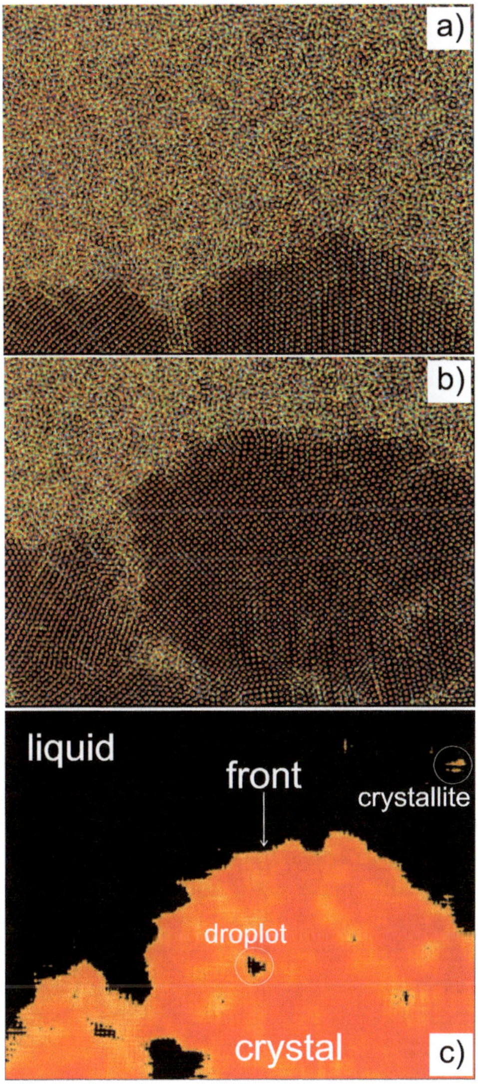

FIGURE 5.18

(See color insert following page 242). Crystallization front in a 3D complex plasma (Rubin-Zuzic *et al.* 2006). Figures (a) and (b) illustrate the front propagating upwards (images are about 16 s apart from each other). Each figure is a superposition of 10 consecutive video frames (about 0.7s); particle positions are color-coded from green to red, i.e., "caged" particles appear redder, "fluid" particles are multicolored. (c) The local order for figure (b), where red implies high crystalline order, black denotes the fluid phase, and yellow indicates transitional regions. Along with the crystallization front, droplets and crystallites are seen that may grow and then dissolve again. Particles are of 1.28 μm diameter.

to the special kinetic observations are possible with complex plasmas, these features could be resolved down to sizes of a few particles. There are two general features worth noting: (i) The size spectra of both droplets and crystallites are compatible with power laws (Figure 5.19, right column). This suggests that within the observable parameter range (~ 10 to $\sim 10^3$ particles) there is no characteristic length scale that determines either formation or dissolution. (ii) The larger crystallites and droplets tend to live longer (Figure 5.19, left column). By "life time" we mean the growth + dissolution phases, so this result is not too surprising. There is, however, a substantial spread in the individual life times.

At first sight, the development of the crystallites can be simply explained in terms of the thermodynamics: If we naturally assume the temperature (both in the liquid and crystalline regimes) below the melting point T_M, then the evolution of seed crystallites (which always form due to random fluctuations) is determined by the competition between a decrease in the bulk free energy and an increase in the surface energy. If the seed crystallite is large enough, the bulk contribution overcomes the surface part and it can grow further.

As for the droplets observed in the crystal regime, the mechanism responsible for their formation should be quite different, because thermodynamically, both the bulk and the surface contributions cause the free energy to increase. It is possible that after the initial solidification, a gradual relaxation from a metastable to a ground state (for example, from hcp to bcc or fcc structure) occurs downstream from the crystallization front. This is naturally accompanied by a release of latent heat. Then the droplets could be a local manifestation of this relaxation. The larger the droplet, the longer it takes to dissipate the released heat, and the longer its lifetime. The existence of interfacial melting between two large domains seen in Figure 5.18 supports this.

Interfacial melting. Regular solids usually exhibit domains of locally ordered regimes (grains), which are separated by domain (grain) boundaries (see, e.g., Gleiter 2000). Thermodynamically, these grain boundaries are different (both in energy and entropy) from the homogeneous crystal regimes. When such a grainy crystal is heated and approaches its melting point, the grain boundaries may play a special role – they can act as "seeds" of pre-melting regions. In a number of experiments using different colloidal suspensions, the effect of grain boundary melting has been demonstrated (Pusey and van Megen 1986; Gasser *et al.* 2001; Alsayed *et al.* 2005).

A kinetic (first principle) theory of melting faces several obstacles – there are long-range many-body interactions to contend with, there is the structural symmetry and periodicity, and universality classes are not known. Experimentally it has been possible to conduct studies with hard sphere colloids, and more recently using special temperature dependent colloidal systems [which contain microgel particles where diameters depend on temperature and therefore allow controlled tuning of the volume packing fraction, see Pusey and van Megen (1986), and Alsayed *et al.* (2005)].

As a result of these experiments (and particularly also studies of water ice), it has become established that crystal surfaces may form melted layers and that similar premelting occurs at defects in crystals, too. This suggests that the less perfect crystal structure and the associated interfacial free energy is the parameter that determines grain boundary melting.

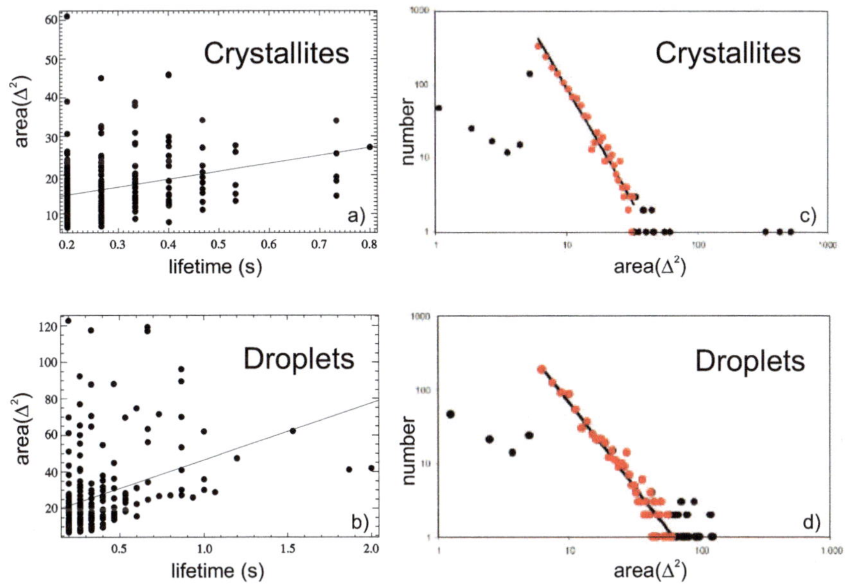

FIGURE 5.19
Characteristics of crystallites and droplets (adapted from Rubin-Zuzic *et al.* 2006). Area of crystallites (a) and droplets (b) measured in units of a single particle cell (squared interparticle distance Δ^2) versus their lifetimes, and histograms showing number of crystallites (c) and droplets (d) versus their areas.

Figure 5.18 shows that in the dynamical 3D crystallization front studies using complex plasmas, interfacial melting can be also observed. This is significant for three reasons:

- The measurement "slice" shown in Figure 5.18 was obtained in a large (millions of particles) complex plasma assembly, many interparticle spacings away from the boundaries. Hence, the measurements confirm that interfacial melting is not necessarily an effect confined to narrow surface regions.
- Damping in strongly coupled complex plasmas is rather weak. This implies that energy transport is to a large extent governed by phonons in the crystalline phase and dust-acoustic (sound) waves in the fluid regime.
- The particle interaction is primarily electrostatic. This implies that the same process – interfacial melting – occurs in different systems with different forms of binary particle interactions. In other words, the process can be generic and is not dependent on peculiarities or special features of the system.

For these reasons we conclude that the complementarity of research between atomic, molecular, colloidal, and complex plasma studies promises to yield much more than just the sum of its parts. A ubiquitous and still poorly understood process – like melting – needs different inputs, different constraints, generalization from different sources and new approaches, so that the principal mechanisms can be identified and combined to a fundamental kinetic theory.

References

Abd el Kader, A. and Earnshaw, J. C. (1999). Shear-induced changes in two-dimensional foam. *Phys. Rev. Lett.*, **82**, 2610–2613.

Abraham, F. F. and Gao, H. (2000). How fast can cracks propagate? *Phys. Rev. Lett.*, **84**, 3113–3116.

Alba-Simionesco, A., Coasne, B., Dosseh, G., Dudziak, G., Gubbins, K. E., Radhakrishnan, R., and Sliwinska-Bartkowiak, M. (2006). Effects of confinement on freezing and melting. *J. Phys. Condens. Matter*, **18**, R15–R68.

Alsayed, A. M., Islam, M. F., Zhang, J., Collings, P. J., and Yodh, A. G. (2005). Premelting at defects within bulk colloidal crystals. *Science*, **309**, 1207–1210.

Arp, O., Block, D., Piel, A., and Melzer, A. (2004). Dust Coulomb balls: Three-dimensional plasma crystals. *Phys. Rev. Lett.*, **93**, 165004/1–4.

Carlson, J. D., Sprecher, A. F., and Conrad, H. (eds.) (1990). *Electrorheological fluids*. Technomic, Lancaster.

Carstensen, J., Greiner, F., Hou, L. J., Maurer, H., and Piel, A. (2009). Effect of neutral gas motion on the rotation of dust clusters in an axial magnetic field. *Phys. Plasmas*, **16**, 013702/1–8.

Chakrabarti, J., Dzubiella, J., and Löwen, H. (2004). Reentrance effect in the lane formation of driven colloids. *Phys. Rev. E*, **70**, 012401/1–4.

Chan, C. L., Woon, W. Y., and I, L. (2004). Shear banding in mesoscopic dusty plasma liquids. *Phys. Rev. Lett.*, **93**, 220602/1–4.

Chandrasekhar, S. (1961). *Hydrodynamic and hydromagnetic stability*. Oxford University, Oxford.

Chen, T., Zitter, R. N., and Tao, R. (1992). Laser diffraction determination of the crystalline structure of an electrorheological fluid. *Phys. Rev. Lett.*, **68**, 2555–2558.

Chu, J. H. and I, L. (1994). Direct observation of Coulomb crystals and liquids in strongly coupled rf dusty plasmas. *Phys. Rev. Lett.*, **72**, 4009–4012.

Chui, S. T. (1982). Grain-boundary theory of melting in two dimensions. *Phys. Rev. Lett.*, **48**, 933–935.

Chui, S. T. (1983). Grain-boundary theory of melting in two dimensions. *Phys. Rev. B*, **28**, 178–194.

Dassanayake, U., Fraden, S., and van Blaaderen, A. (2000). Structure of electrorheological fluids. *J. Chem. Phys.*, **112**, 3851–3858.

de Mello, A. J. (2006). Control and detection of chemical reactions in microfluidic systems. *Nature*, **442**, 394–402.

Donko, Z., Goree, J., Hartmann, P., and Kutasi, K. (2006). Shear viscosity and shear thinning in two-dimensional Yukawa liquids. *Phys. Rev. Lett.*, **96**, 145003/1–4.

Dubin, D. (1990). First-order anharmonic correction to the free energy of Coulomb crystal in periodic boundary conditions. *Phys. Rev. A*, **42**, 4972–4982.

Ernst, M. H., Hauge, E. H., and van Leeuwen, J. M. J. (1970). Asymptotic time behavior of correlation functions. *Phys. Rev. Lett.*, **25**, 1254–1256.

Farouki, R. and Hamaguchi, S. (1993). Thermal energy of crystalline one-component plasma from dynamical simulation. *Phys. Rev. E*, **47**, 4330–4336.

Fink, V. A., Kretschmer, M., Fortov, V., Höfner, H., Konopka, U., Morfill, G. E., Petrov, O., Ratynskaia, S., Usachev, A, and Zobnin, A. (2005). Cooperative phenomena in laminar fluids: Observation of streamlines. *AIP Conf. Proc.*, **799**, 295–298.

Fischer, E. W. (1993). Light scattering and dielectric studies on glass forming liquids. *Physica A*, **201**, 183–206.

Fisher, M. E. (1974). The renormalization group in the theory of critical behavior. *Rev. Mod. Phys.*, **46**, 597–616.

Fortov, V. E., Khrapak, A. G., Khrapak, S. A., Molotkov, V. I., and Petrov, O. F. (2004). Dusty plasmas. *Phys. Usp.*, **47**, 447–492.

Fortov, V. E., Ivlev, A. V., Khrapak, S. A., Khrapak, A. G., and Morfill, G. E. (2005). Complex (dusty) plasmas: Current status, open issues, perspectives. *Phys. Rep.*, **421**, 1–103.

Fortov, V. E., Vaulina, O. S., Petrov, O. F., Vasiliev, M. N., Gavrikov, A. V., Shakova, I. A., Vorona, N. A., Khrustalyov, Y. V., Manohin, A. A., and Chernyshev, A. V. (2007). Experimental study of the heat transport processes in dusty plasma fluid. *Phys. Rev. E*, **75**, 026403/1–8.

Frank, F. C. (1952). Supercooling of liquids. *Proc. R. Soc. London A*, **215**, 43–46.

Frenkel, J. (1955). *Kinetic theory of liquids.* Dover, New York.

Gasser, U., Weeks, E. R., Schofield, A., Pusey, P. N., and Weitz, D. A. (2001). Real-space imaging of nucleation and growth in colloidal crystallization. *Science*, **292**, 258–262.

Gavrikov, A., Shakhova, I., Ivanov, A., Petrov, O., Vorona, N., and Fortov, V. (2005). Experimental study of laminar flow in dusty plasma liquid. *Phys. Lett. A*, **336**, 378–383.

Gleiter, H. (1989). Nanocrystalline materials. *Prog. Mater. Sci.*, **33**, 223–315.

Gleiter, H. (2000). Nanostructured materials: Basic concepts and microstructure. *Acta Materialia*, **48**, 1–29.

Gulley, G. L. and Tao, R. (1997). Structures of an electrorheological fluid. *Phys. Rev. E*, **56**, 4328–4336.

Gumbsch, P. and Gao, H. (1999). Dislocations faster than the speed of sound. *Science*, **283**, 965–968.

Halperin, B. I. and Nelson, D.R. (1978). Theory of two-dimensional melting. *Phys. Rev. Lett.*, **41**, 121–124.

Hamaguchi, S., Farouki, R. T., and Dubin, D. H. E. (1997). Triple point of Yukawa systems. *Phys. Rev. E*, **56**, 4671–4682.

Hansen, J. P. and McDonald, I. R. (1986). *Theory of simple liquids*. Academic, New York.

Hansen, J. P. and Schiff, D. (1973). Influence of interatomic repulsion on the structure of liquids at melting. *Mol. Phys.*, **25**, 1281–1290.

Hansen, J. P. and Verlet, L. (1969). Phase transitions of the Lennard-Jones system. *Phys. Rev.*, **184**, 151–161.

Hayashi, Y. (1999). Structure of a three-dimensional Coulomb crystal in a fine-particle plasma. *Phys. Rev. Lett.*, **83**, 4764–4767.

Helbing, D., Farkas, I. J., and Vicsek, T. (2000). Freezing by heating in a driven mesoscopic system. *Phys. Rev. Lett.*, **84**, 1240-1243.

Heller, D. A., Jeng, E. S., Yeung, T. K., Martinez, B. M., Moll, A. E., Gastala, J. B., and Strano, M. S. (2006). Optical detection of DNA conformational polymorphism on single-walled carbon nanotubes. *Science*, **311**, 508–511.

Huang, Y. H. and I, L. (2007). Memory and persistence correlation of microstructural fluctuations in two-dimensional dusty plasma liquids. *Phys. Rev. E*, **76**, 016403/1–6.

Hummer, G., Rasaiah, J. C., and Noworyta, J. P. (2001). Water conduction through the hydrophobic channel of a carbon nanotube. *Nature*, **414**, 188–190.

Hynninen, A. P. and Dijkstra, M. (2005). Phase diagram of dipolar hard and soft spheres: Manipulation of colloidal crystal structures by an external field. *Phys. Rev. Lett.*, **94**, 138303/1–4.

Ichimaru, S. (1982). Strongly coupled plasmas: High-density classical plasmas and degenerate electron fluids. *Rev. Mod. Phys.*, **54**, 1017–1059.

Ikezi, H. (1986). Coulomb solid of small particles in plasmas. *Phys. Fluids*, **29**, 1764–1766.

Ivlev, A. V., Steinberg, V., Kompaneets, R., Höfner, H., Sidorenko, I., and Morfill, G. E. (2007a). Non-Newtonian viscosity of complex-plasma fluids. *Phys. Rev. Lett.*, **94**, 145003/1–4.

Ivlev, A. V., Zhdanov, S. K., and Morfill, G. E. (2007b). Free thermal convection in complex plasma with background-gas friction. *Phys. Rev. Lett.*, **99**, 135004/1–4.

Ivlev, A. V., Morfill, G. E., Thomas, H. M., Räth, C., Joyce, G., Huber, P., Kompaneets, R., Fortov, V. E., Lipaev, A. M., Molotkov, V. I., Reiter, T., Turin, M., and Vinogradov, P. (2008). First observation of electrorheological plasmas. *Phys. Rev. Lett.*, **100**, 095003/1–4.

Ivlev, A. V., Zhdanov, S. K., Thomas, H. M., and Morfill, G. E. (2009). Fluid phase separation in binary complex plasmas. *Europhys. Lett.*, **85**, 45001/1–5.

Jäckle, J. (1986). Models of the glass transition. *Rep. Prog. Phys.*, **49**, 171–231.

Juan, W. T., Chen, M. H., and I, L. (2001). Nonlinear transports and microvortex excitations in sheared quasi-two-dimensional dust Coulomb liquids. *Phys. Rev. E*, **64**, 016402/1–5.

Khrapak, S. A. and Morfill, G. E. (2008). A note on the binary interaction potential in complex (dusty) plasmas. *Phys. Plasmas*, **15**, 084502/1–4.

Khrapak, S. A., Morfill, G. E., Ivlev, A. V., Thomas, H. M., Beysens, D. A., Zappoli, B., Fortov, V. E., Lipaev, A. M., and Molotkov, V. I. (2006). Critical point in complex plasmas. *Phys. Rev. Lett.*, **96**, 015001/1–4.

Khrapak, S. A., Klumov, B. A., and Morfill, G. E. (2008). Electric potential around an absorbing body in plasmas: Effect of ion-neutral collisions. *Phys. Rev. Lett.*, **100**, 225003/1–4.

Kirchner, H. O. K., Michot, G., and Schweizer, J. (2002). Fracture toughness of snow in shear under friction. *Phys. Rev. E*, **66**, 027103/1–3.

Kittel, C. (1961). *Introduction to solid state physics*. Wiley, New York.

Knapek, C. A., Ivlev, A. V., Klumov, B. A., Morfill, G. E., and Samsonov, D. (2007a). Kinetic characterization of strongly coupled systems. *Phys. Rev. Lett.*, **98**, 015001/1–4.

Knapek, C. A., Samsonov, D., Zhdanov, S., Konopka, U., and Morfill, G. E. (2007b). Recrystallization of a 2D plasma crystal. *Phys. Rev. Lett.*, **98**, 015004/1–4.

Kompaneets, R., Morfill, G. E., and Ivlev, A. V. (2009). Design of new binary interaction classes in complex plasmas. *Phys. Plasmas*, **16**, 043705/.1–6.

Kosterlitz, J. M. and Thouless, D. J. (1973). Ordering, metastability and phase transitions in two-dimensional systems. *J. Phys. C*, **6**, 1181–1203.

Kremer, K., Robbins, M. O., and Grest, G. S. (1986). Phase diagram of Yukawa systems: Model for charge-stabilized colloids. *Phys. Rev. Lett.*, **57**, 2694–2697.

Lai, Y. J. and I, L. (2002). Avalanche excitations of fast particles in quasi-2D cold dusty-plasma liquids. *Phys. Rev. Lett.*, **89**, 155002/1–4.

Landau, L. D. and Lifshitz, E. M. (1978). *Statistical physics. Part I.* Pergamon, Oxford.

Landau, L. D. and Lifschitz, E. M. (1986). *Theory of elasticity*. Pergamon, Oxford.

Leunissen, M. E., Christova, C. G., Hynninen, A. P., Royall, C. P., Campbell, A. I., Imhof, A., Dijkstra, M., van Roij, R., and van Blaaderen, A. (2005). Ionic colloidal crystals of oppositely charged particles. *Nature*, **437**, 235–240.

Lindemann, F. A. (1910). Über die Berechnung molekularer Eigenfrequenzen. *Z. Phys.*, **11**, 609–612.

Liu, B. and Goree, J. (2007). Superdiffusion in two-dimensional Yukawa liquids. *Phys. Rev. E*, **75**, 016405/1–5.

Liu, B. and Goree, J. (2008). Superdiffusion and non-gaussian statistics in a driven-dissipative 2D dusty plasma. *Phys. Rev. Lett.*, **100**, 055003/1–4.

Löwen, H. (1996). Dynamical criterion for two-dimensional freezing. *Phys. Rev. E*, **53**, R29–R32.

Löwen, H., Palberg, T., and Simon, R. (1993). Dynamical criterion for freezing of colloidal liquids. *Phys. Rev. Lett.*, **70**, 1557/1–4.

Maitland, G. C., Rigby, M., Smith, E. B., and Wakeham, W. A. (1981). *Intermolecular forces*. Clarendon, Oxford.

March, N. and Tosi, M. P. (2002). *Introduction to liquid state physics*. World Scientific, London.

Marcus, A. H. and Rice, S. A. (1997). Phase transitions in a confined quasi-two-dimensional colloid suspension. *Phys. Rev. E*, **55**, 637–656.

Meijer, E. J. and Frenkel, D. (1991). Melting line of Yukawa system by computer simulation. *J. Chem. Phys.*, **94**, 2269–2271.

Melzer, A., Homann, A., and Piel, A. (1996a). Experimental investigation of the melting transition of the plasma crystal. *Phys. Rev. E*, **53**, 2757–2766.

Melzer, A., Schweigert, V. A., Schweigert, I. V., Homann, A., Peters, S., and Piel, A. (1996b). Structure and stability of the plasma crystal. *Phys. Rev. E*, **54**, 46–49.

Meyers, M. A. and Chawla, K. K. (1998). *Mechanical behavior of materials*. Prentice Hall, New Jersey.

Miller, S. A., Young, V. Y., and Martin, C. R. (2001). Electroosmotic flow in template-prepared carbon nanotube membranes. *J. Am. Chem. Soc.*, **123**, 12335–12342.

Mitic, S., Sütterlin, R., Ivlev, A. V., Höfner, H., Thoma, M. H., Zhdanov, S., and Morfill, G. E. (2008). Convective dust clouds driven by thermal creep in a complex plasma. *Phys. Rev. Lett.*, **101**, 235001/1–4.

Morfill, G. E., Khrapak, S. A., Ivlev, A. V., Klumov, B. A., Rubin-Zuzic, M., and Thomas, H. M. (2004a). From fluid flows to crystallization: New results from complex plasmas. *Phys. Scripta*, **T107**, 59–64.

Morfill, G. E., Rubin-Zuzic, M., Rothermel, H., Ivlev, A., Klumov, B., Thomas, H., and Konopka, U. (2004b). Highly resolved fluid flows: "Liquid plasmas" at the kinetic level. *Phys. Rev. Lett.*, **92**, 175004/1–4.

Morfill, G. E., Konopka, U., Kretschmer, M., Rubin-Zuzic, M., Thomas, H. M., Zhdanov, S. K., and Tsytovich, V. (2006). The "classical tunnelling effect" – observations and theory. *New J. Phys.*, **8**, 7/1-20.

Murayama, M., Howe, J. M., Hidaka, H., and Takaki, S. (2002). Atomic-level observation of disclination dipoles in mechanically milled, nanocrystalline Fe. *Science*, **295**, 2433-2435.

Murray, C. H. and Winkle, D. H. V. (1987). Experimental observation of two-stage melting in a classical two-dimensional screened Coulomb system. *Phys. Rev. Lett.*, **58**, 1200-1203.

Nefedov, A. P., Morfill, G. E., Fortov, V. E., Thomas, H. M., Rothermel, H., Hagl, T., Ivlev, A. V., Zuzic, M., Klumov, B. A., Lipaev, A. M., Molotkov, V. I., Petrov, O. F., Gidzenko, Y. P., Krikalev, S. K, Shepherd, W., Ivanov, A. I., Roth, M., Binnenbruck, H., Goree, J. A., and Semenov, Y. P. (2003a). PKE-Nefedov: Plasma crystal experiments on the International Space Station. *New J. Phys.*, **5**, 33/1–10.

Nelson, D. R. (2002). *Defects and geometry in condensed matter physics*. Cambridge University, Cambridge.

Nelson, D. R. and Halperin, B. I. (1979). Dislocation-mediated melting in two dimensions. *Phys. Rev. B*, **19**, 2457–2484.

Nosenko, V. and Goree, J. (2004). Shear flows and shear viscosity in a two-dimensional Yukawa system (dusty plasma). *Phys. Rev. Lett.*, **93**, 155004/1–4.

Nosenko, V., Zhdanov, S., and Morfill, G. (2007). Supersonic dislocations observed in a plasma crystal. *Phys. Rev. Lett.*, **99**, 025002/1–4.

Nosenko, V., Zhdanov, S., Ivlev, A. V., Morfill, G., Goree, J., and Piel, A. (2008). Heat transport in a two-dimensional complex (dusty) plasma at melting conditions. *Phys. Rev. Lett.*, **100**, 025003/1–4.

Nosenko, V., Zhdanov, S. K., Ivlev, A. V., Knapek, C. A., and Morfill, G. E. (2009). 2D Melting of plasma crystals: Equilibrium and nonequilibrium regimes. *Phys. Rev. Lett.*, **103**, 015001/1–4.

Nunomura, S., Samsonov, D., Zhdanov, S., and Morfill, G. (2006). Self-diffusion in a liquid complex plasma. *Phys. Rev. Lett.*, **96**, 015003/1–4.

Ohta, H. and Hamaguchi, S. (2000). Molecular dynamics evaluation of self-diffusion in Yukawa systems. *Phys. Plasmas*, **7**, 4506–4514.

Phillpot, S. R. and Wolf, D. (1990). Grain boundaries in silicon from zero temperature through melting. *J. Amer. Ceramic Soc.*, **73**, 933–937.

Pusey, P. N. (2005). Freezing and melting: Action at grain boundaries. *Science*, **309**, 1198–1199.

Pusey, P. N. and van Megen, W. (1986). Phase behaviour of concentrated suspensions of nearly hard colloidal spheres. *Nature*, **320**, 340–342.

Qian, T., Wang, X. P., and Sheng, P. (2006). A variational approach to moving contact line hydrodynamics. *J. Fluid Mech.*, **564**, 333–360.

Quinn, R. A., Cui, C., Goree, J., Pieper, J. B., Thomas, H., and Morfill, G. E. (1996). Structural analysis of a Coulomb lattice in a dusty plasma. *Phys. Rev. E*, **53**, R2049–R2052.

Räth, C., Bunk, W., Huber, M., Morfill, G., Retzlaff, J., and Schuecker, P. (2002). Analysing large-scale structure I. Weighted scaling indices and constrained randomization. *Month. Not. Roy. Astr. Soc.*, **337**, 413–421.

Ratynskaia, S., Rypdal, K., Knapek, C., Khrapak, S., Milovanov, A. V., Ivlev, A., Rasmussen, J. J., and Morfill, G. E. (2006). Superdiffusion and viscoelastic vortex flows in a two-dimensional complex plasma. *Phys. Rev. Lett.*, **96**, 105010/1–4.

Raveche, H. J., Mountain, R. D., and Streett, W. B. (1974). Freezing and melting properties of the Lennard-Jones system. *J. Chem. Phys.*, **61**, 1970–1984.

Reichert, H., Klein, O., Dosch, H., Denk, M., Honkimäki, V., Lippmann, T., and Reiter, G. Observation of five-fold local symmetry in liquid lead. (2000). *Nature*, **408**, 839–841.

Reichman, D. R., Rabani, E., and Geissler, P. L. (2005). Comparison of dynamical heterogeneity in hard-sphere and attractive glass formers. *J. Phys. Chem. B*, **109**, 14654–14658.

Robbins, M. O., Kremer, K., and Grest, G. S. (1988). Phase diagram and dynamics of Yukawa systems. *J. Chem. Phys.*, **88**, 3286–3312.

Rosakis, A. J., Samudrala, O., and Coker, D. (1999). Cracks faster than the shear wave speed. *Science*, **284**, 1337–1340.

Rubin-Zuzic, M., Morfill, G. E., Ivlev, A. V., Pompl, R., Klumov, B. A., Bunk, W., Thomas, H. M., Rothermel, H., Havnes, O., and Fouquet, A. (2006). Kinetic development of crystallization fronts in complex plasma. *Nature Physics*, **2**, 181–185.

Saigo, T. and Hamaguchi, S. (2002). Shear viscosity of strongly coupled Yukawa systems. *Phys. Plasmas*, **9**, 1210–1216.

Salin, G. and Caillol, J. M. (2002). Transport coefficients of the Yukawa one-component plasma. *Phys. Rev. Lett.*, **88**, 065002/1–4.

Salmon, J. B., Colin, A., Manneville, S., and Molino, F. (2003). Velocity profiles in shear-banding wormlike micelles. *Phys. Rev. Lett.*, **90**, 228303/1–4.

Saltzman, E. J. and Schweizer, K. S. (2006). Activated hopping and dynamical fluctuation effects in hard sphere suspensions and fluids. *J. Chem. Phys.*, **125**, 044509/1–19.

Schall, P., Cohen, J., Weitz, D. A., and Spaepen, F. (2004). Visualization of dislocation dynamics in colloidal crystals. *Science*, **305**, 1944–1948.

Shimada, T., Murakami, T., Yukawa, S., Saito, K., and Ito, N. (2000). Simulational study on dimensionality dependence of heat conduction. *J. Phys. Soc. Jpn.*, **69**, 3150–3153.

Sillescu, H. (1999). Heterogeneity at the glass transition: A review. *J. Non-Cryst. Solids*, **243**, 81–108.

Stangroom, J. E. (1983). Electrorheological fluids. *Phys. Technol.*, **14**, 290–296.

Stevens, M. J. and Robbins, M. O. (1993). Melting of Yukawa systems: A test of phenomenological melting criteria. *J. Chem. Phys.*, **98**, 2319–2324.

Strandburg, K. J. (1988). Two-dimensional melting. *Rev. Mod. Phys.*, **60**, 161–207.

Sütterlin, K. R., Wysocki, A., Ivlev, A. V., Räth, C., Thomas, H. M., Rubin-Zuzic, M., Goedheer, W. J., Fortov, V. E., Lipaev, A. M., Molotkov, V. I., Petrov, O. F., Morfill, G. E., and Löwen, H. (2009). Dynamics of lane formation in driven binary complex plasmas. *Phys. Rev. Lett.*, **102**, 085003/1–4.

Tao, R. (1993). Electric-field-induced phase transition in electrorheological fluids. *Phys. Rev. E*, **47**, 423–426.

Teng, L. W., Tu, P. S., and I, L. (2003). Microscopic observation of confinement-induced layering and slow dynamics of dusty-plasma liquids in narrow channels. *Phys. Rev. Lett.*, **90**, 245004/1–4.

Tennekes, H. and Lumley, J. L. (1972). *A first course in turbulence*. MIT Press, Cambridge.

Thomas, H. M. and Morfill, G. E. (1996). Melting dynamics of a plasma crystal. *Nature*, **379**, 806–809.

Thomas, H. M., Morfill, G. E., Fortov, V. E., Ivlev, A. V.,Molotkov, V. I., Lipaev, A. M., Hagl, T., Rothermel, H., Khrapak, S. A., Suetterlin, R. K., Rubin-Zuzic, M., Petrov, O. F., Tokarev, V. I., and Krikalev, S. K. (2008). Complex plasma laboratory PK-3 plus on the international space station. *New J. Phys.*, **10**, 033036/1–14.

Vaitheeswaran, S., Rasaiah, J. C., and Hummer, G. (2004). Electric field and temperature effects on water in the narrow nonpolar pores of carbon nanotubes. *J. Chem. Phys.*, **121**, 7955–7965.

van Kampen, N. G. (1981). *Stochastic processes in physics and chemistry*. Elsevier, Amsterdam.

van Zanten, J. and Rufener, K. P. (2000). Brownian motion in a single relaxation time Maxwell fluid. *Phys. Rev. E*, **62**, 5389–5396.

Vaulina, O. S. and Dranzhevskii, I. E. (2007). Transport properties of quasi-two-dimensional dissipative systems with a screened Coulomb potential. *Plasma Phys. Rep.*, **33**, 494–502.

Vaulina, O. S. and Khrapak, S. A. (2000). Scaling law for the fluid-solid phase transition in Yukawa systems (dusty plasmas). *JETP*, **90**, 287–289.

Vaulina, O. S. and Khrapak, S. A. (2001). Simulation of the dynamics of strongly interacting macroparticles in a weakly ionized plasma. *JETP*, **92**, 228–234.

Vaulina, O., Khrapak, S., and Morfill, G. (2002). Universal scaling in complex (dusty) plasmas. *Phys. Rev. E*, **66**, 016404/1–5.

Vlasov, Y. A., Bo, X. Z., Sturm, J. C., and Norris, D. J. (2001). On-chip natural assembly of silicon photonic band gap crystals. *Nature*, **414**, 289–293.

Whitby, M. and Quirke, N. (2007). Fluid flow in carbon nanotubes and nanopipes. *Nature Nanotech.*, **2**, 87–94.

Woon, W. Y. and I, L. (2004). Defect turbulence in quasi-2D creeping dusty-plasma liquids. *Phys. Rev. Lett.*, **92**, 065003/1–4.

Wysocki, A. and Löwen, H. (2004). Instability of a fluid-fluid interface in driven colloidal mixtures. *J. Phys. Condens. Matter*, **16**, 7209–7224.

Yethiraj, A. and van Blaaderen, A. (2003). A colloidal model system with an interaction tunable from hard sphere to soft and dipolar. *Nature*, **421**, 513–517.

Yethiraj, A., Thijssen, J. H. J., Wouterse, A., and van Blaaderen, A. (2004). Large-area electric-field-induced colloidal single crystals for photonic applications. *Adv. Mater.*, **16**, 596–600.

Young, A. P. (1979). Melting and the vector Coulomb gas in two dimensions. *Phys. Rev. B*, **19**, 1855–1866.

Zahn, K. and Maret, G. (2000). Dynamic criteria for melting in two dimensions. *Phys. Rev. Lett.*, **85**, 3656–3659.

Zhdanov, S., Nunomura, S., Samsonov, D., and Morfill, G. (2003). Polarization of wave modes in a two-dimensional hexagonal lattice using a complex (dusty) plasma. *Phys. Rev. E*, **68**, 035401/1–4.

Zuzic, M., Ivlev, A. V., Goree, J., Morfill, G. E., Thomas, H. M., Rothermel, H., Konopka, U., Sütterlin, R., and Goldbeck, D. D. (2000). Three-dimensional strongly coupled plasma crystal under gravity conditions. *Phys. Rev. Lett.*, **85**, 4064–4067.

6

Dusty plasmas in the solar system

Mihály Horányi, Ove Havnes, Gregor E. Morfill

6.1 Introduction

Dust particles immersed in plasmas and UV radiation collect electrostatic charges and respond to electromagnetic forces in addition to all the other forces acting on uncharged grains. Simultaneously they can alter their plasma environment. Dusty plasmas represent the most general form of space plasmas. The interplanetary space, comets, planetary rings, the dusty surfaces of airless celestial objects, Noctilucent clouds are all examples where electrons, ions and dust particles coexist. As the subject is rapidly expanding, there are several reviews that cover different aspects of dusty plasma phenomena in the solar system (Axford and Mendis 1974; Grün *et al.* 1984; Mendis *et al.* 1984; Goertz 1989; Hartquist *et al.* 1992; Mendis and Rosenberg 1994; Bliokh *et al.* 1995; Horányi 1996; Cho and Röttger 1997; Horányi *et al.* 2004). In this chapter we discuss a selected set of examples, including a) dusty plasmas in the Earth's mesosphere, where remote sensing observations using radars were recently recognized as a possible diagnostic tool of the dusty plasmas in this region; b) planetary rings, including the spokes in Saturn's B-ring that played an important role in the early development of dusty plasma studies; and c) the lunar surface.

6.2 Noctilucent clouds

The Earth's mesosphere, $\sim$ 50 to $\sim$ 90 km above the surface, is the boundary region between outer space and the denser lower atmosphere. In the mesosphere most of the incoming precipitation of energetic electrons and ions is stopped and most of the meteoric particles burn up. Since there is a comparatively low energy input from solar radiation into the mesosphere, its temperature on the average falls off from its highest value just above the stratosphere to a minimum at a height $\sim$ 85–90 km, the mesopause. Above this, the temperature increases with height due to UV absorption by molecular oxygen and nitrogen, while below the mesosphere in the stratosphere there is also a positive temperature gradient with height due to absorption of UV by ozone. The global mesospheric air circulation system also has a profound influence on the mesospheric temperatures. On the northern hemisphere from mid-May to late August the circulation pattern is such that there is a net upward wind draught of a few cm s^{-1} in the polar region. Due to adiabatic cooling of the expanding upward wind draught the polar mesospheric summer temperature is lowered and the minimum at the mesopause is generally in the range of 110–150 K, then the coldest region on the Earth (von Zahn and Meyer 1989; Lübken 1999). The wind draught is downward in winter and the temperature in the polar mesopause region is then around 200–220 K (Lübken *et al.* 2006).

The plasma environment in the mesosphere is one with a very low fractional ionization. The neutral density in the upper part of the mesosphere around 85 km is 10^{20} m^{-3} while the electron density can vary from as low as 10^8 m^{-3} or less during night-time conditions with no ionizing particle precipitation, to several times 10^9 m^{-3} during quiet day-light conditions and in excess of 10^{10} m^{-3} during disturbed conditions with ionizing precipitation. Also, during quiet night-time conditions the majority of the negative charges can be on heavy negative ions, creating a charging environment where the positive ions may be lighter than the negative charge carriers, possibly creating positively charged dust of low charges (Rapp *et al.* 2005). It is probable that there always is a population of small meteoritic so-called "smoke particles" present in the upper parts of the mesosphere as a result of a process where meteor particles burn up, mainly in the height region 75–110 km, as they enter the atmosphere. Part of the evaporated gas re-condenses to form the smoke-particles (Rosinsky and Snow 1961; Hunten *et al.* 1980; Kalashnikova *et al.* 2000; Megner *et al.* 2006). The smoke particles most likely consist of metals and silicate compounds (Plane 2004) and have radii probably less than a few nanometers. The material evaporated off the meteoric particles as they burn up in the atmosphere can also lead to the formation of atomic metallic layers in the upper parts of and above the mesosphere. Observed anticorrelations between the strength of these metallic layers and the presence of dust particles in noctilucent clouds (NLC) (Gadsden and Schröder, 1989) and in the radar scattering dusty Polar Mesospheric Summer Echoes (PMSE) layers (Cho and Röttger 1997; Rapp and Lübken 2004) show that metals can be deposited on the dust particles (Plane 2004).

In the summer, when the mesospheric temperature falls to below the dew point for water vapor, larger particles are formed when water condenses, most likely on meteoric smoke particles, to form icy dust particles which may be observed visually as NLC if they are large enough. Von Cossart *et al.* (1999) found that in NLC observed with lidars, the average dust radius is $\sim$50 nm with an average number density density of $\sim$ 100 cm^{-3}. The dust particles formed in the summer mesosphere, whether visual or sub-visual will normally be electrically charged and thereby affect the electron density in their vicinity. Due to the low plasma temperature the dust charges are low, from –1e to several times this, depending on the dust size. Density gradients in the dust gas, most likely caused by neutral air turbulence (Rapp and Lübken 2004), will cause density gradients also in the electron (and ion) gas which will lead to radar backscatter from the dust layers. That the dust controls the electron gas in the PMSE layers has been shown in a number of rocket experiments which find a strong

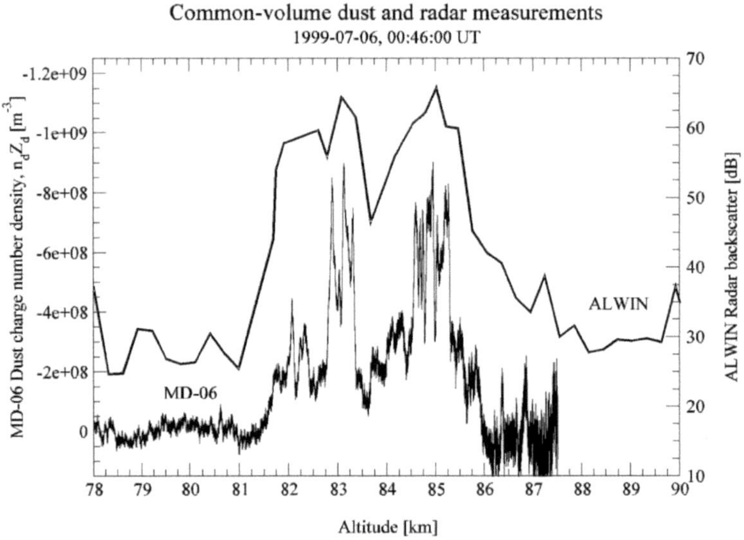

FIGURE 6.1
The dust charge number density as measured with the dust probe on the dust payload MiniDusty-06 (MD-06) and the PMSE radar backscatter measured with a 53.5 MHz radar (Havnes *et al.* 2001)

correlation between dust charge density and PMSE backscatter strength (Havnes *et al.* 1996, 2001; Mitchell *et al.* 2001, 2003; Smiley *et al.* 2003). Figure 6.1 shows the strong correlation between dust charge density in a PMSE layer and the radar backscatter from the same layer, for a case where the rocket payload passed through the PMSE layer along the radar beam (Havnes *et al.* 2001).

The very first direct detection of charged mesospheric dust particles was done by Havnes *et al.* (1996) with their bucket shaped DUSTY probe. At the opening of the bucket (Faraday cup), facing in the ram direction, a grid at +6.2 V closed the interior of the probe to the thermal ambient plasma which has a temperature of $\sim$ 150 K, or $\sim$ 0.01 eV. Dust particles with sizes from a few nm in radius and more will easily penetrate the grid since they have a velocity relative to the grid on the order of 1 km s^{-1}, the rocket payload velocity, leading to relative kinetic energies $\geq$ 10 eV for ice particles of radius $>$ 1 nm. For dust particles of around 1 to 2 nm and less the airflow around the payload will most likely deflect most of the dust particles away from the probe. The detection of such small particles, with new types of dust detectors, has been at the focus of rocket in situ investigations of the mesosphere the last several years (Rapp *et al.* 2005; Amyx *et al.* 2008; Rapp and Strelnikova 2008; Strelnikova *et al.* 2009). Observations indicate that the winter mesosphere from 60–90 km is populated by very small nanometer meteor smoke particles of densities varying from 10^9 m^{-3} at 65 km to more than 10^{11} m^{-3} at 85 km. These densities are a factor of 5 above model calculations (Megner *et al.* 2006). During the summer the mesosphere below around 80 km apparently does not contain many meteoric smoke particles, presumably because they have been swept out of this height region by the upward wind draught. In the majority of launches with different dust probes, the observations show that the dust particles most often are negatively charged. Also, in cases with high dust densities, the majority of the electrons resided on the dust particles (Havnes *et al.* 1996; Lübken *et al.* 1998) creating the so-called electron bite-outs (Pedersen *et al.* 1969; Ulwick *et al.* 1988) with few free electrons.

Estimates of the dust charge and dust sizes is also dependent on the ion–electron pair production Q_I due to neutral gas ionization by radiation or particle precipitation. Rapp and Lübken (2001) considered the charging for various dust densities and sizes, and plasma densities or Q_I values. For typical quiet conditions with little or no particle precipitation, they find that the charges on the dust can vary from $-1e$ for radius below 10 nm with a linear increase to $\sim -2e$ for 30 nm and $-3e$ for 50 nm particles. However, dust density effects may change this since an increase in dust density will lower the average charge (Havnes *et al.* 1990; Rapp and Lübken 2001). A direct computation of dust charges based on dust charge densities and plasma density measurements with rocket probes, combined with lidar measurements of dust sizes and densities (e.g., von Cossart *et al.* 1999), is not always possible because there may be a large population of dust particles which are too small ($\leq$ 15–20 nm) to be detected by the lidars. A population of small but numerous dust particles may easily be carrying most of the dust space charge and will be the ones mainly detected by the dust probes. For the launch rocket flight ECT02 (Havnes *et al.* 1996, Rapp and Lübken 2004) where the absence of lidar detection of dust indicates small dust particles, the observations are consistent with dust sizes from 20 nm and below with charges of the order of $-1e$. It appears that the production of negative dust charges is well understood and that the main charging effects are due to electron and ion collision with and attachment onto dust particles, including polarization effects during collisions (e.g. Rapp and Lübken 2001). Photodetachment of electrons (Weingartner and Draine 2001; Dimant and Milikh 2004) has been shown by Havnes *et al.* (2004), from measurements of the relaxation rate of dust charges in PMSE radar overshoot experiments (Havnes *et al.* 2003) where the PMSE signal is modulated by high power transmitters (Rietveld *et al.* 1993), to be of minor importance for the PMSE dust charging.

The PMSE radar backscatter modulation is caused by the heating of the electrons during the time the heating transmitters are on, combined with the extra dust charging which the heated electrons cause during the same phase. Havnes (2004) predicted that an overshoot can be produced where the PMSE signal strength, after first being weakened when the heater is switched on (Chilson *et al.* 2000), flares up or overshoots to a strength up to 7 to 8 times that of the undisturbed PMSE signal as the heating is switched off (Figure 6.2). This effect is only apparent in heating cycles where the heater off time is long enough to allow the dust charges to relax back to their equilibrium charges in the unheated electron gas. The overshoot effect, which has been observed in several PMSE overshoot campaigns (Havnes *et al.* 2003; Kassa *et al.* 2005; Biebricher *et al.* 2006; Naesheim *et al.* 2008), shows promise of being a useful diagnostic tool. It has also been observed for Polar Mesospheric Winter Echoes (PMWE) (La Hoz and Havnes 2008) where it is very weak in spite of a con-

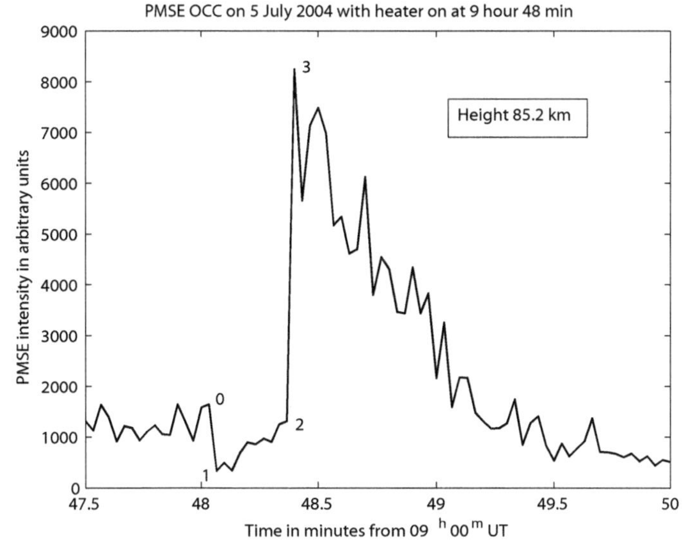

FIGURE 6.2
The radar PMSE backscatter during a heater cycling where the heater transmitters were on for 20 sec (from 1 to 2) and off for 160 sec (from 3 to 0 in the next heating cycle). We see the immediate weakening from 0 to 1 as the heater is switched on and the strong overshoot from 2 to 3 as the heater is switched off (from Kassa *et al.* 2005).

siderable electron heating effect from around 200 K to around 1000 K indicating the presence of very small dust particles with around 3 nm radius (Havnes and Kassa 2009).

A number of rocket dust charge measurements have indicated that positive dust particles may also be present in the summer mesosphere (Havnes *et al.* 1996; Gelinas *et al.* 1998; Horányi *et al.* 2000; Rapp *et al.* 2005; Amyx *et al.* 2008). However, in several cases this is clearly the effect of secondary charge production during dust impacts where charges may be rubbed off the impact point during glancing impacts (Tomsic 2001). For one rocket probe experiment DUSTY 2 (Havnes *et al.* 1996), positive currents were measured on the front grid of the probe and at first interpreted as due to impacts of positive dust particles. A later detailed modeling of the effect of payload rotation and grid structure on the impact rate, including fragmentation of impacting dust particles and secondary charge production, showed that the observed variation of the positive current with the payload rotation, could only be explained by electrons being rubbed off from the front grid by the impact of the dust particles (Havnes and Naesheim 2007). Similar effects can probably explain other reported observations of positively charged dust particles (e.g., Amyx *et al.* 2008) but there may be cases where the observed positive charges on the mesospheric dust particles are real (Rapp *et al.* 2005).

6.3 Planetary rings

The motion of small charged grains making up the faint, diffuse rings of Jupiter and Saturn can be surprisingly complex, as this is often determined by electromagnetic forces in addition to gravity, drag and radiation pressure. There are many exciting phenomena associated with the interaction of magnetospheric fields and plasmas with the embedded dust grains. Lorentz resonances (Schaffer and Burns 1987), gyrophase-drifts due to compositional and/or plasma density and/or plasma temperature gradients (Northrop *et al.* 1989), transport due to charge (Morfill *et al.* 1980) or magnetic field fluctuations (Consolmagno 1980), shadow resonances (Horányi and Burns 1991), and the coupling between radiation pressure and electrodynamic forces (Horányi *et al.* 1992), for example, might all play a role in shaping the distribution of small charged grains in planetary rings. The dust becomes an integral component of the magnetosphere since it acts as a source/sink of the plasma. The produced low energy photo and secondary electrons or the sputtered off ions might significantly alter the magnetospheric plasma distribution, for example. Though many of these processes are now recognized, dusty planetary magnetospheres still hold surprises, as they can result in possible capture of interplanetary dust, the transport of ring material across vast distances, and even the ejection of small charged grains from the magnetosphere, for example (Horányi 1996).

6.3.1 Simplified dynamics

The equation of motion of a charged dust grain (of mass m and charge Q), as written in Gaussian units in an inertial coordinate system fixed to the planet's center, is

$$\ddot{\mathbf{r}} = \frac{Q}{m}\left(\frac{\dot{\mathbf{r}}}{c} \times \mathbf{B} + \mathbf{E_c}\right) - \frac{\mu}{r^3}\mathbf{r}\,, \tag{6.1}$$

where $\mathbf{r}$ is the grain's position vector, c is the speed of light, $\mathbf{B}$ is the magnetic field, and μ equals the gravitational constant times the planet's mass. For an infinite conductivity magnetosphere that rigidly co-rotates with the planet with a rotation rate of Ω, $\mathbf{E_c} = (\mathbf{r} \times \Omega) \times \mathbf{B}/c$ is the co-rotational electric field. We have neglected the planet's oblateness, as well as the forces due to radiation pressure and the plasma and neutral drags.

To make a connection to the familiar Kepler problem, let us assume that the magnetic field can be given as that of a simple dipole located at the center with its magnetic moment aligned with the rotation axis of the planet. This picture is a reasonable first approximation for the magnetic fields at Jupiter (J) and Saturn (S). Now, in the equatorial plane $B = B_0(R/r)^3$, where B_0 is the magnetic field at the surface and R is the radius of the planet ($B_0^J = 4.2$, $B_0^S = 0.21$ Gauss and $R_J = 7.1 \times 10^4, R_S = 6 \times 10^4$ km). The magnetic field lines pierce the equatorial plane at right angles pointing anti-parallel to Ω. The resulting co-rotating electric field points radially outward with an amplitude of $E_c = E_0(R/r)^2$ ($E_0^J = 5.8 \times 10^{-5}, E_0^S = 2.4 \times 10^{-6}$ V m^{-1}). Naturally, the force acting on dust particles associated with the co-rotational electric field also depends on their charge, $F_{\rm el} = QE_c$. The charge in turn, as discussed above, is a function of the plasma environment, the material properties of the grain, the charging history, relative velocity, etc. Generally the grain's charge can have a complicated history, but for the moment let us assume it remains a constant.

Assuming an average density of $\rho = 1$ g cm^{-3}, the ratio of the electrostatic force to gravity is

$$F_{\rm el}/F_{\rm G} = \alpha \phi_v a_\mu^{-2} \tag{6.2}$$

($\alpha^J = 5.7 \times 10^{-3}$, $\alpha^S = 5.3 \times 10^{-4}$). In the typical range of $-50 \leq \phi_v \leq 10$, for particles with the radius in units of microns $a_\mu \gg 1$, electrostatic forces represent a perturbation only and these grains follow approximate Kepler orbits. On the other hand, particles with $a_\mu \ll 1$ can be dominated by electromagnetic forces and gravity becomes a perturbation.

Let us return to Equation (6.1) and rewrite it in the equatorial plane of a planet using polar coordinates

$$\ddot{r} = r\dot{\varphi}^2 + \frac{q}{r^2}(\dot{\varphi} - \Omega) - \frac{\mu}{r^2}, \tag{6.3}$$

$$\ddot{\varphi} = -\frac{\dot{r}}{r}\left(\frac{q}{r^3} + 2\dot{\varphi}\right), \tag{6.4}$$

introducing $q \equiv QB_0R^3/mc$, so that the combination $q/r^3 = \omega_g$ becomes the local gyrofrequency (the angular rate dust particles circle about magnetic field lines).

On a circular equilibrium orbit, where the the sum of the radial components of all the forces is zero ($\ddot{\varphi} = \ddot{r} = \dot{r} = 0$ and $\dot{\varphi} = \text{constant} = \psi$), Equation (6.3) yields an algebraic equation for the angular velocity, ψ

$$\psi^2 + \omega_g \psi - \omega_g \Omega - \omega_k^2 = 0, \tag{6.5}$$

where $\omega_k = (\mu/r^3)^{1/2}$ is the Kepler angular velocity. For big particles, terms that contain ω_g can be dropped and we recover $\psi = \pm\omega_k$. For very small particles, terms that are not multiplied with ω_g are to be dropped and $\psi = \Omega$. Very small grains are picked up by the magnetic field and co-rotate with the planet.

Equations (6.3) and (6.4) can be integrated to yield constants of the motion

$$\mathscr{E} = \frac{1}{2}(\dot{r}^2 + r^2\dot{\varphi}^2) - \frac{\mu + q\Omega}{r}, \tag{6.6}$$

$$J = r^2\dot{\varphi} - \frac{q}{r}. \tag{6.7}$$

For large particles ($q \to 0$) these constants become the Kepler energy and angular momentum. The Jacobi constant, $H = \mathscr{E} - \Omega J$, remains a constant even if Q changes with time (Northrop and Hill 1983; Schaffer and Burns 1987).

The right-hand side of Equation (6.3) can be written solely as a function of r using Equation (6.7) to replace $\dot{\varphi}$, and we can express $\ddot{r} = -\frac{\partial U}{\partial r}$, where the effective

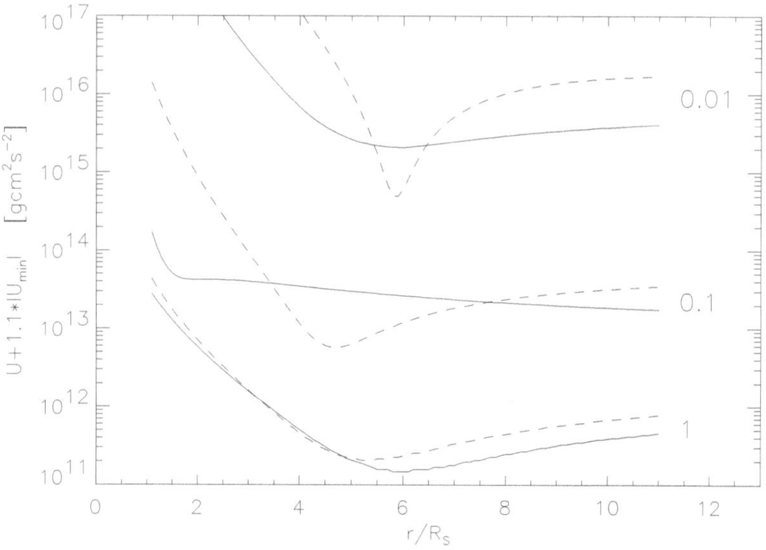

FIGURE 6.3
The effective potential for dust grains with a_μ = 0.01, 0.1, and 1, started from Io on circular Kepler orbits with $\phi_V = -30$ (*dashed lines*) and +3 (*continuous lines*). To avoid the overlap of these curves, since only their shape is important, we have shifted them apart by plotting U+ 1.1$\times|\min(U)|$ instead of U itself.

potential

$$U(r) = -\frac{\mu + q\Omega}{r} + \frac{J^2}{2r^2} + \frac{qJ}{r^3} + \frac{q^2}{2r^4}\,. \tag{6.8}$$

The equilibrium orbit with a given J can be found from here by solving the equation $\frac{\partial U}{\partial r} = 0$. For any r, the initial condition $\dot{r} = 0$ and $\dot{\varphi} = \psi$ [that is, the solution to Equation (6.5)] satisfies $\frac{\partial U}{\partial r} = 0$.

Small grains are constantly generated by micro-meteoroid bombardment of the moons, or in the case of Jupiter, probably also by the volcanoes on Io. Generally, their escape velocity is small that $\dot{\varphi}(t=0) = \omega_k$. For the initial Kepler orbit $J = r^2(\omega_k - \omega_g)$.

Figure 6.3 shows $U(r)$ for particles started from Io ($r_0 = 5.9R_J$) at Jupiter for the two typical values $(-30, +3)$ for ϕ_v (Horányi *et al.* 1993a,b). Particles with negative surface potentials remain confined in the vicinity of r_0. However, grains in a certain size range with positive charges are not confined [$U(r)$ shows no minima]. What sets this size range? In the case of positively charged grains the force due to the co-rotational electric field points radially out, opposing gravity. The upper limit in size for ejection, $a_\mu^{\max}$ is set by the condition $F_{\mathrm{el}}/F_{\mathrm{G}} > 1$.

The lower limit in dust size for ejection is due to the fact that very small grains behave like ions or electrons circling magnetic field lines. The radius of their trajectory is the gyroradius $r_g = | wmc/QB | = | w/\omega_g |$, where w is the relative velocity between the co-rotating magnetic fields and the particle. For Kepler initial conditions, $w = r(\Omega - \omega_k)$. The motion of these grains is well described by the guiding center approximation if the size of their orbit is smaller than the characteristic length scale for variations in the magnetic fields, $| r_g \nabla B/B | < 0.1$ ($| \nabla B/B | = 3/r$ in the equatorial plane of an aligned centered dipole). The upper limit in grain size satisfying this condition (i.e., the smallest grains that will be ejected) is

$$a_\mu^* = \left(\frac{10^{-3} B_0 R^3 \phi}{4\pi r^2 \rho w c} \right)^{1/2} . \tag{6.9}$$

Grains in the range $a_\mu^* < a_\mu < a_\mu^{\max}$ will be ejected from the magnetosphere. As these positively charged grains move outward, they gain energy from the co-rotational electric field

$$W = \int_{r_0}^{r_1} EQdr = E_0 R^2 Q \left(\frac{1}{r_0} - \frac{1}{r_1} \right) , \tag{6.10}$$

where the upper limit of the integration, r_1, is the characteristic size of the magnetosphere. This mechanism explains the ejection of small grains from a planetary magnetosphere that was first observed by Ulysses during its close encounter with Jupiter in 1992 (Grün *et al.* 1993). The process was also identified at Saturn by Cassini measurements in 2004 (Kempf *et al.* 2005).

Substituting the approximate values for Jupiter ($r_0 = 6R_J$, the location of Io, and $r_1 = 50R_J$, the characteristic size of the magnetosphere) and Saturn ($r_0 = 6.3R_S$, the location of Dione, and $r_1 = 20R_S$) in Equation (6.10) yields an estimate for the escape speeds $v_{escape} = 3/a_\mu$ for Jupiter, and $v_{escape} = 0.6/a_\mu$ for Saturn (Horányi 2000).

The typically nanometer-sized grains expected to be ejected from Jupiter and Saturn are below the calibration threshold of the dust instruments; however, due to their large predicted speeds, they can generate sufficiently large signals that are consistent with the observations.

6.3.2 Saturn's E-ring

The in situ dust measurements by the Cassini Dust Analyzer (Srama *et al.* 2006), in combination with imaging (Porco *et al.* 2006) and plasma measurements (Wang *et al.* 2006; Kurth *et al.* 2006), provide an unprecedented opportunity to study the size and spatial distributions of dust in Saturn's E-ring. One of the major discoveries of Cassini to date has been that the south-polar region of the moon Enceladus is geologically active, and it is the source of most of the E-ring particles (Spahn *et al.* 2006a,b). Enceladus has been long thought to be the source of the E-ring grains as the optical depth sharply peaks just outside the orbit of this moon. Another surprising discovery is the extent of the E-ring material much beyond its earlier recognized boundaries of 3–8 R_S, ($R_S \simeq 6 \times 10^4$ km, the radius of Saturn), perhaps reaching even the orbit of Titan at $R_{Ti} = 20.3R_S$ (Srama *et al.* 2006). Dust particles can be transported outwards from Enceladus due to plasma drag (Morfill *et al.* 1983; Havnes *et al.* 1992; Dikarev, 1999), but along the way they also lose mass due to sputtering (Jurac *et al.* 2001a). The competition between these sets the range particles with a given initial size can reach.

Plasma drag acting alone results only in a slow enough adiabatic orbital expansion; hence, dust grains approximately follow slowly expanding circular Kepler orbits. The plasma drag is dominated by the co-rotating oxygen ions. In this initial analytic model we ignore the deviations from co-rotation due to mass-loading and use a simple approximation for the density of $n_{\mathrm{O}^+} \simeq 100(R_E/r_a)^4$, where $R_E = 4R_S$, and r_a is the semi-major axis. The characteristic temperature $T_{\mathrm{O}^+} \simeq 100$ eV, and the thermal speed of $v^{th}_{\mathrm{O}^+} \simeq 25$ km s^{-1}.

Already at Enceladus the relative velocity between dust particles and the co-rotating O^+ ions $v_{rel} = r_a(\Omega - \omega_K) > v^{th}_{\mathrm{O}^+}$, where $\Omega = 1.64 \times 10^{-4}$ s^{-1} is the rotation velocity of Saturn and $\omega_K = \sqrt{\mu/r_a^3}$ is the Kepler angular velocity, with $\mu = 3.8 \times 10^{22}$ cm^{3}s^2, the product of Saturn mass and the gravitational constant. Hence, the drag force acting on a dust particle is dominated by direct collisions as opposed to distant Coulomb interactions (Morfill and Grün, 1979; Morfill *et al.* 1993).

$$F_{\mathrm{d}} = n_p m_i \pi a^2 v_{rel}^2 = n_{\mathrm{O}^+} m_{\mathrm{O}^+} \pi a^2 r_a^2 (\Omega - \omega_K)^2 \,, \tag{6.11}$$

where $m_{\mathrm{O}^+} \simeq 16$ AMU is the mass of the O^+ ions.

The drag force acting on the grain increases the orbital energy; hence, the semi major axis changes at a rate

$$\begin{aligned} \frac{dr_a}{dt} &= \frac{2r_a^2}{\mu}\dot{E} = \frac{2r_a^2}{m\mu} F_{\mathrm{d}} r_a \omega_K 2\pi n_{\mathrm{O}^+} m_{\mathrm{O}^+} \frac{r_a^5}{m\mu} a^2 \omega_K (\Omega - \omega_K)^2 \\ &= \frac{3 n_{\mathrm{O}^+} m_{\mathrm{O}^+}}{2a} \left(\Omega^2 \mu^{-1/2} r_a^{7/2} - 2\Omega r_a^2 + \mu^{1/2} r_a^{1/2} \right), \end{aligned} \tag{6.12}$$

where $E = -\mu/(2r_a)$ is the orbital energy per unit mass, and $m(t) = (4\pi/3)a^3(t)$ is the mass of the dust grain. Substituting the density of O^+ ions, measuring the semi-major axis in units of Saturn's radius, the grain radius a_μ in units of microns, and the time in years Eg.(6.12) can be rewritten as

$$\frac{dr_{as}}{dt_y} = \frac{1}{a_\mu} \left(0.12 r_{as}{}^{-1/2} - 0.61 r_{as}{}^{-2} + 0.78 r_{as}{}^{-7/2} \right) \simeq \frac{0.03}{a_\mu} \,, \tag{6.13}$$

approximately a constant for fixed a_μ in the range of $4 < r_{as} < 20$.

The drag time from Enceladus to Titan is $\tau_D \simeq 500 a_\mu$ year. If the characteristic lifetime of a grain due to sputtering in the E-ring is $\tau_s = \tau_0 a_\mu$, where $\tau_0 \simeq 50$ years is the lifetime of a grain with $a_\mu = 1$ (Jurac *et al.* 2001a), no particle could make this trip. However, as particles move outwards, the plasma density drops and their sputtering lifetime gets prolonged; simultaneously their drag rate increases due to their diminishing size.

Due to sputtering, the radius of a dust particles decreases at a rate, independently of its size, as

$$\frac{da_\mu}{dt} = -\frac{1}{\tau_0}. \tag{6.14}$$

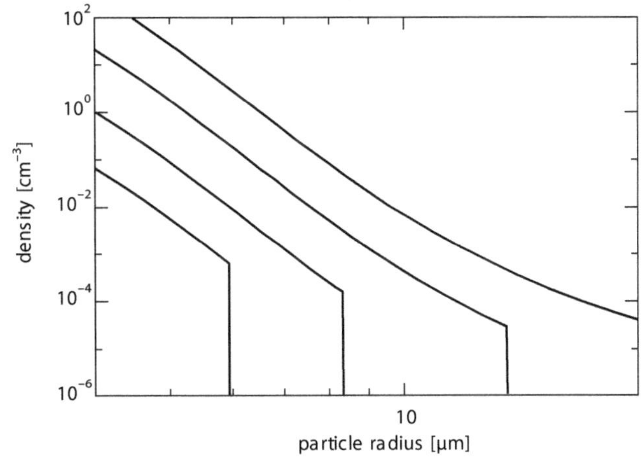

FIGURE 6.4
The number density of grains with a^* = 0.1, 0.3, 1, and 3 μm (from top to bottom, respectively), based on Equation 6.16, assuming an initial power-law size distribution $n(a) \sim a^{-p}$ with p = 2.5 in the size range of $0.1 < a < 10$ μm, normalized to $n(a^* = 1\ \mu\text{m}) = 1$ at the orbit of Enceladus (Horányi *et al.* 2008)

We have used energy-dependent sputtering yields (Jurac *et al.* 2001a,b) to estimate that the lifetime of a micron sized particle is expected to increase from $\simeq 50$ at $r_{as} = 4$ to $\simeq 600$ years at $r_{as} = 20$. Combining Equations (6.13) and (6.14), and using a simple fit for $\tau_0 \simeq 78 - 15r_{as} + 2r_{as}{}^2$, results in

$$\frac{da_\mu}{dr_{as}} = -\frac{a_\mu}{0.06r_{as}{}^2 - 0.45r_{as} + 2.34}, \tag{6.15}$$

and the radius of the eroding particle as function of distance for grains originating from Enceladus becomes

$$\frac{a}{a_0} = 1.18\exp\left[-3.34\arctan(0.2r_{as} - 0.75)\right], \tag{6.16}$$

indicating that the radius of a grain drops to $< 5\%$ of its initial value a_0 when reaching $r_{as} = 20$. Equation (6.16) also indicates that the shape of any initial size distribution remains fixed as particles drift and erode, as the entire distribution shifts towards smaller sizes with increasing r_{as}.

From the conservation of particle flux, the number density of grains with an initial a_0 drops as $n_{a_0}(r_{as})/n_{a_0}(r_{as} = R_E) = (R_E \dot{r}_{aR_E})/(r_{as}\dot{r}_{as})$, which remains independent of a_0; however, the coupled Equations (6.12) and (6.14) have to be integrated numerically (for any a_0) to obtain the universal $(R_E \dot{r}_{aR_E})/(r_{as}\dot{r}_{as})$ function.

The density for a fixed grain size a^* can be calculated for an initial size distribution by first using Equation (6.16) to get the original size $a_0(a^*, r_{as})$ the grain started with at Enceladus. For example, Figure 6.4 shows the spatial distribution of grains for various a^*, assuming a power-law size distribution with an index of –3.5 for the production rate of grains at Enceladus (Juhász and Horányi, 2002; Juhász *et al.* 2007; Kempf *et al.* 2008). Due to plasma drag, the density becomes $n(a_0) \sim a_0^{-2.5}$, and we set the size range of $0.1 < a_0 < 10\ \mu$m. The upper limit of an initial size distribution determines the spatial range of particles with a given size. In the example shown in Figure (6.4), there are no particles with $a_\mu > 3$ beyond $r_{as} \simeq 6$, or particles with $a_\mu > 1$ beyond $r_{as} \simeq 8.2$.

A continuous scan of the dust densities has been derived from the impact rates measured by Cassini during the ring-plane crossing in orbit 7. The estimated densities drop from a maximum of $\simeq 0.5$ m^{-3} near Enceladus in the ring-plane to $\simeq 10^{-4}$ m^{-3} at a distance of about 15 R_S at a height of about 2 R_S (Srama *et al.* 2006). The $\simeq$ 4 orders of magnitude drop seems to be reasonably well matched by the simple analytical estimates of the combined effects of drag and sputtering of small grain. Due to plasma drag, particles can be transported from 4 to 20 R_S and will arrive there with a few % of their original radius due to sputtering losses. If grains with $a_\mu \gg 0.5$ are found at this distance, they would have to originate from one of the more distant moons. However, to date no dust density enhancements have been noticed while crossing the ring-plane near the orbits of any other satellites. Alternatively, if found, these grains could also be signatures of dust impacts on unseen small moonlets. These calculations also indicate that the geysers on Enceladus have been supplying the E-ring material at an approximately constant rate at least for several hundreds of years.

6.3.3 Spokes

The intermittently appearing, approximately radial markings on Saturn's B-ring (see Figure 6.5) are thought to be caused by small charged dust particles that are lofted from their parent bodies due to electrostatic forces. They were first recognized in measurements made by the Voyager spacecraft (Smith *et al.* 1981, 1982), though they were possibly seen earlier in ground-based observations (Robinson 1980). The Hubble Space Telescope (HST) monitored the activity of the spokes starting shortly before the ring-plane crossing in 1995 until October 1998, when spokes were no longer seen by HST (McGhee *et al.* 2005). This was thought to be due to the changing illumination of the rings, as the solar elevation angle (measured from the ring-plane) increased due to the orbital motion of Saturn. Contrary to expectations, Cassini did not find spokes either on its approach or after its orbital insertion in July 2004 until September of 2005, indicating that spokes are a seasonal phenomenon and their formation can be suspended.

The seasonal variation of spoke activity has been suggested to be a consequence of the variable plasma density near the ring. The plasma density is a function of the solar elevation angle, since it is generated mainly from the rings by photoelectron production and by photo-sputtering of neutrals that are subsequently ionized (Farrell *et al.* 2006; Mitchell *et al.* 2006). While this seems to be a reasonable explanation for the seasonal variability of spoke activity after their formation, we are still lacking a generally accepted explanation for their triggering mechanism. It is generally

FIGURE 6.5

Top: Voyager 2 images of spokes in the B-ring (Smith *et al.* 1981, 1982). The image on the left was captured in back-scattered light before closest encounter, the spokes appear as dark radial features across the ring plane. The image on the right was taken in forward-scattered light after Voyager crossed the ring plane, looking back towards the Sun, and the spokes appear as bright markings. Typical dimension of these spokes are 10,000 km in length and 2000 km in width. The changing brightness indicates that spokes consist of small grains with radii comparable to the wavelength of the visible light ($< 1\mu$m). At the time these images were taken, the rings' opening angle to the sun was 8°. Bottom: The first set of spoke observations by Cassini taken on September 5, 2005 (opening angle is 20.4°), over a span of 27 minutes. These faint and narrow spokes were seen from the unilluminated side of the B-ring. These spokes are approximately 3,500 kilometers long and 100 kilometers wide, much smaller than the average spokes seen by Voyager. These images were taken with a resolution of 17 km per pixel at a phase angle of 145° when Cassini was 13.5° above the dark side of the rings as the spokes were about to enter Saturn's shadow. Courtesy NASA/JPL-Caltech and NASA/JPL/SSI.

believed that spoke formation involves charging and electric fields acting on small grains, but this process requires a much higher plasma density than is commonly expected in the near-environment of the rings (Hill and Mendis 1982; Goertz and Morfill, 1983).

Spokes are composed of dust particles in a narrow size distribution centered at about $a_\mu \simeq 0.6$, where a_μ is the radius in units of μm (McGhee *et al.* 2005). Spokes initially cover an approximately radial strip with an area of $A \simeq 10^3 \times 10^4$ km^2, with a characteristic optical depth of $\tau \simeq 0.01$. The total number of elevated grains can be estimated to be on the order of $N_d \simeq A\tau/(\pi r_a^2) \simeq 10^{23}$. If the grains are released simultaneously and carry just a single electron when released from their parent bodies, the formation of the spoke cloud requires a minimum surface charge density (associated with the spoke grains), measured in units of electron charges $\sigma_e^* = N_d/A \simeq 10^6$ cm^{-2}, orders of magnitude higher than the charge density, σ_e^0, expected from the nominal plasma conditions in the B-ring.

The nominal plasma environment near the optically thick B-ring is set by the competing electron and ion fluxes to and from the ring due to photoelectron production from the ring, as well as the ionosphere, and the photo-ionization of the ring's neutral atmosphere that is maintained by photo-sputtering. All of these are expected to show a seasonal modulation with the ring's opening angle with respect to the Sun. The characteristic electron energy is $T_e \simeq 2$ eV, and the plasma density is expected to be on the order of 0.1–1 cm^{-3} (Waite *et al.* 2005). The characteristic shielding distance is $\lambda_D = 740\sqrt{T_e/n_p} = 1 - 3 \times 10^3$ cm, larger than the average distance between the cm – m sized objects in the B-ring, which has a comparable vertical thickness, $h \simeq 10^3$ cm. Hence, it is reasonable to treat the B-ring as a simple sheet of material (Goertz and Morfill, 1983). The nominal surface potential, including its possible seasonal variations, is expected to be in the range of $-5 < \phi_R < 5$ V. The surface charge density can be estimated from Gauss's law, $\sigma_e^0 = E_\perp/(4\pi e) \simeq (\phi_R^V/300)/(4\pi e\lambda_D) \simeq 750\phi_R^V (n_p/T)^{1/2} < 1 - 3 \times 10^3$ cm^{-2}, where $E_\perp$ is the electric field normal to the ring, ϕ_R^V is the electrostatic potential measured in volts, and e is the charge of an electron. The fact that $\sigma_e^0 \ll \sigma_e^*$ indicates that the formation of a spoke requires a higher than normal plasma density, or requires increasing the ring potential directly as we are suggesting. The expected charge, in units of e, on a grain resting on the surface is $Q_d = 10^{-8}\sigma_e^0 \pi a_\mu^2 \ll 1$; hence, the vast majority of micron sized grains remain uncharged, and the fractional Q_d can be interpreted as the probability of a grain to have a single charge (Goertz and Morfill, 1983):

$$P_e = 2.4 \times 10^{-5} a_\mu^2 \phi_R^V \left(\frac{n_p}{T_e}\right)^{1/2}. \tag{6.17}$$

Neglecting cohesive/adhesive forces, a small grain with a single e charge can be lifted from the ring-plane if the electrostatic force acting on it exceeds gravity, $eE_\perp > 2\pi\sigma_m G m$, where $\sigma_m = 100$ g cm^{-2} is the average surface mass density of the B-ring (Cuzzi *et al.* 1984), and m is the grain's mass. This is true for both positive or negative ϕ_R. The radius of the largest grain that can be lifted is

$$a_\mu^{\max} \simeq 2\left(\phi_R^V\right)^{1/3}\left(\frac{n_p}{T_e}\right)^{1/6} . \tag{6.18}$$

Even in the low plasma density of $n_p = 0.1$ cm^{-3}, $a_\mu^{\max} \simeq 2$ due to the very small gravitational acceleration normal to the ring-plane.

Once a grain is lifted its charge could change due to the currents collected in the sheath above the ring-plane, where the plasma is dominated by charge carriers with the sign of charge opposite to ϕ_R. For example, when charging of the ring is dominated by photoelectron production from the ring itself, there is a sheath of electrons above the positively charged ring. Based on numerical examples, if a grain can travel with its original charge to a height of $\simeq 0.1\lambda_D$, it will probably get through the sheath, even if it subsequently loses its charge or changes its polarity (Morfill *et al.* 1983; Mitchell *et al.* 2006). The time it takes to travel across $0.1\lambda_D$ is $t_D = \lambda_D\sqrt{0.2m/e\phi_R} \simeq 5\times 10^2\sqrt{T_e/(\phi_R^V n_p)}\, a_\mu^{3/2}$ s. Spokes seem to form quickly, as they were noticed fully developed on images taken five minutes after a previous frame with no sign of spoke activity (Smith *et al.* 1982). Requiring, for example, that $t_D < 30$ s for $a_\mu = 0.6$ necessitates a plasma density $n_p > 10$ cm^{-3}, higher than its normally expected value.

The time to neutralize a grain is approximately $t_Q = \left(n_p v_{th}\pi a^2\right)^{-1}$, where v_{th} is the thermal speed of electrons (for positive grain potential, $\phi_R > 0$) or ions (for negative grain potential, $\phi_R < 0$), and the probability of crossing the sheath with its charge intact is $P_c = e^{-t_D/t_Q}$, with the exponent $t_D/t_Q = \alpha a_\mu^{3.5}$, where $\alpha \simeq 10^3 T_e(n_p/\phi_R^V)^{1/2}$ for $\phi_R^V > 0$, or $10^3 T_i(n_p/\phi_R^V)^{1/2}(m_e/m_i)^{1/2}$ for $\phi_R < 0$.

The net probability for a grain to leave the ring-plane is the product of the probabilities of having a charge on the surface and the probability of crossing the plasma sheath above it, $P = P_e P_c \sim a^2 e^{-\alpha a^{3.5}}$. This suggests a narrow size distribution centered on $a_\mu^* = \beta\left(\phi_R^V/n_p\right)^{1/7} T^{-2/7}$, with $\beta \simeq 1.2\times 10^{-1}$ and 0.6, for electrons and O_2^+ ions, respectively. Figure 6.6 shows P as function of grain size and plasma density. These consideration indicate that spoke formation involves intermittent, short lived periods of $\phi_R^V < 0$ with a plasma density of $10 < n_p < 100$ cm^{-3}. The lower limit is set by the requirement that spokes form in less than 5 minutes, and the upper limit and that $\phi_R^V < 0$ are set to support a narrow size distribution of the spoke particles near 0.6 μm in radius.

It was suggested that the spoke initiation conditions described above could be produced by a meteorite impact–produced plasma cloud (Goertz and Morfill, 1983). Such a cloud was shown to expand, cool and recombine, as it rapidly propagated in the radial direction, explaining many of the observed spoke characteristics. The proposed propagation speed of such a cloud was recently reexamined and awaits further observational constraints (Farmer and Goldreich, 2005; Morfill and Thomas 2005).

During the first four years (2004–2008) of the Cassini missions, spokes observations remained a high priority. After almost a year of no spoke activity, they were

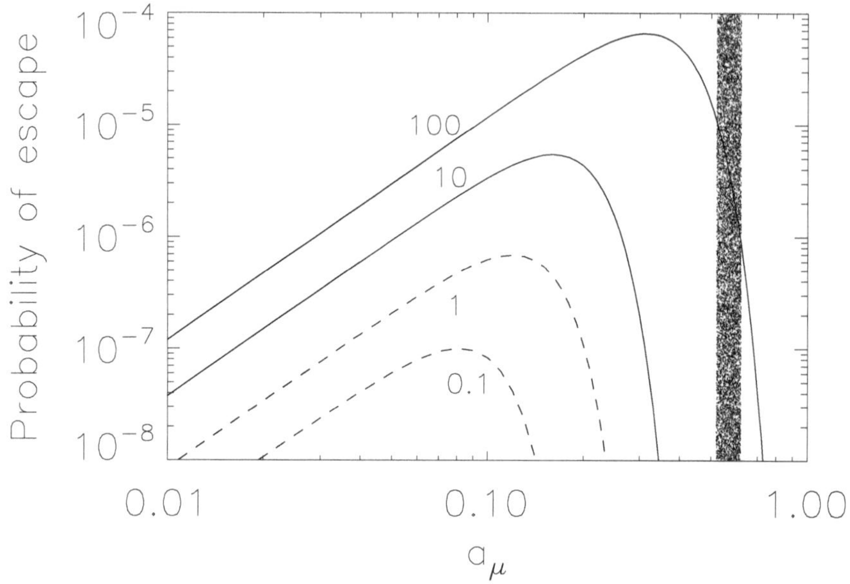

FIGURE 6.6
The probability for a dust grain to leave the ring-plane as function of its size for low plasma densities assuming $\phi_R^V = 5$ V (dashed lines) and high plasma densities with $\phi_R^V = -5$ V. The curves are labeled with the value of the plasma density, n_p. The vertical dark band marks the characteristic size range of spoke particles indicated by HST observations (McGhee *et al.* 2005).

seen to reappear in the fall of 2005, indicating the governing role of the solar elevation angle with respect to the rings, as predicted (Nitter *et al.* 1998; Mitchell *et al.* 2006). The new observations show spokes that are much fainter than those seen by the Voyagers though the solar elevation angle by now (2008) has a similar value. This could be a result of the solar cycle variability of the UV flux reaching Saturn, as indicated in Figure 6.7, that is responsible for plasma production in both the atmosphere and over the rings. The average solar ionizing UV flux can typically change over a factor of two between solar maximum and minimum conditions, and its effects on spokes are yet to be investigated.

6.4 Lunar surface

The electrostatic levitation and transport of lunar dust remains an interesting and controversial science issue since the Apollo era of the 1970s. This issue is also of great engineering importance in designing human habitats and protecting optical and me-

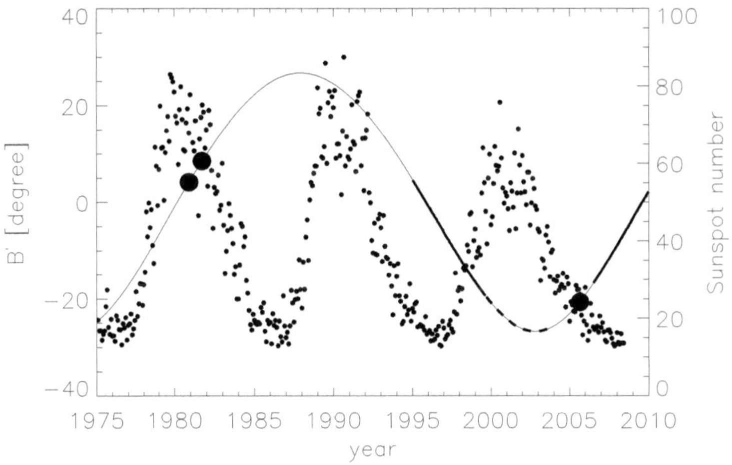

FIGURE 6.7
The approximately sinusoidal variation of the solar elevation angle B' as a function of time. The first two dots on this curve indicate the Voyager encounters in 1980 and 81, and the third dot indicates the first spoke siting by Cassini in September 2005. The thick continuous (dashed) segments (1995–2004) indicate a period of (un)successful spoke observations by HST (McGhee *et al.* 2005). The thick segment starting in 2006 is indicating the observations of spokes by Cassini. The small dots in the background show the number of sunspots as a proxy for solar UV variability.

chanical devices (Colwell *et al.* 2007). As a function of time and location, the lunar surface is exposed to solar wind plasma, UV radiation, and/or the plasma environment of our magnetosphere. Dust grains on the lunar surface collect an electrostatic charge and contribute to the large-scale surface charge density distribution. They emit and absorb plasma particles and solar UV photons, and represent an electromagnetic interface to the lunar interior. There are several in situ and remote sensing observations that indicate that dusty plasma processes could be responsible for the mobilization and transport of lunar soil. These include a) imaging by the TV cameras of Surveyor 5, 6 and 7; b) the fields and particles measurements by the Suprathermal Ion Detector Experiment (SIDE) of Apollo 12, 14 & 15, and the Charged Particle Lunar Environment Experiment (CPLEE) of Apollo 14; and c) the dust measurements by the Lunar Ejecta and Meteorite Experiment (LEAM) of Apollo 17.

6.4.1 Imaging

Images taken by the television cameras on Surveyors 5, 6 and 7 gave the first indication of dust transport on the airless surface of the Moon (Criswell 1973; Rennilson and Criswell 1974). These TV cameras consisted of a vidicon tube, 25 and 100 mm focal length lenses, shutters, and color filters surmounted by a mirror that could be

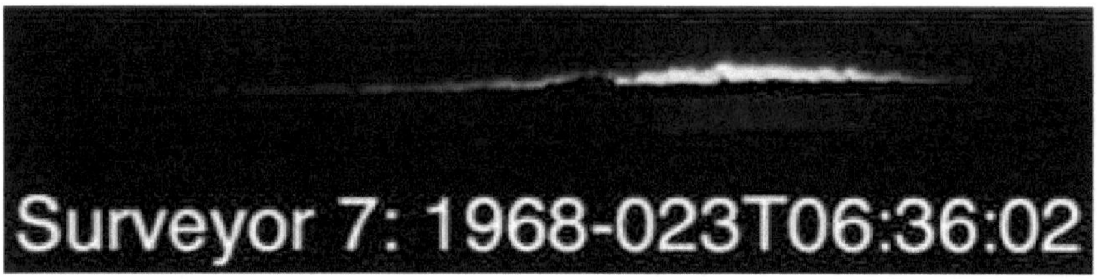

FIGURE 6.8
An unprocessed image of the lunar horizon glow from Surveyor 7 (National Space Science Data Center).

adjusted by stepping motors to move in both azimuth and elevations. Images taken of the western horizon shortly after sunset showed a distinct glow just above the lunar horizon dubbed horizon glow (HG). This light was interpreted to be forward-scattered sunlight from a cloud of dust particles < 1 m above the surface near the terminator. The HG had a horizontal extent of about 3 degrees on each side of the direction to the Sun (Figure 6.8).

Assuming that the observed signal is dominated by diffraction of sunlight, this horizontal extent corresponds to spheres of radius $\simeq 5\ \mu$m for observations at visible wavelengths. The observed intensity of the signal, its duration (up to 2.5 hours), and its vertical and horizontal extent rule out micrometeoroid ejecta, scattering off surface grains, and reflections involving glints off the spacecraft (Rennilson and Criswell 1974). However, it is difficult to analyze these images. In order to determine the physical dimensions of the bright cloud, the determination of the distance to the cloud is needed. By analyzing the shape of the lower boundary of the Surveyor 7 HG cloud and matching it to the local topography from orbital photographs of the Surveyor 7 landing site, Rennilson and Criswell (1974) placed the cloud at the visible horizon, or approximately 150 m from the camera. The vertical extent of the cloud is 1.9 mrad or about 30 cm at that distance. Its horizontal extent of 100 mrad makes the observed cloud 14 m wide, though this dimension may be a result of the light scattering properties of the cloud: it could be much larger with the parts of the cloud further from the Sun line not scattering sufficient light into the cameras.

The astrophotometer on the Lunokhod-2 rover also reported excess brightness, most likely due to HG (Severny *et al.* 1975). An independent set of observations related to dust levitation/transport phenomena is the description of the visual observations of the Apollo-17 crew during sunrise as it was seen from lunar orbit. They reported the appearance of bright streamers with fast temporal brightness changes (seconds to minutes) extending in excess of 100 km above the lunar surface. McCoy and Criswell (1974) argued for the existence of a significant population of lunar particles scattering the solar light. The rough estimates indicated that the scatterers are submicron ($\simeq 0.1\,\mu$m) sized grains. These drawings were analyzed again (Zook and McCoy 1991) and most of the earlier conclusions were verified. This study also estimated the scale height of this "dusty-exosphere" $H \simeq 10$ km, and suggested that dust levitation could be observed using ground based telescopes. A new simple theoretical model suggests that these could be particles with radii < 10 nm lofted from the lunar surface by electrostatic forces (Stubbs *et al.* 2006).

The last set of observations consist of the images taken of the lunar limb by the star-tracker camera of the the Clementine spacecraft, which showed a faint glow along the lunar surface, stunningly similar to the sketches of the Apollo 17 astronauts (Science News 3/26/94, H. Zook, private communications, 1994). The interpretation of these images was complicated by the presence of the scattered light from zodiacal dust particles, and it was never completed due to the untimely death of H. Zook in 2001.

6.4.2 Plasma and electric field measurements

The Moon is exposed to a variety of plasma conditions along its orbit about the Earth. Outside the Earth's magnetosphere, it is immersed in the solar wind. On the lit side of the surface, solar UV radiation dominates charging, resulting in a few volts positive surface potential (Manka 1973). In the absence of solar illumination, the night-side is expected to charge negatively (Figure 6.9). However, the night-side surface potential is determined by a complex set of plasma conditions.

An early model of the lunar day-side plasma environment (Walbridge 1973) indicated that at local noon the photoelectron flux is $\simeq 10^{11}$ cm^{-2}s^{-1} with an average energy of $\simeq 2$ eV, giving an electron density of $\simeq 4.5 \times 10^3$ cm^{-3} and a surface potential of +3.5 V. However, measurements of photoelectron yield from lunar soil samples (Willis *et al.* 1973) found a lower level of emission, 3×10^9 cm^{-2}s^{-1} or 4.5 μA m^{-2}. This emission level gives an electron density of only 130 cm^{-3} and a Debye length of ~ 1 m.

The typical flow speed is $\simeq 400$ km s^{-1}, and the characteristic temperature of the solar wind plasma is $kT \simeq 10$ eV. The thermal speed of the solar wind protons is on the order of 40 km s^{-1}, much below the flow speed; hence, they represent a supersonic flow. On the contrary, the electron thermal speed is close to 2000 km s^{-1}, much faster than the bulk speed; hence, they remain subsonic, and – to a good approximation – the bulk speed can be neglected. Consequently, a void in the solar wind protons behind the Moon could form. However, as electrons separate from the protons, a polarization electric field builds up, accelerating the ions and slowing the electrons, resulting in a filling of the plasma void behind the Moon. The lunar wake is often modeled as a plasma expansion into vacuum (Samir *et al.* 1983). This expansion leads to enhanced electron temperatures and energetic streaming ion beams towards the surface (Halekas *et al.* 2005). Measurements of electrons on the lunar night-side by the Lunar Prospector spacecraft support this simple model and suggest a night-side lunar surface potential of at least –35 V and more likely near –100 V (Halekas *et al.* 2002).

In the Earth's magnetotail the potential may exceed 500 V negative at times, due to high energy electron fluxes to the surface, usually within the Earth's plasma sheet (Halekas *et al.* 2005). These large negative surface potentials, inferred from Lunar Prospector data, most frequently occurred on the lunar night-side but were also observed on the day-side when the photoelectron current would be expected to prevent

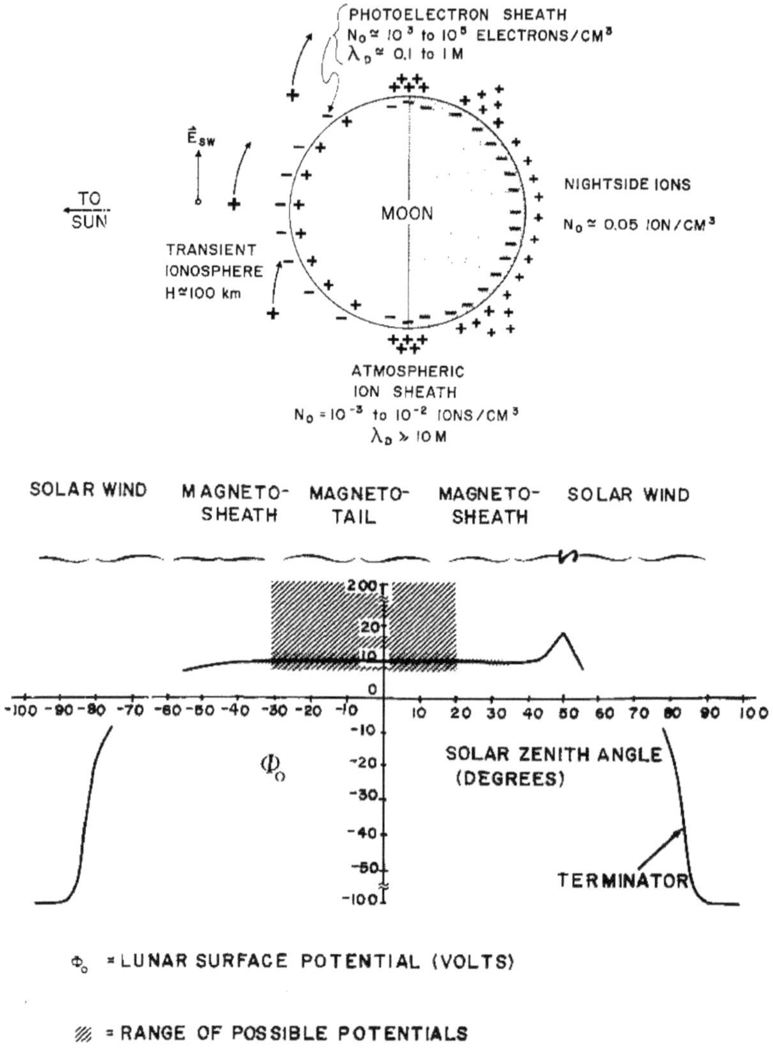

FIGURE 6.9
Top: The predicted electrostatic charge density distribution; Bottom: the measured lunar surface potential as a function of the solar zenith angle. The range of uncertainty is indicated by the shaded box in the figure (Freeman and Ibrahim 1975).

such large negative potentials, possibly indicating a strong variability in the photoelectron current from the lunar surface. This in turn suggests the possibility of small scale spatial and temporal variations in the day-side lunar surface potential.

We are aware of only two sets of in situ measurements that gave estimates of the lunar surface potential. The Apollo SIDE measured the energy of ions at the lunar surface (Freeman *et al.* 1973). The instrument was isolated from the lunar surface by an adjustable voltage biased electrical mesh – working as an artificial ground. This made it possible to vary the electrostatic potential between the SIDE instrument and the lunar surface. By biasing SIDE to negative potentials, it collected ions that were generated in the tenuous lunar atmosphere by solar wind electron impact or by photo-ionization. As the initial energies of these ions are negligibly small, their measured energies (per charge) can be used to estimate the accelerating electric field. Figure 6.9 shows the estimated lunar surface potential as a function of solar zenith angle. However, there are large uncertainties in deriving the surface potential, ϕ_0, from the electric field, E, estimates. This is due to the plasma distribution above the surface that determines a characteristic shielding distance, the Debye length. In units of cm, $\lambda_D \simeq 700(kT/n)^{0.5}$, where kT is the plasma temperature in units of eV, n is the plasma density in units of cm^{-3}, and the electric field $E \simeq \phi_0/\lambda_D$. However, λ_D is expected to change from $\simeq$ 1m in the relatively dense day-side plasma to $\gg$ 100 m in the dilute night-side plasma environment. The SIDE ion detector was approximately 50 cm above the lunar surface, a height that is comparable to λ_D on the lit side, but much shorter than that on the night-side, or when the Moon is shadowed by the Earth in the magneto-tail. Hence, the measured ion energies at these times could be due to only a fraction of the surface electric field, explaining the large uncertainties indicated in Figure 6.9.

The CPLEE had similar science goals to SIDE (O'Brien and Reasoner 1971; Burke and Reasoner 1972). In addition to ions, it also measured the energy (up to 200 eV) of electrons at a height of 26 cm above the surface, in two independent collectors pointing up and 60° from vertical toward lunar west. Due to the geometry of this setup, CPLEE could only measure electrons moving towards the lunar surface. The measurements indicated electrons with energies up to 200 eV on occasion. However, these energetic electron fluxes vanished during eclipse by the Earth, indicating that they are most likely electrons produced from the lunar surface by solar UV. As photoelectrons are generated with a few eV energy, the measurements indicated a large electric field pointing away from the surface, or a surface potential that from time to time is +200 V (Reasoner and Burke 1973). These high potentials were reached when the Moon was in the Earth's magnetotail and shielded from the solar wind.

The terminator region must have a complex history as the generally strong electric fields on the lit side pointing away from the surface are replaced by weaker fields pointing towards the Moon. The continued reduction of photo-emission at diminishing solar zenith angles on large scales are accompanied with the presence of shadows adjacent to more directly illuminated and photo-emitting surfaces on the small spatial scale of boulders, rocks, and crater rims. This could produce strong localized electric fields due to the variation in the photoelectron current from shadowed to illuminated regions. The time dependent changes of the areas of neighboring lit/dark patches have been suggested to possibly lead to "supercharging" and lift-off of lunar fines. (De and Criswell 1977; Pellizari and Criswell, 1978a,b).

6.4.3 Dust measurements

The LEAM Experiment was deployed by the Apollo-17 astronauts on December 11, 1972. It started measurements after the return of the landing module and continued to make observations for about 3 years. The science objectives of LEAM were (1) to investigate the interplanetary dust flux (primary particles) bombarding the lunar surface, (2) to investigate the properties of the lunar ejecta (secondary) particles, (3) to follow the temporal variability of these dust fluxes along the lunar orbit, and (4) to observe interstellar particles. The design and the expected performance of the LEAM experiment were similar to those for the dust experiments onboard the Pioneer 8 and 9 spacecraft that were launched into heliocentric orbits in 1967 and 1968, respectively (Berg *et al.* 1973, 1974).

The LEAM instrument consisted of three sensor systems. The EAST sensor was pointed 25 degrees north of east, so that once per lunation its field of view swept into the direction of the interstellar dust flow. The WEST sensor was pointing in the opposite direction as a control for the EAST sensor, while the UP sensor was parallel to the lunar surface and viewed particles coming from above. Each of these systems was composed of two sets (front and back) of 4 x 4 basic sensor elements to determine the impacting particle's mass, m, and velocity vector, $\mathbf{v}$. The sensors used a combination of thin plastic films and grids to measure the current from the plasma cloud generated as the dust particles penetrated the film, a signal pulse with amplitude proportional to $m \times v^{2.6}$. The two groups of sensors in a system were placed 5 cm apart, and a time-of-flight setup was used to determine the speed of an impacting dust particle. All of the 16 front sensors were enabled to provide a start signal, and all of the 16 back sensors were designed to provide a stop signal for a total of 256 different combinations, enabling the determination of the velocity vector of the penetrating dust particles. In addition, the back film was attached to a microphone with an acoustic signal proportional to the momentum of the grain. The only exception for this redundant arrangement was the WEST sensor, which lacked a front film. This sensor was designed to identify low-speed ejecta impacts that were expected not to penetrate the front film. Hence, the WEST sensor could not measure particle speed. Extensive laboratory calibrations were performed on these sensors using a 2 MeV electrostatic accelerator with particle masses in the range of $10^{-13} < m < 10^{-9}$ g, and velocities in the range of $1 < v < 25$ km s^{-1}. The pulse height amplitudes (PHA) from the film-grid sensors were sorted in the range from 0 to 7, and in the preflight calibration the front film rarely registered a PHA greater than 3. Most preflight calibration dust impacts gave rise to signals from both films, indicating that the particles had penetrated the front film and passed on to the rear film.

Once LEAM started to operate it became clear that its observations contradicted expectations. Based on previous measurements in interplanetary space by Pioneer 8 and 9, for example, the expected impact rate of interplanetary dust particles was a few impact detections per day. Instead, LEAM registered up to hundreds of impacts per day. Most puzzling was the fact that these events registered in the front film only, but with the maximum possible PHA of 7. Additionally, the LEAM op-

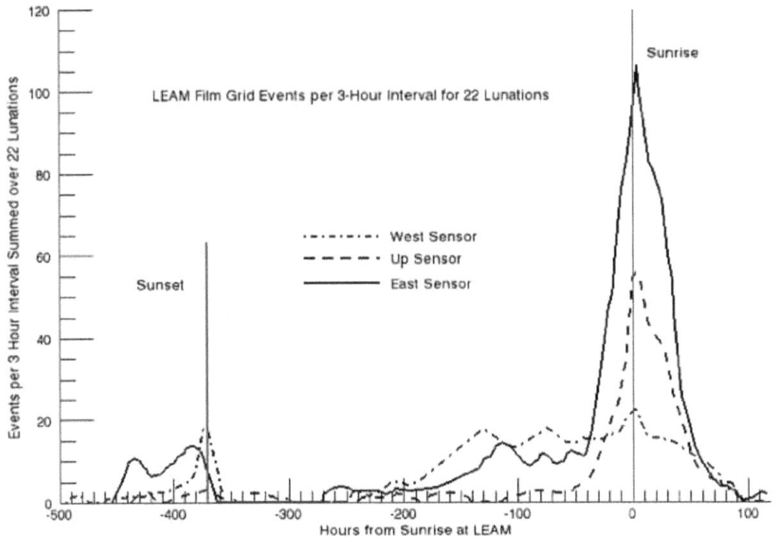

FIGURE 6.10
The number of dust impacts per 3 hour intervals integrated over 22 lunations (Berg *et al.* 1976). The EAST and WEST sensors showed an approximately unchanged behavior with constant PHA, while the UP sensor registered a declining rate with suddenly dropping PHA after about 20 months on the lunar surface (O. Berg, personal communication, 2006).

erating temperature exceeded its predicted maximum value of 146° F at lunar noon, indicating possible thermal problems that were initially believed to be responsible for generating noise in the electronics and possibly responsible for the elevated impact rates. This was supported by the correlation of the elevated impact rates with the passage of the terminator, both at sunrise and at sunset. As data accumulated, a systematic behavior was recognized. The terminator event rate started to increase up to 60 hours before the sunrise at the site, and persisted for a period of approximately 30–60 hours. In this period the rates were up to 100 times higher than the normal background rates (Berg *et al.* 1976). The rates dropped two orders of magnitude during local noon. Interestingly, no increased rates were observed during lunar eclipses. Figure 6.10 shows the number of dust impacts onto Lunar Ejecta and Meteorite Experiment (LEAM) per 3-hour period, integrated over 22 lunar days.

A new picture emerged to replace the high temperature electronics explanation: LEAM was registering slow-moving, highly charged lunar dust particles. There were two subsequent studies done to verify this point: a theoretical work to model the response of the electronics (Perkins 1976) and an experimental study of the LEAM flight spare (Bailey and Frantsvog 1977). The results of the sensor modeling and circuit analysis showed that charged particles moving at velocities $< 1\ \mathrm{km\,s^{-1}}$ do produce large PHA responses via induced voltages on the entry grids, as opposed to signals from impact generated plasmas. This explains why the rear films remained silent even though the front sensor was thought to be hit by an energetic dust grain. The experimental study had a similar conclusion: extremely slow moving particles ($v < 100\ \mathrm{m\,s^{-1}}$) generate a LEAM response up to and including the maximum PHA of 7 if the particles carry a positive charge $Q > 10^{-12}$ C. Both of these studies suggest that the LEAM events are consistent with the sunrise/sunset–triggered levitation and transport of slow moving, highly charged lunar dust particles. Assuming a day-time surface potential of +5 V, the LEAM measurements indicate grain sizes on the order of millimeters in radius! The entire LEAM data set is shown in Figure 6.10 for all three sensor surfaces. While the daily average PHA remained relatively constant for the EAST and WEST sensors, it exhibited a rapid decline for the UP sensor after 20 months, perhaps indicating dust accumulation on the topside of LEAM (O. Berg, personal communications, 2006).

6.4.4 The lunar dust environment

The dust environment of the Moon has remained a controversial issue since the Apollo era. Visual observations and photographic images from the Apollo command modules and images from the Clementine mission have been used to indicate the presence of dust at high altitudes above the lunar surface. The in situ and remote sensing observations on the lunar surface, as discussed above, indicate that dusty plasma processes are responsible for the mobilization and transport of lunar soil. However, there is a lack of both observations and theoretical understanding to directly relate the existing surface and high altitude phenomena, and the existence of a significant "dusty-exosphere" remains a question. In addition to the dusty plasma processes acting on the lunar surface, a lunar dust exosphere could also be maintained by the continual bombardment by interplanetary dust particles. These can produce secondary ejecta grains with sufficient speeds to reach tens of kilometers in altitude, expected to form a "permanently" present, approximately spherically symmetric dust exosphere.

Dust charging, mobilization, and transport is expected to be most efficient near the terminators, where large surface electric fields can be generated due to differential charging across lit and dark regions. Hence, future observations from a satellite in orbit around the Moon at an altitude of $\sim$ tens of kilometers could be used to measure the variability of size and spatial distribution of small grains and gain insight into the efficiency of dusty plasma processes acting on the lunar surface. There is also a need to develop dusty plasma diagnostic packages for future surface landers that could directly measure the plasma parameters, as well as the mass, charge, and velocity distributions of lofted dust particles.

6.5 Summary

We used three examples to demonstrate that dusty plasma processes can be responsible for shaping the characteristics of the plasma environment, as well as the size and spatial distributions of the small embedded dust grains. In Noctilucent clouds these processes are thought to be responsible for generating electron density bite-outs and unexpected charge states of small solid particles, for example. Future combination of in situ rocket measurements with improved sensors, ground based lidar and radar measurements, and satellite observations are expected to bring about major advances in our understanding of the physics of this region. This will be of great significance, as Noctilucent clouds are thought to be indicators of the global changes in our climate. Planetary rings are an excellent laboratory to study dusty plasma processes acting on large scales. The combination of imaging, with in situ measurements of the plasma parameters, the electric and magnetic fields, and dust provide a rich complementary data set. The extension of the Cassini mission beyond its prime (2004–2008) for an additional four years could provide an unprecedented series of observations that cover a full period of seasons. The dusty plasma processes acting on the lunar surface are of great scientific interest, and their understanding has great importance in the design of future lunar habitats, to ensure the safety of the astronauts, and the long-term operations of all mechanical and optical devices.

References

Amyx, K., Sternovsky, Z., Knappmiller, S., Robertson, S., Horányi, M., and Gumbel, J. (2008). In-situ measurement of smoke particles in the wintertime polar mesosphere between 80 and 85 km altitude. *J. Atmos. Solar-Terrestrial Phys.*, **70**, 61–70.

Axford, W. I. and Mendis, D. A. (1974). Satellites and magnetospheres of the outer planets. *Annu. Rev. Earth and Planet. Sci.*, **2**, 419–474.

Bailey, C. L. and Frantsvog, D. J. (1977). Response of the LEAM detector to positively charged microparticles. NASA Contract No. NAS5-23557. Concordia College.

Berg, O. E., Richardson, F. F., and Burton, H. (1973). Apollo 17 preliminary science report. NASA SP-330, 16.

Berg, O. E., Richardson, F. F., Rhee, J. W., and Auer S. (1974). Preliminary results of a cosmic dust experiment on the Moon. *Geophys. Res. Lett.*, **1**, 289–290 .

Berg, O. E., Wolf, H., and Rhee, J. W. (1976). Lunar soil movement registered by the Apollo 17 cosmic dust experiment. In *Interplanetary dust and zodiacal light*, Elsasser, H. and Fechtig, H. (eds.), pp. 233–237. Springer-Verlag, New York.

Biebricher, A., Havnes, O., Hartquist, T. W., and La Hoz, C. (2006). On the influence of plasma absorption by dust on the PMSE overshoot effect. *Adv. Space Res.*, **38**, 2541–2550.

Bliokh, P., Sinitsin, V., and Yaroshenko, V. (1995). *Dusty and self-gravitational plasmas in space*, Kluwer Academic Publishers, Dordrecht, Boston .

Burke, W. J. and Reasoner, D. L. (1972). Absence of the plasma sheet at lunar distance during geomagnetically quiet times. *Planet. Space Sci.*, **20**, 429–436 .

Chilson, P. B., Belova, E., Rietveld, M. T., Kirkwood, S., and Hoppe, U.-P. (2000). First artificially induced modulation of PMSE using the EISCAT heating facility. *Geophys. Res. Lett.*, **27**, 3801–3804.

Cho, J. Y. N. and Röttger, J. (1997). An updated review of polar mesosphere summer echoes: Observations, theory, and their relationship to noctilucent clouds and subvisible aerosols. *J. Geophys. Res.*, **102**, 2001–2020.

Colwell, J., Batiste, S., Horányi, M., Robertson, S., and Sture, S. (2007). Lunar surface: Dust dynamics and regolith mechanics. *Rev. Geophys*, **45**, RG 2006.

Consolmagno, G. J. (1980). Electromagnetic scattering lifetimes for dust in Jupiter's ring. *Nature*, **285**, 557–558.

Cuzzi, J. N., Lissauer, J. J., Esposito, L. W., Holberg J. B., Marouf, E. A., Tyler, G. L., and Boischot, A. (1984). Saturn's rings: Properties and processes. In *Planetary rings*, Greenberg, R. and Brahic, A. (eds.), pp. 73–199. Univ. Arizona Press, Tucson.

Criswell, D. R. (1973). Horizon-glow and the motion of lunar dust. In *Photon and particle interaction in space*, Grard, R. J. L. (ed.), pp. 545–556. Reidel, Dordrecht.

De, B. R. and Criswell, D. R. (1977). Intense localized charging in the lunar sunset terminator region: Development of potentials and fields. *J. Geophys. Res.*, **82**, 999–1004.

Dikarev, V. V. (1999). Dynamics of particles in Saturn's E ring: effects of charge variations and the plasma drag force. *Astron. Astrophys.*, **346**, 1011–1019.

Dimant, Y. S. and Milikh, G. M. (2004). Effect of radio wave heating on polar mesospheric clouds. *Adv. Space. Res.*, **34**, 2413–2421.

Farrell, W. M., Desch, M. D., Kaiser, M. L., Kurth, W. S., and Gurnett, D. A. (2006). Changing electrical nature of Saturn's rings: Implications for spoke formation. *Geophys. Res. Lett.*, **33**, L07203, doi: 10.1029/2005GL024922.

Farmer, A. J. and Goldreich, P. (2005). Spoke formation under moving plasma clouds. *Icarus*, **179**, 535–538.

Freeman, J. W. and Ibrahim, M. (1975). Lunar electric fields, surface potential and associated plasma sheaths. *The Moon*, **8**, 103–114.

Freeman, J. W., Fenner, M. A., and Hills, H. K. (1973). The electric potential of the Moon in the solar wind. In *Photon and particle interaction in space*, Grard, R. J. L. (ed.), pp. 363–368. Reidel, Dordrecht.

Gadsden, M. and Schröder, W. (1989). *Noctilucent clouds*. Springer-Verlag, New York.

Gelinas, L. J., Lynch, K. A., Kelly, M. C., Collins, S., Baker, S., Zhou., Q., and Friedman, J. S. (1998). First observation of meteoritic charged dust in the tropical mesosphere. *Geophys. Res. Lett.*, **25**, 4047–4050.

Goertz, C. K. (1989). Dusty plasmas in the solar system. *Rev. Geophys.*, **27**, 271–292.

Goertz, C. K. and Morfill, G. (1983). A model for the formation of spokes in Saturn's ring. *Icarus*, **53**, 219–229.

Grün, E., Morfill, G. E., and Mendis, D. A. (1984). Dust-magnetosphere interactions. In *Planetary rings*, Greenberg, R. and Brahic, A. (eds.), pp. 275–332. University of Arizona Press, Tucson.

Grün, E., Zook, H. A., Baguhl, M., Balogh, A., Bame, S. J., Fechtig, H., Forsyth, R., Hanner, M. S., Horányi, M., Kissel, J., Lindblad, B.-A., Linkert, D., Linkert, G., Mann, I., McDonnell, J. A. M., Morfill, G. E., Phillips, J. L., Polanskey, C., Schwehm, G., Siddique, N., Staubach, P., Svetska, J., and A. Taylor. (1993). Ulysses discovers dust emissions from Jupiter and probable interstellar grains. *Nature*, **362**, 428–430.

Halekas, J. S., Mitchell, D. L., Lin, R. P., Hood, L. L., Acuña, M. H., and Binder, A. B. (2002). Evidence for negative charging of the lunar surface in shadow. *Geophys. Res. Lett.*, **29**, 1435, doi:10.1029/2001GL014428.

Halekas, J. S., Lin, R. P., and Mitchell, D. L. (2005). Large negative lunar surface potentials in sunlight and shadow. *Geophys. Res. Lett.*, **32**, L09102, doi:10.1029/2005GL022627.

Hartquist, T. W., Havnes, O., and Morfill, G. E. (1992). The effects of dust on the dynamics of astronomical and space plasmas. *Fund. Cosmic Phys.*, **15**, 107–142.

Havnes, O. (2004). Polar mesospheric summer echoes (PMSE) overshoot effect due to cycling of artificial electron heating. *J. Geophys. Res.*, **109**, A02309, doi:10.1029/2003JA010159.

Havnes, O. and Kassa, M. (2009). On the sizes and observable effects of dust particles in polar mesospheric winter echoes. *J. Geophys. Res.*, **114**, D09209, doi:10.1029/2008JD011276.

Havnes, O. and Naesheim, L. I. (2007). On the secondary charging effects and structure of mesospheric dust particles impacting on rocket probes. *Ann. Geophys.*, **25**, 623–637.

Havnes, O., Aanesen, T. K., and Melandsø, F. (1990). On dust charges and plasma potentials in a dusty plasma with dust size distribution. *J. Geophys. Res.*, **95**, 6581–6585.

Havnes, O., Morfill, G., and Melandsø, F. (1992). Effects of electromagnetic and plasma drag forces on the orbit evolution of dust in planetary magnetospheres. *Icarus*, **98** , 141–150.

Havnes, O., Trøim, J., Blix, T., Mortensen, W., Nsheim, L. I., Thrane, E., and Tønnesen, T. (1996). First detection of charged dust particles in the Earth's mesosphere. *J. Geophys. Res.*, **101**, 10839–10847.

Havnes, O., Brattli, A., Aslaksen, T., Singer, W., Latteck, R., Blix, T., Thrane, E., and Trøim, J. (2001). First common volume observations of layered plasma structures and polar mesospheric summer echoes by rocket and radar. *Geophys. Res. Lett.*, **28**, 1419–1422.

Havnes, O., La Hoz, C., Naesheim, L. I., and Rietveld, M. T. (2003). First observations of the PMSE overshoot effect and its use for investigating the conditions in the summer mesosphere. *Geophys. Res. Lett.*, **30**, 2229, doi:10.1029/2003GL018429.

Havnes, O., La Hoz, C., Biebricher, A., Kassa, M., Meseret, T., Naesheim, L. I., and Zivkovic, T. (2004). Investigation of the mesosspheric PMSE conditions by use of the new overshoot effect. *Phys. Scr.*, **T107**, 70–78.

Hill, J. R. and Mendis, D. A. (1982). The dynamical evolution of the Saturnian ring spokes. *J. Geophys. Res.*, **87**, 7413-7420.

Horányi, M. (1996). Charged dust dynamics in the solar system. *Annu. Rev. Astron. Asrophys.*, **34**, 383–418.

Horányi, M. (2000). Dust streams from Jupiter and Saturn. *Phys. Plasmas*, **7**, 3847–3850.

Horányi, M. and Burns, J. A. (1991). Charged dust dynamics: Orbital resonances due to planetary shadows. *J. Geophys. Res.*, **96**, 19283–19289.

Horányi, M., Burns, J. A., and Hamilton, D. (1992). The dynamics and origin of Saturn's E ring. *Icarus* , **97**, 248–259.

Horányi, M., Grün, E., and Morfill, G. (1993a). The dusty ballerina skirt of Jupiter. *J. Geophys. Res.*, **98**, 21245–21251.

Horányi, M., Morfill, G., and Grün, E. (1993b). Mechanism for the acceleration and ejection of dust grains from Jupiter's magnetosphere. *Nature*, **363**, 144–146.

Horányi, M., Robertson, S., Smiley, B., Gumbel, J., Witt, G., and Walch, B. (2000). Rocket-borne mesospheric measurement of heavy ($m = 10$ amu) charge carriers. *Geophys. Res. Lett.*, **27**, 3825–3828.

Horányi, M., Hartquist, T. W., Havnes, O., Mendis, D. A., and Morfill, G. E. (2004). Dusty plasma effects in Saturn's magnetosphere. *Rev. Geophys.*, **42**, RG4002, doi:10.1029/2004RG000151.

Horányi, M., Juhász, A., and Morfill, G. E. (2008). The large scale structure of Saturn's E-ring. *Geophys. Res. Lett.*, **35**, L04203, doi: 10.1029/2007GL032726.

Hunten, D. M., Turco, R. P., and Toon, O. B. (1980). Smoke and dust particles of meteoric origin in the mesosphere and stratosphere. *J. Atmos. Sci.*, **37**, 1342–1357.

Juhász, A. and Horányi, M. (2002). Saturn's E ring: A dynamical approach. *J. Geophys. Res.*, **107**, A6, doi:10.1029/2001JA000182.

Juhász, A., Horányi, M., and Morfill, G. E. (2007). Signatures of Enceladus in Saturn's E-ring. *Geophys. Res. Lett.*, **34**, L09104. doi: 10.1029/2006GL029120.

Jurac, S., Johnson, R. E., and Richardson, J. D. (2001a). Saturn's E ring and the production of the neutral torus. *Icarus*, **149**, 384–396.

Jurac, S., Johnson, R. E., Richardson, J. D., and Paranicas, C. (2001b). Satellite sputtering in Saturn's magnetosphere. *Planet. Space Sci.*, **49**, 319–326.

Kalashnikova, O., Horányi, M., Thomas, G. E., and Toon, O. B. (2000). Meteoric smoke production in the atmosphere. *Geophys. Res. Lett.*, **27**, 3293–3296.

Kassa, M., Havnes, O., and Belova, E. (2005). The effect of electron bite-outs on artificial electron heating and the PMSE overshoot. *Ann. Geophys.*, **23**, 1–11.

Kempf, S., Srama, R., Horányi, M., Burton, M., Helfert, S., Moragas-Klostermeyer, G., Roy, M., and Grün, E. (2005). High-velocity streams of dust originating from Saturn. *Nature*, **433**, 289–291.

Kempf, S., Beckmann, U., Moragas-Klostermeyer, G., Postberg, F., Srama, R., Economou, T., Schmidt, J., Spahn, F., and Grün, E. (2008). The E ring in the vicinity of Enceladus. I. Spatial distribution and properties of the ring particles. *Icarus*, **193**, 420–437.

Kurth, W. S., Averkamp, T. F., Gurnett, D. A., and Wang, Z. (2006). Cassini RPWS observations of dust in Saturn's E ring. *Planet. Space Sci.*, **54**, 988–998.

La Hoz, C. and Havnes, O. (2008). Artificial modification of Polar Mesospheric Winter Echoes (PMWE) with an rf heater: Do charged dust particles play an active role? *Geophys. Res. Lett.* **113**, D19205, doi: 10.1029/2008JD010460.

Lübken, F.-J. (1999). Thermal structure of the Arctic summer mesosphere. *J. Geophys. Res.*, **104**, 9135–9149.

Lübken, F.-J., Rapp, M., Blix, T., and Thrane, E. (1998). Microphysical and turbulent measurements of the Schmidt number in the vicinity of polar mesosphere summer echoes. *Geophys. Res. Let.*, **25**, 893–896.

Lübken, F.-J., Strelnikov, B., Rapp, M., Singer, W., Latteck, R., Brattli, A., Hoppe, U.-P., and Friedrich, M. (2006). The thermal and dynamical state of the atmosphere during polar mesosphere winter echoes. *Atmos. Chem. Phys.*, **6**, 13–24.

Manka, R. H. (1973). Plasma and potential at the lunar surface. In *Photon and particle interaction in space*, Grard R. J. L. (ed.), pp. 347–361. Reidel, Dordrecht.

McCoy, J. E. and Criswell, D. R. (1974). Evidence for a high latitude distribution of lunar dust. *Proc. 5th Lunar Conf.*, **3**, 2991–3005.

McGhee, C. A., French, R. G., Dones, L., Cuzzi, J. N., Salo, H. J., and Danos, R. (2005). HST observations of spokes in Saturn's B ring. *Icarus*, **173**, 508–521.

Mendis, D. A. and Rosenberg, M. (1994). Cosmic dusty plasmas. *Annu. Rev. Astr. Astrophys.*, **32**, 419–463.

Mendis, D. A., Hill, J. R., Ip, W.-H., Goertz, C. K., and Grün, E. (1984). Electrodynamic processes in the ring system of Saturn. In *Saturn*, (A85-33976), pp. 15–91. Univ. Arizona Press, Tucson.

Megner, L., Rapp, M., and Gumbel, J. (2006). Distribution of meteoric smoke sensitive to microphysical properties and atmospheric conditions. *Atmos. Chem. Phys.*, **6**, 4415–4426.

Mitchell, J. D., Croskey, C. L., and Goldberg, R. A. (2001). Evidence for charged aerosols and associated meter-scale structure in identified PMSE/NLC regions. *Geophys. Res. Lett.*, **28**, 1423–1426.

Mitchell, J, D., Croskey, C. L., Goldberg, R. A., and Friedrich, M. (2003). Charged particles in the polar mesospheric region: probe measurements from the MaCWAVE and DROPPS programs. *Proceedings of the 16th ESA symposium on European Rocket and Bolloon Programmes and related research*, St. Gallen, Switzerland (ESA SP-530), 351–356.

Mitchell, C. J., Horányi, M., Havnes, O., and Porco, C. C. (2006). Saturn's spokes: Lost and found. *Science*, **311**, 1587–1589.

Morfill, G. E. and Grün, E. (1979). The motion of charged dust particles in interplanetary space. I - The zodiacal dust cloud. *Planet. Space Sci.*, **27**, 1269–1292.

Morfill, G. E. and Thomas, H. M. (2005). Spoke formation under moving plasma clouds – The Goertz-Morfill model revisited. *Icarus*, **179**, 539–542.

Morfill, G. E., Grün, E., and Johnson, T. V. (1980). Dust in Jupiter's magnetosphere. I – Physical processes. II – Origin of the ring. III – Time variations. IV – Effect on magnetospheric electrons and ions. *Planetary. Space Sci.*, **28**, 1087–1100 .

Morfill, G. E., Grün, E., and Johnson, T. V. (1983). Saturn's E, G, and F rings – Modulated by the plasma sheet? *J. Geophys. Res.*, **88**, 5573–5579.

Morfill, G. E., Havnes, O., and Goertz, C. K. (1993). Origin and maintenance of the oxygen torus in Saturn's magnetosphere. *J. Geophys. Res.*, **98**, 11285–11297.

Naesheim, L. I., Havnes, O., and La Hoz, C. (2008). A comparison of polar mesosphere summer echo at VHF (224 MHz) and UHF (930 MHz) and the effects of artificial electron heating. *J. Geophys. Res.*, **113**, D08205/1–11.

Nitter, T., Havnes, O., and Melandsø, F. (1998). Levitation of charged dust in the photoelectron sheath above surfaces in space. *J. Geophys. Res.*, **103**, 6605–6620.

Northrop, T. H. and Hill, J. R. (1983). The adiabatic motion of charged dust grains in rotating magnetospheres. *J. Geophys. Res.*, **88**, 1–11.

Northrop, T. G., Mendis, D. A., and Schaffer, L. (1989). Gyrophase drifts and the orbital evolution of dust at Jupiter's gossamer ring. *Icarus*, **79**, 101–115.

O'Brien, J. and Reasoner, D. L. (1971). Charge particle lunar environment experiment. In *Apollo 14 preliminary science report*, NASA SP-272, 193–213.

Pedersen, A., Trøim, J., and Kane, J. (1969). Rocket measurement showing removal of electrons above the mesopause in summer at high latitudes. *Planet. Space Sci.*, **18**, 945–947.

Pelizzari, M. A. and Criswell, D. R. (1978a). Differential charging of nonconducting surfaces in space. *J. Geophys. Res.*, **83**, 5233–5244.

Pelizzari, M. A. and Criswell, D. R. (1978b). Lunar dust transport by photoelectric charging at sunset. *Proc. 9th Lunar Planet. Sci. Conf.*, 3225–3237.

Perkins, D. (1976). Analysis of the LEAM experiment response to charged particles, Bendix Aerospace Systems Division, BSR 4233 (NASA Contract No. NAS9-14751).

Plane, J. M. C. (2004). A time-resolved model of the mesospheric Na layer constraints on the meteoric input function. *Atmos. Chem. Phys.*, **4**, 627–638.

Porco, C. C., Helfenstein, P., Thomas, P. C., Ingersoll, A. P., Wisdom, J., West, R., Neukum, G., Denk, T., Wagner, R., Roatsch, T., Kieffer, S., Turtle, E., McEwen, A., Johnson, T. V., Rathbun, J., Veverka, J., Wilson, D., Perry, J., Spitale, J., Brahic, A., Burns, J. A., DelGenio, A. D., Dones, L., Murray, C. D., and Squyres S. (2006). Cassini observes the active south pole of Enceladus. *Science*, **311**, 1393–1401.

Rapp, M. and Lübken, F.-J. (2001). Modelling of particle charging in the polar summer mesosphere: Part 1 – General results. *J. Atmos. Sol. Terr. Phys.*, **63**, 759–770.

Rapp, M. and Lübken, F.-J. (2004). Polar mesosphere summer echoes (PMSE): Review of observations and current understanding. *Atmospheric Chem. Phys.*, **4**, 2601–2633.

Rapp, M. and Strelnikova, I. (2008). Measurements of meteor smoke particles during the ECOMA-2006 campaign: 1. Particle detection by active photoionization. (2008). *J. Atmos. Sol.–Terr. Phys.*, **71**, 477–485.

Rapp, M., Hedin, J., Strelnikova, I., Friedrich, M., Gumbel, J., and Lübken, F.-J. (2005). Observations of positively charged nanoparticles in the night-time polar mesosphere. *Geophys. Res. Lett.*, **32**, L23821, doi:10.1029/2005GL024676.

Reasoner, D. L. and Burke, W. J. (1973). Measurement of the lunar photoelectron layer in the geomagnetic tail. In *Photon and particle interaction in space*, Grard, R. J. L. (ed.), pp. 369–387. Reidel, Dordrecht.

Rennilson, J. J. and Criswell, D. R. (1974). Surveyor observations of lunar horizon glow. *The Moon*, **10,** 121–142.

Rietveld, M. T., Kohl, H., and Kopka, H. (1993). Introduction to ionospheric heating at Tromsø. Part I: Experimental overview. *J. Atmos. Terr. Phys.*, **55**, 577–599.

Robinson, L. J. (1980). Closing in on Saturn. *Sky and Telescope*, **60**, 481.

Rosinski, J. and Snow, R. H. (1961). Secondary particulate matter from meteor vapors. *J. Met.*, **18**, 736–745.

Samir, U., Wright, K. H., Jr., and Stone, N. H. (1983). The expansion of a plasma into a vacuum: Basic phenomena and processes and applications to space plasma physics. *Rev. Geophys. Space Phys.*, **21**, 1631–1646.

Schaffer, L. and Burns, J. A. (1987). The dynamics of weakly charged dust – Motion through Jupiter's gravitational and magnetic fields. *J. Geophys. Res.*, **92**, 2264–2280.

Severny, A. B., Terez, E. I., and Zvereva, A. M. (1975). The measurements of sky brightness on Lunokhod-2. *The Moon*, **14**, 123–128.

Smiley, B., Robertson, S., Horányi, M., Blix, T., Rapp, M., Latteck, R., and Gumbel, J. (2003). Measurement of positively and negatively charged particles inside PMSE during MIDAS SOLSTICE 2001. *J. Geophys. Res. Atmospheres*, **108**, 8444, doi:10.1029/2002JD002425.

Smith, B. A., Soderblom, L., Beebe, R. F., Boyce, J. M., Briggs, G., Bunker, A., Collins, S. A., Hansen, C., Johnson, T. V., Mitchell, J. L., Terrile, R. J., Carr, M. H., Cook, A. F., Cuzzi, J. N., Pollack, J. B., Danielson, G. E., Ingersoll, A. P., Davies, M. E., Hunt, G. E., Masursky, H., Shoemaker, E. M., Morrison, D., Owen, T., Sagan, C., Veverka, J., Strom, R., and Suomi, V. E. (1981). Encounter with Saturn: Voyager 1 imaging science results. *Science*, **212**, 163–191.

Smith, B. A., Soderblom, L., Batson, R. M., Bridges, P. M., Inge, J. L., Masursky, H., Shoemaker, E., Beebe, R. F., Boyce, J., Briggs, G., Bunker, A., Collins, S. A., Hansen, C., Johnson, T. V., Mitchell, J. L., Terrile, R. J., Cook, A. F., Cuzzi, J. N., Pollack, J. B., Danielson, G. E., Ingersoll, A. P., Davies, M. E., Hunt, G. E., Morrison, D., Owen, T., Sagan, C., Veverka, J., Strom, R., and Suomi, V. E. (1982). A new look at the Saturnian system: The Voyager 2 images. *Science*, **215**, 504–536.

Spahn, F., Albers, N., Hörning, M., Kempf, S., Krivov, A.V., Makuch, M., Schmidt, J., Seiss, M., and Sremcević, M. (2006a). E ring dust sources: Implications from Cassini's dust measurements. *Planet. Space Sci.*, **54**, 1024–1032.

Spahn, F., Schmidt, J., Albers, N., Hörning, M., Makuch, M., Seiss, M., Kempf, S., Srama, R., Dikarev, V., Helfert, S., Moragas-Klostermeyer, G., Krivov, A. V., Sremcević, M., Tuzzolino, A. J., Economou, T., and Grün, E. (2006b). Cassini dust measurements at Enceladus and implications for the origin of the E ring. *Science*, **311**, 1416–1418.

Srama, R., Kempf, S., Moragas-Klostermeyer, G., Helfert, S., Ahrens, T. J., Altobelli, N., Auer, S., Beckmann, U., Bradley, J. G., Burton, M., Dikarev, V. V., Economou, T., Fechtig, H., Green, S. F., Grande, M., Havnes, O., Hillier, J. K., Horanyi, M., Igenbergs, E., Jessberger, E. K., Johnson, T. V., Kruger, H., Matt, G., McBride, N., Mocker, A., Lamy, P., Linkert, D., Linkert, G., Lura, F., McDonnell, J. A. M., Mohlmann, D., Morfill, G. E., Postberg, F., Roy, M., Schwehm, G. H., Spahn, F., Svestka, J., Tschernjawski, V., Tuzzolino, A. J., Wasch, R., and Grun, E. (2006). In situ dust measurements in the inner Saturnian system. *Planet. Space Sci.*, **54**, 967–987.

Strelnikova, I., Rapp, M., Strelnikov, B., Brattli, A., Svenes, K., Hoppe, U.-P., Friedrich, M., Gumbel, J., and Williams, B. P. (2009). Measurements of meteor smoke particles during the ECOMA-2006 campaign: 2. Results. *J. Atmos. Sol.–Terr. Physics*, **71**, 486–496.

Stubbs, T. J., Vondrak, R. R., and Farrell, W. M. (2006). A dynamic fountain model for lunar dust. *Adv. Space Res.*, **37**, 59–66.

Tomsic, A. (2001). *Collisions between water clusters and surfaces, PhD thesis.* Göthenborg University, Göthenborg.

Ulwick, J. C., Baker, K. D., Kelley, M. C., Balsley, B. B., and Ecklund, W. L. (1988). Comparison of simultaneous MST radar and electron density probe measurements during STATE. *J. Geophys. Res.*, **93**, 6989–7000.

von Cossart, G., Fiedler, J., and von Zahn, U. (1999). Size distributions of NLC particles as determined from 3-color observations of NLC by ground-based lidar. *Geophys. Res. Lett.*, **26**, 1513–15161.

von Zahn, U. and Meyer, W. (1989). Mesopause temperatures in polar summer. *J. Geophys. Res.*, **94**, 14647–14650.

Waite, J. H., Jr., Cravens, T. E., Ip, W.-H., Kasprzak, W. T., Luhmann, J. G., McNutt, R. L., Niemann, H. B., Yelle, R. V., Mueller-Wodarg, I., Ledvina, S. A., and Scherer, S. (2005). Oxygen ions observed near Saturn's A ring. *Science*, **307**, 1260–1262.

Walbridge, E. (1973). Lunar photoelectron layer. *J. Geophys. Res.*, **78**, 3668–3687.

Wang, Z., Gurnett, D. A., Averkamp, T. F., Persoon, A. M., and Kurth, W. S. (2006). Characteristics of dust particles detected near Saturn's ring plane with the Cassini Radio and Plasma Wave instrument. *Planet. Space Sci.*, **54**, 957–966.

Weingartner, J. C. and Draine, B. T. (2001). Electron-ion recombination on grains and poycyclic aromatic hydrocarbons. *Astrophys. J.*, **563**, 842–852.

Willis, R. F., Anderegg, M., Feuerbacher, B., and Fitton, B. (1973). Horizon-glow and the motion of lunar dust. In *Photon and particle interaction in space*, Grard, R. J. L. (ed.), pp. 389–401. Reidel, Dordrecht.

Zook, H. A. and McCoy, E. (1991). Large scale lunar horizon glow and a high altitude lunar dust exosphere. *Geophys. Res. Lett.*, **18**, 2117–2120 .

7

Numerical simulation of complex plasmas

Olga S. Vaulina and Boris A. Klumov

7.1 Molecular dynamics simulations of complex plasmas: Basic concepts

7.1.1 Methods of simulation of the dynamics of dust particles

The major problem encountered in studying the physical properties of non-ideal systems is associated with the absence of an analytical theory of liquid that would be capable of explaining its thermodynamic properties, giving the equation of state, describing the effects of heat and mass transfer, etc. Two basic approaches are employed in the development of approximate models for the description of the liquid state. The first approach involves a semiempirical method of determining the correlation of the parameters of liquid with one another and with the properties of initial crystals, which is based on analogies between the crystalline and liquid states of matter (Balescu 1975; Frenkel 1976; March and Tosi 2002). The second approach is based on a complete statistical calculation of the properties of non-ideal media by the method of molecular dynamics using model data on the energy of particle interaction (Young and Alder 1971; March and Tosi 2002). This simulation procedure enables one to study diverse physical phenomena (phase transitions, thermal diffusion of particles, viscosity and thermal conductivity, the dynamics of the system approaching the equilibrium state, etc.). Numerical simulations of the dynamical properties of non-ideal systems are of great importance from the standpoint of the theory of liquids because, due to strong interparticle interaction, no small parameter is present in such systems which could be used for analytical description of the state of the system and thermodynamic characteristics, as it is possible in the case of gas.

Two well-known numerical algorithms are usually employed for analyzing the transport characteristics of systems of interacting particles, namely, the Monte Carlo method and the method of molecular dynamics (MMD). As distinct from the Monte Carlo method, which was developed for the calculation of equilibrium values, the MMD enables one to describe the approach of the system under investigation to the state of equilibrium. Therefore, the MMD is an indispensable tool for use in studying the processes of heat and mass transfer and wave propagation, and the dynamics of formation of instabilities. This method is based on the solution of a set of ordinary differential equations, i.e., equations of motion of particles in the field of various forces. Within this approach, we can identify the method of molecular dynamics based on the integration of invertible equations of motion of particles (MIM, the method of invertible motion) and the method of Brownian dynamics (or Langevin dynamics) (MBD) based on the solution of Langevin equations and taking into account the irreversibility of the processes under investigation. In the former case (MIM), only the elastic interactions of particles are taken into account ignoring the dissipation (friction) and other processes of energy exchange between particles and surrounding medium (thermostat). The motion of particles in such a system is not stable, and procedures of re-normalization of the calculation data are employed after a certain number of integration steps for maintaining the equilibrium temperature of the particles. This approach is adequate to simulate processes in atomic systems, but cannot be employed for analyzing the motion of macroparticles in a laboratory plasma where the dissipation due to collisions with gas atoms or molecules plays an important role.

Unlike the MIM, the MBD takes into account the loss of kinetic energy of particles due to friction forces, and the equilibrium state of a system with constant temperature is maintained due to its exchange of energy with the thermostat. This exchange is modeled by a random force F_{ran} correlated with the friction forces in the system using the fluctuation-dissipation theorem (Landau and Lifshitz 1980; Lichtenberg and Lieberman 1992). The special importance of the MBD in simulating the dynamics of particles in dusty plasma is based on the fact that the Langevin equation enables one to take into account the interactions between dust particles and "thermostat particles", which maintain statistical equilibrium in the system. This equilibrium is observed in numerous experimental situations, where the Maxwellian distributions of the velocities of dust particles are reported. In so doing, the MBD enables one to take into account the processes of energy exchange between the particles and surrounding medium both due to their collisions with molecules of surrounding gas and due to other stochastic processes, for example, fluctuations of the particle charges, which cause an increase in the kinetic temperature of the particle system relative to the gas temperature (Zhakhovskii *et al.* 1997; Vaulina *et al.* 1999).

7.1.2 Equations of motion of dust particles

Along with random forces F_{ran}, which are the sources of thermal motion of particles, the forces of pair interparticle interaction F_{int} are taken into account in a set of N_d equations of motion (where N_d is the number of particles) for simulating equilibrium microscopic processes in uniform extended clouds of interacting particles,

$$m_d \frac{d^2\mathbf{r}_k}{dt^2} = \sum_j F_{int}(r)|_{r=|\mathbf{r}_k-\mathbf{r}_j|} \frac{\mathbf{r}_k - \mathbf{r}_j}{|\mathbf{r}_k - \mathbf{r}_j|} - m_d \nu_{fr} \frac{d\mathbf{r}_k}{dt} + \mathbf{F}_{ran}. \tag{7.1}$$

Here, $F_{int}(r) = -\partial U/\partial r$, $r = |\mathbf{r}_k - \mathbf{r}_j|$ is the interparticle spacing, U is the potential energy of pair interaction, m_d is the particle mass, and ν_{fr} is the coefficient of friction of dust particles due to their collisions with surrounding medium (mostly with neutrals). Under conditions of the local thermodynamic equilibrium, the average value of the random force is $< F_{ran} >= 0$, and the autocorrelation func-

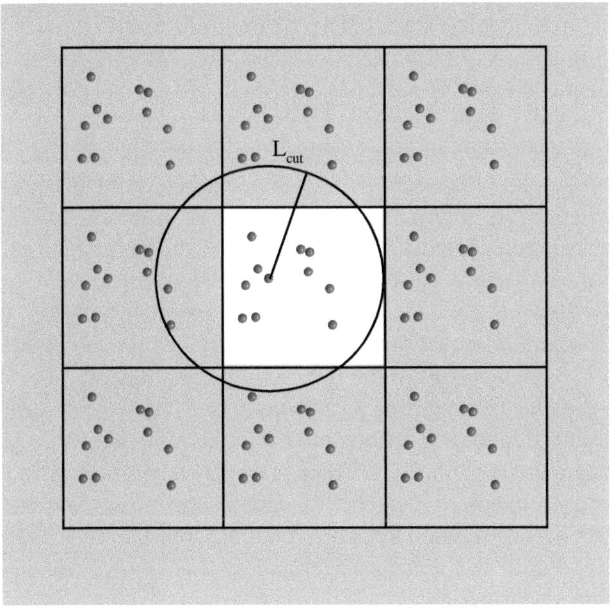

FIGURE 7.1
Sketch of numerical simulation procedure using periodic boundary conditions.

tion $< F_{ran}(0)F_{ran}(t) >= 6Tm_d\nu_{fr}\delta(t)$ describes the delta-correlated Gaussian process (Cummins and Pike 1974; Ovchinnikov *et al.* 1989). Such stochastic processes can be simulated using random increments of momentum of particles $p^x_{ran} = (2T\nu_{fr}\Delta t m_d)^{1/2}\psi$, where p^x_{ran} is the momentum increment by one degree of freedom, ψ is some random quantity distributed by normal law with mean-square deviation equal to unity, and Δt is the time integration step for Equations (7.1). For correct integration of random forces, the integration step Δt has to be short enough. Periodic boundary conditions in directions *x*, *y* and *z*, which enable one to maintain constant the number of particles and their average kinetic energy, are usually used for studying equilibrium processes in extended three-dimensional dust systems. Such conditions can be realized by way of simulating 27 identical cubic cells, the space positions of particles in which are maintained similar to their position in the central cell at each instant of count time (see Figure 7.1). Upon crossing any boundary of the central cell, the particle comes back at the velocity it escapes from the cell, but from the opposite side. For simulating the dynamics of particles in extended dust layers formed, for example, in the sheath region of an rf discharge, periodic boundary conditions in the two selected directions (nine computational cells) are used. In the remaining direction the effect of balanced external forces is usually used.

Numerical experiment usually proceeds as follows: Initially the particles are located randomly within the central cell. Then, due to interaction between them, the process of self-organization starts. After the equilibrium (for the given system parameters) configuration of particles is reached, the dynamic characteristics (the velocities and displacements of particles) are analyzed for the particles in the central cell.

In calculating the forces of interparticle interaction, all particles of the complete system of 27 (or nine, in the case of two-dimensional problem) cells are taken into account. However, the interparticle interaction is often cut-off at some distance $r = L_{cut}$. For example, for the Yukawa potential the computational cell size L is defined by the condition $L \gg \lambda_D$ (Farouki and Hamaguchi 1992) and the cut-off length usually does not exceed several interparticle distances L_{cut} = (4–8)Δ. This corresponds to the number of "independent" particles (in the central cell) ranging from 64 to 512. Such a cut-off of the interaction potential does not result in a significant error in the case of screening parameter $\kappa = \Delta/\lambda_D > 1$. In simulating systems with $\kappa < 1$, longer-range interactions must be taken into account. This can be realized using an appropriate algorithm which is based on constructing a large number of translation cells (Hamaguchi *et al.* 1997). This method, however, was developed for studying the properties of crystals and is far from always acceptable for simulating the dynamics of liquid systems characterized by the absence of long-range order in the arrangement of particles.

In order to simulate the dynamics of a finite system of particles contained in a trap formed, for example, by electric fields $\mathbf{E}_{ext}$ of a gas-discharge chamber, the force term describing this electric confinement as well as other forces which can be present in the system (e.g., gravity, ion drag force, thermophoretic force, etc.) should be added to Equation (7.1).

To conclude this section let us derive the scaling parameters of the equations of motion. Dimensionless equations of motion are often employed for simulating the dynamics of particles; these equations enable one to avoid repeated calculations of the properties of systems with the same self-similar properties. By way of example, consider the re-normalization of equations of motion (7.1) for a system of particles interacting via screened Coulomb (Yukawa) pair potential. Using the mean interparticle spacing $\Delta = n_d^{-1/3}$ as a unit of distance, the inverse dust plasma frequency $\omega_{pd}^{-1} = (4\pi Q^2 n_d/m_d)^{-1/2}$ as a unit of time, and the thermal velocity of particles $v_{Td} = (T_d/m_d)^{1/2}$ as a velocity unit, the normalized equations of motion of particles in projection into Cartesian coordinate axis OX have the form

$$\dot{X}_k = (4\pi\Gamma)^{-1/2} V_k, \tag{7.2}$$

$$\dot{V}_k = \sum_{j \neq k} (4\pi\Gamma)^{-1/2} \frac{1 + \kappa R_{kj}}{R_{kj}^3} \exp(-\kappa R_{kj})(\mathbf{R}_{kj}\mathbf{e}_X) - \xi V_k + \sqrt{2\xi} f(\tau), \tag{7.3}$$

where X_k and V_k denote dimensionless coordinate and velocity of the k-th particle, τ is dimensionless time, $\mathbf{R}_{kj} = \mathbf{R}_k - \mathbf{R}_j$, $\mathbf{e}_X$ is the unit vector in the direction of the axis OX, $\xi = v_{fr}/\omega_{pd}$, and $f(\tau)$ is the delta-correlated white noise: $< f(\tau) > = 0$ and $< f(0) f(\tau) > = \delta(\tau)$. Therefore, the behavior of the system under consideration is defined by three dimensionless parameters Γ, κ, and ξ. In the frictionless ($\xi = 0$)

limit, the method of molecular dynamics (MIM) is adequate and the system is characterized only by the non-ideality parameter Γ and the screening parameter κ. As discussed above, in this case the periodic re-normalization of particle velocities must be employed for maintaining a constant temperature of the system because of free exchange between the potential and kinetic energies. For finite ξ, modified MBD can be used. In this case the system temperature is defined by the parameters of the Langevin force. It is maintained constant automatically and requires no correction during simulation.

7.2 Numerical simulation of spatial correlations between dust particles

7.2.1 Pair and three-particle correlation functions

Laboratory dusty plasma, which exhibits a number of unique properties, is a good experimental model both for studying the properties of highly non-ideal plasma and for better understanding the phenomena of self-organization of matter in nature. Experimental investigations of dusty plasmas can play an important role in verifying the existing models and developing new phenomenological models for highly non-ideal liquid systems. Such models are of great importance because, due to strong interparticle interaction, no small parameter is present in the theory of liquid which could be used for analytical description of the state of the liquid system and thermodynamic characteristics, as is possible for gases (Balescu 1975; Frenkel 1976; March and Tosi 2002).

The equilibrium properties of liquid are fully described by a set of probability density functions $g_s(\mathbf{r}_1, \mathbf{r}_2, \ldots \mathbf{r}_s)$ of the location of particles at points $\mathbf{r}_1$, $\mathbf{r}_2$, $\ldots \mathbf{r}_s$. In the case of isotropic pair interaction, the physical properties of liquid (such as pressure, energy density, and compressibility) are fully defined by the pair correlation function $g(r) = g_2(|\mathbf{r}_1 - \mathbf{r}_2|)$ (Ailawadi 1980) which, in turn, depends on the type of interaction potential between the particles and on the particles' temperature (Ichimaru 1982),

$$g(r) = \exp\{-[U(r)/T_d] + N(r) + B(r)\}, \tag{7.4}$$

where r is the distance between two particles, $U(r)$ is the potential energy of pair interaction, T_d is the kinetic energy of random (thermal) motion of particles, $N(r)$ is defined by the functions $g_1(\mathbf{r}_1)$ and $g(r)$, and $B(r)$ takes into account higher order correlations and is a complex integral function of $g_s(\mathbf{r}_1, \mathbf{r}_2, \ldots, \mathbf{r}_s)$ at $s > 2$. In the hypernetted-chain approximation it is assumed that $B(r) = 0$, and the Ornstein–Zernike relation is used to determine $N(r)$ (Ailawadi 1980; Ichimaru 1982). However, the available results of numerical investigations demonstrate the inadequacy of the hypernetted-chain approximation even in the case of weak non-ideality of the systems under investigation (Raverche and Mountain 1972; Raverche *et al.* 1972; Wang and Crumhansr 1972; Ailawadi 1980; Ichimaru 1982). It is only the inclusion of higher order correlations $[B(r) \neq 0]$ that produces results which coincide with the data of numerical simulation (Ichimaru 1982). Therefore, determining the form of $g(r)$ in the general case requires information both about the type of the pair interaction potential $U(r)$ and about the behavior of correlation functions $g_s(\mathbf{r}_1, \mathbf{r}_2, \ldots, \mathbf{r}_s)$ at $s > 2$, or it requires the use of some approximations for these functions. Within the model of one-component plasma, the approximating functions for deviation of the effective energy of interacting particles from the energy of their Coulomb interaction $U(r) = Q^2/r$ at distances shorter than the mean interparticle spacing $\Delta = n_d^{-1/3}$ can be written in the form (Ichimaru 1982)

$$[N(r) + B(r)]T = Q^2(1.25 - 0.39\frac{r}{r_s})/r_s \quad \text{at} \quad 0.4 < r/r_s < 1.8 \tag{7.5}$$

$$[N(r) + B(r)]T = Q^2[1.057 - 0.25(\frac{r}{r_s})^2]/r_s \quad \text{at} \quad r/r_s < 0.2, \tag{7.6}$$

where $r_s = (4\pi n_d/3)^{-1/3}$ is the Wigner-Seitz radius. Therefore, even in the approximation of pair interparticle interaction, correlation functions of higher order ($s > 2$) are of significant importance. Information about the three-particle correlation function $g_3(\mathbf{r}_1, \mathbf{r}_2, \mathbf{r}_3)$ is of importance in calculating the physical characteristics of the medium, which depend on derivatives of the pair correlation function $g(r)$ with respect to temperature $\partial g(r)/\partial T$ or density $\partial g(r)/\partial \rho$ (entropy, coefficients of thermal expansion, etc.). The three-particle correlation function $g_3(\mathbf{r}_1, \mathbf{r}_2, \mathbf{r}_3)$ defines the probability of simultaneous detection of three particles in the vicinity of points $\mathbf{r}_1$, $\mathbf{r}_2$, and $\mathbf{r}_3$. The superposition approximation (Kirkwood relation) is most frequently used for the approximation of the three-particle correlation function,

$$g_3(\mathbf{r}_1, \mathbf{r}_2, \mathbf{r}_3) \simeq g_3^{sp}(\mathbf{r}_1, \mathbf{r}_2, \mathbf{r}_3) = g(\mathbf{r}_1 - \mathbf{r}_2)g(\mathbf{r}_2 - \mathbf{r}_3)g(\mathbf{r}_3 - \mathbf{r}_1). \tag{7.7}$$

Relation (7.7) is often employed for calculating integral equations in the kinetics of interacting particles, as well as in recovering interparticle interaction potentials, by methods based on the use of the hypernetted-chain approximation or the Percus–Yevick equation (Raverche and Mountain 1972; Raverche *et al.* 1972; Wang and Crumhansr 1972; Ailawadi 1980; Ichimaru 1982). Nevertheless, the available results of numerical investigations performed for the case of hard sphere interaction and for particles interacting with potentials of the Lennard–Jones type demonstrate the inadequacy of the superposition approximation even in the case of low densities of particles (Raverche and Mountain 1972; Raverche *et al.* 1972; Wang and Crumhansr 1972; Ailawadi 1980; Ichimaru 1982). As the non-ideality of the liquid systems under investigation increases, the difference of the approximation $g_3^{sp}(\mathbf{r}_1, \mathbf{r}_2, \mathbf{r}_3)$ from the exact value $g_3(\mathbf{r}_1, \mathbf{r}_2, \mathbf{r}_3)$ can be as significant as $\sim$100%. The experimental verification of superposition approximation for real liquids and gases is difficult because no direct determination of the three-particle correlation function is possible in the absence of information about coordinates of individual particles. Analysis of three-particle correlations in real liquids involves the use of indirect diagnostic methods,

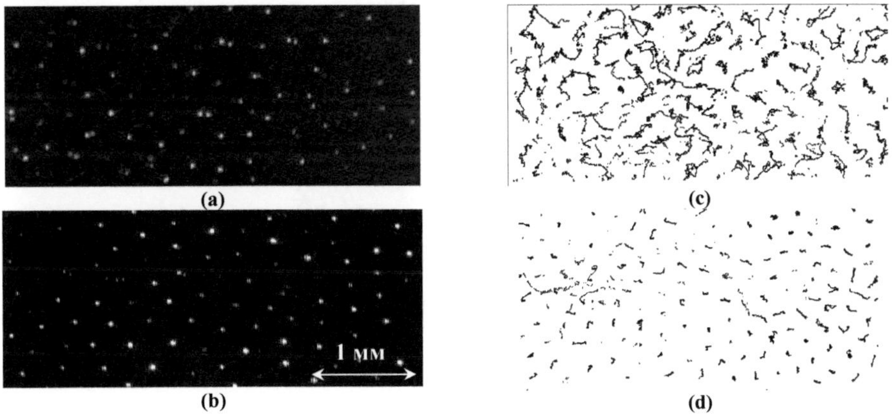

FIGURE 7.2
Video images of illuminated particle layer (a, b) and trajectories of motion of particles (c, d) for different experiments: (a, c) p = 3 Pa, W = 10 W; (b, d) p = 7 Pa, W = 10 W.

for example, measurements of the structure factor $S(k)$ for several values of pressure of the medium at a constant temperature, whence $\partial g(r)/\partial n$ is recovered; this latter derivative, in turn, contains information about three-particle correlation function (Raverche and Mountain 1972; Raverche *et al.* 1972). The extraction of such information calls for additional data on isothermal compressibility of the medium under investigation.

Unlike real liquids and gases, laboratory dusty plasma is a good experimental model for studying the physical properties of non-ideal systems because, due to the size of dust particles, their coordinates can be videofilmed which greatly simplifies the use of direct contactless methods for their diagnostics. An experimental investigation of three-particle correlations in a complex plasma was performed by Vaulina *et al.* (2004a) in an rf discharge with a setup shown schematically in Figure 1.3. This experiment was performed in argon gas at a pressure between $p = 2$ Pa and $p = 10$ Pa and discharge power between $W \simeq 2$ and $W \simeq 10$. The particles used were monodisperse latex spheres of radius $a \simeq 1.7$ μm and mass density of $\rho \simeq 1.5$ g cm^{-3}. Under the experimental conditions, the particles formed four to ten flat horizontal layers in a sheath region above the lower discharge electrode. The particle cloud was illuminated by a laser knife ($\sim$200–300 μm thick) and the scattered light was recorded by a video camera. Fragments of video images of the particles registered in the illuminated particle layer are shown in Figures 7.2a and 7.2b for different experimental conditions. The observed particle structures were of the liquid-like type with the mean interparticle spacing Δ ranging from 260 to 350 μm. The processing of video records produced pair correlation functions $g(r)$ and three-particle correlation functions $g_3(r_{12}, r_{23}, r_{31})$ ($r_{ij} = |\mathbf{r}_i - \mathbf{r}_j|$) averaged over a period of $\sim$1–2.5 s under

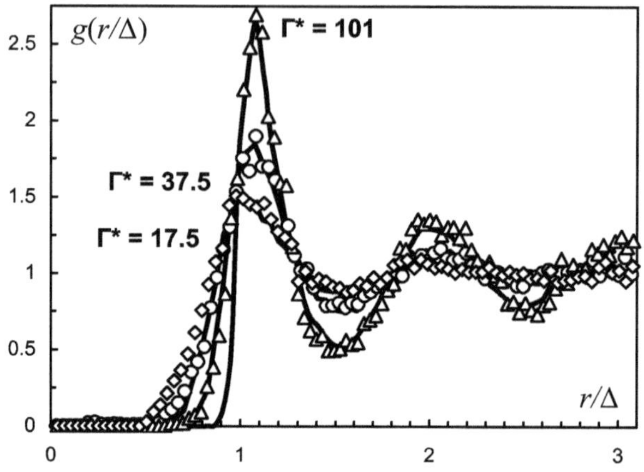

FIGURE 7.3
The pair correlation functions of particles in an rf discharge plasma $g(r/\Delta)$ measured experimentally for different plasma parameters: $p = 3$ Pa, $W = 10$ W (diamonds); $p = 3$ Pa, $W = 2$ W (circles); $p = 7$ Pa, $W = 10$ W (triangles). Solid curves correspond to numerical simulations for different values of the modified coupling parameter Γ^* (the values are shown in the figure).

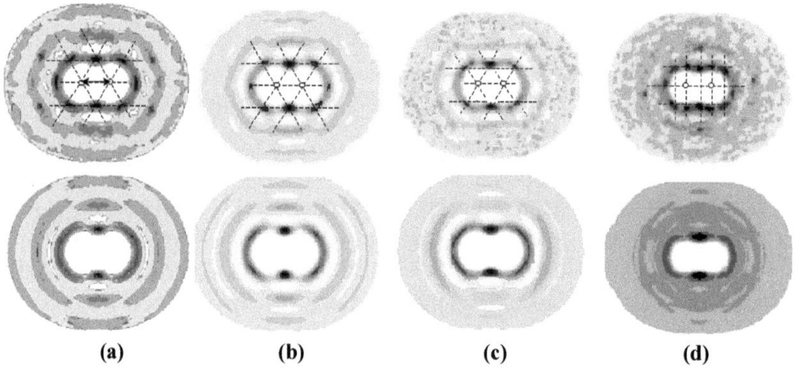

FIGURE 7.4
The measured correlation functions g_3 (top row) and superposition approximation g_3^{sp} (bottom row) for different experimental conditions: (a) $p = 7$ Pa, $W = 10$ W, $\delta = 0.61$; (b) $p = 3$ Pa, $W = 2$ W, $\delta = 0.28$; (c) $p = 3$ Pa, $W = 10$ W, $\delta = 0.2$; (d) $p = 5$ Pa, $W = 9$ W, $\delta = 0.3$.

constant experimental conditions. The trajectories of motion of the particles during the time of averaging of pair correlation functions are shown in Figures 7.2c and 7.2d.

The pair correlation functions $g(r/\Delta)$ are given in Figure 7.3 for different values of gas pressure p and discharge power W. The cross sections of the obtained three-particle correlation functions $g_3(r_{12}, r_{23}, r_{31})$ ($r_{ij} = |\mathbf{r}_i - \mathbf{r}_j|$) with a fixed value of r_{12}, which is equal to that of the most probable interparticle spacing r^{max} ($r_{12} = r^{max}$) determined by the position of the maximum of the pair correlation function $g(r)$, are shown in Figure 7.4 for the same discharge parameters. Also shown in this figure are the results of calculation of the three-particle function $g_3^{sp}(r_{12}, r_{23}, r_{31})$ within superposition approximation (7.7). In order to represent these functions $g_3(r_{12}, r_{23}, r_{31})$ and $g_3^{sp}(r_{12}, r_{23}, r_{31})$ in a descriptive "two-dimensional" form convenient for comparison, they were normalized to the value of the maximum of $g_3(r_{12}, r_{23}, r_{31})$: black color corresponds to unity, and white color corresponds to $g_3 = g_3^{sp} \equiv 0$. The deviation of the function $g_3^{sp}(r_{12}, r_{23}, r_{31})$ from the results of calculation of $g_3(r_{12}, r_{23}, r_{31})$, which is given in the caption of Figure 7.4, was calculated proceeding from the relative mean-square error of superposition approximation,

$$\delta = \frac{1}{N^{1/2}} \left[\sum_{i=1}^{N} \left(g_3(r_{12}, r_{2i}, r_{i1}) - \frac{g_3^{sp}(r_{12}, r_{2i}, r_{i1})}{g_3(r_{12}, r_{2i}, r_{i1})} \right)^2 \right]^{1/2}, \tag{7.8}$$

where N is the total number of space elements $d\mathbf{r}_i$ in the neighborhood of a point with coordinate $\mathbf{r}_i$, into which the dust layer being analyzed was broken up.

The pair correlation functions obtained as a result of numerical simulation of systems of particles interacting via the screened Coulomb (Yukawa) potential are shown in Figure 7.3 for different values of the effective non-ideality parameter

$$\Gamma^* = Q^2/(T\Delta)(1 + \kappa + \kappa^2/2)\exp(-\kappa), \tag{7.9}$$

which defines the phase state of Yukawa systems (Vaulina and Vladimirov 2002; Vaulina *et al.* 2002; Vaulina and Petrov 2004), as discussed in Section 5.1. The results of measurements of $g(r)$ agree well with the systems being simulated with $\Gamma^* \sim 100$, 38, and 18. The calculation of three-particle correlation functions and their superposition approximations for some non-ideality parameters are illustrated in Figure 7.5. Note the good agreement between the cross sections of three-particle correlation functions for numerical and experimental data in the cases where similar agreement is observed between the shapes of the pair correlation functions.

Visual analysis of the results reveals that the recorded particle structures exhibit short-range orientational order. This is reflected in the emergence of maxima of $g_3(r_{12}, r_{23}, r_{31})$ in the nodes of hexagonal clusters shown by dashed lines in Figures 7.4a–c. As the maximum of the pair correlation function increases (see Figure 7.3), the magnitude of these maxima located at distances r close to r^{max} increases, and new maxima arise at distances $r \simeq 2r^{max}$. This effect does not show up when the superposition approximation $g_3^{sp}(r_{12}, r_{23}, r_{31})$ is analyzed. The formation of regular dust clusters is observed as a result of the increase in the effective coupling parameter (at $\Gamma^* > 25$) when the dynamics of the liquid system become similar to those of a solid and can be treated within the "theory of jumps" developed for describing strongly correlated molecular liquids (Frenkel 1976; Vaulina and Vladimirov 2002).

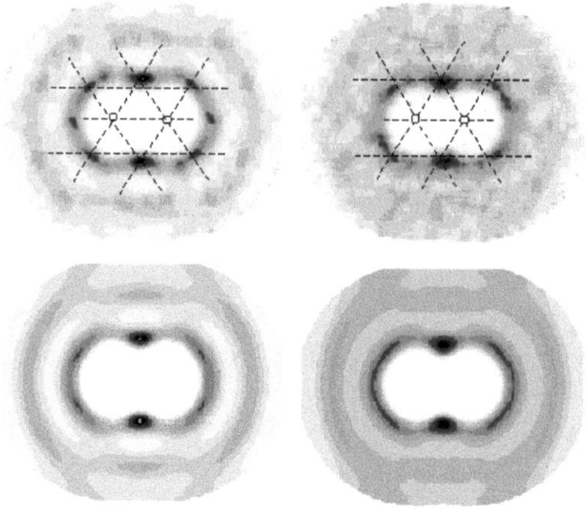

(a) (b)

FIGURE 7.5
Three-particle correlation functions g_3 (top) and superposition approximation g_3^{sp} (bottom) obtained in numerical simulations for different values of the effective coupling parameter Γ^*: (a) $\Gamma^* = 37.5$, $\delta = 0.62$; (b) $\Gamma^* = 17.5$, $\delta = 0.6$.

In so doing, the orientational number (which defines the number of nearest neighbors) becomes equal to the number of nearest neighbors in lattices of the crystal type. Therefore, the particle subsystem melts with the type of its packing retained. The majority of metals melt in a similar manner.

The results of experimental investigations of three-particle correlations can be used for qualitative structure analysis of non-ideal systems and enable one to readily detect the presence of clusters of different shapes (similar to the cross sections of cubic lattices of different types) in the dust layer being analyzed (see Figure 7.4d). The cross sections of different crystal lattices (on the face of elementary cubic cell), as well as the functions $g_3(r^{max}, r_{12}, r_{23}, r_{31})$ and $g_3^{sp}(r^{max}, r_{12}, r_{23}, r_{31})$ for $\Gamma^* \sim 400$ are illustrated in Figure 7.6.

7.2.2 Pair correlation functions and phase states of the particle subsystems

Pair correlation functions $g(r)$ are most frequently employed for quantitative analysis of the phase state and space order of a particle system in plasmas. Numerical simulation by Vaulina and Petrov (2004) reveals that, for a wide range of isotropic pair potentials $U(r)$, the space correlation of particles in non-ideal systems is defined by the ratio of the second derivative U'' of potential at a point of mean interparticle spacing Δ to the particle temperature T and does not depend on friction. In systems of particles interacting via the screened Coulomb (Yukawa) potential with $\kappa \lesssim 6$, the

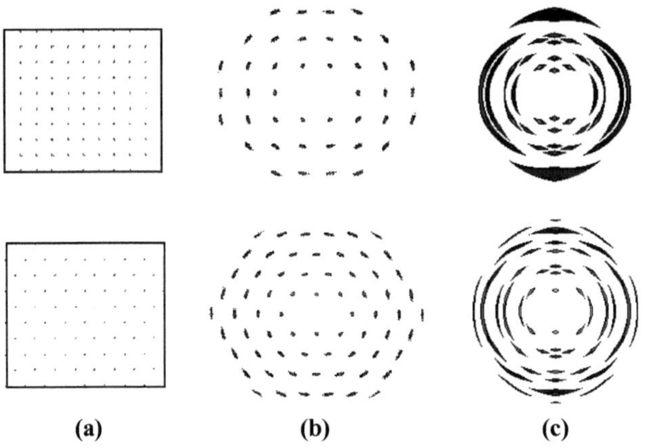

FIGURE 7.6
Cross sections of (a) crystal cubic lattice and of functions (b) g_3 and (c) g_3^{sp} calculated for these cross sections. Top row – the cross section of a bcc lattice; bottom row – the cross section of the fcc lattice (δ = 0.4), which corresponds to hexagonal arrangement of particles (δ = 0.8).

effective coupling parameter Γ^* fully defines the shape of pair correlation function $g(r)$ from $\Gamma^* \sim 1$ to the crystallization point of the system, where a body centered cubic (bcc) crystal structure is formed at $\Gamma^* \simeq 106$ (Vaulina and Vladimirov 2002; Vaulina *et al.* 2002; Vaulina and Petrov 2004). An example of pair correlation functions for such systems with different parameters Γ^* is shown in Figure 7.3. The dependencies of the first maxima (g_1, S_1) of the pair correlation functions $g(r)$ and the structure factor $S(q)$ and the positions of these maxima ($r = d_{g1}$, $k = d_{S1}$), as well as of the ratio g_1/g_{min} (here, $g_{\min} = \min[g(r)]$ at $r \neq 0$), on the parameter Γ^* are shown in Figures 7.7–7.9 for different system parameters. An abrupt increase (jumps) in the values of the first maxima of correlation functions $g(r)$ and $S(q)$ (from ~ 2.65 to ~ 3.1, Figure 7.7), as well as of the ratio g_1/g_{min} (from ~ 5 to ~ 7, Figure 7.9), is observed between $\Gamma^* \simeq 100$ and $\Gamma^* \simeq 110$, where crystallization occurs. Such behavior is in good agreement with the well-known criterion of crystallization suggested by Hansen and Verlet (1969), which defines the value of the first maximum S_1 for the liquid structure factor as being lower than 2.85, as well as with another empirical criterion according to which the ratio of the maximal value of pair correlation function to its value at the first minimum on the crystallization line of different systems is close to $\simeq 5$. Because this value does not depend on the viscosity of surrounding gas, this criterion agrees with the results of calculation of crystallization of non-dissipative systems by methods of molecular dynamics, which do not take into account the friction of particles (coefficient of friction of particles $\nu_{fr} = 0$) (Van Horn 1969; Robbins *et al.* 1988; Meijer and Frenkel 1991; Hamaguchi *et al.* 1997).

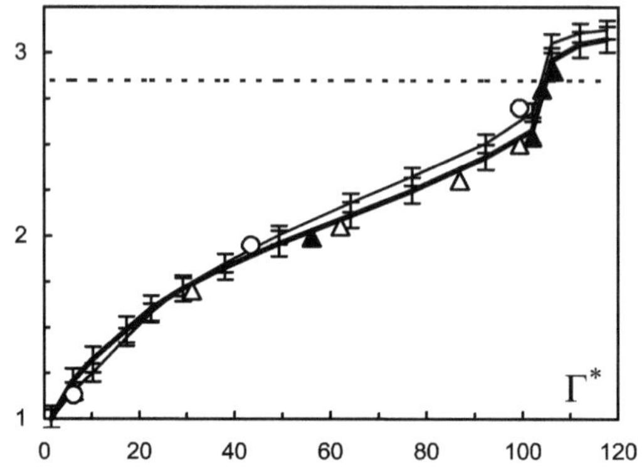

FIGURE 7.7
The maximums of the structure factor (S_1, fine line) and the pair correlation function (g_1, bold line) as a function of the effective coupling parameter Γ^* for dissipative Yukawa systems (Vaulina and Vladimirov 2002; Vaulina *et al.* 2002; Vaulina and Petrov 2004), as well as for non-dissipative systems (ν_{fr} = 0): solid triangles indicate the values of g_1 for Yukawa systems (Robbins *et al.* 1988); open triangles correspond to the values of g_1 in the one-component plasma (OCP) model (Ichimaru 1982); circles show the values of S_1 in the OCP model (Ichimaru 1982). Horizontal dashed line corresponds to S_1 = 2.85.

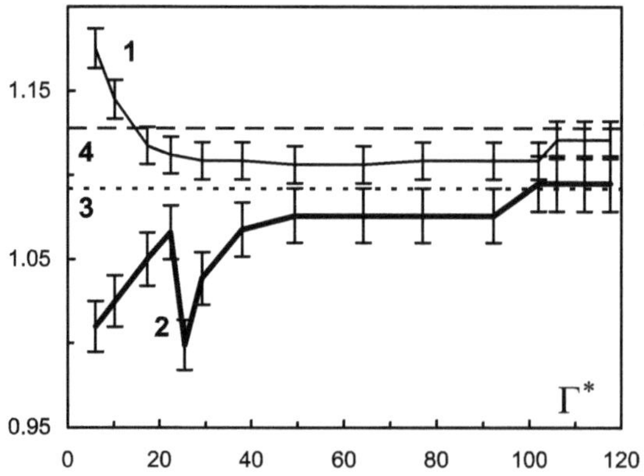

FIGURE 7.8
The relative position k_{S1}/k_Δ (1, fine line) of S_1 and the relative position d_{g1}/Δ (2, bold line) of g_1 as functions of the effective coupling parameter Γ^* (here $k_\Delta = 2\pi/\Delta$). Dashed lines indicate the positions of maximum of correlation functions for a bcc lattice: 3, $d_{g1} = (3\sqrt{3}/(4n_d))^{1/3}$; 4, $k_{S1} = 2\pi(\sqrt{2}n_d)^{1/3}$.

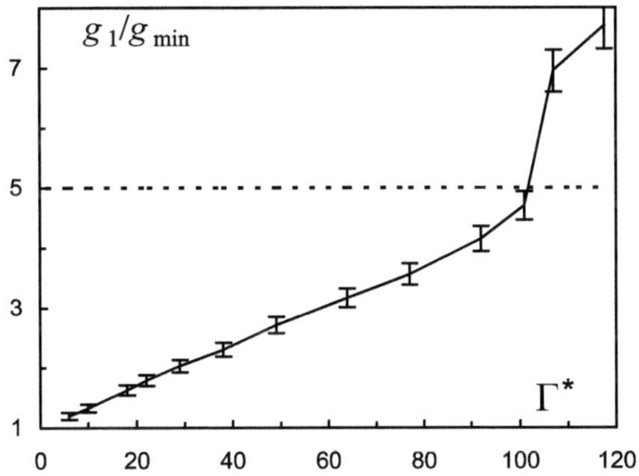

FIGURE 7.9
The ratio g_1/g_{min} as a function of the effective coupling parameter Γ^*.

7.3 Transport properties of complex plasma: Numerical study

7.3.1 Transport of particles in non-ideal media

The problems associated with transport processes in dissipative systems of interacting particles are of significant interest in various fields of science and engineering (like hydrodynamics, plasma physics, medicine, physics and chemistry of polymers, etc.) (Cummins and Pike 1974; Balescu 1975; Frenkel 1976; Ovchinnikov *et al.* 1989; March and Tosi 2002). The major problem encountered in studying the physical properties of such systems is associated with the absence of an analytical theory of liquid that would be capable of explaining its thermodynamic properties, giving the equation of state, describing the effects of heat and mass transfer, etc. As was already mentioned, two basic approaches are employed at present in developing the theory of highly non-ideal systems and approximate models for the description of the liquid state of matter. The first approach involves the complete statistical calculation of the properties of non-ideal media using model data on the energy of particle interaction (Balescu 1975; Frenkel 1976; March and Tosi 2002). The second approach is the semiempirical method of determining the correlation of the parameters of liquid with one another and with the properties of crystals, which is based on analogies between the crystalline and liquid states of matter (Young and Alder 1971; March and Tosi 2002).

The semiempirical method is based on the so-called "theory of jumps". This theory is based on the assumption that the molecules of liquid are in an equilibrium ("settled") state for a period of time required to acquire an energy (activation energy) sufficient to overcome the potential barriers related to the the presence of the neighboring molecules and to move ("jump") to a new "settled" state. Therefore, one can assume that a diffusing (active) particle is capable to migrate by performing a jump to one of the equivalent positions in an imaginary lattice. The random walk of an active particle over the "lattice" sites after a large number of jumps is described by equations of macroscopic diffusion (with some time-independent coefficient D), i.e., Fick's laws are valid (Cummins and Pike 1974; Ovchinnikov *et al.* 1989). However, the existing level of experimental physics dictates the need for going beyond the limits of diffusion approximation. In particular, the description within macroscopic kinetics may turn out to be insufficient for analyzing the processes occurring on short time scales. The investigation of processes of mass transfer over short times of observation is of special importance from the standpoint of studying fast processes, such as the propagation of shock waves, pulsed action, or the motion of the front of chemical transformations in condensed media (Cummins and Pike 1974; Ovchinnikov *et al.* 1989), as well as from the standpoint of analyzing the transport characteristics of highly dissipative ($\xi \gg 1$) media, such as colloidal solutions, plasma of combustion products, nuclear-induced plasma at atmospheric pressures (Fortov *et al.* 1996, 1999), where the correct measurement of diffusion coefficients of particles calls for long-term experiments.

The use of hydrodynamic approaches results in a successful description of particle transport only in the case of short-range interactions. When the interparticle interactions are not as weak as in the case of gases, one fails to construct a correct kinetic equation. The fundamental theories of transport are based on the fact that the particle number density of each component of the system under consideration is a hydrodynamic variable which slowly varies in space and time. Such systems are in the state of statistical equilibrium and can be characterized by a certain set of physical parameters, for example, concentration, kinetic temperature, and pressure, which can experience only slight fluctuations around their average equilibrium values. In statistical physics, such a state is described using various Gibbs distributions depending on the type of contact of the system with environment (thermostat), which prohibits or allows the exchange of energy or particles with this environment; Nyquist formulas, Green functions, and the fluctuation-dissipative theorem are used for analyzing equilibrium fluctuations and transport coefficients (Young and Alder 1971; Cummins and Pike 1974; Balescu 1975; Frenkel 1976; Landau and Lifshitz 1980; Ovchinnikov *et al.* 1989; Lichtenberg and Lieberman 1992; Zhakhovskii *et al.* 1997; Vaulina *et al.* 1999; March and Tosi 2002).

The transport coefficients (such as the coefficients of diffusion D, thermal conductivity χ, viscosity η) characterize the thermodynamic state of the system and reflect the nature of interparticle interaction. For gases the coefficients of self-diffusion, kinematical viscosity $\nu = \eta/\rho$, and thermal diffusivity $\theta = \chi/(\rho c_p)$ are on the same order of magnitude and can be approximated by analytical expressions (Frenkel 1976; March and Tosi 2002). Here $\rho = m_d n_d$, and c_p is the specific heat capacity under constant pressure. The presence of such relations for the liquid state of matter enables one to use the known hydrodynamic models for analyzing the propagation of waves, shear flows, formation of vortexes and various instabilities in highly non-ideal media.

The numerical simulation of transfer processes in simple monatomic liquids with a wide range of interaction potentials reveals that, when the particle mean free path l_{d-d} is comparable to the mean interparticle distance Δ, the transport coefficients can be approximated by the following relations (March and Tosi 2002):

$$D \simeq 0.6\Delta v_{T_d} \exp(-0.8s), \tag{7.10}$$

$$\eta \simeq 0.2\rho\Delta v_{T_d} \exp(0.8s), \tag{7.11}$$

$$\chi \simeq (1.5k_B/m_d)\Delta v_{T_d} \exp(0.5s). \tag{7.12}$$

Here, $v_{T_d} = (T_d/m_d)^{1/2}$, and s is proportional to configurational entropy.

For highly correlated liquids ($l_{d-d} \ll \Delta$), the coefficients $D \propto \exp(-W/T_d)$ and $\eta \propto \exp(W/T_d)$ acquire an exponential dependence on inverse temperature. Here, W is the activation energy (Young and Alder 1971; Frenkel 1976; March and Tosi 2002). Simulations of transport processes in systems with isotropic pair potentials reveal that the correlation between the coefficients D and η is independent of the degree of correlation between the particles and obeys the Stokes formula, $\eta \simeq T_d/(8\Delta D)$ (Young and Alder 1971; March and Tosi 2002). One can assume that this agreement is observed for the coefficients of diffusion and thermal conductivity of such liquids. Then, in view of Equations (7.10) and (7.12), the coefficient $\theta = \chi/(\rho c_p)$ can be written as (Fortov *et al.* 2005b)

$$\theta \simeq \frac{1.5k_B}{m_d c_p}\Delta V_t \left(\frac{0.6\Delta v_{Td}}{D}\right)^{5/8}, \tag{7.13}$$

where k_B is the Boltzmann constant.

The main difference between the properties of simple liquids and complex plasmas is associated with the presence of dissipation, mostly due to particle–neutral collisions. In the limit of very low dissipation the results of simulations of the dynamics of particles in plasmas and simple liquids should coincide. The normalized coefficients of diffusion $D^*(\Gamma^*) = D/(\omega^*\Delta^2)$ and viscosity $\eta^*(\Gamma^*) = \eta/(\omega^*\Delta^2)$ for this case are shown in Figure 7.10 for the simulated systems of particles interacting via the screened Coulomb (Yukawa) potential. Here $\omega^* = \sqrt{T_d\Gamma^*/\pi\Delta^2 m_d}$ is the effective frequency associated with the interacting particle component. Also shown in Figure 7.10 is the normalized coefficient of thermal diffusivity $\theta^* = \theta/(\omega^*\Delta^2)$ determined by the data of simulation of particle diffusion using relation (7.14) of the following section for $c_p = 2.5k_B/m_d$.

7.3.2 Diffusivity

The diffusion of particles is the main process of mass transfer, which defines the energy loss (dissipation) in systems under investigation. In the case of small deviations of such a system from statistical equilibrium, the diffusion coefficient D of particles is described by the relation which is a particular case of Green–Kubo formulas (Young and Alder 1971; Cummins and Pike 1974; March and Tosi 2002),

$$D = \frac{1}{3}\int_0^\infty \langle V(0)V(t)\rangle dt, \tag{7.14}$$

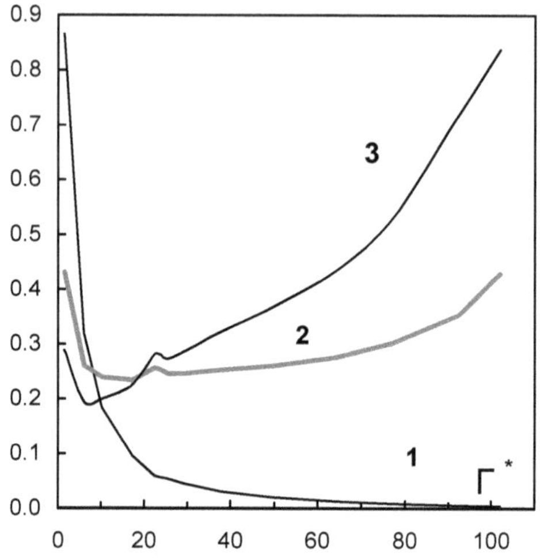

FIGURE 7.10
Normalized transport coefficients D^* (1) obtained by Ohta and Hamaguchi (2000), θ^* (2) calculated from Equation (7.13), and ν^* (3) obtained by Saigo and Hamaguchi (2002) as functions of the effective coupling parameter Γ^* for non-dissipative Yukawa systems.

where $\langle V(0)V(t)\rangle$ is the autocorrelation function of particle velocities V, and t is time. For investigating the evolution of the process of mass transfer $D(t)$, the autocorrelator of velocities $\langle V(0)V(t)\rangle$ is integrated for a finite interval of time,

$$D(t) = \frac{1}{3}\int_0^t \langle V(0)V(t)\rangle dt. \tag{7.15}$$

The diffusion coefficient of interacting particles can be further obtained from particle displacements at $t \to \infty$,

$$D_L = \lim_{t\to\infty} D(t), \tag{7.16}$$

where the function $D(t)$ can be written as

$$D(t) = \langle\langle(\Delta r)^2\rangle_N\rangle_t. \tag{7.17}$$

Here, $\Delta r = \Delta r(t)$ is the displacement of an individual particle during time t, $\langle\ldots\rangle_N$ denotes the averaging over an ensemble consisting of N particles, and $\langle\ldots\rangle_t$ denotes the averaging over all periods of time of duration t during the total time of measurements. The need for the last is defined by the requirement of correct determination of the mean characteristics of strongly correlated liquid systems which are not ergodic (in accordance with the "theory of jumps"). The dynamical behavior of systems of particles interacting via screened Coulomb (Yukawa) potential was numerically investigated by Vaulina and Vladimirov (2002). It was observed that $\langle[\Delta r(t)]^2\rangle_N \simeq \langle\langle[\Delta r(t)]^2\rangle_N\rangle_t$ only for the case of weakly correlated systems with the effective coupling parameter $\Gamma^* \lesssim 40$.

The "jumps" observed in the systems simulated are illustrated in Figure 7.11 which shows the difference between the ensemble averaging $\Delta_N = \langle[\Delta r(t)]^2\rangle_N/\Delta$ and the time averaging $\Delta_N^t = \langle\langle[\Delta r(t)]^2\rangle_N\rangle_t/\Delta$ in the vicinity of the crystallization line.

Note that in determining the diffusion coefficient of dust particles in Equation (7.16), the limit $t \to \infty$ is understood in the sense that the time t is long compared to other microscopic time scales of the system, but it is short compared to the characteristic time of diffusion through a distance of the order of the system size or to a time during which the parameters of dust plasma can significantly vary in the experiment. Because of interaction between particles, the value of D turns out to be lower than that of the Brownian coefficient of diffusion for the same particles in the absence of interaction, $D_0 = T_d/(m_d \nu_{fr})$. In the limiting case of crystal structure we have $D \to 0$, because the displacements of particles located in the crystal lattice sites are limited. Therefore, the ratio D/D_0 in dissipative systems of interacting particles is largely reflective of the nature and force of interaction between particles.

Because no assumptions of the pattern of thermal motion are made in deriving relations (7.14)–(7.17), these relations are valid for gases, liquids and solids. However, in the majority of cases, the calculation of diffusion coefficient using these relations

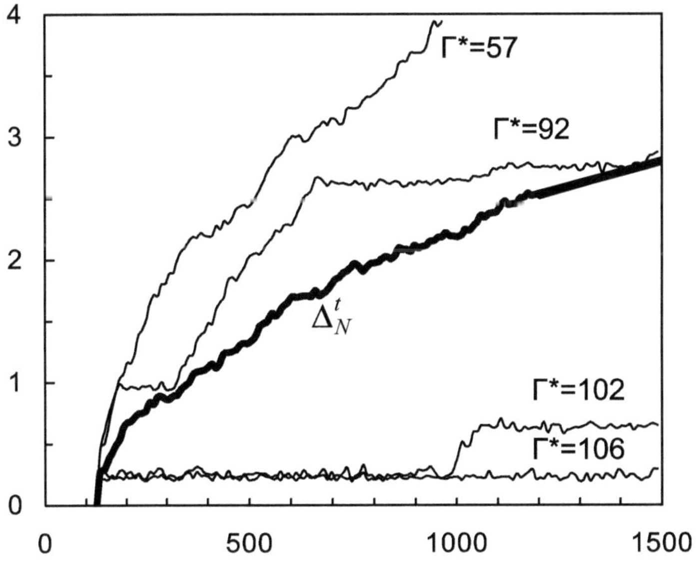

FIGURE 7.11
Illustration of "jumps" in simulations of Yukawa systems (bold: Δ_N^t line for Γ^* = 92).

does not permit analytical solutions. The simple relation $D \equiv D_0 = T_d/(m_d \nu_{fr})$, which is known as the Einstein relation, can be derived only for non-interacting ("Brownian") particles. Virial expansions or relations based on analogies with critical phenomena in liquids were used to include the forces of interparticle interaction in processes of thermal diffusion of particles (Cummins and Pike 1974; Donko and Nyiri 2000; Hofman *et al.* 2000; Ohta and Hamaguchi 2000). Within the semiempirical "jumps theory" the analytical relation for the coefficient of diffusion of molecules in liquid can be written as (Frenkel 1976; March and Tosi 2002)

$$D = \frac{\Delta^2}{2m\tau_0} \exp(-W/T_d), \tag{7.18}$$

where Δ is the average spacing between molecules, m is the molecular mass, τ_0 is the characteristic time defining the frequency of transitions of molecules from one "settled" state to another, and W is the energy barrier the particle has to overcome during these transitions. The exponential dependence of D on temperature T of molecular liquids is supported by experimental results.

Vaulina and Vladimirov (2002), Vaulina and Petrov (2004), and Vaulina *et al.* (2002) proposed that a similar temperature dependence for the coefficient of thermal diffusion D of interacting particles in plasmas can hold. They suggested that the behavior of the diffusion coefficient of particles forming a dissipative Yukawa system is governed by the effective coupling parameter Γ^* and the scaling parameter $\zeta = \omega^*/\nu_{fr}$. For highly non-ideal liquid structures ($\Gamma^* > 50$), Vaulina and Vladimirov (2002) and Vaulina and Petrov (2004) proposed the following approximate expression

$$D = \frac{T_d \Gamma^*}{12\pi m_d \nu_{fr}(1+\zeta)} \exp(-c\Gamma^*/\Gamma^*_M), \tag{7.19}$$

where $\Gamma^*_M \simeq 106$ is the melting (crystallization) point; $c = 2.9$ for $\zeta \geq 0.41$ and $c = 3.15$ for $\zeta \leq 0.14$.

For systems with screening parameter $\kappa \lesssim 6$, the error of approximation of the results of calculation of D using formula (7.19) does not exceed 3% for $\Gamma^* \gtrsim 50$. The error increases up to 7–13% when Γ^* decreases to ~ 40. For $\Gamma^* \simeq 30$, this error is approximately 25–30%. The results of simulation of the dynamics of particles in dissipative ($\nu_{fr} \neq 0$) and non-dissipative ($\nu_{fr} \equiv 0$, $\zeta \to \infty$) systems are shown in Figure 7.12 (Ohta and Hamaguchi 2000; Vaulina and Vladimirov 2002). These results indicate that the normalized value of the diffusion coefficient $D^* \equiv D(\nu_{fr} + \omega^*)m_d/T_d$ is a universal function of the effective coupling parameter Γ^* in the range $5 \lesssim \Gamma^* \lesssim 106$, provided $\kappa \lesssim 6$ (i.e., when bcc lattice is formed). As the value of Γ^* approaches that at the melting point of the system ($\Gamma^* \to \Gamma^*_M \simeq 106$), the value of diffusion coefficient experiences a jump (decreases by several orders of magnitude in the narrow range of $102 \leq \Gamma^* \leq 106$). This jump is an indicator of the first-order phase transition (crystallization). In addition, it follows from Figure 7.12 that the ratio D/D_0 is fully defined by the parameters Γ^* and ζ and weakly depends on κ alone as long as $\kappa \lesssim 6$. Moreover, if the dissipation is high enough ($\zeta \ll 1$), D/D_0 does not depend on ζ either. On the other hand, for $\zeta \gg 1$, the numerical results tend

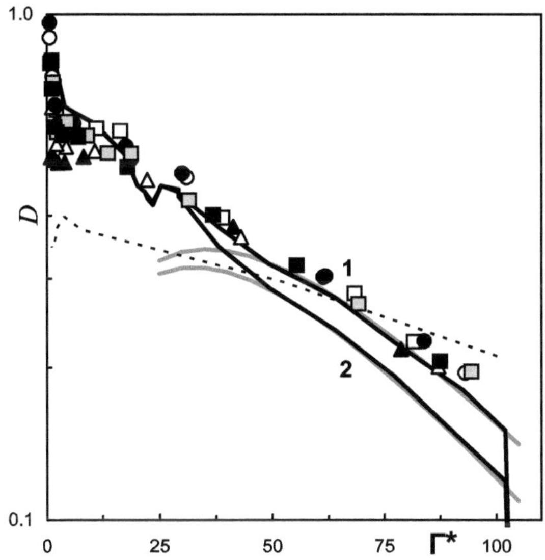

FIGURE 7.12
The normalized function D^* (black lines), averaged over the results of numerical simulation of Vaulina and Vladimirov (2002), as a function of parameter Γ^*, and the approximation of this dependence (gray lines) using relation (7.19) for (1) $\zeta \geq 0.41$ and (2) $\zeta \geq 0.14$; and the value of D^* for Yukawa non-dissipative systems ($\zeta \to \infty$) (Ohta and Hamaguchi 2000): open circles, $\kappa = 0.16$; solid circles, $\kappa = 0.48$; open squares, $\kappa = 0.97$; grey squares, $\kappa = 1.61$; solid squares, $\kappa = 2.26$; open triangles, $\kappa = 3.2$; solid triangles, $\kappa = 4.8$; dotted line, $\kappa = 8$.

to the results obtained for non-dissipative Yukawa systems by Ohta and Hamaguchi (2000). Therefore, the dynamic criterion of melting (Löwen *et al.* 1993), according to which $D/D_0 \simeq 0.1$ at crystallization, turns out to be valid for strongly dissipative systems. However, it finds no confirmation in the case of weakly dissipative systems ($\zeta \gg 1$). As the screening parameter increases to $\kappa \gtrsim 6$, the type of lattice of the system in its solid state changes to the fcc lattice (Vaulina *et al.* 2002), and significant differences are observed in the behavior of the diffusion coefficient compared to the case $\kappa \lesssim 6$ (see Figure 7.12).

Relation (7.19) enables one to determine the effective coupling parameter Γ^* from the results of measurements of the mean interparticle spacing, temperature, and diffusion coefficient of dust particles in liquid systems. Numerical simulation data can be used for the diagnostics of weakly correlated structures with $\Gamma^* \lesssim 50$ (see Figure 7.12). Agreement between the transport characteristics of the simulated liquid-like Yukawa systems and experimentally obtained plasma–particle structures was experimentally verified in a number of complex plasma studies in gas discharges of dif-

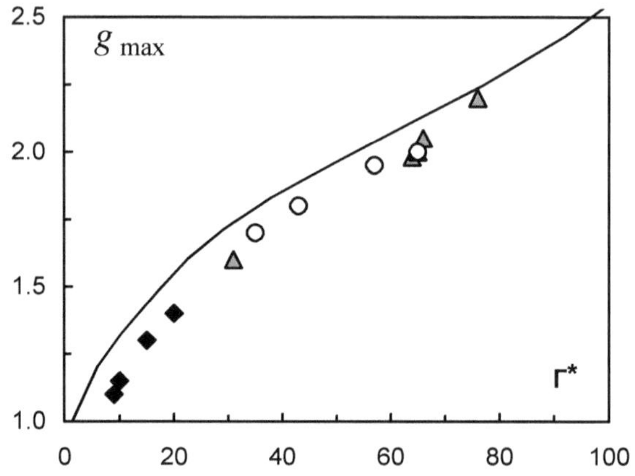

FIGURE 7.13
Experimentally obtained by Vaulina *et al.* (2003) values of g_{max} as a function of the effective coupling parameter Γ^* calculated from the results of measurements of diffusion of dust particles in a dc discharge: Diamonds correspond to ground-based conditions, triangles, microgravity experiments onboard Mir station. In addition, results obtained in an rf discharge in ground-based conditions are shown by circles. The solid line indicates numerical calculation of the function $g_{max}(\Gamma^*)$ for Yukawa systems.

ferent types (Vaulina *et al.* 2003). The dependence of the maximum g_{max} of the measured pair correlation functions $g(r)$ on the value of the effective coupling parameter obtained using the results of measurements of the diffusion coefficients and temperatures of particle systems is shown in Figure 7.13. Reasonable agreement is observed in all cases investigated. However, it must be emphasized once again that, for a number of reasons (such as plasma anisotropy, long-range interactions including the shadowing effects, the effect of external forces and/or boundary conditions, variability of the particle charge), the direct application of results of numerical simulation of three-dimensional Yukawa systems for analyzing the experimental results is mainly limited by the conditions of microgravitation or by rather small particle sizes in the laboratory conditions.

7.3.3 Viscosity

Viscosity is a transport phenomenon which defines the dissipation of energy upon deformation of the medium. The factor of proportionality between the rate of shear strain and the arising shear stress (shear viscosity coefficient η) can be obtained using the Green–Kubo relation

$$\eta = \frac{n_d}{TN}\int_0^{\infty}\langle J^{xy}(0)J^{xy}(t)\rangle dt, \tag{7.20}$$

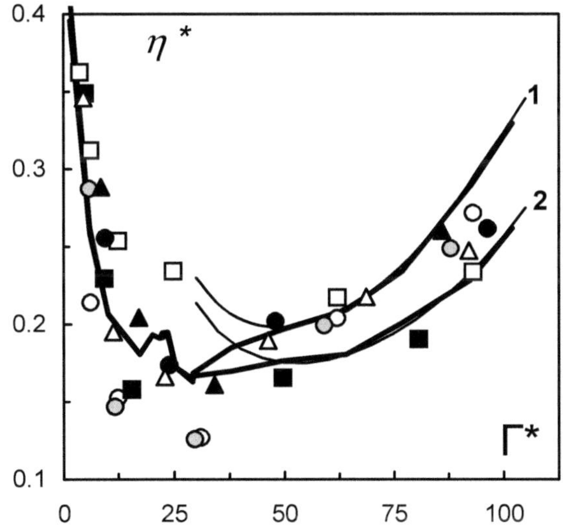

FIGURE 7.14
The normalized viscosity coefficient η^* as a function of the effective coupling parameter Γ^* obtained from Equation (7.22) (bold lines) and from Equation (7.23) (fine lines) for: (1) $\zeta \leq 0.14$, (2) $\zeta \leq 0.41$ in dissipative Yukawa systems. Symbols show the function $\eta^*(\Gamma^*)$ for non-dissipative systems (ν_{fr} = 0): Solid squares (Wallenborn and Baus 1978) and open squares (Donko and Nyiri 2000) represent calculations for the OCP model. Other symbols correspond to the Yukawa model (Saigo and Hamaguchi 2002) with different screening parameters: Open circles, κ = 0.16; grey circles, κ = 0.81; solid circles, κ = 1.61; open triangles, κ = 3.2; solid triangles, κ = 4.8.

where $\langle J^{xy}(0)J^{xy}(t)\rangle$ is the autocorrelation function of momentum fluxes J^{xy}. In the case of highly non-ideal media, this function can be obtained by numerical simulation of the dynamics of interacting particles. The results of numerical calculations of coefficient η in the Yukawa model (for $\zeta = 0.16-4.8$) (Saigo and Hamaguchi 2002) and in the OCP model (Wallenborn and Baus 1978; Ohta and Hamaguchi 2000) for the case of vanishing viscosity ($\zeta \rightarrow \infty$, $\nu_{fr} \rightarrow 0$) are presented in Figure 7.14, which shows the dimensionless coefficient η/η_0, where $\eta_0 = \frac{\Gamma^*}{\Delta^2}\sqrt{\frac{\pi T_d m_d}{\Gamma^*_M}}$. One can readily see that the values of viscosity obtained in these studies are defined by the value of effective coupling parameter within the numerical error estimated by the authors to be 20% or even higher.

The Stokes relation is often employed for analyzing the correlation between the coefficients of viscosity and self-diffusion in liquid metals and dielectrics in the vicinity of their crystallization point (Young and Alder 1971; March and Tosi 2002),

$$\eta = \frac{T_d}{6\pi D a_{\text{eff}}}. \tag{7.21}$$

Here, a_{eff} is some effective radius of a spherical molecule. The complexity of verifying this relation for a wide range of parameters of liquid state of matter largely consists in that the measurements of the coefficient of self-diffusion in liquids often give only the order of its magnitude (in view of the presently employed spectrometric procedures). Another difficulty consists in that a_{eff} in real liquids may appreciably depend on temperature or pressure. Within the theory of "jumps", the simultaneous solution of Equations (7.18) and (7.21) produces the well-known Andrade semiempirical formula $\eta \propto f(T)\exp(W/T)$, where W is the activation energy of jumps and $f(T)$ is some function which exhibits a weaker temperature dependence than the exponent. This formula is the main analytical relation used for the approximation of the temperature dependence of viscosity of liquids over rather short intervals of variation of their temperatures.

A similar correlation between transport coefficients was obtained by way of empirical fitting the numerically data (in a wide range of Γ^* between $\Gamma^* \simeq 1$ and $\Gamma^* \simeq 100$) for the coefficients D and η in Yukawa systems (Vaulina and Petrov 2004),

$$\eta \simeq \frac{T_d}{8D\Delta}. \tag{7.22}$$

In so doing, it was found that a_{eff} in the considered systems is almost constant: $a_{\text{eff}}(T_d) \simeq const$. The normalized value of viscosity $\eta^* = \eta/[\eta_0(1+\zeta^{-1})]$, obtained using formula (7.22) and data of calculation of the diffusion coefficients of particles for weakly dissipative systems ($\zeta \geq 0.41$) is shown in Figure 7.14 (curve **1**). (Note that $\eta^* = \eta/\eta_0$ for $\zeta^{-1} = 0$.) The mean-square deviation of this quantity from the data of direct numerical calculation of coefficients of viscosity (Wallenborn and Baus 1978; Donko and Nyiri 2000; Saigo and Hamaguchi 2002) is within the numerical errors (~20%). Curve **2** in Figure 7.14 corresponds to the case of highly dissipative systems ($\zeta \leq 0.14$), which is often realized in experiments with complex plasmas. An analytical approximation of shear viscosity for strongly correlated particle structures ($\Gamma^* > 50$) can be written as

$$\eta \simeq \frac{4.65(1+\zeta)\nu_{fr} m_d}{\Gamma^* \Delta} \exp\left(c\frac{\Gamma^*}{\Gamma_M^*}\right), \tag{7.23}$$

where $c = 2.9$ for $\zeta \geq 0.41$ and $c = 3.15$ for $\zeta \leq 0.14$ (see Figure 7.14).

Finally we note that experimental analysis of transport coefficients (diffusion and viscosity) in liquid-like complex plasma structures, which are formed in the sheath area of an rf discharge in a wide range of complex plasma parameters, reveals good agreement with the results of numerical investigations by Vaulina *et al.* (2003, 2004b, 2008).

7.4 Complex plasmas in narrow channels

7.4.1 2D complex plasmas in narrow channels

One of the important problems in the physics already mentioned in Chapter 5 is the behavior of micro- and nanoparticles in narrow channels when the interparticle distance is comparable with the channel width; in this case, the effect of the walls on the state of the particle system can be very important (see, e.g., Teng *et al.* 2003). The features of the flow of charged fluids in capillaries (Deegan *et al.* 1997), the investigation of confinement-induced phase transitions (see, e.g., Christenson 2001; Alba-Simionesco 2006), the physics of nanofluids (Pozhar 2000), the penetrability of ion channels in biophysics (Doyle *et al.* 1998), and the effect of the confinement on the state of granulated media (see, e.g., Clerc *et al.* 2008) are all among problems for which investigations of complex plasmas can be very informative.

Due to the fast diffusion of electrons on the walls of the discharge chamber, the central region of the discharge is positively charged and is a potential well for negatively charged particles. The profile of the confining potential (confinement) Φ_c near the center can be considered as parabolic,

$$\Phi_c(x) \propto (x - x_c)^2, \tag{7.24}$$

where x_c is the center of the discharge region. The confinement was measured by Konopka *et al.* (2000) who showed that the confinement is close to the parabolic shape in the central discharge region. The electric field in the near-electrode region of the discharge increases much more strongly than in the bulk, and the confinement is close to the hard wall.

The boundaries-induced effects are usually insignificant in the investigations of a two-dimensional complex plasma, because large particle ensembles ($N_d > 10^4$) with close longitudinal (L_x) and transverse (L_y) spatial scales are usually considered. In this case, the effect of the N_b boundary dust particles is small: $N_b \sim N_d(\Delta/L_{x,y}) \propto \sqrt{N_d} \ll N_d$, where Δ is the mean interparticle distance $[\Delta = \sqrt{(L_x L_y)/N_d}]$. Usually, in the experiments with complex plasmas, $\Delta \sim 10^{-2}$ cm and L_x, $L_y \sim 1$ cm.

Firstly, let's consider crystallization of 2D complex plasmas in narrow channels ($L_y \ll L_x$) whose width is comparable with the interparticle distance ($L_y \sim \Delta$) (Klumov and Morfill 2007). In this case, the effect of the boundary particles can become dominant since $N_b \propto N_d(\Delta/L_y) \sim N_d$.

We discuss here the impact of the confinement potential on the crystallization properties and local order of the particle component. For this reason, we consider a complex plasma which is in the strongly coupled state. This means that the nonideality (coupling) parameter of the system of charged particles in plasmas is

$$\Gamma_s \equiv (Q^2/T_d\Delta)\exp(-\kappa) > 1, \tag{7.25}$$

where T_d is the kinetic temperature of the particle component and $\kappa = \Delta/\lambda_D$ is the screening parameter. Note that here the coupling parameter Γ_s is defined as the actual strength of electric interaction at the mean interparticle separation normalized to the particles kinetic temperature.

It is known that two-dimensional Yukawa systems (e.g., 2D plasma crystals) with $\Gamma_s \gg 1$ and moderate screening parameter values ($\kappa \sim 1$) have a hexagonal crystal lattice. Thus, for 2D Yukawa systems we define the local order as the ratio of the six-fold cells to the total number of particles.

The behavior of an ensemble of the particles is numerically simulated by the molecular dynamics (MD) method. For simplicity, it is assumed that all the particles have the same charge, $|Q| \sim 3 \times 10^3 e$, and the pair interaction between the dust particles is described by the Debye–Hückel (Yukawa) potential. The equations of particle motion

$$m\ddot{\mathbf{r}}_i = -Q\nabla\Phi_c - Q\sum\nabla\phi - m\gamma\dot{\mathbf{r}}_i + \mathbf{L}_i \tag{7.26}$$

are numerically solved for $N = 400$ particles in the two-dimensional geometry (planar monolayer). The terms on the right-hand side of equations (7.26) describe the interaction of the particles with the confinement potential Φ_c, electrostatic interaction between the particles, the drag of the dust particles due to collisions with neutral atoms and molecules of the buffer gas (neutral drag), and the random Langevin force $\mathbf{L}_i$ (thermal noise induced by neutral particles), which is determined from the relation:

$$\langle \mathbf{L}_i(t)\mathbf{L}_j(t+\tau)\rangle = 2\gamma m k_B \delta_{ij}\delta(\tau) \tag{7.27}$$

under the condition

$$\langle \mathbf{L}_i(t)\rangle = 0. \tag{7.28}$$

For 2D systems, the periodic boundary conditions are imposed on the side edges ($x = \{0, L_x\}$). In the transverse direction, the particles are confined by the confinement potential ($y = \{0, L_y\}$, which is either a parabolic potential, $\Phi_c(y) \propto (y - L_y/2)^2$, or a hard wall potential, $\Phi_c(y) \propto \exp[(y - L_y)/\Delta_w]$ at $y > L_y$ and $\Phi_c(y) \propto \exp(-y/\Delta_w)$ for $y < 0$, where Δ_w is a stiffness of the hard wall potential. In the simulations described here a value of $\Delta_w \simeq \Delta/3$) was used.

Figure 7.15 shows the results of the MD simulations of the simplest 1D Yukawa system, a chain of 20 particles for the parabolic and hard wall confinements. The qualitative difference between the equilibrated positions for these types of confinements is clearly seen: for the parabolic potential, the interparticle distance Δ is minimal in the center of the system and increases near the boundaries. In the case of the hard wall, the interparticle distance for a given parameter $\kappa \simeq 2$ is nearly the same in the bulk and decreases noticeably only near the boundaries. Thus, by varying the confinement potential, one can change the particle density near the boundaries and, correspondingly, the density of the defects appearing in the crystal lattice due to the presence of the boundary. The results of the 2D MD simulations of the Yukawa system presented below confirm this expectations.

In the 2D case, initially, the particles with the charge Q are randomly distributed in the two-dimensional space (a rectangle with the sides L_x and L_y) with the initial value $N_6/N_d < 0.5$ at given values of the neutral gas temperature T_n and screening parameter κ, which correspond to a certain coupling parameter Γ_s. When simulating the channels with different thicknesses, the area of the system, $L_x \times L_y$, and the

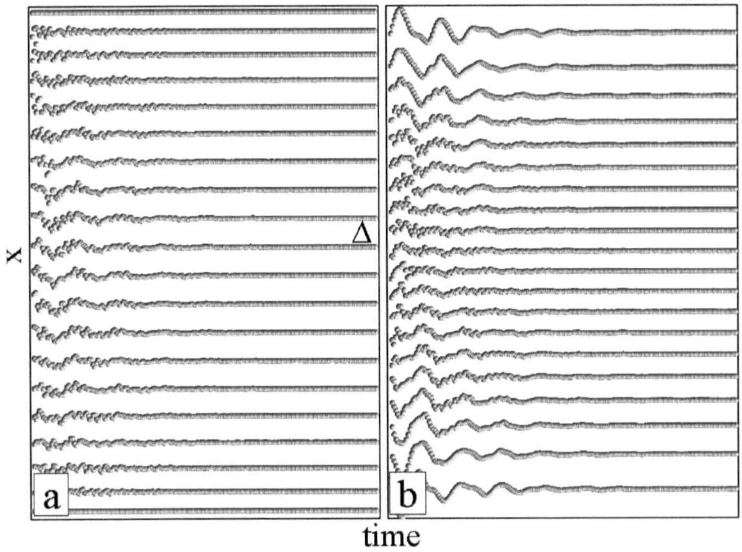

FIGURE 7.15
Impact of the elastic hard wall (a) and parabolic (b) confinements on the evolution of a one-dimensional chain of particles interacting via Yukawa potential. The $x-t$ diagrams for 20 particles with the size $2a = 1$ μm are presented. Initially, the particles are randomly distributed over the space $L = 1$ cm. The neutral gas density is $\rho_g \sim 10^{-7}$ g/cm^3, the particle charge is $|Q| = 3\times10^3 e$, and the screening length is $\lambda_D = 1$ mm. It is seen that the interparticle distance Δ near the system boundaries in the steady-state decreases in the case (a) and increases in the case (b).

number of particles, N_d are conserved. The method of establishing the steady state is used to determine the quasi-equilibrium configuration of the particles, for which the Delaunay triangulation and Voronoi cell method are used to determine the nearest neighbors for each particle. Using these data, the ratio N_6/N_d characterizing the local order is calculated. Some results of the molecular dynamics simulation of such a system of particles for various Γ_s and κ values are presented and discussed below.

The mean value N_6/N_d, characterizing the local order in the system with the parabolic and hard wall confinements is shown in Figure 7.16 as a function of Γ_s at $\kappa = 1$ for different system aspect ratios L_x/L_y. The local order is determined by using 50 sequential snapshots of the particle positions in the steady-state stage in the system ($t \gg \gamma^{-1}$), and the time interval δt between the neighboring frames is chosen from the condition $\delta t \geq \gamma^{-1}$ in order to reduce the influence of the time correlations. Figure 7.16 also shows the snapshot of the particle positions in both confinement potentials for close parameters $\Gamma_s \simeq 100$ and Delaunay triangulations convenient for

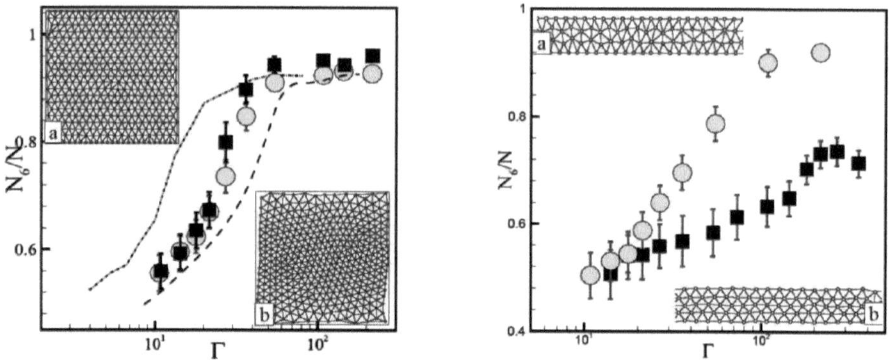

FIGURE 7.16
Relative number of particles having six nearest neighbors N_6/N_d versus the coupling parameter Γ_s for the hard wall (squares) and parabolic confinements (circles) at different aspect ratios of a 2D Yukawa system.

visualizing the nearest neighbors of the particles and the presence of the defects.

In agreement with the expectations for such a system when L_x/Δ, $L_y/\Delta \gg 1$ (left panel of Figure 7.16) the effect of the boundaries is small and both confinements provide close dependencies of N_6/N_d on Γ_s. Note that value of N_6/N_d increases sharply with Γ_s at $\Gamma_s \simeq 30$–40 and is saturated at the level $\langle N_6/N_d \rangle \simeq 1$ for $\Gamma_s \geq 10^2$. The sharp increase of N_6/N_d corresponds to the well-known liquid–solid phase transition (see, e.g., Fortov *et al.* 2005a). Figure 7.16 also shows $\langle N_6/N_d \rangle$ versus Γ_s for the parabolic confinement at $\kappa = 0.5$ (dashed line) and $\kappa = 2$ (dash-dotted line). This plot shows that the liquid–solid transition region is shifted toward smaller Γ_s when the screening parameter κ increases.

Similar dependencies for smaller channel widths $L_y/\Delta \sim 3$ are presented in Figure 7.16 (right panel). In this case, the liquid–solid transition is less pronounced than for the case of $L_y/\Delta \gg 1$. A significant decrease in the fraction of "crystallized" particles $\langle N_6/N_d \rangle$ for $\Gamma_s \gg 1$ is observed for the hard wall confinement. Such a result can be explained by the fact that fast crystallization of the particles occurs near the wall. In this case, the density of the particles near the wall is higher and, correspondingly, the interparticle distance is smaller than that in the bulk. For this reason, the rigid wall induces defects, which are usually pairs consisting of the particles with five and seven nearest neighbors (five- and seven-fold cells). This effect disappears in the case of the parabolic confinement, because the particle density near the boundary is lower than that in the bulk; hence, the defect number is much smaller than this number for the hard wall confinement for the same values of Γ_s and κ.

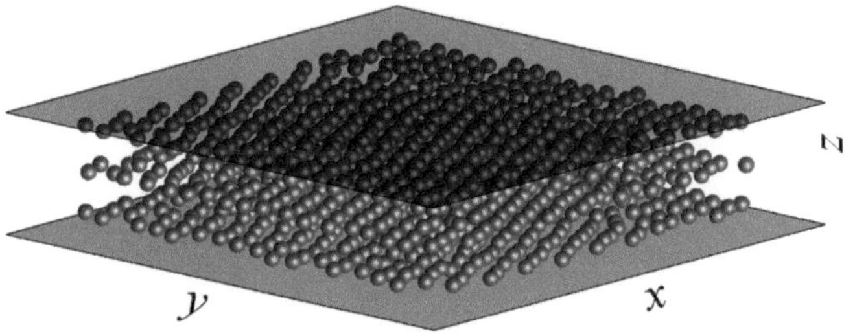

FIGURE 7.17
Geometry of the considered 3D problem. Initially N_d = 16,000 particles with the charge Q and interacting via Yukawa potential are randomly distributed in the space between two plates. The confinement limits the z coordinate of the particle ($0 \leq z \leq L_z$), whereas in the (x, y) plane, the particles are located in the region $L_x \times L_y$: $0 \leq x,\ y \leq L_{x,y}$.

7.4.2 3D complex plasmas in narrow channels

The 3D case was investigated by Klumov and Morfill (2008) for $N_d = 16{,}000$ particles, which are randomly distributed in a narrow channel at the initial time. Figure 7.17 shows the geometry of the 3D problem. The confinement limits the position of the particles along the z axis ($0 \leq z \leq L_z$), whereas in the (x,y) plane, the particles are located in the region ($0 \leq x, y \leq L_{x,y}$). The periodic boundary conditions are used on the lateral edges: ($x = \{0, L_x\}$, $y = \{0, L_y\}$).

As discussed in Section 5.1, in an infinite system of particles interacting via the Yukawa potential, the solid state in equilibrium can have only two types of crystal lattices: face-centered cubic (fcc) for large values of κ and body-centered cubic (bcc) for small κ. The hexagonal close-packed (hcp) phase can be formed in non-equilibrium Yukawa systems (see, e.g., Klumov *et al.* 2006; Rubin-Zuzic *et al.* 2006). This is due to the closeness of the energies required for the formation of the hcp and fcc phases (Hamaguchi *et al.* 1997). To identify the fcc/hcp/bcc lattice types, it is sufficient to know the positions of three nearest crystal layers A, B, and C. In this case, the type of the appearing crystal lattice can be determined visually. For this reason, we consider the behavior of a Yukawa system consisting of three layers.

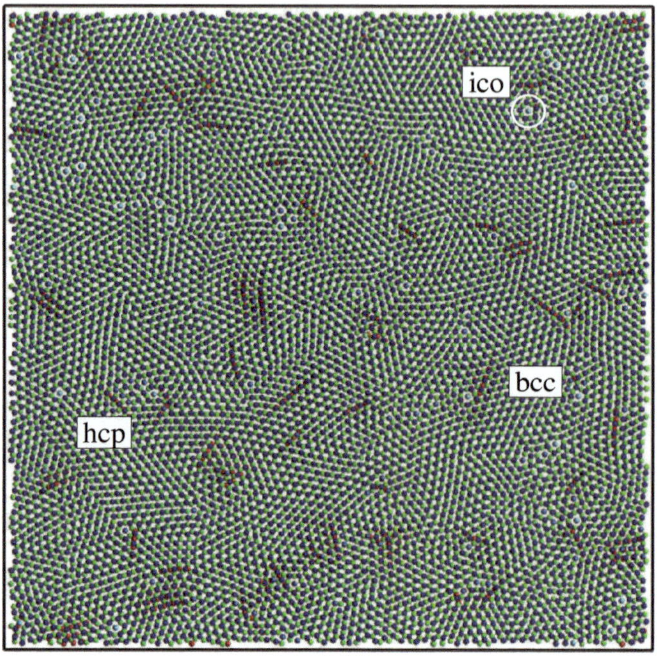

FIGURE 7.18
(See color insert following page 242). Crystallized Yukawa system in the narrow channel for the parabolic confinement. The particles form three layers A, B, and C. The particles of layers A, B, and C are marked by blue, green, and red, respectively. It is seen that layer C is almost completely screened by layer A, which means that the hcp lattice is dominant in that case. The bcc phase is also seen. A small number of clusters (∼1%) have icosahedral-like (fivefold) symmetry.

Some simulation results for the Yukawa system are presented in Figures 7.18 and 7.19 for the parabolic and hard wall confinements, respectively. Here the positions of the particles in the (x, y) plane are shown (all three layers of the particles are given). The calculations were performed with the following system parameters: the size and charge of the particles were $a \simeq 1\ \mu\text{m}$ and $|Q| = 3 \times 10^3 e$, respectively; $\kappa = 2$–3; and the neutral gas density was $\rho_g \sim 10^{-7}\ \text{g cm}^{-3}$.

Both figures demonstrate the steady-state crystalline phase of the Yukawa system at $\Gamma_s \sim 10^4$. In both figures, the particles of a certain layer are shown by a certain color (particles are color-coded by z-coordinate). It is seen that the parabolic confinement leads primarily to the formation of crystallites with the ABA layer arrangement (the third layer is screened by the first layer), which is typical for the hcp or bcc phases. A relatively small number of clusters ($\sim 1\%$) have icosahedral (ico)

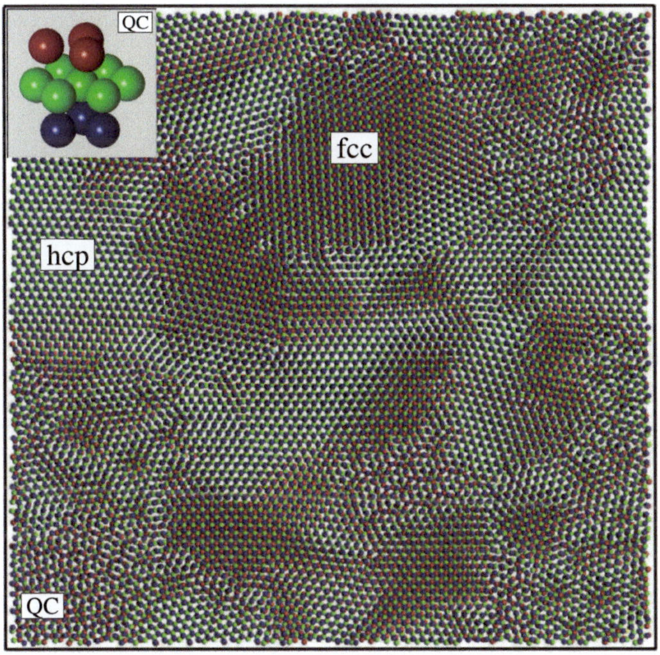

FIGURE 7.19
(See color insert following page 242). Crystallization of the particles in the narrow channel for the hard wall confinement. The domains having hcp and fcc lattice types are clearly seen. A significant number of clusters have a quasi-crystalline (QC) phase. The inset shows the unit cell of the QC phase, which is a distorted hcp/fcc unit cell.

symmetry. Note that this lattice type is induced by the boundaries of the system. The regions with the ABA (hcp/bcc phases) and ABC (particles of three layers are seen) layer arrangements are clearly seen in Figure 7.19 for the hard wall confinement. A significant number of dust particles have fcc and hcp lattice types.

The appearance of a new quasi-crystalline (QC) phase for the hard wall confinement is very interesting. The unit cell of this phase is shown in the inset in Figure 7.19 and is the hcp/fcc phase distorted by the rotation of the upper and lower layers with respect to the middle layer.

The compression of the considered three-layer system along the z-axis (or the decrease the screening parameter κ) results in bifurcation of the system to two-layer systems at definite "wall separation". Similarly, the increase of "wall separation" results in bifurcation to stable four-layer system for both types of confinement. So the stable three-layer system can exist only in a rather narrow range of parameters (particle charge, screening length, screening parameter). Figure 7.20 shows stable

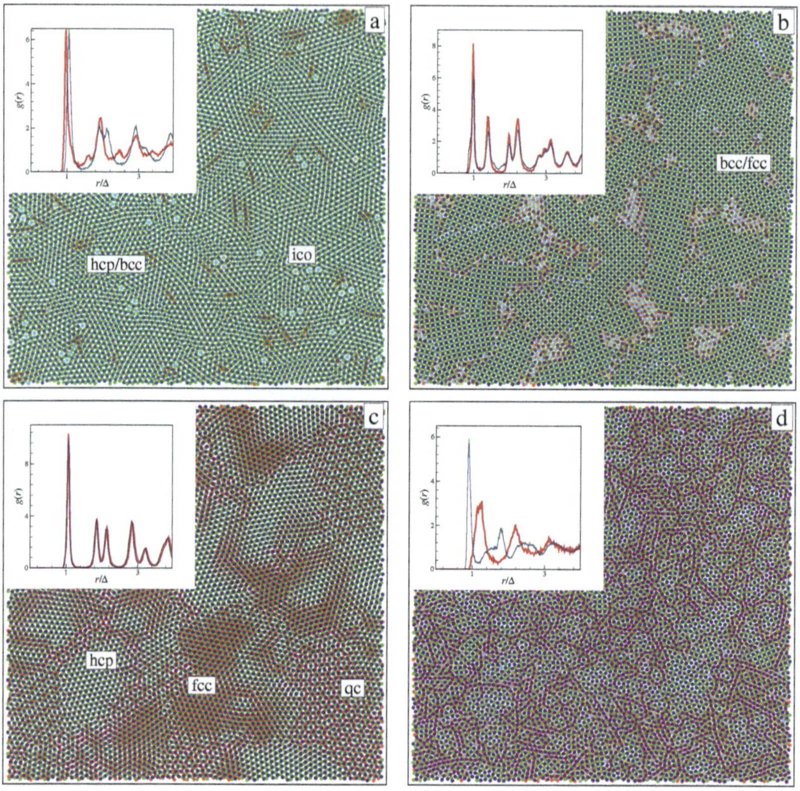

FIGURE 7.20
(See color insert following page 242). Crystallization of the particles in the narrow channel for the parabolic (a, b) and hard wall (c, d) confinements. Particles are color-coded by z-coordinate. Stable three-layers configurations of the Yukawa system (a, c) are presented. These systems close to the two-layer configuration (b, d) are also shown. The insets show pair correlation function $g(r/\Delta)$ for each layer, including the central one (solid line).

configurations of the confined Yukawa systems at different "wall separations". It is clearly seen that the phase state of the crystallized systems changes significantly with "wall separation", which means that we can control phase state of such systems via decrease or increase of the κ value or the "wall separation". Figure 7.21 presents the evolution of Yukawa systems at different "wall separations". Relative density of particles is presented for both parabolic (top panel) and hard wall (bottom panel) types of confinement.

Figure 7.22 shows the particle density distribution and the total number of particles in each layer for two types of confinement. It is seen that, as in the two-dimensional case, the density of the particles for the hard wall confinement near the boundaries is

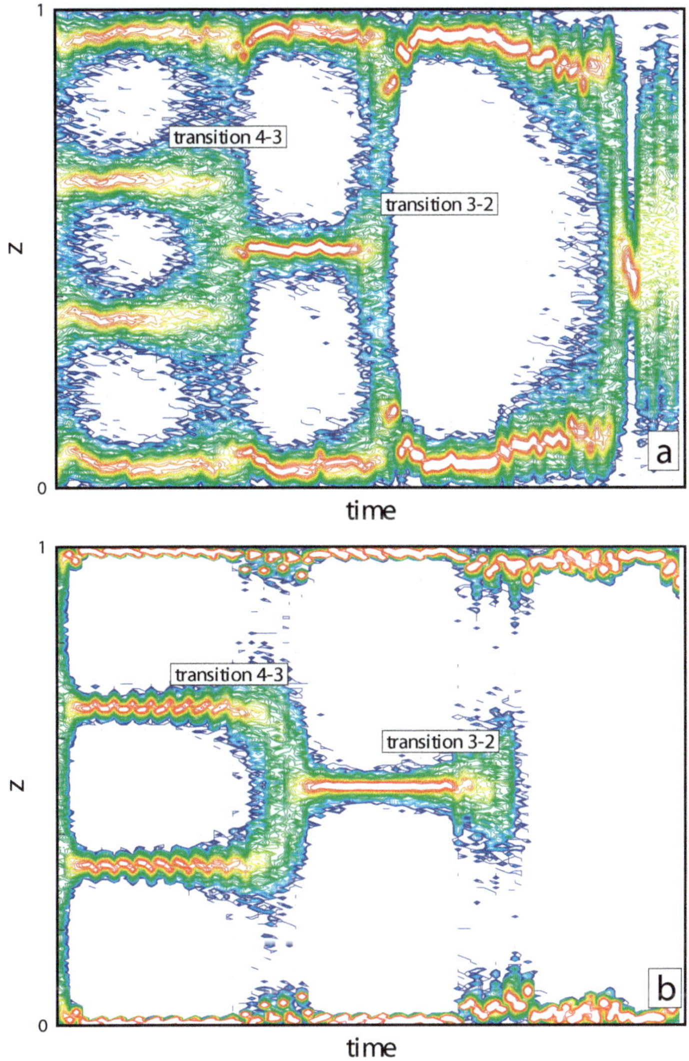

FIGURE 7.21
Evolution of the Yukawa system at different "wall separations". Relative density of particles is shown for both parabolic (top panel) and hard wall (bottom panel) types of confinement.

higher than that in the center. The relation is opposite for the parabolic confinement. Such a density distribution is the main cause of the indicated features of the Yukawa system crystallization in narrow channels.

For the case of the hard wall confinement, the particles are first crystallized near

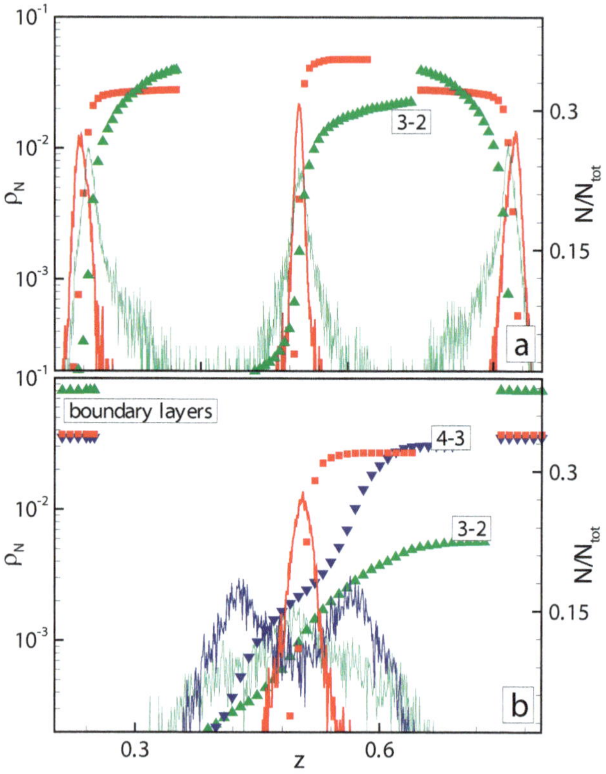

FIGURE 7.22
Relative density of the particles ρ_N versus dimensionless coordinate $\tilde{z} \equiv z/L_z$ for parabolic (*a*) and hard wall (*b*) confinements for different "wall separations". The total number of particles in each layer $N/N_{tot}(\tilde{z})$ as a function of $\tilde{z}$ are also shown for stable three-layers systems (squares) and systems close to bifurcation points: transition (3-2) (up triangles) and (4-3) (down triangles), respectively.

the boundary ($z = 0,\ L_z$), whereas all three layers are formed almost simultaneously for the parabolic confinement. Thus, the confinement type significantly affects the local order and the crystal–lattice type of the system of particles interacting via the Yukawa potential in narrow channels. Note that the confinement leads to the appearance of the new crystal–lattice types (quasi-crystalline and icosahedral phases), which are absent on the phase diagram of the 3D infinite Yukawa systems.

Thus, the effect of confinement on the behavior/crystallization of the Yukawa systems in both two-dimensional and three-dimensional narrow channels has been numerically investigated using the molecular dynamics simulations. The parabolic and hard wall confinements, which are the "soft" and "hard" confinements, have been considered. These types of confinement lead to quite different behaviors of the particle density near the boundaries. This behavior, in turn, strongly affects the local order and the type of the crystal lattice, in particular, a new stable quasi-crystalline phase appears for the hard wall confinement. Thus, the principle possibility of controlling the behavior/flow of complex plasmas in narrow channels is confirmed, which can be very important for the investigations and applications of micro- and nanofluids and nanomaterials.

7.5 Crystallization waves in complex plasmas

We note that the crystalline state (plasma crystal) of the complex plasma was experimentally discovered in 1994 and was theoretically predicted as early as in 1986. However, crystallization waves in the complex plasma were experimentally observed only recently by Rubin-Zuzic *et al.* (2006).

Some of the related results have been already discussed in Section 5.3.4. Below we describe numerical simulations which are able to reproduce the observations. In the experiment, an extended plasma crystal consisting of about 10^7 polymer 1.3 μm particles (with a mean interparticle distance of $\Delta \simeq 80$ μm) was first created in an rf discharge plasma in argon (at a pressure of $p \approx 0.23$ mbar and a frequency of 13.6 MHz). The plasma crystal was melted by sharply reducing the discharge power (followed by a sharp increase to the initial value). The decrease in the power likely led to a decrease in the magnitude of the particle charge Q and, correspondingly, to a decrease in the coupling strength (i.e., in the Γ_s parameter). As a result, the particle component passed from the strongly coupled to weakly coupled state. The further recrystallization process of the particle component was observed by a high-resolution video camera (1028$\times$ 772 pixels, 15 fps).

As revealed, recrystallization in such a system is of the wave character: crystallization begins with the lower boundary of the melted crystal and has the form of a wave with a pronounced front, which propagates upwards in the direction opposite to the gravitational force. Figure 7.23 shows photographs (obtained by superposition of ten sequential frames) demonstrating the propagation of the crystallization wave in the system at various times. The interface between the crystal and liquid phases is clearly seen; the characteristic front thickness is on the order of the interparticle distance. The characteristic front-propagation velocity is $v_{cf} \simeq 100$ μm/s, which corresponds to the formation of approximately one ordered layer of particles per second. As the crystallization wave propagates, the front velocity decreases slightly. The mean interparticle distance is equal to 80 μm and depends slightly on the form of the phase.

It is worth noting the complex fractal structure of the crystallization wave front with the minimum inhomogeneity scale of about Δ. In the process of the crystallization wave propagation, small inhomogeneities on the wave front disappear: their coalescence occurs with the formation of larger inhomogeneities. It is important to

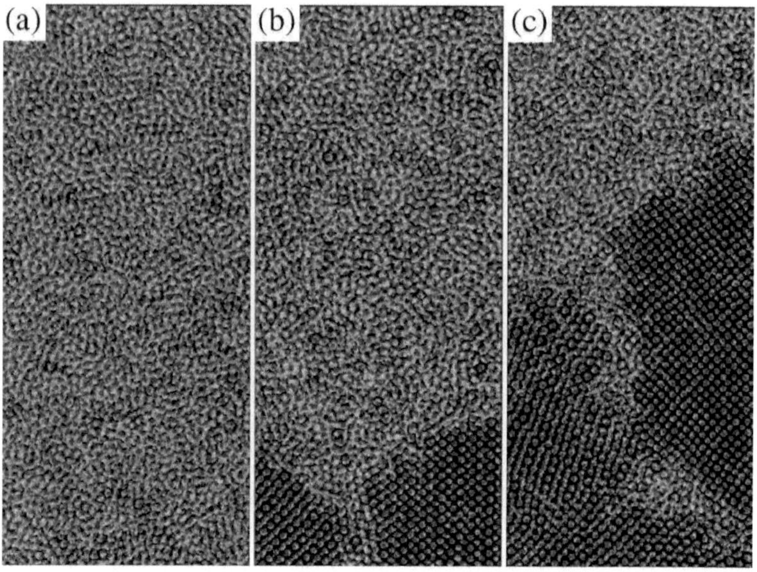

FIGURE 7.23
Photographs (obtained by superposition of ten sequential frames) demonstrating the evolution of the crystallization wave propagating through the complex plasma at $t = 0$ (*a*), $t = 10$ (*b*) and $t = 25$ sec (*c*) after the beginning of the crystallization process. In the beginning the particles are in a disordered state (the crystal is melted). In the process of system recrystallization, the order wave (crystallization that travels from the bottom to the top and has a pronounced front whose width is of the order of the interparticle distance Δ is seen. The particles behind the crystallization wave front form two regions with different structures. Note that the kinetic temperature of the particles behind the crystallization wave front is noticeably lower.

note that Figure 7.23 also shows the relative kinetic temperature of the particles: the temperature of the particles behind the crystallization wave front is noticeably lower. According to frame-by-frame analysis of the motion of particles, the ratio of the temperatures $T_{\rm liq}$ and $T_{\rm cr}$ ahead of and behind the crystallization wave front is estimated as $T_{\rm liq}/T_{\rm cr} \simeq 2$. The presented experimental results demonstrate the propagation of the crystallization wave in a complex plasma at the kinetic level.

Let us try to numerically reproduce the observation data by simulating the behavior of an ensemble of particles by the molecular dynamics simulations. For simplicity, we assume again that all the particles have the same charge $|Q| \sim 3 \times 10^3 e$, where e is the electron charge, and the pair interaction between the dust particles is described by a screened Coulomb (Yukawa) potential, $U(r) = (Q^2/r)\exp(-r/\lambda_{\rm D})$, where $\lambda_{\rm D}$ is the screening length. Other parameters of the complex plasma (particle size, neutral gas density, etc.) are the same as in the experiment. We use the follow-

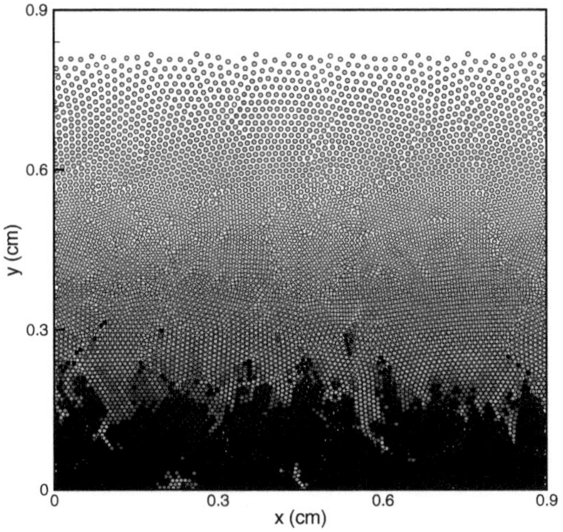

FIGURE 7.24
Recrystallization in a 2D Yukawa system. The temperature field and positions of the particles at a certain time after the start of recrystallization are presented. The darker color corresponds to the lower kinetic temperature of the particles.

ing boundary conditions in the simulations: the lower boundary of the calculation region is a potential well of hard wall type and simulates the near-electrode discharge region, whereas the upper boundary is free. The periodic boundary conditions on the lateral edges are used for both 2D and 3D geometries. Initially, the particles in the preliminarily created crystalline state are instantaneously heated and the plasma crystal is rapidly melted. Some results of the molecular dynamics simulation of the further recrystallization of the particle system are presented below.

Figure 7.24 shows the positions of 900 particles and their energy at a certain moment of time for a 2D plasma crystal. It is seen that a particle cooling wave with a complex fractal structure is formed and propagates from the bottom to the top. Such a (2D) system is rapidly crystallized, and this crystallization of the particle system is of the bulk type, rather than wave. Nevertheless, such a simulation reproduces well a number of the features of experiments such as the cooling wave and its structure, as well as the sedimentation of particles in the crystallization process and gravitation compression of the plasma crystal.

Figure 7.25 shows the initial stage of the formation of the crystallization wave according to the 3D molecular dynamics (N_d = 27,000) simulation. In this case, the crystallization-wave velocity is close to the experimental values and depends slightly on the screening length and effective gravity force. Figure 7.26 shows the dependence of the crystallization wave velocity on the dimensionality (1D, 2D, and 3D) of the particles system.

The strong decrease in the crystallization wave velocity with increasing of the di-

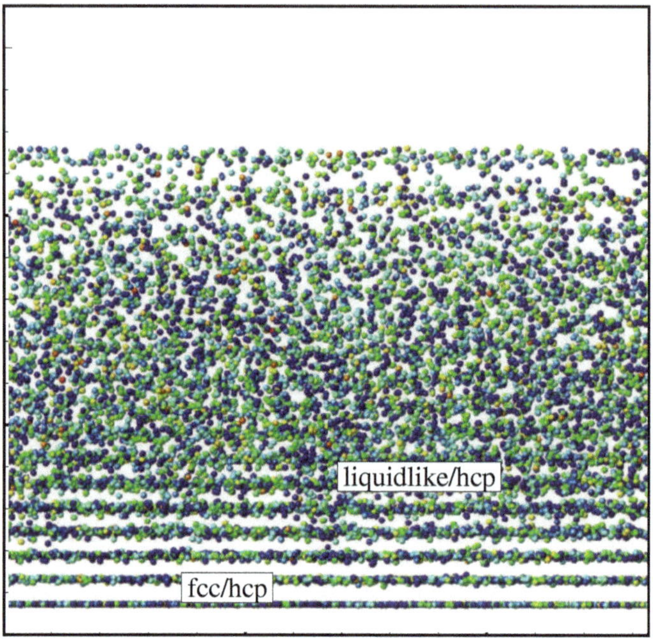

FIGURE 7.25
Crystallization wave in a 3D Yukawa system. The temperature field and positions of the particles at a certain moment of time after the plasma crystal melting are presented. The darker color corresponds to the lower kinetic temperature of the particles. The local order ahead of/behind the crystallization wave front is determined. The coexistence of different lattice types, including the metastable phase (hcp), is shown.

mensionality of the plasma crystal can be explained by the increase in the number N_b of the nearest neighbors with which each single particle interacts. Indeed, the inequality $\lambda_D/\Delta \leq 1$ is often satisfied in experiments with complex plasmas. For this reason, the contribution of long-range interactions ($r \geq \Delta$) is exponentially small. For the linear particle chain, $N_b = 2$; for the 2D system, $N_b = 6$; for the 3D case $N_b = 12$ [$N_b = 12$ for the perfect face centered cubic (fcc) and hexagonal close packing (hcp) lattice types and $N_b = 8$ for the body centered cubic (bcc) lattice]. According to the thermodynamic representations, $N_c \propto \exp(N_b)$ combinations are possible for the formation of an ordered structure consisting of N_b elements; therefore, the crystallization wave front velocity is expected to be $v_{\text{cf}} \propto \exp(-N_b)$. Figure 7.26 also shows the values calculated by the expression $\exp(-N_b)$. It reproduces well the results of molecular dynamics simulation.

We emphasize that the mean interparticle distance Δ, which depends on the charge and density of the particles, on the effective gravity force, and is determined in the simulation, is one of the key parameters in the molecular dynamics calculations re-

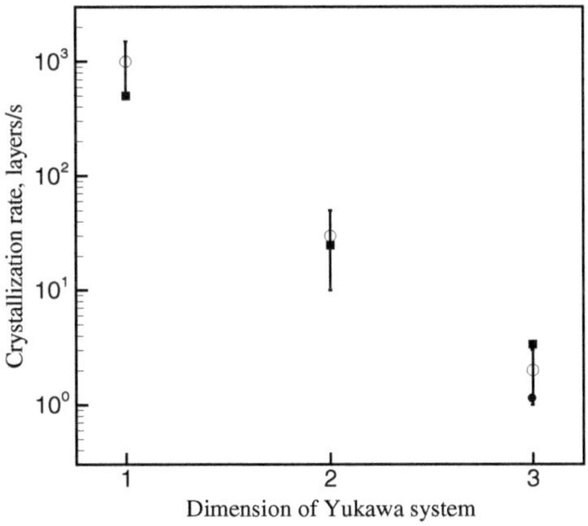

FIGURE 7.26
Crystallization wave velocity versus the dimensionality of the Yukawa system according to the molecular dynamics simulation (open circles) and experiment (solid circle). The values calculated by the expression $\exp(-N_b)$, where $N_b = 2$, 6, and 12 for the 1D, 2D, and 3D geometries, respectively, are also shown (solid squares).

ported here. This circumstance makes it possible to estimate the charge Q of the particles using the crystallization wave propagation velocity and the interparticle distance Δ. For the considered experiment, $|Q| \sim (3 \div 5) \times 10^3 e$.

The 3D molecular dynamics simulation allows the determination of an important characteristic of the crystallization process, the local order of dust particles behind the crystallization wave front (see next section). It is well known that the infinite crystallized 3D Yukawa systems in equilibrium can form only two lattice types: bcc (for small κ values) and fcc (for larger κ) (see, e.g., Fortov *et al.* 2005a).

Finally, in this section, the behavior of the ensemble of charged particles interacting via the Yukawa potential has been investigated by the molecular dynamics method. The complex plasma parameters were taken to be close to the experimental values. The systems of the particles were considered in the 1D (linear particle chain), 2D (plane monolayer), and 3D geometries. The local order of the particle system ahead of/behind the crystallization wave front has been determined for the 3D case. The coexistence of different types of the crystal lattice including the metastable hcp phase has been observed behind the crystallization wave front, which indicates the presence of the strong non-equilibrium conditions behind the front. Although the results of 2D simulation are in qualitative agreement with the experiment and, in particular, reproduce the temperature field of the particles, the sharp temperature front, and the particle cooling wave propagating from the bottom to the top, demonstrate the complex structure of the crystallization wave front, only 3D simulation results are in quantitative agreement with the crystallization wave propagation velocity. The visualization of the particles in experiments with complex plasmas is usually performed in the 2D geometry (side view image). The results of this work allow an important conclusion that the effect of the real geometry (dimensionality) can be decisive in experiments with complex plasmas.

7.5.1 Local order analysis of 3D data

To determine the local order of the particles a bond order parameter method is used (Steinhardt *et al.* 1983). In the framework of this method, the local rotational invariants for each particle are calculated and compared with those for perfect lattice types like fcc/hcp/bcc/ico (see Figure 7.27).

Local rotational invariants of second order $q_l(i)$ and third order $w_l(i)$ are calculated for each particle i by using M nearest neighbors $N_b(i)$:

$$q_l(i) = \left(\frac{4\pi}{(2l+1)} \sum_{m=-l}^{m=l} | q_{lm}(i)|^2 \right)^{1/2} \tag{7.29}$$

$$w_l(i) = \sum_{\substack{m_1,m_2,m_3 \\ m_1+m_2+m_3=0}} \begin{bmatrix} l & l & l \\ m_1 & m_2 & m_3 \end{bmatrix} q_{lm_1}(i) q_{lm_2}(i) q_{lm_3}(i), \tag{7.30}$$

where

$$q_{lm}(i) = \frac{1}{N_b(i)} \sum_{j=1}^{N_b(i)} Y_{lm}(r_{ij}) \tag{7.31}$$

and Y_{lm} are the spherical harmonics, $r_{ij} = r_i - r_j$, where r_i are the coordinates of i-th particle. In Equation (7.30) $\begin{bmatrix} l & l & l \\ m_1 & m_2 & m_3 \end{bmatrix}$ are the Wigner $3j$-symbols, and the summation in the latter expression is performed over all indices $m_i = -l, ..., l$, that satisfy the condition $m_1 + m_2 + m_3 = 0$.

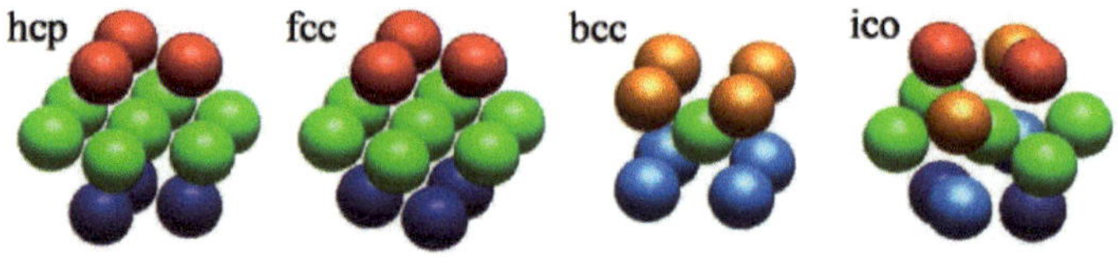

FIGURE 7.27
(See color insert following page 242). The lattice types we try to identify: hexagonal close packing (hcp), face centered cubic (fcc), body centered cubic (bcc) and icosahedron (ico) (from left to right).

It is important to stress that each lattice type has its own unique set of q_l and w_l rotational invariants. This gives us the possibility to identify observed lattice types, by comparing the observed q_l, w_l values with those types $q_l^{\rm id}$, $w_l^{\rm id}$ for perfect lattice.

To define the local order around a particle we used q_4, q_6, w_4, and w_6 rotational invariants. The rotational invariants can be easily calculated for perfect fcc/hcp/ico/bcc. For fcc/hcp/ico the number of nearest neighbors $N_b = 12$ and we have for fcc: $q_4^{\rm fcc} = 0.1909$, $q_6^{\rm fcc} = 0.5745$, $w_4^{\rm fcc} = -0.1593$, $w_6^{\rm fcc} = -0.01316$; for hcp: $q_4^{\rm hcp} = 0.0972$, $q_6^{\rm hcp} = 0.4847$, $w_4^{\rm hcp} = 0.1341$, $w_6^{\rm hcp} = -0.01244$; and for icosahedral lattice type (ico): $q_4^{\rm ico} = 0$, $q_6^{\rm ico} = 0.6633$, $w_4^{\rm ico} = -0.1593$, $w_6^{\rm ico} = -0.1697$.

High values of $q_6^{\rm cr}$ for all cited lattice types can be used to study early stage of nucleation/crystallization in different systems (see, e.g., Auer and Frenkel 2004). We note, that for uncorrelated system, mean value $\langle q_6 \rangle \simeq N_b^{-\frac{1}{2}}$ is significantly smaller than $q_6^{\rm cr}$. For instance, if $N_b = 12$, the value $\langle q_6 \rangle \approx 0.29$.

For the bcc case $N_b = 8$ and $q_4^{\rm sc} = 0.5092$, $q_6^{\rm sc} = 0.6285$, $w_4^{\rm sc} = -0.1593$, $w_6^{\rm sc} = 0.1316$. Sometimes, to identify bcc clusters it is important to know the position of the second shell having 6 particles [the distance between the nearest-neighbor and the second shell in the bcc lattice is relatively small, $(2/\sqrt{3}-1)\Delta \simeq 0.15\Delta$]. Consequently, the thermal motion of the second shell particles can easily distort the rotational invariants and they can be identified as nearest neighbors. For $N_b = 14$ we have $q_4^{\rm bcc} = 0.0363$, $q_6^{\rm bcc} = 0.510$, $w_4^{\rm bcc} = 0.1593$, $w_6^{\rm bcc} = 0.01316$.

The lattice of a real plasma crystal is always distorted because of various factors. For example, a difference in slow particle drifts between different regions of the crystal can give rise to shear stress and, therefore, to its local distortion. Rotational motion can give rise to torsional defects on scales comparable to Δ. Crystalline structure can also be distorted by short-wavelength acoustic disturbances, etc. Therefore, of interest are the changes in q_l and w_l due to various distortions of fcc/hcp/ico/bcc lattices. Data of these kinds are shown in Figure 7.28, where the variations of q_l and w_l due to weak shear, compression/dilation, and torsion of these lattices (without change in the nearest neighbors) are shown on the q_4–q_6 plane. Note that these distortions generally reduce the value of q_6 for all of the lattice types. Note also that local invariants are more sensitive to torsion and less sensitive to shear and compression. The data shown in Figure 7.28 can be used to find dilated hcp/fcc lattices and quasi-crystalline phase regions (torsional defects of fcc/hcp lattice).

We identified the hcp/fcc clusters behind the crystallization front (see Figure 7.25) in our 3D molecular dynamics simulations of a Yukawa system. Below we present the local order analysis of some recent three-dimensional experimental and simulation data.

7.5.1.1 Complex plasmas in a homogeneous dc discharge tube

Here we present a full 3D reconstruction of a particle cloud in a dc discharge plasma (Mitic *et al.* 2008). In contrast to typical complex plasma experiments in dc discharge tubes, the particles in the experiment described here were not levitated in striations, where strong variations in the electric field lead to inhomogeneities in

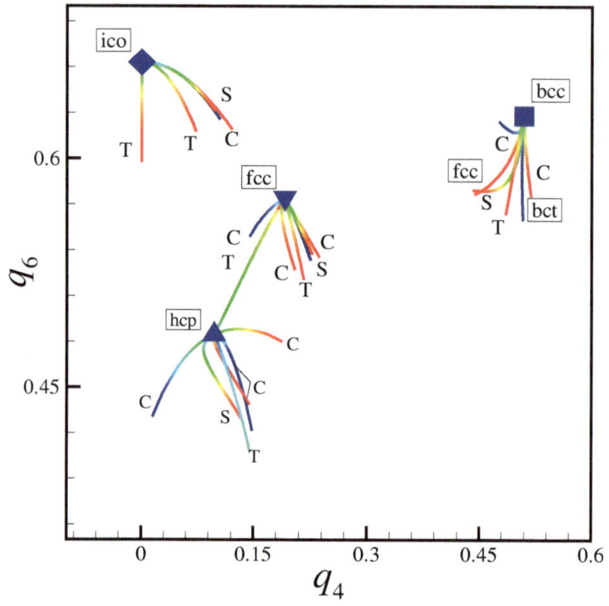

FIGURE 7.28
(See color insert following page 242). Variations of the rotational invariants q_4 and q_6 for different lattice types at different distortions: compression/extraction (C), shear (S) and torsion (T). The curves are color-coded by relative deformation values (S, T). We also plotted here the invariants for the fcc lattice (calculated by using 8 nearest neighbors) and body centered tetragonal (bct) – the compressional modification of bcc.

particle clouds even on small scales. Instead, the particles were confined in a horizontally mounted discharge tube by a weak radial electric field of the discharge. The discharge conditions were selected in such a way that no striations were present in the positive column. In this way large homogeneous particle clouds could be established.

The three-dimensional positions of the particles in the observed part of the cloud are presented in Figure 7.29. Particle positions are overlain for all microspheres in the cloud showing the projection of the system in the side view and top view (images in Figure 7.29 upper part and lower part, respectively). As can be clearly seen, the particles in the bulk of the cloud represent a kind of vertically oriented structure with a single distinct outer particle layer which represents the structure of the radial confinement across the discharge tube.

Local order analysis of the data presented in Figure 7.29 reveals liquid-like be-

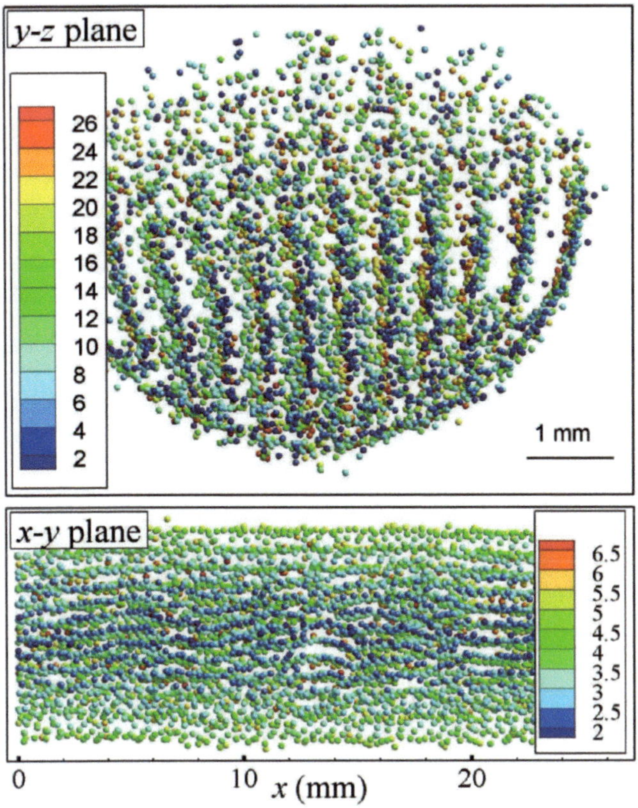

FIGURE 7.29
(See color insert following page 242). Experimentally recorded particle positions in the y–z (top) and x–y (bottom) planes. Particles are color-coded by the corresponding third coordinate presented in mm. About 6000 particles were detected.

havior of the particles. Figure 7.30 shows the simulated distributions of a Yukawa system in q_4–q_6 plane for both a liquid-like system ($\Gamma_s \sim 1$) (b) and a crystallized one ($\Gamma_s \simeq 10^4$) (a). Also, results of the experiment are plotted revealing liquid-like behavior of the observed data (c). To obtain the crystallized Yukawa system we performed 3D MD simulations of 6000 particles in a box with external confinement of a hard wall type in vertical direction (z) with periodic boundary conditions in horizontal (x, y) directions.

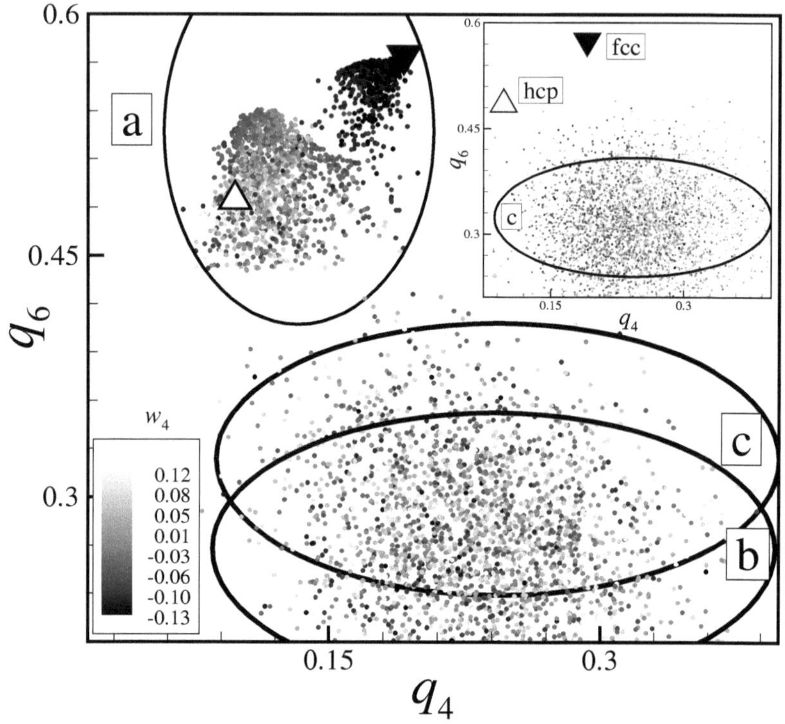

FIGURE 7.30
Distribution of dust particles at different values of Γ_s in the plane of local order parameters q_4–q_6 (calculated by using 12 nearest neighbors) as seen from MD simulations of Yukawa systems along with experimental data. Scattered data are color-coded by third order rotational invariant w_4 value. Data for perfect hcp (Δ) and fcc (∇) are also plotted. Distribution (b) shows a liquid-like system with $\Gamma_s \sim 1$, while case (a) corresponds to crystallized Yukawa system with $\Gamma_s \simeq 10^4$. Experimental data are scattered within the area marked with (c) and presented in detail in the insert.

7.5.1.2 3D Complex plasmas onboard International Space Station

The three-dimensional positions of the particles observed in an experiment onboard the International Space Station (ISS) are presented in Figure 7.31. The figure shows one of the best plasma crystal ever created. As local order analysis of the 3D data shows, more than 90% of the particles are in the crystalline state. Only a small part of the plasma crystal close to the void is melted (liquid-like state).

Figures 7.32 and 7.33 show particle distribution over q_4–q_6 plane for both experimental data and results of molecular dynamics simulations of the crystallized Yukawa system. The results are in favor of the Yukawa crystallization onboard ISS.

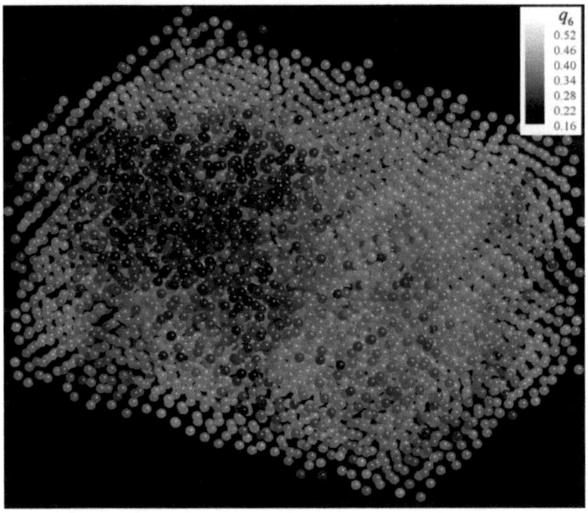

FIGURE 7.31
Recorded particle positions in an experiment onboard ISS. Particle are color-coded by q_6 value. About 10^4 particles were detected.

7.6 On the role of dust in cometary plasma

The cometary coma consists of neutral gas, plasma, and dust grains. Dust grains can influence both the neutral and charged coma's constituents. Usually, the presence of dust particles in a plasma results in additional losses of both electrons and ions due to the plasma recombination on the particle surfaces. Solar radiation makes the impact of dust even more complicated: It now depends on the solar flux, the dust number density, photoelectric properties of the dust particles, the dust particle composition, distribution of sizes, etc. Here we discuss a simple kinetic model developed to evaluate the role of dust particles in the coma plasma chemistry. This model demonstrates that this role can be crucial, resulting in a non-trivial behavior of both electron and ion densities in the plasma. It turns out that the coma's dust particles can be negatively as well as positively charged depending on their composition. These opposite charges can result in fast coagulation of the particles forming complex aggregate shapes of cometary grains.

Let us analyze the impact of dust grains on the plasma chemistry of inner cometary comae. As we are mostly interested in qualitative effects of the grain presence in the coma, the simplified model of the cometary coma is used. The comet is located at heliocentric distance of 1 AU (1 AU $\approx 1.5 \times 10^{13}$ cm is the average Earth–Sun distance). Main ionization source for such a coma is photoionization of water molecules by extreme solar UV light ($\lambda < 98.4$ nm). Typical plasma densities are n_e, $n_i \sim 10^2$–10^4 cm^{-3}; the density of neutrals is in the range of $n_n \sim 10^{10}$–10^{13} cm^{-3}. As density decays with the distance as r^{-2} from the nucleus, the typical size of the region of interest is on the order of 10^3 km.

In this study, we use the gas-to-dust mass ratio equal to unity. A typical size a of dust particles is believed to be 0.3–3 μm, so the typical dust number density is $n_d \sim 0.1 - -10^2$ cm^{-3} for the neutral density $n_n \sim 10^{13}$ cm^{-3} of interest here (we assume the mass density of a grain to be 1 g cm^{-3}). These parameters of the coma are quite realistic (Altwegg *et al.* 1999; Combi *et al.* 2004; Rodgers *et al.* 2004; Haider and Bhardwaj 2005).

Microparticles can affect the coma plasma composition via different ways. First, plasma recombination on the surfaces of dust particles can constitute a significant sink of the plasma. This effect, depending on the dust number density, often results in strong depletion of the plasma. Second, the dust particle surfaces can be the source of electrons due to the photoelectric effect. We note that dependence of the efficiency of the radiation absorption on the refractive index, size of a dust particle, and the radiation wavelength is quite complex. Figure 7.34 shows the absorption (scattering) efficiencies $\sigma_{abs/sca}$ calculated for 121-nm radiation (hydrogen Lyman-α radiation) according to the Mie theory as functions of the particle size a. We consider here the Lyman$-\alpha$ line because it is the most important for the photoelectric effect on dust

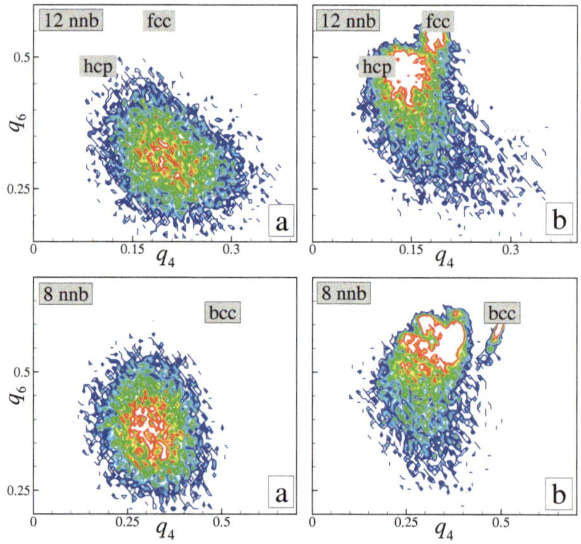

FIGURE 7.32
Local order analysis of data from an experiment onboard ISS. The figures show initial liquid-like (a) and crystallized (b) state of the complex plasma on the plane of rotational invariants q_4–q_6, calculated by using 12 nearest neighbors (top panels) and 8 nearest neighbors (bottom panels). Additionally rotational invariants for perfect lattice types (hcp, fcc, bcc) are plotted.

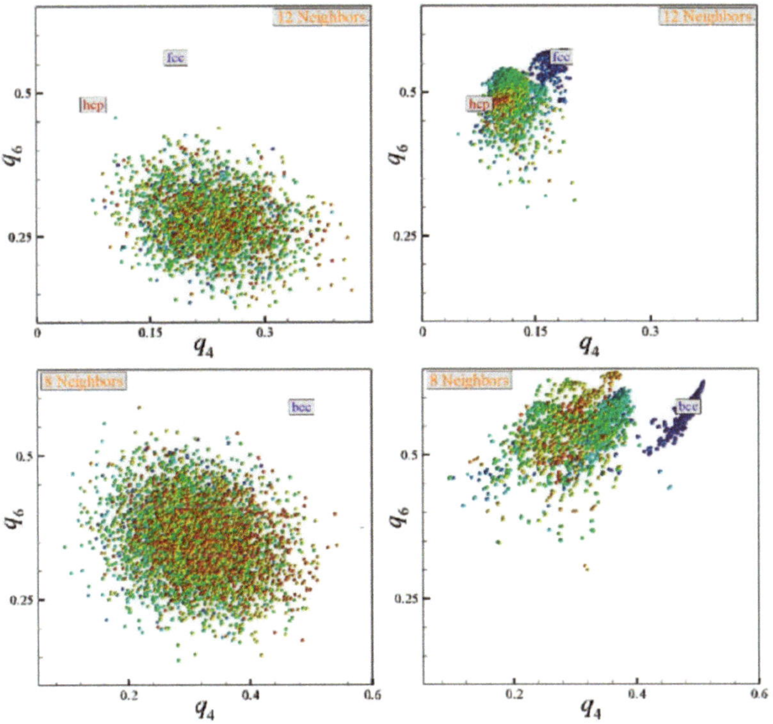

FIGURE 7.33
Local order analysis of MD simulations of Yukawa systems with hard wall confinement.

particles, see Equation (7.34) below.

The inset of Figure 7.34 shows the complex refractive index as a function of the photon energy for several different materials (Palik 1998). We point out a very small imaginary part of the refractive index m (which determines the absorption of radiation by a particle) in the optical range for ice particles. The values of m for other considered materials are relatively large. We see that the absorption efficiencies of the radiation are close to 1 for the grain sizes of interest ($a \sim 1$ μm).

The effect of dust particles on the ionization balance in the cometary coma can be estimated from the system of the continuity equations for the number densities of electrons n_e and positive ions n_i (H_2O^+, OH^+, H^+, O^+, etc), as well as for the grain charge $Q = Ze$. In the local approximation, the set of equations is given by (Klumov

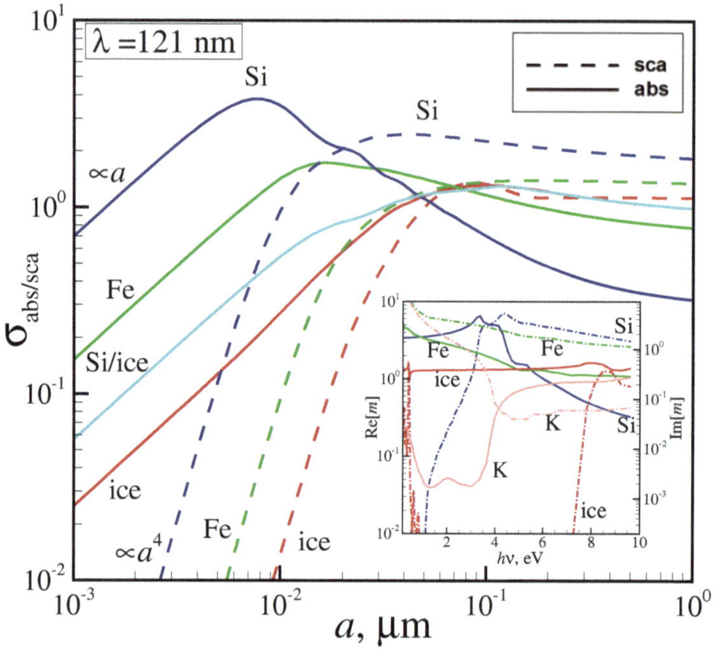

FIGURE 7.34
Optical properties of various-composition micro-particles as calculated using the Mie theory. The efficiencies of scattering σ_{sca} and absorption σ_{abs} of 121-nm radiation versus the particle size a for various micro-particle materials: ice, iron, silicon, and ice-coated silicon (at the corresponding radius ratio 1:2). The known scaling is shown for small a values ($\sigma_{sca} \propto a^4$ and $\sigma_{abs} \propto a$). The inset shows both the (solid lines) real and (dash-dotted) imaginary parts of the complex refractive index m versus the photon energy $h\nu$ for several materials which can be present in a dust particle.

et al. 2000, 2005a,b):

$$\begin{aligned}
\frac{\partial n_e}{\partial t} &= q_e - \alpha_m^{\mathrm{rec}} n_e n_i^m + \pi \langle \nu_{ed}^{\mathrm{photo}} a^2 n_d \rangle - \pi \langle \nu_{ed} a^2 n_d \rangle, \\
\frac{\partial n_i^a}{\partial t} &= q_e^a - \beta_a n_i^a - \pi \langle \nu_{ad} a^2 n_d \rangle, \\
\frac{\partial n_i^m}{\partial t} &= q_e^m + \beta_a n_i^a - \alpha_m^{\mathrm{rec}} n_e n_i^m - \pi \langle \nu_{md} a^2 n_d \rangle, \\
\frac{\partial Z^a}{\partial t} &= \nu_d^{\mathrm{photo}} + \nu_{ad} + \nu_{md} - \nu_{ed}.
\end{aligned} \tag{7.32}$$

Here, we divide positive ions into two distinct groups: the atomic (superscript a) (O^+, H^+, etc.) and molecular (superscript m) ions (H_2O^+, OH^+, CO_2^+, etc.). For atomic ions recombination in three-body collisions can be neglected in comparison with other processes. The recombination coefficient for molecular ions is $\alpha_m^{\mathrm{rec}} \sim$

10^{-7} cm^3 s^{-1}. The introduced groups reflect the presence of atomic and molecular positive ions in the inner coma. More detailed plasma chemistry of the cometary coma is not necessary for the present purpose. The terms ν_{ed} and ν_{jd} ($j = a, m$) describe the losses of electrons and ions on dust particles; β_a is the conversion rate of atomic ions to molecular ones due to the charge transfer reaction; $q_e = q_e^a + q_e^m$ is the ionization rate due to primary sources of ionization (solar UV radiation, photoelectron, fast particles, etc.) that provide a steady state plasma density in the dust-free coma at the level of $n_e = n_i \approx \sqrt{q_e/\alpha^{\text{rec}}} \simeq 10^2 \div 10^4$ cm^{-3}. Furthermore, in Equations (7.32) $\langle ... \rangle$ means averaging over the particle size distribution and ν_d^{photo} describes creation of photoelectrons when solar radiation is absorbed by a single dust particle with the radius a. The coefficients ν_{ed} and ν_{jd} can be evaluated using the orbit motion limited (OML) approach (see Section 2.1.1.2).

The dust induced photoelectric rate ν_d^{photo} can be evaluated from the expression

$$\nu_{ed}^{\text{photo}} = \int_0^{\lambda_W} \sigma_{\text{abs}}(\lambda, a, m) \Phi_\lambda Y(\lambda, m, a) d\lambda, \tag{7.33}$$

where Φ_λ is the spectral solar flux and λ_W is the maximum (threshold) radiation wavelength for the photoelectric effect ($hc/\lambda_W = W$, where W is the work function for a given grain material). For example, λ_W for ice, iron, sodium, potassium, aluminum, and silicon corresponds to the photon energy of 8.7, 4.7, 2.4, 2.3, 4.1, and 4.85 eV, respectively. The photoelectron yield $Y(\lambda, m, a)$ entering into the expression for ν_{ed}^{photo} increases rapidly with the photon energy in the above-threshold region ($|\lambda/\lambda_W| \leq 1$) and is often estimated by the expression $Y(\lambda) = Y_\infty(1 - \lambda/\lambda^*)$ (Draine 1978), which interpolates experimental data. The characteristic values are $Y_\infty \sim 10^{-2}$–10^{-1} and $\lambda^* \simeq \lambda_W$. It is worth noting that Y increases when the grain size a decreases. In the considered model, the yield Y appears to be an important parameter. We stress that λ_W depends on the particle charge. If the particle is positively charged, only photons with energies $2\pi\hbar c/\lambda - W - Qe/a$ create photoelectrons.

In the absence of the photoelectric effect, the OML currents lead to the net negative charge on dust due to higher mobility of plasma electrons as compared to that of ions. It should be noted that for typical coma conditions the effect of ion–neutral collisions on the particle charging can be important. It has been shown (see, e.g., Fortov *et al.* 2005a; Khrapak *et al.* 2005) that even rare ion–neutral collisions increase total ion current to an individual particle and, hence, the absolute magnitude of the (negative) charge decreases. In collision-dominated regime, the ion current decreases due to decrease of the ion mobility, so that the particle charge increases in the absolute magnitude (see, e.g., Khrapak *et al.* 2007). Within the range of plasma coma parameters (of interest here) both decrease as well as increase of the particle charge are possible, see sketch in Figure 2.7. However, the associated differences in the charges in comparison with the (collisionless) OML approach are not too large. Moreover, the presence of other dust grains also leads to a change in the particle charges. In fact, this effect can completely mask the effect of ion–neutral collisions on particle charging, and hence, the collisionless OML model can be employed in the first approximation.

The ionization rate q_d^{photo} induced by the photoelectric effect on dust particles can be roughly estimated from

$$q_d^{\text{photo}} \simeq \pi a^2 n_d \sigma_{\text{abs}} F(\text{Ly}-\alpha) Y(\text{Ly}-\alpha, m, a). \tag{7.34}$$

For the assumed parameters of cometary coma, viz., the dust number density $n_d \simeq 0.1 \div 10^2$ cm^{-3} and the grain size $a \simeq 1$ μm, the value of q_d^{photo} can achieve $10^2 \div 10^4$ cm^{-3} s^{-1}, which can be much higher than the steady state ionization rate $q_e \simeq \alpha^{\text{rec}} n_e^2 \simeq 0.1 \div 10\,\text{cm}^{-3}\text{s}^{-1}$ for a dust-free coma.

Figure 7.35 shows results of kinetic simulations of dusty coma. The set of equations (7.32) is solved numerically for the assumed realistic conditions. The case study presented in Figure 7.35 shows the dust particle charge number Z (bottom panel), the electron density n_e (top panel), and the molecular ion density n_m (middle panel) versus the grain number density n_d and the effective (resulting in the photoelectric effect) cumulative solar flux. We can clearly see that strong influence of dust particles on the plasma composition occurs at fairly moderate values of dust number densities $n_d \sim 1$ cm^{-3}. As a result, both depletion and increase of the electron density can be observed in the dusty coma (depending on the grain properties and number density) complemented by complicated behavior of the molecular ion component. This complex behavior can be explained by the competition of ionization/recombination processes.

The particle charge numbers can be rather large ($|Z| \gg 1$), and this can drastically change the transport properties of such a plasma. This means that any quantitative analysis of cometary plasma environment [e.g., by using the multiscale MHD model (Haberli *et al.* 1997)] should take into account the dust-related effects.

We stress that the charge on dust particles can be positive as well as negative, depending on the grains photoelectric properties and their concentration. The effect is shown in more detail in Figure 7.36.

Figure 7.36 shows case study of the impact of particle number density n_d and composition (ice/iron) on plasma parameters. Top panel (a) shows the variations of charge numbers of both icy (Z^{ice}) and iron (Z^{iron}) micrometer size particles versus n_d at different ice/iron particles mixing ratio γ_{ii} ($\gamma_{\text{ii}} \equiv n_d^{\text{ice}}/n_d^{\text{iron}}$, $n_d^{\text{ice}} + n_d^{\text{iron}} = n_d$). Bottom panel (b) shows electron n_e and ion n_i densities at different dust number density n_d and different ice/iron particles mixing ratio. The value of ice/iron mixing ratio γ_{ii} varies from 0 to 1. It is clearly seen from Figure 7.36 that both positively and negatively charged particles are present in the coma in wide ranges of n_d and γ_{ii}.

The effect of bipolar charges on dust particles can result in coagulation of the oppositely charged grains, which in turn strongly influences the composition of the coma plasma. Because of relatively strong electric forces, the characteristic times of the coagulation are relatively short. As a result of coagulation, larger particles form, with more probable negative charge residing on them due to plasma absorption (with higher electron mobility). The formed aggregate particles can be of complex fractal shape. It is important that these aggregates are formed directly in coma bulk and not

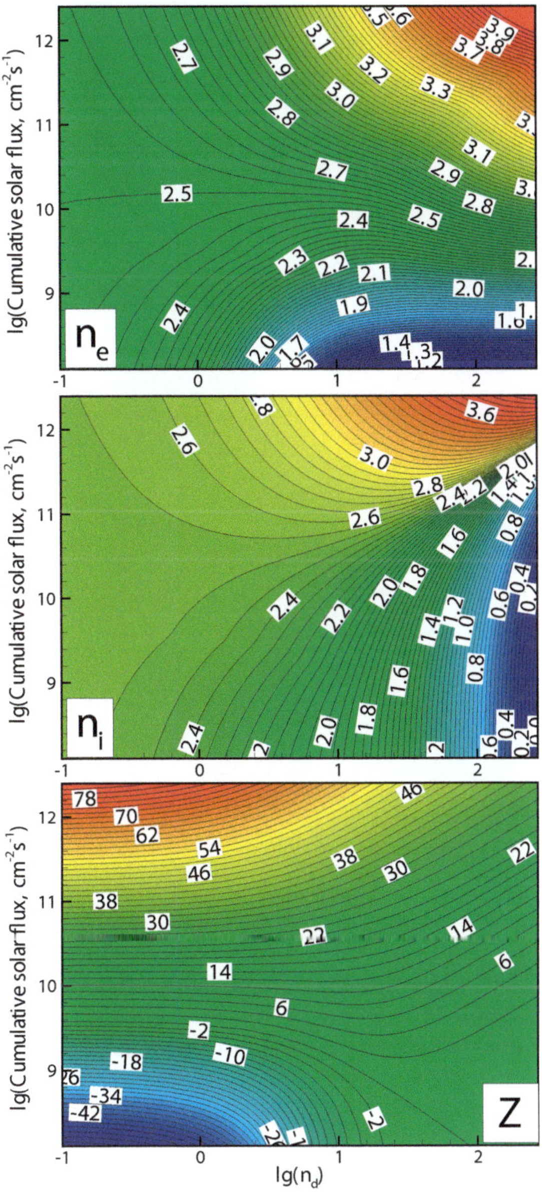

FIGURE 7.35

Plasma composition of a cometary coma from numerical solution of Equations (7.32). The dust particle charge number Z (bottom panel), logarithm of the electron density $\lg(n_e)$ (top panel) and logarithm of the molecular ion density $\lg(n_i^m)$ (middle panel) in the cometary coma at 1 AU are presented versus the effective solar flux and the dust particle number density n_d. The neutral gas number density is about 10^{12} cm^{-3}; the dust particle size is 1 μm; the ionization rate is $q_e \simeq 1$ cm^{-3} s^{-1}. Strong influence of grains on the plasma composition occurs at fairly moderate values of the dust number density n_d.

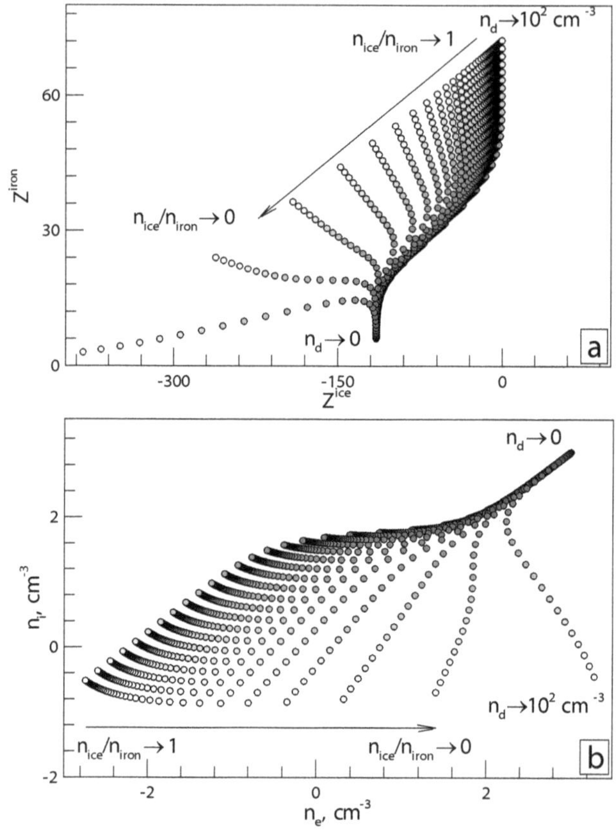

FIGURE 7.36
Impact of the particle number density n_d and composition (ice/iron) on plasma parameters. Top panel (a) shows the variations of charge numbers of both icy (Z^{ice}) and iron (Z^{iron}) micrometer size particles versus n_d at different ice/iron particles mixing ratio. Bottom panel (b) shows electron n_e and ion n_i densities at different dust number density n_d and different ice/iron particles mixing ratio.

introduced there by expanding neutral gas drag from evaporating nucleus' surface. Recent optical observations by Kolokolova *et al.* (2004) provide some support to the assumption about complex shapes of coma grains. In future modeling of cometary environment, the effects of time-dependent distribution of grain sizes (and their influence on the coma plasma) should be taken into account.

To conclude, we have discussed here a kinetic model describing the impact of dust particles on cometary comae. We have shown that dust particles, depending on their sizes, number density and photoelectric properties, can strongly affect plasma composition of the dusty cometary coma. It is important that positively as well as negatively charged dust particles can appear in the coma. These opposite charges result in fast coagulation of particles. The shape of the resulting aggregated grains can be complex.

7.7 Electronegative complex plasmas

Addition of an electronegative gas leads to the appearance of negative ions in the discharge. This can strongly affect the plasma parameters and particle charges and, hence, the phase state of the particle component of complex plasmas. The presence of negative ions also affects forces acting on the particles which, in turn, can affect the equilibrium properties of particle structures, e.g. location, shape of the structure, void size.

Here we discuss a change in the state of complex plasma upon the addition of molecular oxygen O_2 to the rf argon discharge. To begin with, we estimate the influence of molecular oxygen on the plasma composition. For this purpose, we will use the plasma chemical model of discharge in the Ar/O_2 mixture (all plasma parameters are averaged over the discharge volume). For a quasi-neutral complex plasma, the corresponding set of equations has the form:

$$\frac{\partial n_j(t)}{\partial t} = R_j^{\text{prod}} - R_j^{\text{loss}} - R_j^d - n_j/\tau_j, \tag{7.35}$$

$$\frac{\partial Z}{\partial t} = \nu_{i+} - \nu_{i-} - \nu_e, \tag{7.36}$$

$$n_{i+} + Zn_d = n_e + n_{i-}. \tag{7.37}$$

Here, n_j describes the concentrations of electrons and all sorts of ions (positive and negative) in the discharge and also the concentrations of metastable argon and oxygen atoms and molecules which can affect the plasma parameters. R_j^{prod} and R^{loss} describe the photochemical sources and sinks of the j-th component; R_j^d describes the loss of j-th component at the surface of dusty particles; and $\tau_j \simeq L^2/D_j$ is the diffusional lifetime of the j-th component in the discharge where D_j is the corresponding diffusion coefficient. We note that the diffusion coefficient strongly depends on the plasma composition; e.g., for the positive ions, D_j ranges from the ambipolar (at low concentrations of negative ions) to unipolar (for ion-ion plasma) type. In the charging equation, the terms ν_e, ν_{i-} and ν_{i+} describe the electron, negative-ion and positive-ion fluxes to a particle, respectively. The last can be evaluated using OML approximation.

Since $T_{i+} \approx T_{i-} \ll T_e$, the contribution of the negative ions to the charging flux balance can be ignored. The equilibrium concentrations of the charged components and the charge of a micron-sized particle in the considered discharge are presented in Figure 7.37 as functions of the partial concentration ([O_2]/[Ar]) of molecular oxygen

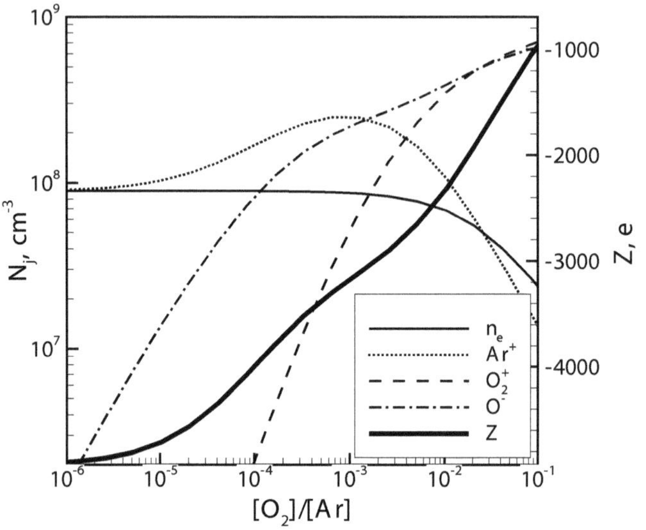

FIGURE 7.37
The composition of an rf discharge plasma in O_2/Ar mixture and the charge number *Z* of an individual particle as functions of the partial concentration $[O_2]/[Ar]$ of molecular oxygen.

in the mixture. One can see that, starting with even negligible concentrations of molecular oxygen ($O_2/Ar \geq 10^{-6}$), the plasma composition changes considerably to transform from the electron–ion plasma (*e* and Ar^+) to the ion–ion plasma, in which O_2^+ and O^- are the major ions while the electron density is strongly suppressed. Such a drastic transformation of the plasma composition is caused by the fast charge-transfer reaction $Ar^+ + O_2 \rightarrow Ar + O_2^+$ of argon ions on oxygen molecules (the back reaction is almost fully inhibited at room temperature due to a large difference in the ionization potentials of argon (15.75 eV) and molecular oxygen (12.2 eV) and by the electron dissociative attachment to the oxygen molecule: $O_2 + e \rightarrow O^- + O + e$.

The latter reaction produces negative ions, which are accumulated in the discharge due to the trapping electric field configuration for negatively charged particles. This effect results in a significant decrease in the absolute magnitude of the particle charge, as compared with the pure argon plasma.

It should be emphasized that such a change in plasma composition can have an appreciable effect on the processes of ion transport in the discharge, because the resonance charge–exchange Ar^+–Ar cross section exceeds, by approximately one order of magnitude, the polarization scattering cross section of the O_2^+ ion in argon. This effect can also be important for the momentum transfer from ions to particles, i.e. for the determination of the ion drag force. The dimensionless charge (potential)

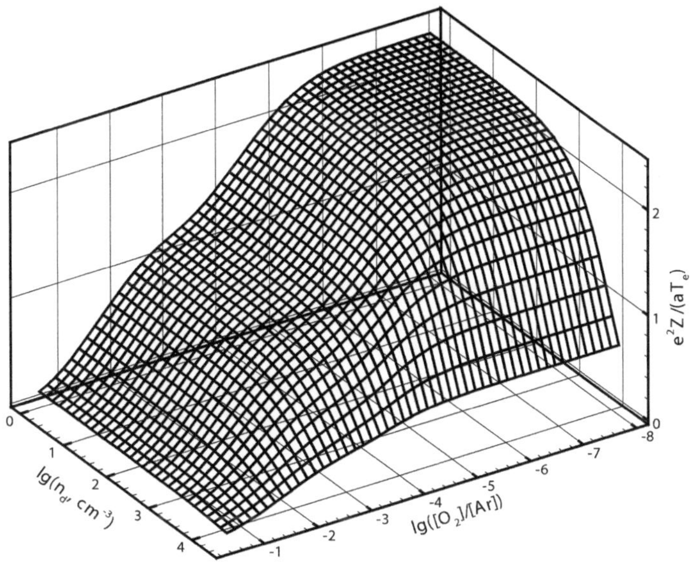

FIGURE 7.38
Dimensionless charge of micrometer particle in an rf discharge in O_2/Ar mixture versus the partial concentration ([O_2]/[Ar]) of molecular oxygen in argon and the particle concentration.

$|Q|e/aT_e$ of a dust particle is shown in Figure 7.38 as a function of particle concentration n_d and partial concentration of molecular oxygen $[O_2]/[Ar]$. One can see that, for $n_d \sim 10^3$–10^5 cm^{-3}, i.e., for experimental conditions typical for complex plasmas, a considerable decrease in the particle charge can be caused by the O_2 impurity. Note that the effects considered, likely show little dependence on the type of electronegative gas M. This is because the main processes inducing these effects – the formation and accumulation of negative ions in plasma and the fast charge exchange of argon ions on impurity species ($Ar^+ + M \rightarrow Ar + M^+$; the back reaction is inhibited because the ionization potential of the argon atom is greater than the ionization potential of M) – are efficient for any electronegative gas.

The results presented qualitatively describe changes in the charge composition of complex plasmas in the presence of the O_2 admixture. However, to determine the forces acting on the particles in a real discharge, it is necessary to know the spatial distribution of the complex plasma parameters in the discharge. Let us consider a one-dimensional discharge geometry (the coordinate $x = \mp L/2$ corresponds to electrodes, and $x = 0$ corresponds to the discharge center). The ion and electron spatiotemporal distributions can be determined from the set of balance equations after adding to Equations (7.35) the term describing the j-th component transport in the drift–diffusion approximation:

$$\frac{\partial n_j(x,t)}{\partial t} + \nabla J_j(x,t) = R_j^{\text{prod}}(x,t) - R_j^{\text{loss}}(x,t) - R_j^d(x,t), \tag{7.38}$$

$$J_j(x,t) = \mu_j n_j(x,t)E(x,t) - D_j \nabla n_j(x,t), \tag{7.39}$$

$$\frac{\partial Z}{\partial t} = \nu_p - \nu_n - \nu_e. \tag{7.40}$$

Here, J_j is the j-th charged component flux, μ_j is the mobility coefficient, and E is the electric field. As long as the mean free path of ions (both positive and negative) is much smaller than L, the drift–diffusion approximation is quite justified for the ions. To determine the electron spatial distribution, the Boltzmann distribution $D_e \nabla n_e(x,t) + \mu_e n_e E(x,t) \approx 0$ can be used. For the metastable species, only the diffusion term is taken into account in the transport equation. The boundary conditions for the set of equations (7.38)–(7.40) are the following: from the symmetry considerations, $J_j = 0$, $E = 0$ in the discharge center, and $n_j = 0$ in the near electrode region. The solution of the system of Equations (7.38)–(7.40) is shown in Figure 7.39.

The thermophoretic force can also make a contribution to the balance of forces acting on the particles. One can easily show that the pure argon discharge does not induce any noticeable thermophoretic force, because the temperature inhomogeneity of neutrals is very small for the typical discharge parameters. Indeed, in pure argon, the quenching $\text{Ar} + \text{Ar}^* \to 2\text{Ar}$ of metastable argon atoms with the rate constant k_q^{Ar} is the main source of heating neutrals. In this case, the neutral gas is heated up to $\Delta T_n \simeq k_q^{\text{Ar}} n^* \varepsilon^* \tau^* \leq 0.01$ K, where $\varepsilon^* \sim 10$ eV is the energy released by quenching and τ^* is the diffusion lifetime of a metastable atom in the discharge. Hence, this process cannot lead to a noticeable heating of neutrals. As a result, the thermophoretic force induced in a pure argon discharge is considerably smaller than the ion drag and electric forces.

In contrast, in the Ar/O_2 mixture, the situation with heating the neutral gas becomes quite different, because the metastable argon atoms can be efficiently quenched in the reaction $\text{O}_2 + \text{Ar}^* \to 2\text{O} + \text{Ar}$ with the rate constant $k_q^{\text{O}_2} \sim 10^{-10}$ cm^3s^{-1} resulting in heating of the neutral gas by $\Delta T_n \simeq 1$ K.

Even more efficient heating is induced by the metastable oxygen ion $\text{O}(^1\text{D})$. This effect is caused by a high concentration of $\text{O}(^1\text{D})$ in the discharge (because of the low excitation energy) and controlled by the rate of the quenching reaction $\text{Ar} + \text{O}(^1\text{D}) \to \text{Ar} + \text{O}$. In the general case, a change ΔT_N in the temperature of neutrals in their reactions of quenching metastable oxygen species can be estimated by the formula

$$\Delta T_n \simeq \sum_i k_e^i n_e \frac{[\text{O}_2]}{[\text{Ar}]} \tau_D \varepsilon_i, \quad \tau_D = L^2/D_n, \tag{7.41}$$

where k_e^i is the rate of the formation of the i-th metastable species by electron impact and ε_i is the corresponding transition energy. For $[\text{O}_2]/[\text{Ar}] = 10^{-2}$ the heating is mainly due to $\text{O}(^1\text{D})$, and ΔT_n can reach a few K. In this case the thermophoretic force can be comparable to the ion drag force for particles with sizes $a \simeq 3\mu$m. Note

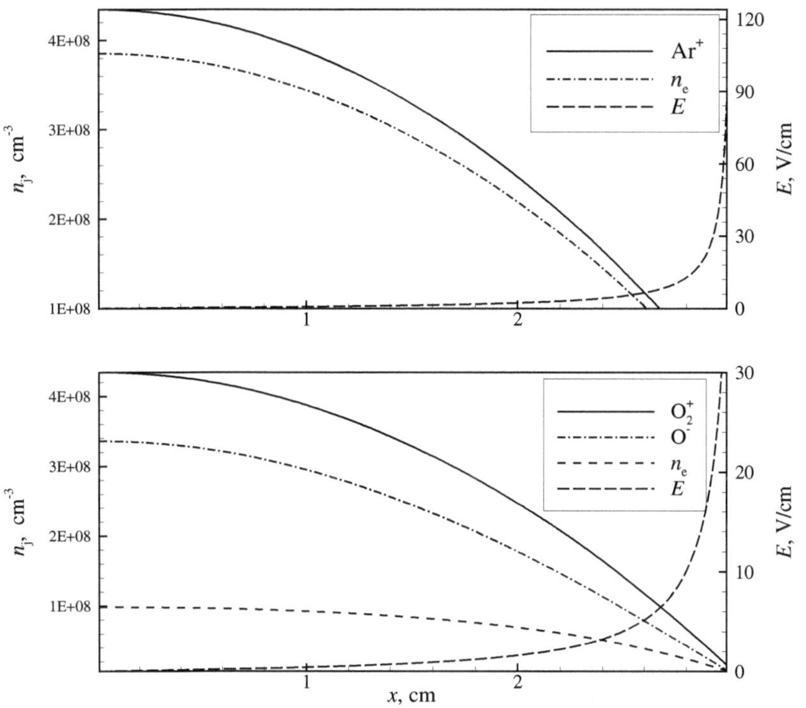

FIGURE 7.39
Spatial dependency of the electric field and ion and electron densities in an rf discharge in O_2/Ar mixture. Pure argon discharge (top panel) and $[O_2]/[Ar] = 10^{-2}$ (bottom panel).

that the central zone of the discharge is heated stronger than its periphery, so that the thermophoretic force expels particles from the discharge. Therefore, the considered mechanism contributes to the void formation in electronegative plasmas.

It is worth noting that the neutrals in argon plasma are heated by practically any impurity gas M. This heating is caused by the quenching reaction: $\mathrm{Ar}^* + \mathrm{M} \rightarrow \mathrm{Ar} + \mathrm{M}$ (as a rule, the rate constant k_q^{M} for such reactions considerably exceeds k_q^{Ar}). The efficiency of this process depends on the impurity concentration in the discharge, on the rate constant k_q^{M} and the fraction of the excited metastable state energy released to heat the neutrals (the rest of the energy is lost in the inelastic processes: ionization, dissociation, and excitation of the radiating M levels).

Thus, the addition of molecular oxygen to argon plasma induces a number of important effects. The composition and transport properties of plasma change substantially: electron–ion plasma transforms to the ion–ion plasma. The appreciable decrease in the particle charge can also change the phase state (e.g., melt plasma crystal) and the configuration of the particle structures. The electric field also decreases in the discharge (see Figure 7.39), thereby changing the force balance for the particles. In addition, metastable argon and oxygen states initiate heating of the neutral gas, and the corresponding induced thermophoretic force makes a considerable contribution to the balance of forces acting on the particles.

References

Ailawadi, N. K. (1980). Equilibrium theory of simple liquids. *Phys. Rep.*, **57**, 241–306.

Alba-Simionesco, C., Coasne, B., Dosseh, G., Dudziak, G., Gubbins, K. E., Radhakrishnan, R., and Sliwinska-Bartkowiak, M. (2006). Effects of confinement on freezing and melting. *J. Phys.: Condens. Matter*, **18**, R15–R68.

Altwegg, K., Balsiger, H., and Geiss, J. (1999). Composition of the volatile material in Halley's coma from in situ measurements. *Space Sci. Rev.*, **90**, 3–18.

Auer, S. and Frenkel, D. (2004). Numerical prediction of absolute crystallization rates in hard-sphere colloids. *J. Chem. Phys.*, **120**, 3015–3029.

Balescu, R. (1975). *Equilibrium and nonequilibrium statistical mechanics.* Wiley Interscience, Chichester.

Christenson, H. K. (2001). Confinement effects on freezing and melting. *J. Phys.: Condens. Matter*, **13**, R95–R133.

Clerc, M. G., Cordero, P., Dunstan, J., Huff, K., Mujica, N., Risso, D., and Varas G. (2008). Liquid–solid-like transition in quasi-one-dimensional driven granular media. *Nature Physics*, **4**, 249–254.

Combi, M. R., Harris, W. M., and Smyth, W. H. (2004). Gas dynamics and kinetics in the cometary coma: Theory and observations. In *Comets II*, Festou, M. C., Keller, H. U., and Weaver, H. A. (eds.), pp. 523–552. Univ. Arizona Press, Tuscon.

Cummins H. Z. and Pike E. R. (1974). *Photon correlation and light beating spectroscopy.* Plenum, New York.

Deegan, R. D., Bakajin, O., Dupont, T. F., Huber, G., Nagel, S. R., and Witten T. A. (1997). Capillary flow as the cause of ring stains from dried liquid drops. *Nature*, **389**, 827–829.

Donko, Z. and Nyiri, B. (2000). Molecular dynamics calculation of the thermal conductivity and shear viscosity of the classical one-component plasma. *Phys. Plasmas*, **7**, 45–50.

Doyle, D. A., Cabral, J. M., Pfuetzner, R. A., Kuo, A. L., Gulbis, J. M., Cohen, S. L., Chait, B. T., and MacKinnon, R. (1998). The structure of the potassium channel: Molecular basis of K^+ conduction and selectivity. *Science*, **280**, 5360, 69–77.

Draine, B. T. (1978). Photo-electric heating of inter-stellar gas. *Astron. J. Suppl.*, **36**, 595–619.

Farouki, R. T. and Hamaguchi, S. (1992). Phase transition of dense systems of charged "dust" grains in plasmas. *Appl. Phys. Lett.* **61**, 2973–2975.

Fortov, V. E., Nefedov, A. P., Petrov, O. F., Samarian, A. A., and Chernyschev, A. V. (1996). Particle ordered structures in a strongly coupled classical thermal plasma. *Phys. Rev. E*, **54**, 2236–2239.

Fortov, V. E., Nefedov, A. P., Vladimirov, V. I., Deputatova, L. V., Molotkov, V. I., Rykov, V. A., and Khudyakov, A. V. (1999). Dust particles in a nuclear-induced plasma. *Phys. Lett. A*, **258**, 305–311.

Fortov, V. E., Ivlev, A. V., Khrapak, S. A., Khrapak, A. G., and Morfill, G. E. (2005a). Complex (dusty) plasmas: Current status, open issues, perspectives. *Phys. Rep.*, **421**, 1–103.

Fortov, V. E., Vaulina, O. S., Petrov, O. F., Shakhova, I. A., Gavrikov, A. V., and Khrustalyov, Y. V. (2005b). Experimental study of heat transfer processes for macroparticles in a dusty plasma. *JETP Lett.*, **82**, 492–497.

Frenkel, Y. I. (1976). *Kinetic theory of liquid.* Oxford University Press, Oxford.

Haberli, R. M., Combi, M. R., Gombosi, T. I., De Zeeuw, D. L., and Powell, K. G. (1997). Quantitative analysis of $H2O^+$ coma images using a multiscale MHD model with detailed ion chemistry. *Icarus*, **130**, 373–386.

Haider, S. A. and Bhardwaj, A. (2005). Radial distribution of production rates, loss rates and densities corresponding to ion masses $\lesssim 40$ amu in the inner coma of Comet Halley: Composition and chemistry. *Icarus*, **175**, 196–216.

Hamaguchi, S., Farouki, R. T., and Dubin, D. H. E. (1997). Triple point of Yukawa systems. *Phys. Rev. E*, **56**, 4671–4681.

Hansen, J. P. and Verlet, L. (1969). Phase transitions of the Lennard–Jones system. *Phys. Rev.*, **184**, 151–161.

Hofman, J. M. A., Clercx, Y. J. H., and Schram, P. J. M. (2000). Effective viscosity of dense colloidal crystal. *Phys. Rev. E*, **62**, 8212–8233.

Ichimaru, S. (1982). Strongly coupled plasmas: high-density classical plasmas and degenerate electron liquids. *Rev. Mod. Phys.*, **54**, 1017–1059.

Khrapak, S. A., Ratynskaia, S. V., Zobnin, A. V., Usachev, A. D., Yaroshenko, V. V., Thoma, M. H., Kretschmer, M., Hofner, H., Morfill, G. E., Petrov, O. F., and Fortov, V. E. (2005). Particle charge in the bulk of gas discharges. *Phys. Rev. E*, **72**, 016406/1–10.

Khrapak, S. A., Klumov, B. A., and Morfill, G. E. (2007). Ion collection by a sphere in a flowing highly collisional plasma. *Phys. Plasmas*, **14**, 034502/1–4.

Klumov, B. A. and Morfill, G. E. (2007). Complex plasma in narrow channels: Impact of confinement on the local order. *JETP Lett.*, **85**, 498–502.

Klumov, B. A. and Morfill, G. E. (2008). Effect of confinement on the crystallization of a dusty plasma in narrow channels. *JETP Lett.*, **87**, 409–413.

Klumov, B. A., Popel, S. I., and Bingham, R. (2000). Dust particle charging and formation of dust structures in the upper atmosphere. *JETP Lett.*, **72**, 364–368.

Klumov, B. A., Morfill, G. E., and Popel, S. I. (2005a). Formation of structures in a dusty ionosphere. *JETP*, **100**, 152–164.

Klumov, B. A., Vladimirov, S. V., and Morfill, G. E. (2005b). Features of dusty structures in the upper Earth's atmosphere. *JETP Lett.*, **82**, 632–637.

Klumov, B. A., Rubin-Zuzic, M., and Morfill, G. E. (2006). Crystallization waves in a dusty plasma. *JETP Lett.*, **84**, 542–546.

Kolokolova, L., Hanner, M. S., Levasseur-Regourd, A., and Gustafson, B. (2004). Physical properties of cometary dust from light scattering and thermal emission. In *Comets II*, Festou, M. C., Keller, H. U., and Weaver, H. A. (eds.), pp. 577–604. Univ. Arizona Press, Tuscon.

Konopka, U., Morfill, G., and Ratke, L. (2000). Measurement of the interaction potential of microspheres in the sheath of a rf discharge. *Phys. Rev. Lett.*, **84**, 891–894.

Landau, L. D. and Lifshitz, E. M. (1980). *Statistical physics.* Pergamon, Oxford.

Lichtenberg, A. J. and Lieberman, M. A. (1992). *Regular and chaotic dynamics.* Springer, New York.

Löwen, H., Palberg, T., and Simon R. (1993). Dynamical criterion for freezing of colloidal liquids. *Phys. Rev. Lett.*, **70**, 1557–1560.

March, N. H. and Tosi, M. P. (2002). *Introduction to liquid state physics.* World Scientific, Singapore.

Meijer, E. J. and Frenkel, D. (1991). Melting line of Yukawa system by computer simulation. *J. Chem. Phys.* **94**, 2269–2271.

Mitic, S., Klumov, B., Konopka, U., and Morfill, G. (2008). Structural properties of complex plasmas in a homogeneous dc discharge. *Phys. Rev. Lett.*, **101**, 12502/1–4.

Ohta, H. and Hamaguchi, S. (2000). Molecular dynamics evaluation of self-diffusion in Yuakawa systems. *Phys. Plasmas* **7**, 4506–4514.

Ovchinnikov, A. A., Timashev, S. F., and Belyy, A. A. (1989). *Kinetics of diffusion controlled chemical processes.* Nova Science Publishers, Commack, New York.

Palik, E. D. (ed.). (1998). *Handbook of optical constants of solids*. Academic Press, New York.

Pozhar, L. A. (2000). Structure and dynamics of nanofluids: Theory and simulations to calculate viscosity. *Phys. Rev. E*, **61**, 1432–1446.

Raverche, H. J. and Mountain, R. D. (1972). Three atom correlation in liquid neon. *J. Chem. Phys.*, **57**, 3987–3992.

Raverche, H. J., Mountain, R. D., and Streett, W. B. (1972). Three atom correlation in the Lennard-Jones fluid. *J. Chem. Phys.*, **57**, 4999–5006.

Robbins, M. O., Kremer, K., and Grest, G. S. (1988). Phase diagram and dynamics of Yukawa systems. *J. Chem. Phys.*, **88**, 3286–3312.

Rodgers, R., Charnley, S. B., Huebner, W. F., and Boice, D. C. (2004). Physical processes and chemical reactions in cometary comae. In *Comets II*, Festou, M. C., Keller, H. U., and Weaver, H. A. (eds.), p. 505. Univ. Arizona Press, Tucson.

Rubin-Zuzic, M, Morfill, G. E., Ivlev, A. V., Pompl, R., Klumov, B. A., Bunk, W., Thomas, H. M., Rothermel, H., Havnes, O., and Fouquet, A. (2006). Kinetic development of crystallization fronts in complex plasmas. *Nature Physics*, **2**, 181–185.

Saigo, T. and Hamaguchi, S. (2002). Shear viscosity of strongly coupled Yukawa systems. *Phys. Plasmas*, **9**, 1210–1216.

Steinhardt, P. J., Nelson, D. R., and Ronchetti, M. (1983). Bond-orientational order in liquids and glasses. *Phys. Rev. B.*, **28**, 784–805.

Teng, L. W., Tu, P. S., and I, L. (2003). Microscopic observation of confinement-induced layering and slow dynamics of dusty-plasma liquids in narrow channels. *Phys. Rev. Lett.*, **90**, 245004/1-4.

Van Horn, H. M. (1969). Crystallization of a classical, one-component Coulomb plasma. *Phys. Lett. A.*, **28**, 706–707.

Vaulina, O. S. and Petrov, O. F. (2004). Simulation of mass transfer in systems with isotropic pair correlation of particles. *JETP*, **99**, 510–521.

Vaulina, O. S. and Vladimirov, S. V. (2002). Diffusion and dynamics of macro-particles in a complex plasma. *Phys. Plasmas*, **9**, 835–841.

Vaulina, O. S., Khrapak, S. A., Petrov, O. F., and Nefedov, A. P. (1999). Charge fluctuations induced heating of dust particles in a plasma. *Phys. Rev. E*, **60**, 5959–5965.

Vaulina, O. S., Vladimirov, S. V., Petrov, O. F., and Fortov, V. E. (2002). Criteria of phase transitions in Yukawa systems (complex plasma). *Phys. Rev. Lett.*, **88**, 245002/1–4.

Vaulina, O. S., Petrov, O. F., Fortov, V. E., Chernyshev, A. V., Gavrikov, A. V., Shakhova, I. A., and Semenov, Y. P. (2003). Experimental studies of the dynamics of dust grains in gas-discharge plasmas. *Plasma Phys. Rep.*, **29**, 606–620.

Vaulina, O. S., Petrov, O. F., Fortov, V. E., Chernyshev, A. V., Gavrikov, A. V., and Shakhova, I. A. (2004a). Tree-particle correlations in non-ideal dusty plasma. *Phys. Rev. Lett.*, **93**, 035004/1–4.

Vaulina, O. S., Petrov, O. F., Fortov, V. E., Morfill, G. E., Thomas, H. M., Semenov, Y. P., Krikalev, S. K., and Gidzenko, Y. P. (2004b). Analysis of dust vortex dynamics in gas discharge plasma. *Phys. Scr.*, **107**, 224–229.

Vaulina, O. S., Petrov, O. F., Gavrikov, A. V., Adamovich, X. G., and Fortov V. E. (2008). Experimental study of transport of macroparticles in plasma rf-discharge. *Phys. Lett. A*, **372**, 1096–1100.

Wallenborn, J. and Baus, M. (1978). Kinetic theory of the shear viscosity of a strongly coupled classical one-component plasma. *Phys. Rev. A.*, **18**, 1737–1747.

Wang, S. and Crumhansr, J .A. (1972). Superposition assumption. II. High density fluid argon. *J. Chem. Phys.*, **56**, 4287–4290.

Young, D. A. and Alder, B. J. (1971). Critical point of metals from the van der Waals model. *Phys. Rev. A*, **3**, 364–371.

Zhakhovskii, V. V., Molotkov, V. I., Nefedov, A. P., Torchinskii, V. M., Khrapak, A. G., and Fortov, V. E. (1997). Anomalous heating of a system of dust particles in a gas-discharge plasma. *JETP Lett.*, **66**, 419–425.

8

Diagnostics of complex plasma

Oleg F. Petrov and Olga S. Vaulina

8.1 Introduction

In studying dusty plasmas in addition to the diagnostics of the gas phase, one needs to determine the basic parameters of particles. These parameters, along with the parameters of plasma (densities and temperatures of electrons, ions, and neutrals), define the basic properties of plasma (electrophysical, optical, and thermodynamic). While the parameters of the gas phase can be determined by methods which were previously successfully used in studying gas plasma (in so doing, it is necessary to take into account the possible perturbation influence of particles on the measurement results), the diagnostics of particles requires the development and application of special methods. This chapter deals with the methods of diagnostics of parameters of particles such as their sizes, concentration, refractive index, and surface temperature. Most attention is given to the optical methods of diagnostics, because they offer a number of advantages. These advantages include high accuracy, absence of action on the object subjected to measurements, high-speed response, and the possibility of automatic data processing and acquisition of data in real time. It is further demonstrated how to take into account the effect of particles in measuring the parameters of gas phase by conventional methods such as the generalized reversal method and the total absorption method.

8.2 Light scattering and absorption measurements

Contactless diagnostic methods are based on measurements of attenuation, scattering, or emission of light by the particles. Problems of two basic types exist in the theory of scattering and absorption of light by small particles. The direct problem is that of calculation of scattered field provided that the basic characteristics of particle are preassigned, namely, the size $d = 2a$, the shape, and the absolute refractive index $m = n - ik$ of the particle material. The inverse problem consists in recovering these parameters from the characteristics of scattered field. It is the inverse problem that is of prime interest in the case of experimental investigations (Cummins and Pike

1974; Dubnishchev and Rinkevicius 1982; Bohren and Huffman 1983Klochkov *et al.* 1985).

In practice one often has to deal with aggregates of large numbers of particles. A system of particles is characterized by the extinction coefficient K_{ext}, scattering coefficient K_{sca}, absorption coefficient K_{abs}, and scattering indicatrix p. The coefficients K_{ext}, K_{sca}, and K_{abs} have the meaning of probability of respective process related to a unit volume. The indicatrix p defines the probability that a photon propagating in a certain direction will be scattered in some other direction. The correlation between the coefficients K_{ext}, K_{sca}, and K_{abs} is defined by $K_{ext} = K_{sca} + K_{abs}$.

In the general case, the theoretical approach to light scattering by numerous particles is a very complicated problem. However, simple relations can be obtained under certain conditions, which relate the parameters of dispersed medium with the optical parameters of a single particle,

$$K_{ext} = n_d \sigma_{ext}, \qquad K_{sca} = n_d \sigma_{sca}, \qquad K_{abs} = n_d \sigma_{abs}, \qquad p = p_0, \tag{8.1}$$

where σ_{ext}, σ_{sca}, and σ_{abs} denote the cross sections of extinction, scattering, and absorption of light by a single particle, respectively; p_0 is the indicatrix of light scattering by a single particle; and n_d is the number of particles per unit volume (particle density). Also used for characterizing the optical properties of particles are the efficiencies of extinction Q_{ext}, scattering Q_{sca}, and absorption Q_{abs},

$$Q_{ext} = \sigma_{ext}/S, \qquad Q_{sca} = \sigma_{sca}/S, \qquad Q_{abs} = \sigma_{abs}/S, \tag{8.2}$$

where S is the area of projection of particle onto a plane perpendicular to the incident beam.

The validity of relations (8.2) is defined by the conditions of single and incoherent (independent) scattering. The single scattering implies that the field scattered by all particles is small in the neighborhood of each particle compared to the field generated by electromagnetic wave incident on the particle. The total scattered field is the sum of fields scattered by single particles. The incoherence of scattering implies that the phases of waves scattered by single particles are not related by any relations. The total intensity of scattering by a system of particles in this case is equal to the sum of intensities of scattering by single particles.

At present, the problem involving a sphere of arbitrary radius and refractive index is exactly solvable in the theory of interaction between electromagnetic wave and small particle. This problem is known as Mie theory (Bohren and Huffman 1983).

8.2.1 Mie theory

The Mie theory deals with scattering and absorption of light by single particles. The problem is formulated as follows. A plane-polarized wave is incident on a homogeneous spherical particle of certain size and composition placed in a linear, isotropic, and homogeneous medium. It is necessary to determine the electromagnetic field at all points of the particle and at all points of the medium. As a result of solution of this problem, expressions are derived for the intensities of radiation scattering, as well as expressions for the efficiencies of extinction, scattering, and absorption, and for scattering indicatrix, as functions of complex refractive index $m = n - ik$ and diffraction parameter $\rho = \pi d/\lambda$ (λ is the wavelength of the radiation).

In addition to being used for describing the optical properties of a sphere, the Mie theory can be used in a first approximation for describing the optical properties of non-spherical particles as well. It was found that in the regime of small angles, particles, irrespective of their shape, scatter much like spheres of the same cross section. As the scattering angle θ increases, the differences increase in such a manner that the pattern of scattering for $\theta > 10°$–$15°$ becomes close to the pattern of diffraction on a sphere of the same volume.

A similar effect is observed for a polydisperse system of spheres. The effective dimension for such particles is usually provided by the Sauter diameter D_{32} which is defined by the ratio of average volume to average cross-sectional area and plays a leading role as regards small-angle scattering and cross sections of light scattering and absorption. The simulation of scattering particles of complex geometry or inhomogeneous structure offers no marked advantages, because the inversion of their optical characteristics enables one to form an opinion of only some equivalent parameters.

The diagnostics of particles in an optically dense medium calls for additional calculations which present no special difficulties given the present-day level of development of mathematical apparatus and computer equipment. The estimates of the effect of multiple scattering on radiation transfer in dusty plasma demonstrate that the single scattering approximation is valid in the case of the majority of laboratory investigations, for example, for measuring the attenuation at optical density of the medium τ <5–6 or for registering the scattering in the region of small angles $\theta < 10°$ at τ <1–2.

The processes of condensation or chemical reactions on the surface of particles can cause the inhomogeneity of the material of the particles. A model of a sphere in a shell is used for analyzing the scattering from a nuclei condensation coated with a liquid shell or from chemically reacting particles. The thickness of a nonuniform layer is not known in the majority of practical cases and is defined by the processes of conversion of particles; the equivalent refractive index of a uniform sphere is used to advance in studying these processes. The regular shape of the object is directly confirmed by the results of measurements of the total scattering indicatrix. Information about the uniformity of the sample can be obtained only for narrow-disperse distributions of weakly absorbing particles by way of analyzing the special features of their optical characteristics. These special features (lobes of indicatrix, oscillation of spectral dependence of optical cross sections) are due to the interference pattern of scattering by the particles. Therefore, any non-uniformities (of size, shape, or material) of single scatterers or of their ensemble smooth out the oscillation of the optical characteristics. The correspondence of the selected model (Mie theory, single scattering, etc.) to the experimental conditions can be confirmed by agreement between the results of independent measurements of the parameters of particles using different methods.

8.2.2 Determination of the size, concentration, and refractive index of particles

The principal characteristic defining the choice of the method of particles diagnostics is the diffraction parameter, that is, the ratio between the particle radius a and the wavelength λ of the radiation. For example, a combination of the methods of dynamic and static scattering is used for determining the sizes and complex refractive index m of particles whose sizes are small compared to the wavelength $(a|m-1| \ll \lambda)$. Successful solution of the inverse problem in this (Rayleigh) approximation is largely defined by simple analytical relations between the parameters of particles and the optical characteristics being recorded. The methods involving measurements of small-angle scattering enable one to recover the average diameter, concentration, and size distribution function of particles $f(a)$. These methods are limited to the range of validity of Fraunhofer diffraction $(a|m-1| \gg \lambda)$ and cannot be used for determining the refractive index of particles.

The methods involving measurements of attenuation of radiation are relatively simple among the methods employed for recovering the average diameter, refractive index, and density of particles whose sizes are comparable to the wavelength $(a|m-1| \sim \lambda)$. On the contrary, the methods involving measurements of scattering indicatrix are labor-consuming and complicate the measurement procedure.

For example, the *spectral transparency method* is based on measurements of extinction of light radiation at several wavelengths λ_i and on the data for the refractive index of particles. The extinction of a light beam passing through an aggregate of randomly positioned particles of different sizes is due to the absorption and scattering of radiation by the particles. Assuming that the multiple scattering can be ignored, the optical transparency Tr (extinction of radiation) at some wavelength λ_i is found from the correlation between the intensities of incident I_0 and transmitted I radiation,

$$Tr = I/I_0 = \exp[-\tau(\lambda_i)] = \exp[-n_d \overline{\sigma}_{ext}(\lambda_i) l], \tag{8.3}$$

where l is the optical path, $\tau(\lambda_i)$ is the optical density, and $\overline{\sigma}_{ext}(\lambda_i) = \int_0^\infty \sigma_{ext} f(a) da$ is the average extinction cross section for a wavelength λ_i.

The unknown parameters of particles (diameter or refractive index) are recovered from the results of measurements of relative optical density $q_i = \tau(\lambda_i)/\tau(\lambda_1)$ (at three or more wavelengths λ_i) using the procedure of minimization of the mean-square deviation S between the experimental q_i^{meas} and calculated q_i^{calc} data,

$$S = \sum_{i=1}^{N} \left(\frac{q_i^{calc} - q_i^{meas}}{N \delta_i q_i^{meas}} \right)^2, \tag{8.4}$$

where δ_i is the relative value of experimental error, and N is the number of wavelengths at which measurements are performed. Relative measurements in numerous cases enable one to eliminate from calculations additional unknown parameters, for example, the concentration of particles under investigation or geometry of the measuring volume.

Recovering the parameters of particles by the spectral dependencies of the characteristics being registered most often presumes that the dispersion of complex refractive index $m(\lambda)$ is known or that this dispersion can be ignored, $m(\lambda) = const$. The latter assumption holds for numerous weakly absorbing dielectrics in the visible and near IR spectral ranges, for example, aluminum and silicon oxides, as well as for ash and coal of some types.

In determining the parameters of particles, one needs to know the model size distribution of particles in addition to the model which takes into account the dispersion of refractive index (Nefedov *et al.* 1995, 1997). In the majority of cases, exponential-power functions and Gaussian distributions are used in calculations. In so doing, the dependence of spectral transparency Tr on the particle size for polydisperse particles is defined by their average Sauter diameter,

$$D_{32} = \frac{2\int\limits_0^\infty a^3 f(a)da}{\int\limits_0^\infty a^2 f(a)da}. \tag{8.5}$$

The Sauter diameter is further used for determining the mass and volume concentration C_v of particles,

$$C_v = \frac{2\tau D_{32}\overline{S}}{3l\overline{\sigma}_{ext}}, \tag{8.6}$$

where $\overline{S} = \pi\int\limits_0^\infty a^2 f(a)da$ is the average cross-sectional area for polydisperse particles. For large particles, whose average diffraction parameter is $\overline{\rho} = \pi D_{32}/\lambda \gg 1$, the volume concentration for known optical density τ depends only on D_{32}, because $\overline{\sigma}_{ext}/\overline{S} \approx 2$. For particles with $\overline{\rho} = 5 - 40$, the value of $\overline{\sigma}_{ext}/\overline{S}$ is defined by the Sauter diameter and complex refractive index of particles $m = n - ik$.

One can determine the average size and concentration of particles and, at the same time, recover the real part n of complex refractive index of weakly absorbing ($k \le 0.01$) particles using the *aperture transparency method* which involves measurements of the dependence of attenuation on the angular aperture of photodetector (Nefedov *et al.* 1997). Because the attenuation and scattering of monochromatic radiation are measured using this method, the dispersion of refractive index $n(\lambda)$ can be found as well.

A schematic diagram of measurements is given in Figure 8.1. In recovering the parameters of particles, use is made of the fact that, during measurements of extinction at high values of the parameter of diffraction of particles, a part of the radiation scattered by particles is incident on the photodetector because of finiteness of the aperture angle θ_d of photodetector and strong elongation of scattering indicatrix. As a result, the value of extinction cross section $\sigma^*_{ext}(\theta_d)$ turns out to be lower than their actual σ_{ext},

$$\sigma^*_{ext}(\theta_d) = \sigma_{ext} - \Delta\sigma = \int\limits_0^\infty \left(\sigma_{ext} - (1/2)\sigma_{sca} \int\limits_0^{\theta_d} p(\theta)\sin\theta d\theta \right) f(a)da. \tag{8.7}$$

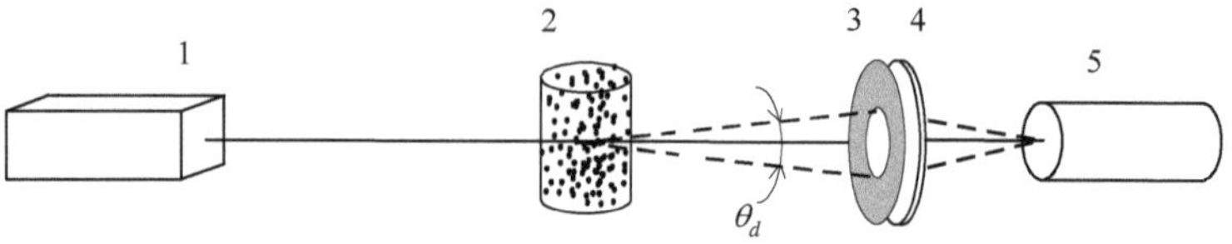

FIGURE 8.1
The scheme of measurements of attenuation of radiation: (1) radiation source, (2) particles, (3) aperture diaphragm, (4) lens, (5) photodetector.

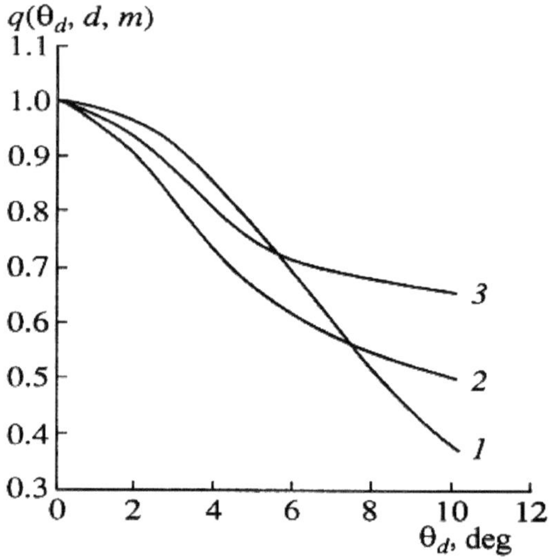

FIGURE 8.2
The angular distribution of relative attenuation $q(\theta_d, d, m)$ for monodisperse latex particles of different diameters d: (1) 2 μm, (2) 4 μm, (3) 3 μm; (λ = 0.633 μm).

The relative value of extinction $q[\theta_d, f(a), m] = \sigma^*_{ext}(\theta_d)/\sigma_{ext}$ will be defined only by the size distribution function of particles $f(a)$ and by the refractive index m. Given by way of illustration in Figure 8.2 is the dependence of $q(\theta_d, d, m)$ on θ_d for monodisperse latex spheres ($m = 1.58$) in water for different particle diameters d.

Note that the effect of multiple scattering on the value of q at $\theta_d < 10°–12°$ is insignificant as regards measurements of particles in plasma media with τ <1–2. The dependence of $q[\theta_d, f(a), m]$ on θ_d for homogeneous spheres is calculated by the Mie theory. The unknown parameters of particles can be found by minimization of mean-square deviation (8.4) between the calculated q_i^{calc} and experimentally obtained q_i^{meas} values for different aperture angles θ_{di} ($i = 1 - N$).

The optical methods are based on the principle of scattering or attenuation of light

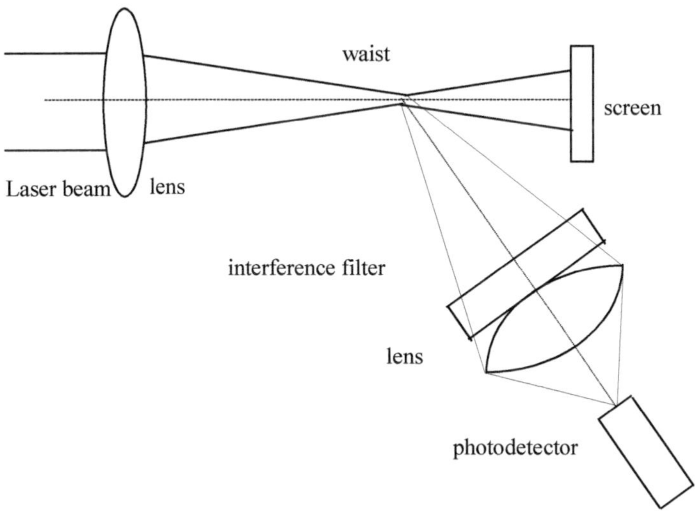

FIGURE 8.3
A typical scheme of optical counter.

by single particles moving in the measuring volume and of registering the intensity of scattered light or its attenuation by a photodetector which transforms light pulses to voltage pulses. Not more than one particle should be located in the measuring volume for correctly determining the size or concentration of particles. Otherwise, an error arises, because two and more particles located in this volume will be registered as one, thus leading to underestimation of the number of particles and to apparent increase in their size.

The devices designed to implement these methods are known as *optical counters*. The particle size is determined by the amplitude of pulsed signal generated by this particle at the photodetector outlet. For comparing the pulse amplitude with the particle size, calculations are performed by the Mie theory and the optical counter is precalibrated against particles of known size.

The advantages offered by optical counters include the high spatial resolution, the possibility of analyzing the parameters of particles in real time, the possibility of determining the particle size distribution function, etc. The most important requirements of optical counters as regards the diagnostics of particles in high-temperature flows include their minimal sensitivity to the complex refractive index of particle material and to the shape and temperature of particles.

A typical optical counter which registers particles with respect to scattering is shown in Figure 8.3. As a rule, the source of radiation is provided by a laser. The laser radiation beam is focused by the first lens for obtaining a waist of certain diameter W_0. The radiation scattered by particles is collected by another lens at an angle θ to the optical axis and directed to the entrance slit of photodetector. The measuring volume is formed by the intersection of the focal region of laser beam with the slit image. Signals from the photodetector are analyzed for obtaining information about the size distribution of particles. Calculations by the Mie theory and calibration measurements using particles of known size (latex particles, as a rule) are performed for determining the dependence of the amplitude of scattering signal on the particle diameter.

The minimal permissible concentration of particles in the flow when determining their size by an optical counter is 10^4–10^5 cm^{-3}. The upper limit of concentration of particles being analyzed is associated with the condition that not more than one particle enters the measuring volume at a time. This limit is inversely proportional to the size of the measuring volume. This in turn restricts the lower limit of the size of particles being analyzed, because a concentration of 10^6–10^7 cm^{-3} can be attained for submicron particles in some technical devices in high-temperature flows. The lower limit for the size of particles is approximately 0.1 μm. The decrease in size of the measuring volume causes a nonuniform distribution of laser radiation density in this volume. In so doing, particles of the same size, some of which fly through the center of measuring volume and the others through its edge, will initiate pulses of different amplitudes at the photodetector outlet. This poses an additional problem of uniqueness of determination of size and restricts the upper limit of the particle size.

8.3 Spectral methods of determination of particle parameters

8.3.1 Particle temperature

The majority of optical methods of determining the temperature of luminous plasma media are based on measurements of the intensity of plasma self-radiation and the attenuation of light from an external source. In the general case, determining the temperature of gas or particles calls for correct calculation of radiation transfer in the plasma medium under investigation and for reliable preliminary data on the emissivity E of the particle layer. The emissivity of particles in plasma is defined by its geometry and by the optical characteristics of single particles. Given the size, refractive index, and concentration of particles, the emissivity can be calculated by means of the scattering theory. However, the optical characteristics of particles are usually not known and need to be studied. Because the particles are exposed to a strong effect of external conditions, it is far from always that preliminary information can be obtained about their optical characteristics. Therefore, for reliable measurements of the temperature of two-phase plasma media, it is necessary that all of the parameters appearing in the equations of radiation transfer should be determined in the course of experimental measurements under the same external conditions.

One of the methods of determining the temperature of particles is by registering the self-radiation of the particles of the medium at different angles or at different optical densities. However, such measurements require *a priori* information either about the quantum survival probability ω or about the scattering indicatrix of particles p.

When the Kirchhoff law is applied, the determination of the temperature of particles reduces to measuring the radiation intensity and the emissivity of the layer of particles (*brightness pyrometry method*). The disadvantage of this method consists in the low accuracy of measurements, because in the general case the equation of radiation transfer must be solved for obtaining the emissivity. With this approach, one needs information about the geometric dimensions of the object under investigation, the scattering indicatrix, the quantum survival probability, the particle size distribution function, and the complex refractive index of the particle material. In addition, it is not always that the results of solution of transfer equation, obtained for infinitely extended media, can be used in application to real objects. Significant indeterminacy with respect to the value of emissivity arises because of inadequate knowledge of complex refractive index of particle material, especially, of its imaginary part (absorption index). The absorption index depends on both the wavelength and temperature.

The accuracy of temperature measurements can be increased by using the *method of color pyrometry*. For example, according to the method of two-color pyrometry, the particle temperature T_d is found from the expression

$$T_d = \left[1/T_c + \frac{\lambda_1\lambda_2}{c_2(\lambda_2-\lambda_1)}\ln\left(\frac{E(\lambda_1,T_d)}{E(\lambda_2,T_d)}\right)\right]^{-1}, \tag{8.8}$$

where $E(\lambda_i,T_d)$ is the emissivity of particles, c_2 is a constant in the Planck equation, and T_c is the color temperature of particles calculated as

$$T_c = \frac{c_2(\lambda_2-\lambda_1)}{\lambda_1\lambda_1}\left[5\ln(\lambda_2/\lambda_1)\ln\left(\frac{I_d(\lambda_2,T_d)}{I_d(\lambda_1,T_d)}\right)\right]^{-1}, \tag{8.9}$$

where $I_d(\lambda_i,T_d)$ is the particle radiation intensity.

The color method is advantageous in that the emissivity ratio $E(\lambda_1,T_d)/E(\lambda_2,T_d)$ used in the method weakly depends on temperature and geometric dimensions of the object. In the case of relative measurements, the effect of the error of measurement of optical density, particle sizes, and other parameters is reduced.

True values of temperature can be measured only if the dependence $E = E(\lambda,T)$ is known. The gray body approximation can be used. In this case, it is assumed that E is independent of λ. The layer of particles radiates as a gray body provided the condition $k\rho \geq 1$ is valid. In the visible region for particles with $k \geq 1$, this condition is valid for particles with radii exceeding 0.5 μm. This condition is less favorable for metal oxides ($k \leq 10^{-2}$), because very large particles are required. The relative error of measurements, caused by the "non-grayness" of radiation, will have the form

$$\frac{\Delta T}{T_c} = -\frac{\lambda_1\lambda_2 T_d}{c_2(\lambda_2-\lambda_1)}\ln\left(\frac{E(\lambda_1,T_d)}{E(\lambda_2,T_d)}\right). \tag{8.10}$$

Real disperse media are characterized by the dependence of temperature on particle radius $T_d = T_d(a)$. In this case the temperature averaged over particle sizes is measured. The exact solution can be obtained only under conditions of single scattering, given the particle size distribution function and the functional dependence $T_d(a)$. Another approach involves the determination of temperature of single particles.

The *radiation-absorption method* uses the transfer equation which can be written as follows for a plane-parallel layer of radiating, absorbing, and scattering medium (a high-temperature gas at temperature T_g containing particles at temperature T_d) assuming local thermodynamic equilibrium:

$$\mu \frac{\partial I_\lambda(x,\mu)}{\partial x} = -(\alpha_g + \alpha_d + \sigma_d) I_\lambda(x,\mu) + \alpha_g I^B(T_g) + \alpha_d I^B(T_g) + j_\lambda^{(s)}, \quad (8.11)$$

where $I_\lambda(x,\mu)$ is the radiation intensity at a point defined by coordinate x and angle ζ ($\cos\zeta = \mu$), α_g is the absorption coefficient of the gas, α_d is the absorption coefficient of the layer of particles, σ_d is the scattering coefficient of the layer of particles, and $I^B(T_g)$ is the Planck function. The term $j_\lambda^{(s)}$ defines the contribution of multiple scattering and can be written as

$$j_\lambda^{(s)} = \frac{\sigma_\lambda}{2\pi} \int_{-1}^{1} p(\mu,\mu') I_\lambda(x,\mu') d\mu', \quad (8.12)$$

where $p(\mu,\mu')$ is the probability of scattering of radiation in the direction μ.

The temperature of "non-gray" particles can be determined using the *spectrometric method* which does not require *a priori* information about the optical properties of particles. The method involves spectral measurements of self-radiation and optical density of a layer of particles and uses an empirical inversion technique (procedure of minimization of mean-square deviation between the experimental and calculated data) (Nefedov *et al.* 1995, 1997).

For plasma with particles, which optical properties at some wavelength λ are characterized by the survival probability of quantum (single scattering albedo) $\omega(\lambda) = \overline{\sigma}_{sca}(\lambda)/\overline{\sigma}_{ext}(\lambda)$, by the optical density $\tau(\lambda)$, by the scattering indicatrix $p(\lambda)$, and by the temperature T_d, the optical characteristics of particles are determined from the results of measurements of three signals, namely, S_P (signal of radiation of particles), S_L (signal of radiation of standard lamp with temperature T_L), and S_{PL} (signal of lamp radiation transmitted through plasma) at wavelength λ,

$$\frac{S_P(\lambda)}{S_L(\lambda)} = \frac{E(\lambda) I^B(T_d,\lambda)}{I^B(T_L,\lambda)}, \quad (8.13)$$

$$\tau(\lambda) = -\ln\left(\frac{S_{PL} - S_P}{S_L}\right), \quad (8.14)$$

$$E(\lambda) = [1 - \omega(\lambda)] \sum_{i=0}^{\infty} \omega(\lambda) \Lambda_\lambda^{(i)}(\tau,p), \quad (8.15)$$

where $\Lambda_\lambda^{(i)}(\tau,p)$ denotes some invariants, the value of which depends on the geometry of the medium and on the direction of radiation. Therefore, the emissivity $E(\lambda)$

is a function of two unknown quantities ω and p which are defined by the size and refractive index of particles.

Measurements of the optical density $\tau(\lambda)$ (8.14) enable one to select the required approximation for solving transfer equations and, consequently, the number of terms of the series $i = N \geq 0$, which must be taken into account in calculating $E(\lambda)$. For measuring the emissivity of the medium with $\tau(\lambda) < 3$, it is sufficient to take into account single scattering ($N = 1$), because exact calculations at $N \geq 2$ give a variation in $E(\lambda)$ of only 1–2%. When multiple scattering can be ignored ($i = 0$), the emissivity of disperse medium can be written as

$$E(\lambda) = [1 - \omega(\lambda)]1 - \exp[-\tau(\lambda)] = W(\lambda)1 - \exp[-\tau(\lambda)]. \tag{8.16}$$

The error in the determination of $E(\lambda)$ due to unaccounted-for multiple scattering for particles with the diffraction parameter $\overline{p} =$ 1–20 and $\omega(\lambda) = 0.6$ is $\lesssim 2\%$ for limited measuring volume at $\tau(\lambda) \leq 1$.

In this case, the problem of determining the temperature reduces to the choice of adequate approximation for spectral dependence of survival probability of quantum albedo $\omega(\lambda)$ [or function $W(\lambda) = 1 - \omega(\lambda)$]. Curves of the following form can be used for spectral approximation of the function $W(\lambda)$:

$$W(\lambda) = c/(\lambda^a \tau^b). \tag{8.17}$$

Here, the parameters a, b, and c are independent of the wavelength. Relative measurements eliminate the parameter c. In the absence of dispersion of refractive index of particles in the working wavelength range, the criterion of the choice of the approximation for $\omega(\lambda)$ can be provided by the parameter $\gamma =< \rho > (n-1)$. The values of a and b are given in Table 8.3.1 for different values of the parameter γ. Note that, for the majority of particles of combustion products, the typical values of refractive index $m = n - ik$ in the optical wavelength range lie in the ranges $n = 1.4 \div 2.4$ and $k = 0 \div 0.1$ or $k = 0.4 \div 1.4$.

The limiting cases 1 and 5 are quite obvious. For example, for Rayleigh particles with $\gamma < 0.1$, the survival probability of quantum albedo is $\omega(\lambda) = const/\lambda^3$, and $\tau(\lambda) \sim 1/\lambda$. In the majority of cases, the value of $\omega(\lambda)$ is less than or approximately equal to 0.01. This means that $W(\lambda) \to 1$, and the emissivity $E(\lambda)$ hardly depends on $\omega(\lambda)$. The plasma medium containing large particles ($\gamma > 40$) is optically gray [$E(\lambda) \sim const$]; $\tau(\lambda)$ and $\omega(\lambda)$ are independent of λ. In other cases, relation (8.17) is used for approximation of spectral dependence $W(\lambda)$. The choice of suitable approximation $E(\lambda)$ is based on the results of analysis of spectral dependence of $\tau(\lambda)$ in the working wavelength range or on additional data on the value of parameter γ obtained, for example, using other diagnostic methods. The parameter a will be the second unknown parameter (T_d, a) in the set of equations (8.13)–(8.15). This problem can be solved using regression analysis techniques by way of ensuring optimal agreement between the measurement and calculation data.

TABLE 8.1
Functions $W(\lambda)$ for spectral approximation of emissivity.

	$\gamma = <\rho> (n-1)$	$W(\lambda)$ $k \leq 0.1$	$W(\lambda)$ $k \geq 0.4$	$\tau(\lambda)$
1	$\gamma > 40/\chi^{(a)}$	$const$	$const$	$const$
2	$\gamma > 3$	$const/(\tau\lambda^a)$; $a = -0.1 \div 0.5$	$const\,\tau$	$d\tau/d\lambda > 0$; $(\pm d\tau/d\lambda)^{(b)}$
3	$3/4\chi < \gamma < 3$	$const/(\tau\lambda^a)$; $a = 1 \div 1.5$	$const/(\tau^{1/3}\lambda^a)$; $a = -0.5 \div 0$	$\pm d\tau/d\lambda$
4	$\gamma < 3/4\chi$	$const/(\tau\lambda^a)$; $a = 0.65 \div 1$	$const/(\tau\lambda^a)$; $a = 0.5 \div 0.6$	$d\tau/d\lambda < 0$; $\tau \sim \lambda^\beta$ $\beta = -4 \div (-1)$
5	$\gamma < \chi/10$	$const \to 1$	$const \to 1$	$\tau \sim 1/\lambda$

Note: $^{(a)}$ $\chi = 1.0$ for particles with $k \leq 0.1$, $\chi = 1.5$ at $k \geq 0.4$; $^{(b)}$ for the case of narrow-disperse distributions of particles with $k \leq 0.1$.

8.3.2 The spectrometric method of the particle size and refractive index determination

If the particle temperature is determined, the emissivity $E(\lambda)$ can be obtained from relation (8.13). On the other hand, $E(\lambda)$ is a function of two unknown quantities $\omega(\lambda)$ and $p(\lambda)$, which are determined by the size and refractive index (material) of particles. By solving Equations (8.13), (8.14), and (8.16), one can obtain the spectral dependence of albedo $\omega(\lambda)$, which contains information about the size and complex refractive index of particles $m = n - ik$. Similar to the spectral transparency method, it is assumed that the dispersion of complex refractive index $m(\lambda)$ is known or it can be ignored $[m(\lambda) = const]$.

For absorbing particles ($k \neq 0$), the spectral dependence of albedo $\omega(\lambda)$ is a function of Sauter diameter D_{32} and refractive index m, provided that the parameter γ meets the condition $\gamma \geq 5$. In this case, the unknown parameters of particles can be determined by the optimal agreement between the experimental $q_i^{meas} = \omega^{meas}(\lambda_i)$ and calculation $q_i^{calc}(D_{32}, m) = \omega^{calc}(\lambda_i)$ data on several wavelengths $\lambda_i (i = 1 \div N)$ using the procedure of minimization of the mean-square deviation (8.4). For particles with $\gamma < 5$, the spectral distribution of $\omega(\lambda)$ can depend on the form of the size distribution function. Then, most suitable model functions $f(a)$ are used.

8.3.3 Simultaneous determination of the particle size, refractive index, and temperature

The main difficulties of the optical diagnostics of particles are associated with the ambiguity of solution of inverse problems of the scattering theory in recovering two or more unknown parameters. Therefore, the majority of existing methods require *a priori* information about the size or refractive index of particles. In a number of cases, a combination of methods enables one to eliminate the ambiguity of inverse solutions in the absence of additional information about the parameters of the particles. For example, a procedure was developed involving measurements of spectral transparency and dynamic scattering for determining the dispersion of optical constants, concentration, and size of Rayleigh particles. The principal difficulties involved in the practical use of this procedure are associated with the application of the Kramers–Kronig relation, because the interpolation of measurement results in a limited spectral region can cause significant errors (from 15% to 50%) in determining $n(\lambda_i)$ and $k(\lambda_i)$ for individual wavelengths λ_i even in numerical calculations.

For particles with sizes comparable with or larger than the wavelength, a combination of the aperture transparency method and the spectrometric method enables one to simultaneously measure average sizes, concentration, and optical constants of particles, as well as to obtain information about the dispersion of complex refractive index (Nefedov *et al.* 1995).

If the dispersion of the refractive index of particles in the working spectral range cannot be ignored, neither the spectral transparency method nor the spectrometric method permits of reliable determination of the sizes in the absence of the information on the function $m(\lambda)$. In this case, an independent method can be employed to measure the average particle diameter D_{32}, and the value of $k(\lambda_i)$ can be determined from the equation $dS(\lambda_i)/dk(\lambda_i) = 0$ [see Equation (8.4)] for each wavelength λ_i.

The average diameter D_{32}, the concentration n_d, and the real part n of refractive index of particles can be obtained using the aperture transparency method. This method enables one to estimate the dispersion of particle sizes. This information can be used for the approximation of polydisperse distributions of particles with $\gamma < 5$ by model functions. Measurements of the spectral dependence of the optical density $\tau(\lambda)$ can be used for correction and elimination of possible ambiguity of the diameter D_{32}. Comparison of particle sizes obtained as a result of independent measurements by the spectrometric and spectral transparency methods serves as indirect verification of the effect of dispersion $m(\lambda)$ on the optical characteristics $E(\lambda)$ and $\tau(\lambda)$.

8.3.4 The effect of particles on the determination of the concentration of alkali metal atoms and the gas temperature

The concentration of alkali metal atoms, along with the temperature of gas and particles, defines the electrophysical properties of a dusty plasma and can have a significant effect on various physicochemical processes. The presence of particles in plasma media causes a variation of their optical and radiative characteristics. Therefore, in measuring the concentration of atoms and the gas temperature by conventional methods of optical diagnostics, one should take into account the effect of particles on the processes of radiation transfer (Samarian *et al.* 2000).

The conventional methods of determining the concentration and temperature of atoms in pure gas include the method of total absorption and the method of spectral line reversal.

Method of total absorption. The measurements of gas temperature and concentration of atoms presume the presence of a standard source of radiation calibrated against the blackbody temperature, with the optical beam $I(\lambda)$ of this source being focused such that scattered radiation would not reach the photodetector. The spectral radiation intensity distribution of the standard source (lamp), which passes in a two-phase medium and reaches the photodetector, can be written as follows, disregarding multiple scattering:

$$I(\lambda) = I^B(T_L,\lambda)\exp(-\tau) + \left[\alpha_g(\lambda)I^B(T_g,\lambda) + \alpha_d(\lambda)I^B(T_P,\lambda)\right][1-\exp(-\tau)]/\tau, \tag{8.18}$$

where τ is the optical density of two-phase plasma medium. Because the optical characteristics of particles (α_d, τ_d) within the narrow spectral range of the radiation line hardly depend on wavelength λ, the integral coefficient of total absorption on the spectral line of gas atoms can be written as

$$A(\lambda) = \frac{[S_L\exp(-\tau_d) + S_P - S_{PL}]\Delta\lambda}{S_L\exp(-\tau)}, \tag{8.19}$$

where S_L is the signal of lamp radiation, S_{PL} is the signal of lamp radiation that passed in the plasma, and S_P is the signal of self-radiation of the plasma. These signals are measured for the total spectral range of absorption with a rectangular instrument function of spectral width $\Delta\lambda$.

One can readily observe that the only difference from the case of "pure" gas consists in the factor $\exp(-\tau_d)$: it is this factor that determines the magnitude of additional attenuation by the particles. Therefore, the measurement of the coefficient $A(\lambda)$ (equivalent width of absorption line) does not present any special difficulties, and the concentration of atoms in a two-phase medium can be determined using conventional computational algorithms developed for "pure" gases.

Conventional reversal method. In the case where the optical density of particles τ_d is much lower than the coefficient of absorption of gas atoms $\alpha_g(\lambda)$ $[\alpha_g(\lambda) \ll \tau_d]$ within the working segment of the spectral line of radiation $\Delta\lambda$ isolated by the photodetector, the gas temperature can be found from the formula

$$I^B(T_g,\lambda) = I^B(T_L,\lambda)S_{PL}/(S_L + S_P - S_{PL}). \tag{8.20}$$

Therefore, if the condition

$$\frac{\tau_d\Delta\lambda}{\int \alpha_g(\lambda)d\lambda} < 0.1 \tag{8.21}$$

is valid at the center of the spectral line of radiation, we can ignore the effect of particles and determine the gas temperature by (8.20). In doing so, the error in determining the temperature T_g due to the effect of particles will not exceed 0.5%.

Generalized reversal method. One of the most optimal methods, which allows the elimination of the effect of particles on the results of measurements of gas temperature, is that of registering three signals (S_L, S_P, and S_{PL}) on two wavelengths λ_i and λ_j,

$$\frac{I^B(\lambda,T_g)}{I^B(\lambda,T_L)} = \frac{S_P(\lambda_i)\tau(\lambda_i)/\Lambda_i - S_P(\lambda_j)\tau(\lambda_j)/\Lambda_j}{[\tau(\lambda_i) - \tau(\lambda_j)]S_L}, \tag{8.22}$$

where $\Lambda_i = 1 - \exp[-\tau(\lambda_i)] = (S_L - S_{PL} + S_P)/S_L$.

The principal advantage in using relation (8.22) is that such measurements do not require information about the optical characteristics of the dispersed phase and, therefore, do not introduce additional errors. In addition, the algorithm described above enables one to partly compensate for the effect of multiple scattering and can be employed for determining the gas temperature in two-phase media with an optical density $\tau_d < 2$.

Note that the determination of T_g and $\alpha_g(\lambda)$ from measuring spectral intensity of lines of radiation of gas atoms calls for a detector device with a high resolution and for significant efforts to compensate errors of measurements. Registering the distribution of intensity over the spectral line profiles under rapidly changing conditions of real flows requires simultaneous spectral measurements on several wavelengths, which involve the use of a corresponding number of detectors. Therefore, the compensation for fluctuations of the optical characteristics of two-phase media results in bulky structures of optical instruments. At present, matrix photodetectors are extensively employed for simultaneous spectral measurements. However, the spectral resolution of such instruments does not always allow reliable measurements of the gas temperature by generalized reversal methods. If the radiation line segment $\Delta\lambda$ isolated by the detector is not narrow enough and the emissivity of the medium varies considerably, such measurements will result in a distortion of the temperature and of the optical characteristics.

References

Bohren C. F. and Huffman D. R. (1983). *Absorption and scattering of light by small particles*. Wiley, New York.

Cummins, H. Z. and Pike, E. R. (1974). *Photon correlation and light beating spectroscopy*. Plenum, New York.

Dubnishchev Y. N. and Rinkevicius B. S. (1982). *Metody lazernoi dopplerovskoi anemometrii (Methods of laser doppler anemometry)*. Nauka, Moscow (in Russian).

Klochkov V. P., Kozlov L. F., Potykevich I. V., and Soskin M. S. (1985). *Lazernaya anemometriya, distantsionnaya spektroskopiya i interferometriya (Laser anemometry, remote spectroscopy, and interferometry)*. Naukova Dumka, Kiev (in Russian).

Nefedov A. P., Petrov O. F., and Vaulina O. S. (1995). Analysis of radiant energy emission from high temperature medium with scattering and absorbing particles. *J. Quant. Spectrosc. Radiat. Transfer*, **54**, 453–470.

Nefedov A. P., Petrov O. F., and Vaulina O. S. (1997). Analysis of particle sizes, concentration, and refractive index in measurement of light transmittance in the forward-scattering-angle range. *Appl. Opt.*, **36**, 1357–1366.

Samarian A. A., Vaulina O. S., Nefedov A. P., and Petrov O. F. (2000). Analysis of the formation of ordered dust-grain structures in a thermal plasma. *Plasma Phys. Rep.*, **26**, 86–591.

9

Applications

Vladimir E. Fortov, Alexey G. Khrapak, Sergey V. Vladimirov

9.1 Technological and industrial aspects

Dusty plasmas have been present in various industrial applications for many decades. These are, e.g., precipitation of aerosol particles in combustion products of electric power stations, plasma spraying, and electrostatic painting. Moreover, powder-contaminated plasmas pose a number of challenges to the microelectronic industry, materials science, and gas discharge research and development areas. For example, particulate powders with the sizes comparable to feature sizes of the semiconductor integrated circuits have become a troublesome factor in the semiconductor micro-manufacturing. Dust in the plasma reactors often causes irrecoverable defects and line shorts in some ultra-large scale integrated circuits, which can totally compromise the entire microchip fabrication process.

In the beginning of the 1990s, it became clear that a large part of contamination found on the surface of silicon wafers after the manufacturing was not because of insufficient cleaning, but in fact was an inevitable consequence of plasma etching and deposition technologies. In most capacitively coupled rf discharge reactors, all particles are charged negatively and levitate close to one of the electrodes. After the discharge is switched off, they are deposited on the wafer surface. Sub-micron particles deposited on the wafer can reduce the working surface, cause dislocations and voids, and reduce adhesion of thin films. Enormous efforts put forth on reduction of the number of undesirable dust particles in industrial plasma reactors have brought positive results (Selwyn *et al.* 1989; Bouchoule 1999; Kersten *et al.* 2001, 2003a,b).

Thus dust obviously plays a role as a contaminant in plasma technological processes. However, it has become obvious in recent years that the presence of dust in plasmas does not necessarily lead to undesirable consequences. Thus, the accents in the dust particle research in technology are gradually shifting from the traditional view of them as unwelcome process "killer" contaminants to often desired elements that can, for example, dramatically affect and even improve the basic properties of plasma-made thin films.

Powders produced by employing plasma technologies can have interesting and useful properties: very small sizes (from a nanometer to micrometer range), monodispersity, and high chemical activity. The size, structure and composition of the powder can be varied easily in compliance with the specific requirements of a certain technology. In this connection, two trends can be distinguished in applied dusty plasma research (Kersten *et al.* 2001, 2003a,b; Vladimirov and Ostrikov 2004; Vladimirov *et al.* 2005). The first one represents a development of well-established technologies of surface modification, with the dust particles now being the subject of treatment. In order to create particles with specific properties, coating, surface activation, etching, modification, or separation of clustered grains in plasmas can be adapted. The second important trend is the creation of new nanostructure materials, like thin films with an inclusion of nanometer-sized particles. The typical size of the elements of integrated circuits in microelectronics is reduced every year and in the nearest future it will likely reach 10 nm. Furthermore, capacitively coupled rf discharges are often replaced by inductively coupled ones: The particle trapping is more difficult in capacitive discharges, which leads to a significant amount of the particles dropping onto the surface of the silicon wafer during plasma processing. Thus, the solution introduced in the 1990's, which was mostly based on dust particle confinement in special traps, does not work for these devices. This poses a serious problem for the production of integrated circuits of the next generation, which demands further applied research of the properties of dusty plasmas.

One of the key modern issues in the industrial applications of the complex plasma systems is in the tailoring of various properties of the plasma-generated micro- and nanoparticles and nanoclusters in the ionized gas phase. Such particles and clusters can be regarded as building units in nanofabrication involving doping, structural incorporation or self-assembly processes. In particular, in the low-energy nanocluster chemical vapor deposition (Jensen 1999), the cluster charge is critical for the growth of various silicon- and carbon-based nanofilms (Hwang *et al.* 2000). To this end, the charge and size of molecular clusters are the crucial parameters for the explanation of the unique architectures of many nano-sized objects. The nanocluster charge appears to be a key reason for the highly anisotropic growth of ordered nanostructures, such as silicon nanowires and carbon nanotubes (Hwang *et al.* 2000).

Apart from the common deleterious aspect, nano- and micron-sized particles have a number of applications in material engineering, optoelectronic, optical, petrochemical, automotive, mineral and several other industries. For example, ultra-fine particles can be efficiently incorporated into polymeric/ceramic materials to synthesize a number of advanced nanostructured materials for the applications as water repellent, protective, fire resistant, functional and other coatings. Furthermore, fine powders of $\sim$10 nm-sized particles have been widely used as catalysts for inorganic manufacturing, ultra-fine UV-absorbing additives for sunscreens and other outdoor applications. Other applications include textiles, wear-resistant ceramics, inks, pigments, toners, cosmetics, advanced nanostructured and bioactive materials, environmental remediation and pollution control, waste management, as well as various colloidal suspensions for mining, metallurgical, chemical, pharmaceutical industries, and food processing.

Nanoparticles have recently emerged as valuable elements of several technologies aiming to tailor the materials properties at nano-scales and manufacture novel nanoparticle-assembled materials with unique optical, thermal, catalytic, mechanical, structural and other properties and featuring nano-scale surface morphologies and architectures (Roco *et al.* 1999). The rapidly emerging applications of nanoparticles include nanopatterned and nanocomposite films, nanocrystalline powders and consolidated structures, and sophisticated nanoparticle assemblies that represent new forms of supramolecular crystalline matter. Further potential applications include but are not limited to nano-scale inorganic synthesis, dispersions and suspensions with the controlled fluid dynamics, and nano-sized single/few-electron data storage units.

In addition, fine particles, advertently injected into the plasma, can be trapped and subsequently coated to enhance their surface properties (e.g., for catalytic applications) (Kersten *et al.* 2003b). Plasmas can also be used to oxidize (and thus eliminate) contaminant particles in a bio-gas. In this way, it is also possible to electrically charge and control nanometer-sized soot particles in diesel engine exhausts. This is an example of the complex plasma treatment of soot and aerosol particles for environmental remediation.

A number of high-pressure discharge systems has recently been used for the synthesis of various nanoparticles. For example, metallic titanium nanoparticles can be generated in an atmospheric pressure plasmas containing argon, hydrogen, and titanium tetrachloride ($TiCl_4$). This is an alternative to conventionally used multi-stage production of metallic titanium from titanium ores such as rutile (Murphy and Bing 1994). Another interesting recent advance is the highly efficient synthesis of silicon nanoparticles in atmospheric pressure microhollow discharges in argon–silane mixtures (Sankaran *et al.* 2003). It is also interesting to note that metal nanoparticles can be synthesized by the plasma enhanced chemical vapor deposition in highly unusual environments such as micron-sized channels of microchannel glass. This can be achieved, e.g., by using a metal–organic precursor ferrocene $(C_5H_5)_2$-Fe and differential pumping across the microchannels (McIlroy *et al.* 2003).

Non-equilibrium atmospheric pressure plasmas in the dielectric barrier discharge configurations can also be used for electrostatic rupture of Gram-negative *E. coli* and Gram-positive *Bacillus subtilis* bacteria (Laroussi *et al.* 2003). The electrostatic disruption mechanism for cell rupture requires that the local electrostatic tension of the cell wall overcome its tensile strength. The electrostatic charge on the surface of bacteria originates due to electron and ion collection currents in a manner similar to solid dust grain charging in a plasma. To this end, airborne bacteria exposed to high-pressure non-equilibrium plasmas can also be regarded as microscopic "dust" particles.

To conclude, fine solid particles in plasma technologies are indeed deleterious process contaminants as, e.g., in microelectronics. To this end, significant progress has been achieved in the development of various methods of removal and suppression of dust in various industrial facilities. Most recently, however, the plasma-grown solid grains have become increasingly attractive for a number of thin-film technologies including but not limited to low-temperature self-assembly of ordered nanoparticle arrays, nanocrystalline and polymorphous silicon–based thin films for optoelectronic functionalities and devices, biocompatible calcium phosphate–based bio-ceramics for dental surgery and orthopedic applications, and hard wear-resistant, self-lubricating, UV-protecting and many other functional coatings. Despite a notable progress in industrial applications of the plasma-grown micro-/nanoparticles, some of them still remain at the research and development stage.

In view of the most recent advances in the science and applications of fine powders, control and manipulation (and eventually the adequate management strategies) of the plasma-grown solid particles is becoming a matter of utmost importance. However, this aim cannot be achieved without proper understanding of the underlying physics of the basics of plasma-fine particle interactions. For this reason, the dynamics and self-organization of the particulate matter, as well as various collective processes in low-temperature complex plasma systems, are crucial (Vladimirov and Ostrikov 2004; Vladimirov *et al.* 2005). Indeed, without proper understanding of the fundamentals of basic dust–plasma (and wherever applicable dust–solid substrate/wall) interactions, it is impossible to adequately and self-consistently describe the real processes in the complex plasma systems with variable-sized particulate matter of complex shapes and internal organization, such as nanoclusters and complex nucleates and agglomerates. On the other hand, knowledge of the collective phenomena involving charged dust particles is important in the studies of self-organization and critical phenomena in gas discharges that affect the origin and growth of fine powder particles. Thus, fundamental and applied aspects of the problems are inseparably associated with each other.

9.2 Dust in fusion reactors

It is well known that dust particles are present in magnetic confinement fusion devices (Winter 2000; Vladimirov and Ostrikov 2004). Their origin is mostly the plasma–surface interaction. The radioactive dust contains large amounts of hydrogen isotopes, with up to 50% in tritium.

Physically, tritium, incorporated into carbonaceous dust, undergoes radioactive decay, and this can lead to dust charging and the formation of the nuclear-induced plasmas. In the plasma, charged dust particles can be transported and levitated. There are thus two major sets of problems related to particulate generation in fusion devices. One of them is related to the safety of operation of the fusion reactor, the other being related to the plasma parameters and stability. Specifically, dust-bound tritium inventory is a major safety concern for future fusion reactors. The main problem in this regard is that dust cannot be reprocessed together with tritium, thus increasing the site inventory. Dust is also a potential carrier of tritium in the case of a severe reactor failure. Furthermore, if the reactor cooling systems are damaged, large amounts of hydrogen can form an explosive mixture with oxygen from the environment.

The key point of another aspect is that large amounts of dust can accumulate at the bottom of the device (which is usually a divertor area in tokamak and stellarator devices). Dust accumulation can impede the heat transfer to the cooling surfaces and also compromise specially designed gaps for electrical insulation or thermal expansion purposes. Such layers can sublimate when exposed to huge heat loads. On the other hand, this can lead to a source of plasma impurities adversely affecting the plasma parameters and stability.

What are the specific sources of origin and formation mechanisms of dust particles in fusion devices? Significant parts of the plasma-exposed surfaces (e.g., limiters, divertors, antennas for rf heating) are often coated with carbon-based materials, such as graphite or carbon fiber composites. However, carbon suffers from high erosion rates from intense physical sputtering and chemical erosion. As a result of the exposure to chemically active hydrogen, several forms of hydrocarbon are released from the surfaces into the plasma edge where they interact with the plasma and could be ionized or dissociated. Edge-localized modes and pressure-driven instabilities or disruptions (quick and uncontrollable discharge quenching leading to deposition of the plasma-stored energy onto the surface) at the plasma edge can lead to excessive heat fluxes onto the divertor surfaces.

It is clear at present that more understanding of the mechanisms responsible for particulate production from plasma–surface interactions in fusion devices is required. Moreover, this area has been highlighted by the U.S. Fusion Safety Program as one of the priority areas of research. Recently developed plasma/fluid and aerosol models of disruption simulation experiments in the SIRENS high heat flux facility integrate the necessary mechanisms of plasma–material interactions, plasma and fluid flow, and particulate generation and transport (Sharpe *et al.* 2001). The model successfully predicts the size distribution of primary particulates generated in SIRENS disruption-induced material immobilization experiments.

The estimated erosion rate of the carbon material can be quite high, up to 2×10^{21} m^2s^{-1} in the TEXTOR (Tokamak EXperiment for Technology-Oriented Research) fusion device. The eroded material is usually redeposited in a form of carbon-based layers in the areas of lower heat fluxes, and contains a large amount of radioactive hydrogen isotopes. The dust thus becomes radioactive and can carry a large proportion of tritium inventory. The main factor to be utilized for successful solution of the problem is the presence of the electrical charge at the dust particles. In these conditions, the equilibrium particle charge is determined by competition of secondary electron emission and electron and ion absorbtion from the ambient plasma.

One can estimate the charge that can be accumulated by a carbon-based particle due to radioactive decay of tritium (half-lifetime of $t_{1/2} = 12.3$ years with a maximum electron energy of 18.6 keV) (Winter 2000). Specifically, the number of β-decays in a 5 μm-sized carbon particle carrying 0.4 hydrogen isotopes (with 50% tritium) per carbon atom, can reach up to 5×10^2 per second. Assuming that all β-electrons leave the particle, the secondary electron emission yield of unity, and mean charge lifetime of 1 s, one can calculate that such a particle can accumulate a positive equilibrium charge of $Q = 5 \times 10^2\ e$. The electric field of 38 V cm^{-1} would be sufficient to confine such a particle near the surface.

The size of particulates in fusion devices varies in broad ranges, from a few tens of nanometers to several millimeters. The estimates of the total amount of redeposited radioactive dust by Winter (2000) show that large amounts up to a few tens of kilograms can be generated in the International Thermonuclear Experimental Reactor (ITER) device. The dust composition is mostly carbon but may also include all other materials used inside the vessel or for wall conditioning purposes (e.g., B, Si). In the TEXTOR experiments, a large number of almost perfect metallic spheres with diameters from 10 μm to 1 mm has been identified. The most likely formation mechanism is the reactor wall flaking (heterogeneous process) with subsequent coagulation of metal atoms on hot and non-wetting graphite surfaces. It is also interesting to note that very small, sub-100 nm carbon particles can be formed in the fusion devices as a result of CVD (Chemical Vapor Deposition) processes in carbon vapor. Formation of small globular clusters, fullerene-like materials, etc., is also possible. In the TEXTOR device, agglomerates of individual particles of about 100 nm in diameter are frequently observed. Another possible mechanism is the dust growth in the scrape-off layers (detached plasmas in the proximity of divertors and limiters), where the conditions are quite similar to those in chemically active low-temperature hydrocarbon plasmas. Under such conditions, the growth will probably proceed via negative hydrocarbon ions and multiple ion–neutral reactions.

Large particles introduced into the plasma can also induce a disruption. However, usually if a discharge is fully developed, their effect on the discharge performance is weak. However, if particles pre-exist in the vessel prior to the plasma start-up, a significant amount of impurities can be released into the plasma volume. Indeed, the intensive impurity radiation is often observed during the start-up phase and may be due to the levitated dust. As was already mentioned, β-decay of tritium may lead to charging of dust and formation of nuclear-induced plasmas, which may affect the initiation phase of a thermonuclear plasma. It is worth noting that the electron number density of nuclear-induced plasmas is typically about 5×10^9 cm^{-3}. When the gas pressure in the vessel increases to about 10^{-3} mbar, the plasma breakdown takes place and the fast β-electrons from T-decay ionize the gas along their track (on the order of 10^3 m at this pressure). In this way, about 500 electron–ion pairs per β-electron can be formed. The plasma induced by radioactive particles can be formed in a simple parallel plate model reactor configuration even without any magnetic field. In this case, dust levitation and formation of ordered dust structures is possible (Fortov *et al.* 2002). Therefore, study of the possible methods to remove the radioactive dust from or minimize its consequences on the operation of the fusion reactors becomes increasingly important. Note that use of the thermophoretic force can be a viable route for the removal of the radioactive dust (Yokomine *et al.* 2001).

In the ITER dust can represent a serious safety hazard. ITER, similar to most of the existing fusion devices, will have wall parts made of graphite and carbon composites. As we have already noticed, the tritium implantation into the carbonaceous dust can result in appearance of dust particles, where for one atom of carbon there are two atoms of tritium. The mass of tritium in large devices like ITER may be as high as dozens of kilograms. Such a high amount raises serious problems related to the safety of the operation. Thus, the problem of dust removal from thermonuclear devices represents one of the most important scientific and technical problems.

9.3 Nuclear photovoltaic electric battery

Application of radioactive materials for electric power production started soon after the discovery of radioactivity. At present, radioactive isotopes are widely used in power engineering. Different methods of energy conversion of the radioactive decay into other kinds of energy (thermal, electric, light) have been developed. Radioisotope thermoelectric power generators have been widely adopted for autonomous hydrometeorlogical stations, for radio and light marine buoys, for medicine (radionuclide cardiostimulators with electric power supply on a basis of ^{238}Pu), and for power supply of spacecrafts. As a main fuel for generators, ^{90}Sr, ^{238}Pu, and ^{210}Po are used. In the foregoing systems, energy needs do not exceed 100 W. When power of the energy-release of 1–10 kW is necessary, the thermionic converters with ^{235}U as the heat source are used. All these energy sources have disadvantages, in particular, very low efficiency. Moreover, a nuclear reactor is very complicated to produce. Recently, a new method of the nuclear-to-electric energy conversion was proposed by Baranov *et al.* (2000) and Filippov *et al.* (2005). The operating principle of the novel atomic battery is as follows: high-energy particles, which are formed during the decay of a radioactive material, ionize an inert gas such as xenon. The dissociative recombination of formed diatomic xenon ions results in the effective excitation of xenon excimers which emit vacuum ultraviolet photons with a wavelength of about 172 nm. These photons are absorbed on a wide band-gap photoconverter and generate electron–hole pairs. Estimates indicate that the total efficiency of a battery utilizing this principle may be as high as 25–35%. A pictorial diagram of the proposed atomic battery is given in Figure 9.1, and the main elements and processes occurring in such a battery are shown in Figure 9.2. Besides small spherical particles, the radio-isotope fuel may have other geometric shapes such as thin wires and foils. The efficiency of the energy yield, averaged over the spectra of β-particles, turns out to be higher than 80% even with a particle diameter (thickness) of 100 microns.

In order to use solid isotopes in the photovoltaic converters, it is necessary to have the isotope surface area as large as possible, because the mean free path of the ionizing particles in the isotope material is very short (e.g., the mean free path of β-radiation with the mean decay energy in ^{90}Sr is about 180 μm). Therefore, a homogeneous mixture of gas and isotope dust is a very good option. Excitation of the gas mixture is performed by α- or β-radiation from the radioactive dust. Estimates show that at a dust size of 1–20 μm and dust number density of 10^5–10^9 cm^{-3}, it is possible to obtain the power density of ~ 1 W m^{-3}. The gas pressure has to be of the order of $1 \div 10$ Bar to ensure effective energy conversion of β- or α-radiation into UV radiation. The main technical problem here is to have a homogeneous gas–dust mixture at high gas pressures. Experiments of Babichev *et al.* (2004), Pal' *et al.* (2005), and Filippov *et al.* (2006) performed in such systems demonstrate that this is possible in principle. Processes of self-organization occurring in nuclear-induced plasmas result in the formation of stationary structures and, hence, provide relatively homogeneous redistribution of particles over the plasma volume.

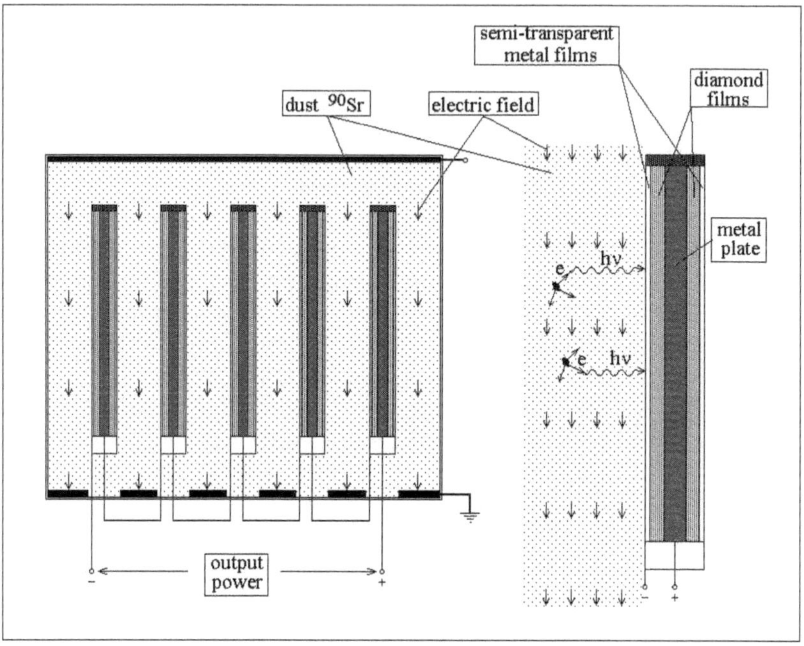

FIGURE 9.1
Schematic diagram of an aerosol photovoltaic source of electric energy.

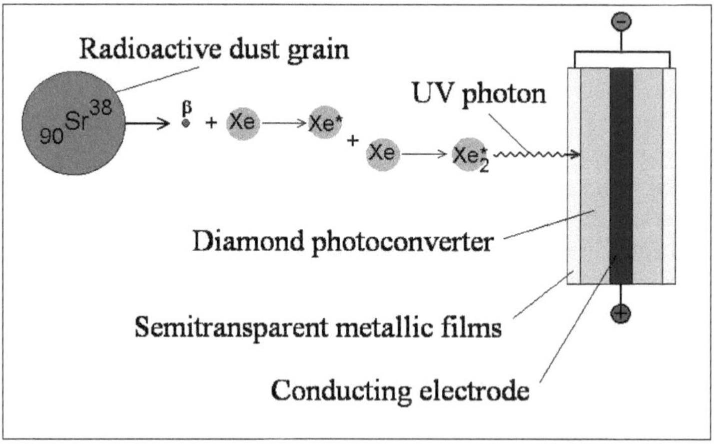

FIGURE 9.2
Scheme of the main components and processes occurring in an aerosol photovoltaic source of electric energy.

Simulation of dusty plasma properties in a photovoltaic electric source was performed by Filippov *et al.* (2005). In noble gases, the condition of locality of the electron energy distribution function (EEDF) is violated. Therefore, Filippov *et al.* (2003) developed a nonlocal model of dust particle charging, which takes into account the nonlocality of the EEDF using the nonlocal method of moments (Ingold 1989). Dust particle charge as a function of radius for different values of the specific power of a photovoltaic source of electric energy was calculated. From this the strength of interparticle coupling can be calculated and predictions can be made whether the particle system forms highly ordered crystal-like or less ordered liquid-like structures.

The plasma created by a beam of high-energy electrons upon ionization of a gaseous medium containing dust particles of micron sizes has physical properties close to the plasma of the aerosol photovoltaic energy source. Investigation of such plasma by Filippov *et al.* (2005) revealed the following:

1. The experimental investigation of the formation of dust particle structures in a nuclear-excited plasma demonstrated the possibility of a gas–dust mixture to organize into a structure levitating in the gravity field.

2. The experiments revealed the possibility of using a photoconverter to generate electrical power upon the excitation of a gas medium by a beam of high-energy electrons which simulate β-particles.

3. β-active isotopes with a half-life of 10 to 30 years are most promising as fuel for an autonomous photovoltaic source of electric energy with a service lifetime of 10 years and longer.

4. The dust particle size was found to have limitations from both below and above. The limitation from below is associated with the fact that, in the case of small particles, their number per unit volume turns out to be enormous (the radius and concentration of dust particles are related by the need to ensure the specific power of the source on a level of 0.1–10 W/l that is acceptable from the practical standpoint). As a result, the main process in the loss of diatomic xenon ions becomes the absorption on the particles. Then, the energy equal to the ionization potential is used to heat dust particles. This causes a reduction of the efficiency of the PSEE. The dust particle size is restricted from above because, first, it is difficult to "suspend" large particles in the gravity field and, second, their concentration turns out to be low; therefore, the mean interparticle distance becomes large, and the particles cease to interact with one another.

5. In order to reduce a loss of energy in the dust component of a gas–dust mixture, it is necessary to raise the pressure as the specific power of a battery increases.

6. A constraint is imposed on the external electric field because of the Joule heating of the gas. The external field is necessary for the formation and prevention

of the sedimentation and deposition of a levitating cloud of dust particles on the construction elements.

To conclude, for the successful development of adequate scientific principles of atomic batteries based on dust-plasma structures, the investigations has to be continued. At present, an experimental setup is available, and the intense research of the possibility to create ordered structures of dusty plasma produced by a stationary beam of high-energy electrons is being performed. The experiments are being continued to determine the efficiency of conversion of the beam energy to the electric power with the aid of a standard photoconverter using xenon, krypton, argon/nitrogen mixtures, and air. The activities which are aimed at developing a wide band-gap diamond photoconverter have been started; after completing these activities, it is planned to perform experiments with xenon.

References

Babichev, V. N., Pal', A. F., Starostin, A. N., Filippov, A. V., and Fortov, V. E. (2004). Stable dust structures in non-self-sustained gas discharge under atmospheric pressure. *JETP Letters*, **80**, 241–245.

Baranov, V. Y., Belov, I. A., Dem'yanov, D. V., Ivanov, A. S., Mazalov, D. A., Pal', A. F., Petrushevich, Y. V., Pichugin, V. V., Starostin, A. N., Filippov, A. V., and Fortov, V. E. (2000). Radioactive isotopes as source of energy in photovoltanic nuclear battery on basis of plasma–dust structures. In *Isotopes: Properties, production, and applications* (in Russian), Baranov, V. Y. (ed.), pp. 626–641. IzdAT, Moscow.

Bouchoule A. (1999). Technological impacts of dusty plasmas. In *Dusty plasmas: Physics, chemistry and technological impacts in plasma processing*, Bouchoule, A. (ed.), pp. 305–396. Wiley, Chichester.

Filippov, A. V., Dyatko, A. V., Pal', A. F., and Starostin, A. N. (2003). Development of a self-consistent model of dust grain charging at elevated pressures using the method of moments. *Plasma Phys. Rep.*, **29**, 190–202.

Filippov, A. V., Pal', A. F., Starostin, A. N., Fortov, V. E., Petrov, O. F., Dyachenko, P. P., and Rykov, V. A. (2005). Atomic battery based on ordered dust–plasma structures. *Ukr. J. Phys.*, **50**, 137–143.

Filippov, A. V., Babichev, V. N., Dyatko, N. A., Pal', A. F., Starostin, A. N., Taran, M. D., and Fortov, V. E. (2006). Formation of plasma dust structures at atmospheric pressure. *JETP*, **102** 342–354.

Fortov, V. E., Vladimirov, V. I., Deputatova, L. V., Nefedov, A. P., Rykov, V. A., and Khudyakov, A. V. (2002). Removal of dust particles from technological plants. *Doklady Physics*, **47**, 367–369.

Hwang, N. M., Cheong, W. S., Yoon, D. Y., and Kim, D. Y. (2000). Growth of silicon nanowires by chemical vapor deposition: Approach by charged cluster model. *J. Cryst. Growth*, **218**, 33–39.

Ingold, J. H. (1989). Nonequilibrium positive column. *Phys. Rev. A*, **40**, 7158–7164.

Jensen, P. (1999). Growth of nanostructures by cluster deposition: Experiments and simple models. *Rev. Mod. Phys.*, **71**, 1695–1735

Kersten, H., Deutsch, H., Stoffels, E., Stoffels, W. W., Kroesen, G. M. W., and Hippler, R. (2001). Micro-disperse particles in plasmas: From disturbing side effects to new applications. *Contrib. Plasma Phys.*, **41**, 598–609.

Kersten, H., Deutsch, H., Stoffels, E., Stoffels, W. W., and Kroesen, G. M. W. (2003a). Plasma-powder interaction: Trends in application and diagnostics. *Int. J. Mass Spectrometry*, **223–224**, 313–325.

Kersten, H., Wiese, R., Thieme, G., Froehlich, M., Kopitov, A., Bojic, D., Scholze, F., Neumann, H., Quaas, M., Wulff, H., and Hippler, R. (2003b). Examples for application and diagnostics in plasma-powder interaction. *New. J. Phys.*, **5**, 93/1–15.

Laroussi, M., Mendis, D. A., and Rosenberg, M. (2003). Plasma interaction with microbes. *New J. Phys.*, **5**, 41/1–10.

McIlroy, D. N., Huso, J., Kranov, Y., Marchinek, J., Ebert, C., Moore, S., Marji, E., Gandi, R., Hong, Y.-K., Grant Norton, M., Cavalieri, E., Benz, R., Lustus, B. L., and Rosenberg, A. (2003). Nanoparticle formation in microchannel glass by plasma enhanced chemical vapor deposition. *J. Appl. Phys.*, **93**, 5643–5649.

Murphy, A. B. and Bing, M. (1994). Equilibrium calculations of the reduction of titanium tetrachloride by aluminium and hydrogen. *High Temp. Chem. Process.*, **3**, 365–373.

Pal', A. F., Filippov, A. V., and Starostin, A. N. (2005). An experimental and theoretical study of the high-pressure dusty plasma created by a stationary e-beam. *Plasma Phys. and Contr. Fusion*, **12B**, B603–B615.

Roco, M. C., Williams, S., and Alivisatos, P. (Eds.) (1999). *Nanotechnology research directions: Vision for nanotechnology research and development in the next decade*. Kluwer Academic, Amsterdam. See also: US National Nanotechnology Initiative, http://www.nano.gov.

Sankaran, R. M., Holunga, D., Flagan, R. C., and Giapis, K. P. (2003). Synthesis of silicon nanoparticles using atmospheric-pressure microdischarges. *Bull. Amer. Phys. Soc.*, **48**, 39.

Selwyn, G. S., Singh, J., and Bennett, R. S. (1989). In situ laser diagnostic studies of plasma-generated particulate contamination. *J. Vac. Sci. Technol. A*, **7**, 2758–2765.

Sharpe, J. P., Merrill, B. J., Petti, D. A., Bourham, M. A., and Gilligan, J. G. (2001). Modeling of particulate production in the SIRENS plasma disruption simulator. *J. Nucl. Mater.*, **290**, 1128–1133.

Vladimirov, S. V. and Ostrikov, K. (2004). Dynamic self-organization phenomena in complex ionized gas systems: New paradigms and technological aspects. *Phys. Rep.*, **393**, 175–380.

Vladimirov, S. V., Ostrikov, K., and Samarian, A. A. (2005). *Physics and applications of complex plasmas*. Imperial College, London.

Winter, J. (2000). Dust: A new challenge in nuclear fusion research? *Phys. Plasmas*, **7**, 3862–3866.

Yokomine, T., Shimizu, A., and Okuzono, M. (2001). The possibility of dust removal in fusion plasma device using thermophoretic force. *Fusion Technol.*, **39**, 1028–1032.

Index